P9-ECZ-802

Volume 3 of *Thunderstorms: A Social, Scientific, and Technological Documentary*

Instruments and Techniques for Thunderstorm Observation and Analysis

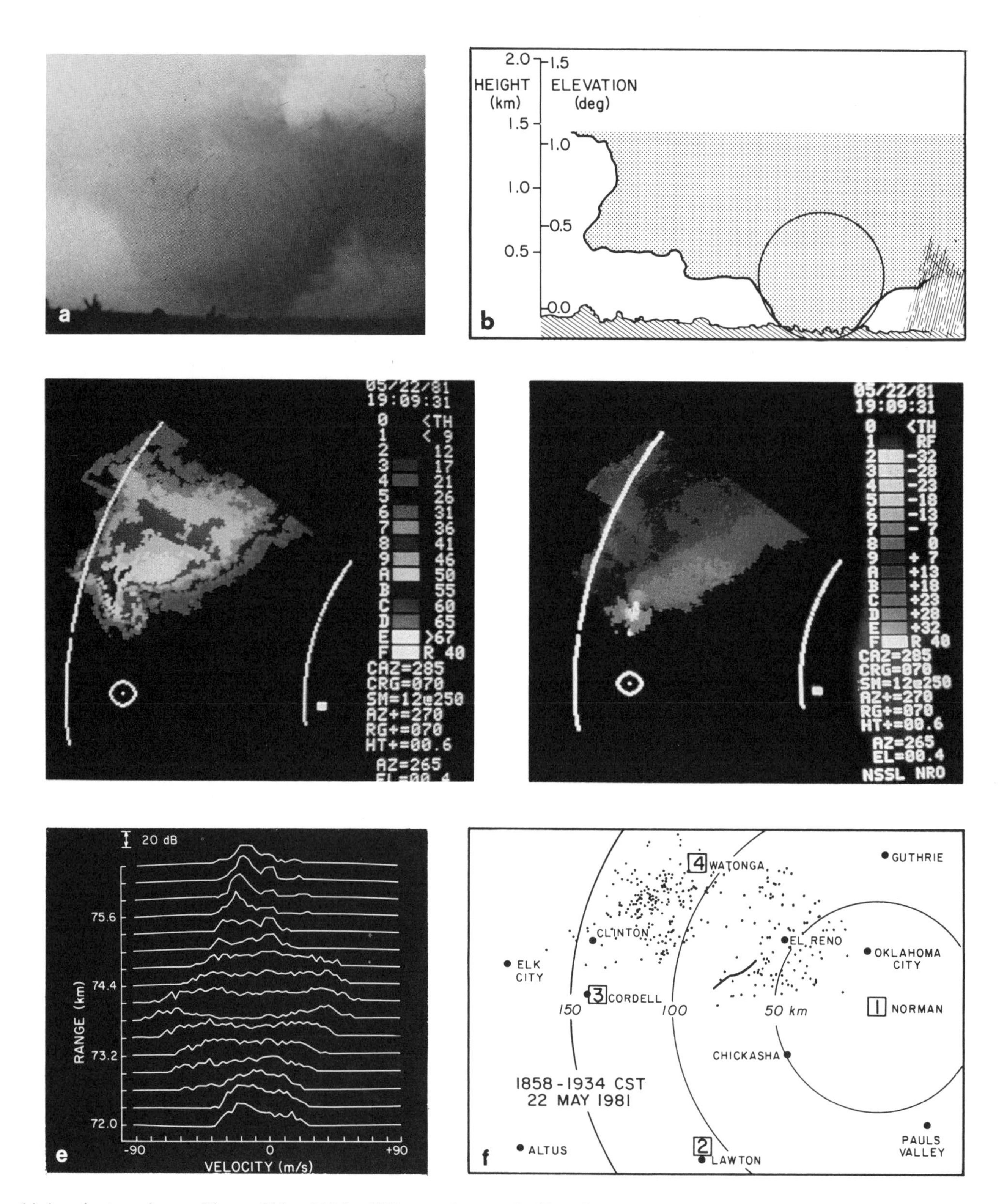

(a) A major tornado near Binger, Okla., 22 May 1981, was photographed by a Storm Intercept team 5 km from the funnel cloud. (b) The circle shows how the Doppler radar at Norman, Okla., illuminated the funnel 75 km away. (c) Plan distribution of reflectivity factor includes some values >60 dBZ and shows a shape characteristic of tornadic storms. (d) Plan distribution of radial velocities—reds denote motion away from the radar; the tornado lies in the strong color-change region. (e) Velocity spectra along beam as in (b) show speeds from −85 to +80 m/s^{-1} within the 1.3-km beam width. (f) Ground strike points of lightning show no clustering along the tornado ground path. See also Chaps. 2, 4, 7, 8, 10, and 11.

Volume 3 of *Thunderstorms: A Social, Scientific, and Technological Documentary*

Instruments and Techniques for Thunderstorm Observation and Analysis

Second Edition, Revised and Enlarged

Edited by Edwin Kessler

University of Oklahoma Press
Norman and London

BY EDWIN KESSLER

On the Distribution and Continuity of Water Substance in Atmospheric Circulations (Boston, 1969)

(Editor) *Thunderstorms: A Social, Scientific, and Technological Documentary,* 1st ed., 3 vols. (Washington, D.C., 1982)

(Editor) *Thunderstorms: A Social, Scientific, and Technological Documentary,* 2d ed., 3 vols. (Norman, 1983–87) (Vol. 1, *The Thunderstorm in Human Affairs,* 1983; Vol. 2, *Thunderstorm Morphology and Dynamics,* 1985; Vol. 3, *Instruments and Techniques for Thunderstorm Observation and Analysis,* 1987)

Library of Congress Cataloging-in-Publication Data

Instruments and techniques for thunderstorm
observation and analysis.

(Thunderstorms—a social, scientific, and
technological documentary; v. 3)
Bibliography: p. 235
Includes index.
1. Thunderstorms—Observations. 2. Thunderstorms—
Measurement. 3. Meteorological instruments.
I. Kessler, Edwin. II. Series: Thunderstorms—a social,
scientific, and technological documentary (2d ed.,
rev. and enl.); v. 3.
QC968.T48 1983 vol. 3 551.5′54 87–40554
ISBN 0–8061–2117–3 [551.5′54′0287]

The paper in this book meets the guidelines for permanence and durability of the Committee on Production Guidelines for Book Longevity of the Council on Library Resources, Inc.

Contents

Preface to the Second Edition

This is the third and final volume of the series of books on thunderstorms, whose preparation and first appearance were a project of the National Severe Storms Laboratory and its parent organization, the Environmental Research Laboratories, in the National Oceanic and Atmospheric Administration, U.S. Department of Commerce.

Our project began in 1976, and significant scientific and technological developments since then are indicated by significant differences between the first and second editions. In this edition of Volume 3, Chapters 2, 10, and 12 have been substantially revised and updated to account for some recent advances. The remaining chapters of the first edition have been reviewed by their authors, partly rewritten for improved clarity of presentation, and somewhat updated. As in all other sciences, progress is uneven, and only a few changes have seemed warranted in Chapters 4, 5, and 6.

Chapters 8 and 13 are wholly new to this edition. Chapter 8 presents a unique gathering together of descriptions of equipment used during the 1980s to observe electrical phenomena that accompany storms. This chapter complements and extends Chapter 7, whose author, Edward T. Pierce, died in 1978. Chapter 7 is virtually unchanged from the first edition; it is a lucid treatment of measurement fundamentals and of equipment used during the 1970s, when the study of electrical phenomena in the atmosphere underwent a significant expansion. Chapter 13 is a pedagogically oriented discussion of methods for using Doppler radar data to learn, in astonishing detail, about coevolving fields of force and motion and all phases of water in thunderstorms.

As with Volumes 1 and 2 in their second editions, this volume is also distinguished from its predecessor by the addition of an index.

We appreciate the support of the National Center for Atmospheric Research, sponsored by the National Science Foundation, whose grant of funds covered a significant portion of the cost of manufacture.

I thank authors and reviewers for attention to many details and substantial improvements. Reviewers who have newly contributed to this edition include Douglas K. Lilly, Tzvi Gal-Chen, Steven Rutledge, Bernard Vonnegut, and others anonymous.

Finally, thank you, Doris Radford Morris, Managing Editor of the University of Oklahoma Press, for your purposeful, dedicated, and persevering work, truly the sine qua non to production of all three of our thunderstorm volumes in their second editions.

EDWIN KESSLER

University of Oklahoma
April, 1987

Preface to the First Edition

This third and last volume of a comprehensive documentation of thunderstorms presents tools and techniques for acquisition and analysis of data on thunderstorms.

The authors have been cooperative and gracious during a process that began in 1976. They have my heartfelt thanks. Reviewers have also contributed substantially. Reviewers of this volume include Sigmund Fritz, Thomas H. Georges, Roland List, Don R. MacGorman, Roddy R. Rogers, W. David Rust, William L. Taylor, and Charles Warner. I particularly thank the staff of the Environmental Research Laboratories, Publication Services Division, Boulder, Colo., for arduous editorial work expertly accomplished. My secretary, Barbara R. Franklin, and my wife, Lottie Menger Kessler, have also been indispensable to completion of this work.

EDWIN KESSLER

Norman, Oklahoma
January, 1982

Instruments and Techniques for Thunderstorm Observation and Analysis

1. Observation and Analysis with Station Networks

Kenneth E. Wilk and Stanley L. Barnes

1. Introduction

The National Weather Service and its predecessor organizations have always maintained and operated networks of meteorological stations for observation at the Earth's surface and in the upper atmosphere (see Vol. 1, Chap. 7). Measurements taken at these stations allow meteorologists to monitor large-scale (larger than 2,000 km) weather systems and to gather climatological records on smaller-scale events, including showers and thunderstorms. In-depth aspects of meteorological measurements are treated by Houghton (1985).

Since individual severe thunderstorms are only 15–30 km across, observations are needed at many points within this narrow region to describe accurately the distributions of storm-associated parameters. Observations by radar, electric field sensors, and weather satellites are sufficiently detailed, but an adequate thermodynamic description of individual local storms also requires observations of the kind provided by the regular network of weather-observing stations. Thus direct measurements of wind velocity, pressure, temperature, and moisture, which are essential for storm research, are provided by a few local mesoscale networks of in situ sensors (*meso* = middle, in this application typically 20 to 200 km, often defined as meso β, as discussed in Vol. 2, pp. 113ff.). This chapter reviews the development of mesoscale research networks and their instrumentation and describes some methods for data acquisition and analysis.

2. Early Mesometeorological Networks for Severe-Storm Research (1946–1966)

Since World War II several mesoscale-observing experiments have provided better definition of patterns of pressure, temperature, wind, and moisture associated with severe local storms. The Thunderstorm Project was the first major national effort to gain comprehensive knowledge of severe thunderstorms (Byers and Braham, 1949). Collection of storm data was undertaken in 1946–47 as a joint project of the University of Chicago, the U.S. Weather Bureau, the National Advisory Committee for Aeronautics, and the U.S. military agencies. The project set a pattern in the use of research stations for observations at the Earth's surface and in the upper air that would continue for two decades.

Table 1.1 lists project instrumentation at 55 network sites, placed first in Florida and later moved to Ohio. The network in Ohio was expanded to cover the somewhat larger scale of thunderstorms and related meteorological fields in the Midwest as compared with those observed in Florida (compare Fig. 1.1a, b). Note also the sites for radar and surrounding rawinsondes, which for the first time played a role in the observation of local storms and their environment. (The development of upper-air sounding capability was detailed by Middleton [1969], and the early history of U.S. aerology was summarized by Ferguson [1933].) During the summers of 1946 and 1947 the networks operated fewer than 300 days yet recorded useful data during 180 rain periods, most of which involved thunderstorms. The composite data set included 1,327 wind soundings and many individual stormtraverses by instrumented aircraft. By comprehensive analysis, case by case, Byers and his associates (Byers and Braham, 1948; Byers and Rodebush, 1948; Braham, 1952; Bleeker and Andre, 1950) produced the first definitive study of meteorological features accompanying thunderstorms in the United States. Using averages from Weather Bureau stations surrounding the network to determine the large-scale patterns of pressure and wind (which were then subtracted from the dense-network data), project meteorologists were able to analyze and track the mesoscale perturbations associated with developing thunderstorms. The high- and low-pressure cells so revealed were on the same scale as the major storm elements and were attributed to vertical motions in the organizing convective-cloud systems. New questions arose with these measurements, with speculation on the origins and implications of gust fronts and local pressure and temperature anomalies. For example, Tepper (1950) suggested that pressure jumps, sometimes observed before squall-line occurrence, might be a basis for advanced warning of tornadoes.

In 1955, the U.S. Army Signal Corps sponsored a reanalysis of the Thunderstorm Project data by staff at the

Table 1.1. Meteorological Instrumentation* in the Thunderstorm Project, 1946–1947

Parameter	Florida	Ohio
Wind direction	Vane with 4 contacts and double-register recorder for 8-point (45°) resolution	Vane with continuous output provided by variable resistor connected to strip chart recorder
Windspeed	3-cup sensor with internal recording of miles of air-flow	3-cup sensor with AC generator providing continuous output from 0 to 75 kn, on strip chart
Temperature-humidity	Bendix hygrothermograph using Bourdon tube for temperature and hair hygrometer for measuring relative humidity	Same
Rainfall	Weighing-bucket rain gage with 20-in-diameter collector	Same
Pressure	Microbarograph using double Sylphon bellows (checked weekly with aneroid barometer)	Same

*Drive gears on most recorders provided charts that could be read to a resolution of 1 min.

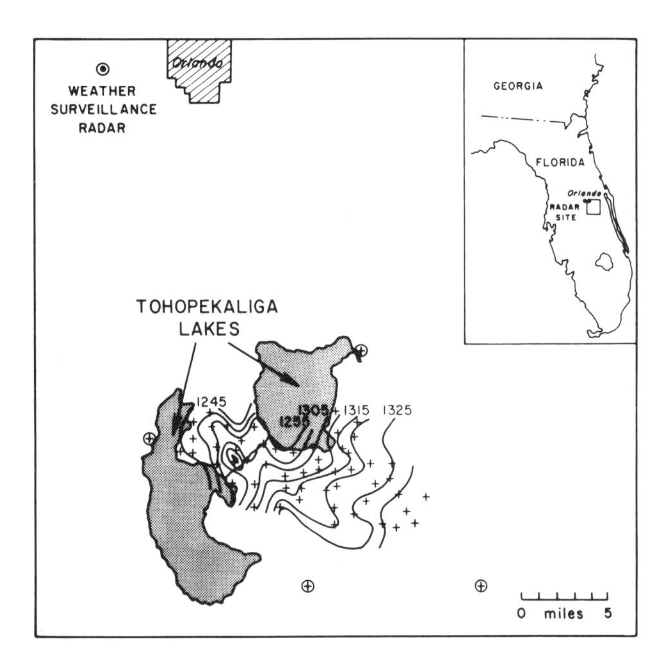

Figure 1.1a. Isochrones of "first gust" 9 July 1946 in the Florida network of the Thunderstorm Project. Such analyses disclosed that new storms often form on the intersection of outflow boundaries from existing storms (Byers and Braham, 1948).

Illinois State Water Survey, searching for repetitive patterns that might improve short-range forecasting. Both organizations then developed local research networks. The Army facility at Fort Huachuca, Ariz., operated a 12-station network for military applications. The Illinois State Water Survey operated several mesoscale rain-gage networks to define thunderstorm rainfall characteristics and later developed the first urban networks designed to improve water resources management (Changnon and Semonin, 1971). During the same period Williams (1953) and Fujita et al. (1956) developed analysis techniques with data collected

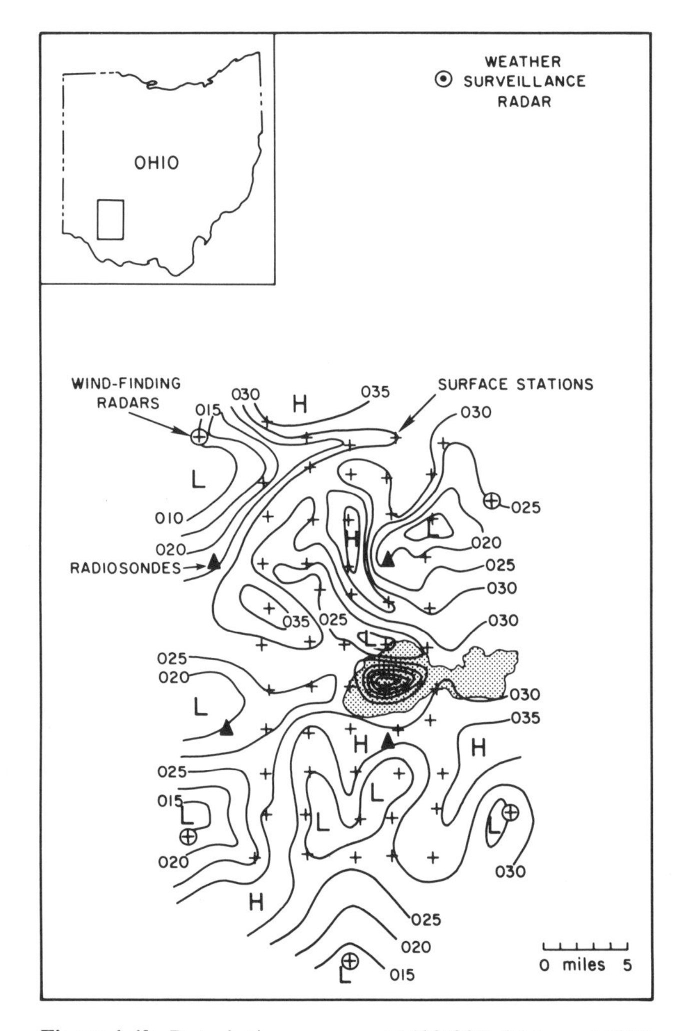

Figure 1.1b. Perturbation pressure at 1430 CST, 25 August 1947, in the Ohio network of the Thunderstorm Project. Shaded area shows radar echo from cumulonimbus cloud producing rain and localized area of high pressure (Bleeker and Andre, 1950).

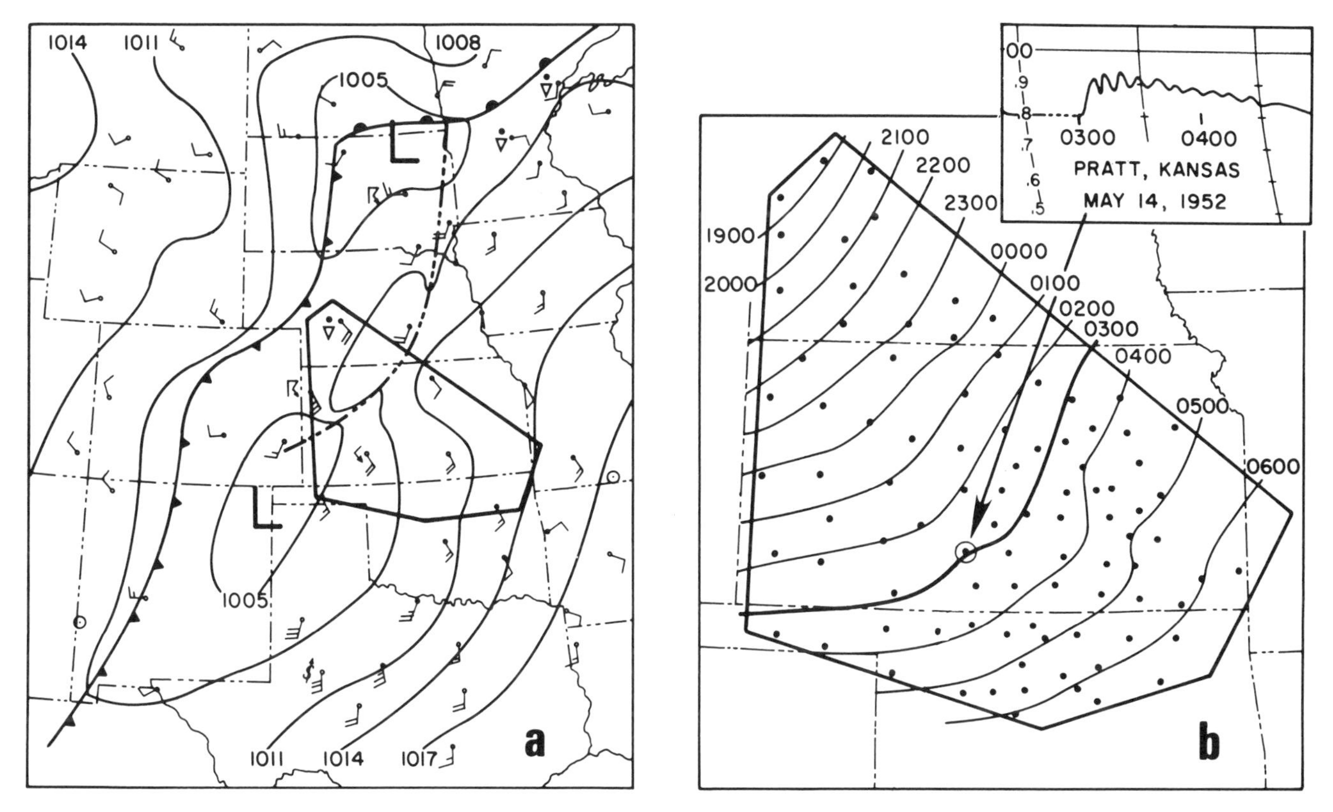

Figure 1.2. (a) Synoptic map for 14 May 1952, showing squall line advancing across mesometeorological network in Nebraska, Kansas, and Oklahoma (Williams, 1953). (b) Isochrone analysis of pressure perturbation associated with the squall line in Fig. 1.2a (Williams, 1953).

from a much larger network in Nebraska, Kansas, and Oklahoma (Fig. 1.2a, b). Such data were used to produce the first views of smaller-scale pressure, wind, and temperature structure in combination with the detailed view of convective precipitation given by weather radar. New terminology, e.g., "wake depression" and "thunderstorm high," described patterns that commonly appeared with severe thunderstorms.

Increases in fatalities from aircraft accidents in thunderstorms and increases in property destruction from tornadoes during the 1950s renewed interest in gathering and analyzing storm data, and the National Severe Storm Project (NSSP), with headquarters in Kansas City, Mo., was created within the U.S. Weather Bureau. The project established a field office in Oklahoma City and designed three networks to collect pressure and rainfall data from squall lines over a three-state area.

The general NSSP research plan, developed in 1961, called for three scales of investigation (NSSP Staff, 1961) using nested networks, shown schematically in Fig. 1.3. The NSSP nested networks (Fig. 1.4) used hygrothermographs, microbarographs, and rain gages extensively—but with stations relatively far apart and over a much larger area than that covered by the Thunderstorm Project. Fujita and Stuhmer (1963) prepared the first extensive mesosynoptic analysis using data from the NSSP interlaced networks, newly installed in 1962 (Fig. 1.5).

As the NSSP program progressed, increased emphasis was placed on the smaller scale (2 to 20 km) to supplement the fine-scale measurements provided by improved radar and aircraft instrumentation. This new mesoscale network (Fujita, 1962) was established in Oklahoma to study surface characteristics of individual severe thunderstorms. Emphasis on storm scale events has continued to the present both at NSSL and at new field projects of the National Center for Atmospheric Research (NCAR), where hailstorm characteristics have been studied (Foote and Knight, 1977).

During 1960–65 the Air Force Weather Service and the Weather Bureau expanded their jointly operated Severe Local Storms Forecast Center at Kansas City to search for better techniques to anticipate squall-line development, and the Federal Aviation Administration's research group (National Aviation Facilities Experiment Center, NAFEC) at Atlantic City, N.J., developed a reporting mesonetwork in the expectation that terminal forecasting might be greatly improved by use of local information in real time. However, the NA-

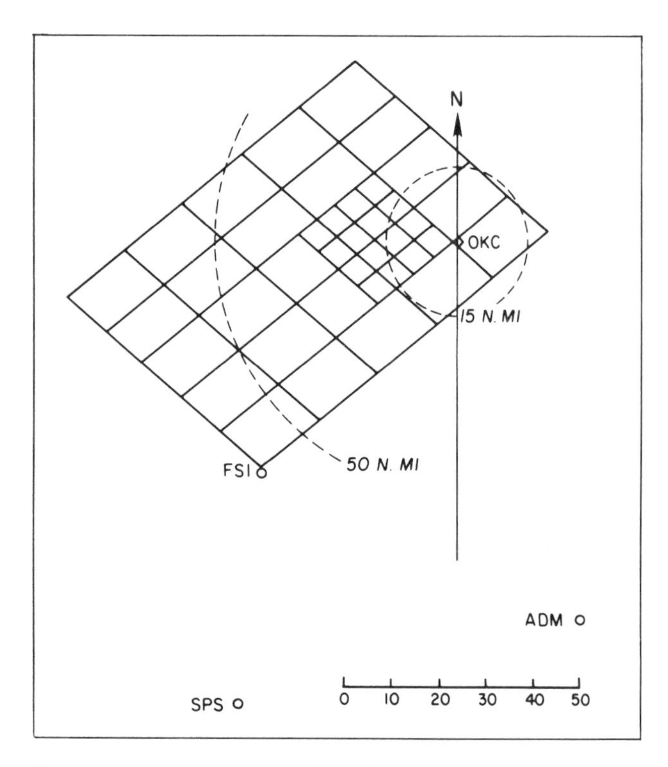

Figure 1.3. Schematic plan for NSSP mesometeorological networks, 1961.

FEC data indicated that there was not yet sufficient knowledge of the behavior of mesoscale weather systems to develop a dependable short-range forecasting system, even with detailed real-time measurements.

In late 1963 the nucleus of the NSSP was moved to Norman, Okla., and reorganized in 1964 as the National Severe Storms Laboratory (NSSL) to become a national focus of observations on severe local storms (Kessler, 1965, 1977). Both NSSP and NSSL gave considerable emphasis to development of weather radar. Radarscope photographs of thunderstorms became commonplace after 1965, as several research groups sought better prediction techniques to anticipate storm formation, movement, and severity.

3. Severe-Storm Networks After 1970

The rapid improvement in research radars in the mid-1960s (Lhermitte and Kessler, 1965) pointed the way to continuous, quantitative observations of storm development. Newton (1963) and Browning and Donaldson (1963) had developed qualitative models of large convective storms; these affirmed a need for three-dimensional mesoscale data to be obtained by combining radar, rawinsonde, and surface network observations. Kreitzberg and Brown (1970) further stressed the value of including serial soundings from special rawinsonde networks by constructing cross sections of wind, temperature, and moisture and tracing the vertical displacement of the air surrounding severe storms.

In 1970, Barnes et al. (1971) led the design and analysis of a major observational program at NSSL to "trap" a tornado-producing thunderstorm within dense, interlaced surface and upper-air networks (Fig. 1.6a). On 29 April 1970, a series of severe storms moved across central Oklahoma (Fig. 1.6b), and after 12 exhausting hours of probing and waiting, the efforts of the 40 scientists and technicians were rewarded with a very detailed data set. (At midnight, the classic tornado-bearing superstorm [Fig. 1.7a, b] was monitored clearly by radar and the surface network as it moved through the Oklahoma City area, causing widespread destruction.) Detailed analyses of these data (Barnes, 1978a, b; Barnes and Nelson, 1978) provided a new perspective of the wind, temperature, and moisture fields measured at all levels from the surface to the tropopause in a severe storm's environment. The studies relating the surface events to the radar echoes were especially valuable in providing clear evidence that the mesocyclone (i.e., the tornado cyclone of Brooks [1949, 1951]) is the parent circulation of tornadoes (Fig. 1.7b). The analyses showed that spacing and recording intervals of the 1970 mesonetwork stations were effective in resolving scales of about 10 km and 5 min.

4. Cooperative Observers and Nested Networks

In the first networks installed by the National Severe Storms Project, sites were secured by contracts with landowners, who also assisted in changing recorder charts, winding instrument clocks, and safeguarding the equipment. With this help a few technicians could oversee many tens of networks sites. Such operations required careful procedures for checking records and weekly contacts with observers to help correct malfunctions. As in any other volunteer program, observer performance varied greatly, and control of quality to a high level was difficult. Later, when the areal coverage of the network was reduced for concentrated study of individual storms, each site was visited frequently to make comparative measurements and properly annotate and replace charts.

Although recent programs place much emphasis on the thunderstorm scale, there has been a continuing need to relate detailed observations to the synoptic scale. At NSSL this link has been forged by "subsynoptic" stations placed between existing NWS and FAA observing facilities. The subsynoptic concept provided a background network having 100-km station spacing that surrounds the mesoscale network having 10-km spacing. Sites for the subsynoptic network in Oklahoma were selected from more than 200 potential locations, which included municipal fire stations, airports, and junior colleges. Staff at many of these facilities needed weather observations for their own operations or research projects and were delighted to share responsibility for locating and operating research weather stations.

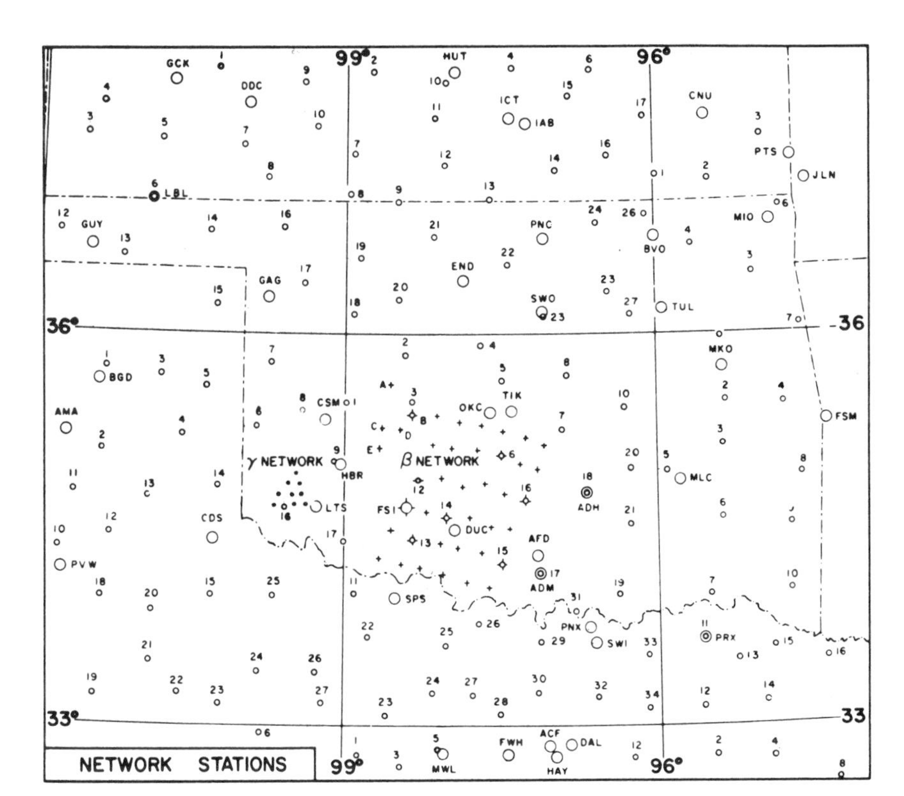

Figure 1.4. NSSP mesometeorological beta and gamma networks, 1963.

Figure 1.5. (below) A mesosynoptic analysis of tornado cyclone case of 26 May 1963.

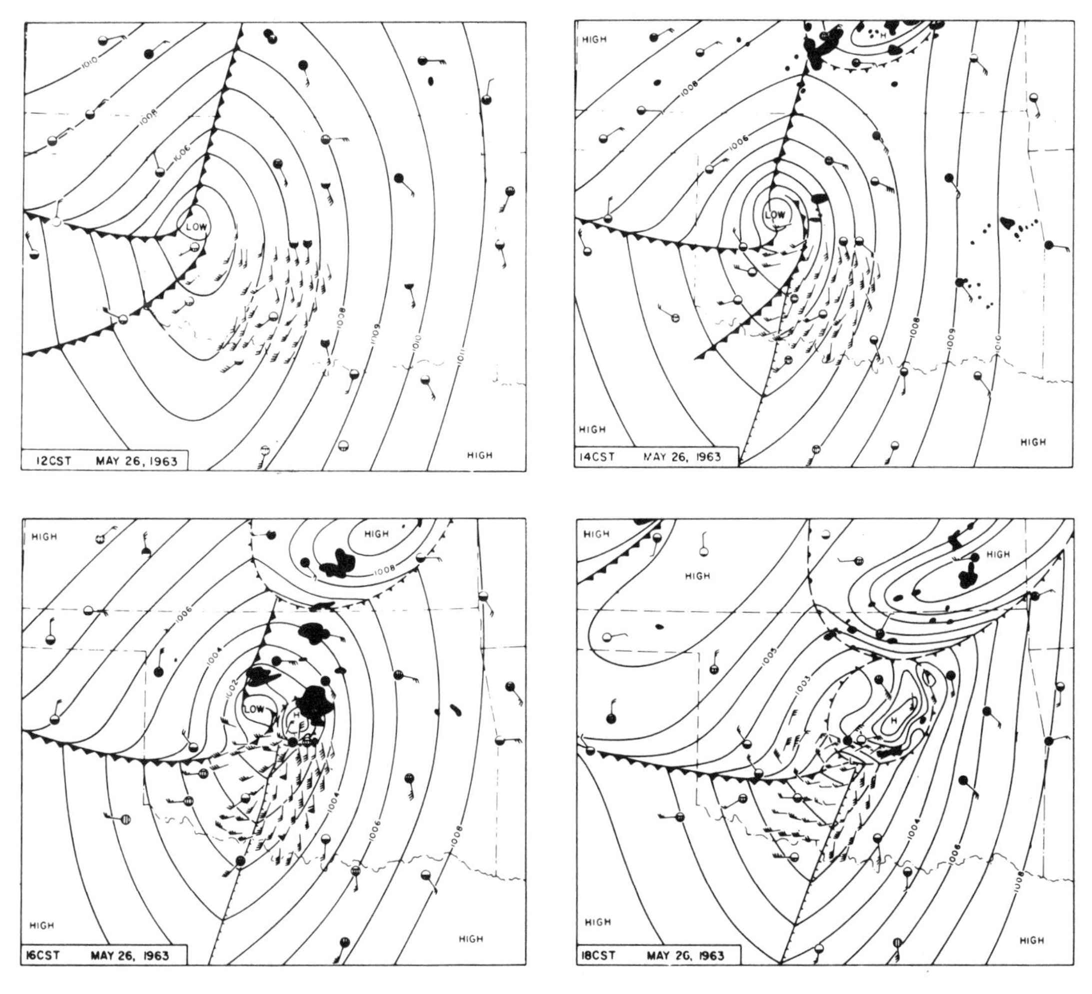

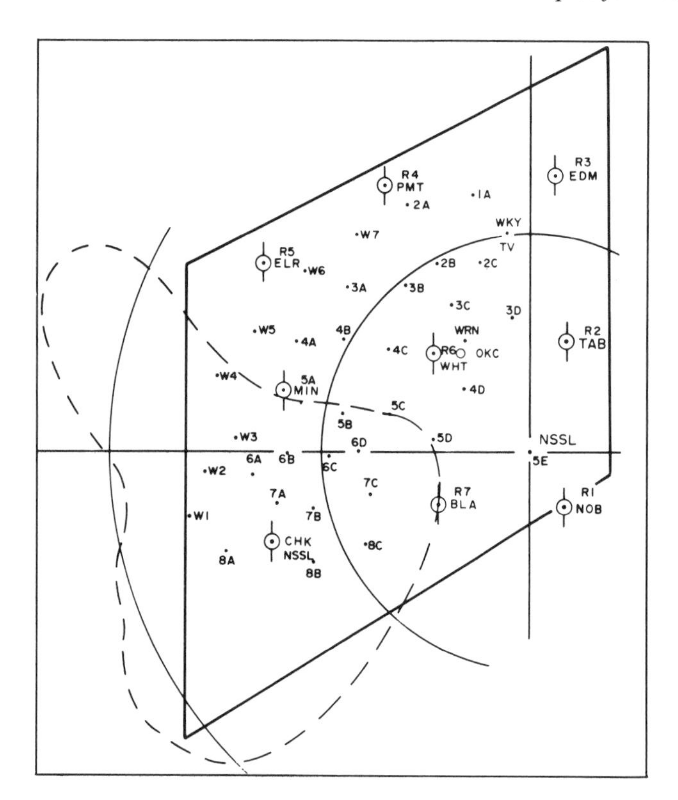

Figure 1.6a. NSSL surface, upper-air, and mesonetwork stations, 1970. Soundings were obtained at approximately 60-min intervals to 400 mb during selected periods, at stations marked by circles with vertical bars. The dashed line marks the boundary of the dense rain-gage network maintained by the Agricultural Research Service. A 20-nmi-range circle intersects the tall, instrumented WKY tower at azimuth 356°.

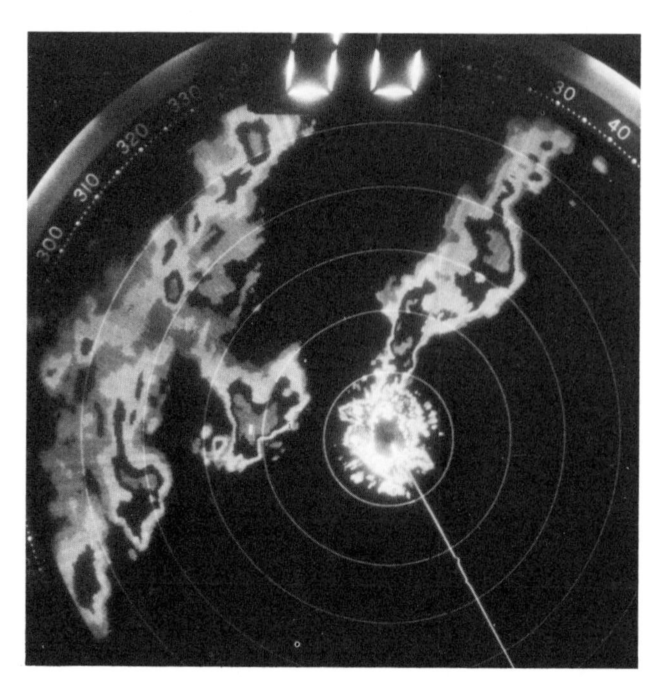

Figure 1.6b. WSR-57 radar display of thunderstorms crossing central Oklahoma on 29 April 1970. The storm 40 mi west of the radar site (second range mark) is also visible in Fig. 1.7a.

Laboratory technicians visited the sites monthly or at the request of the station manager. Since commercial power was available at most subsynoptic stations, multiple-channel strip chart recorders were installed at several locations to record eight channels of information simultaneously (Fig. 1.8). The recording-pen travel was more accurate then the older strip chart records, and the linearity allowed semiautomated data retrieval with x–y readers.

In 1973 the combination of observations from these stations with data from more finely spaced mesoscale sites was a valuable aid to reconstruction of meteorological patterns associated with the first major tornado observed with the NSSL Doppler radar (Brown, 1976). Analysis of wind trajectories located mesocyclogenesis and gust fronts, which have been shown to precede tornado development (Brandes, 1977).

The series of successful field experiments at NSSL and NCAR in the 1970s, with proof of advantages in Doppler radars, led to the design and operation of a national field project in 1979. The SESAME (Severe Environmental Storms and Mesoscale Experiment) observational program was the largest program directed toward severe local storms ever undertaken (Alberty et al., 1979). The SESAME networks (Fig. 1.9) recorded synoptic, subsynoptic, and mesoscale weather patterns using advanced digital recording and Doppler radar systems.

5. Instruments for Mesoscale Meteorological Networks

Until 1979 a typical field station in the NSSL mesometeorological networks (Fig 1.10) consisted of a fenced grassy plot of about 100 m^2, isolated from buildings and trees. Wind-sensing equipment was installed on towers 5 to 10 m high; instruments recording on strip charts were housed in standard wooden shelters used by the Weather Bureau. The hygrothermographs, microbarographs, and rain gages contained precision spring-driven clocks that powered rotating drums (Fig. 1.11). Examples of recordings made by these instruments during severe thunderstorms are shown in Fig. 1.12.

The accuracy of these systems, which are still in occasional use today, depends greatly on precalibrations and adjustments made before field installation. Such procedures are followed carefully by meteorological technicians to ensure that all instruments have similar responses and that their values correspond to preestablished standards, i.e., those set by the calibrated mercurial barometer and aspirated psychrometer.

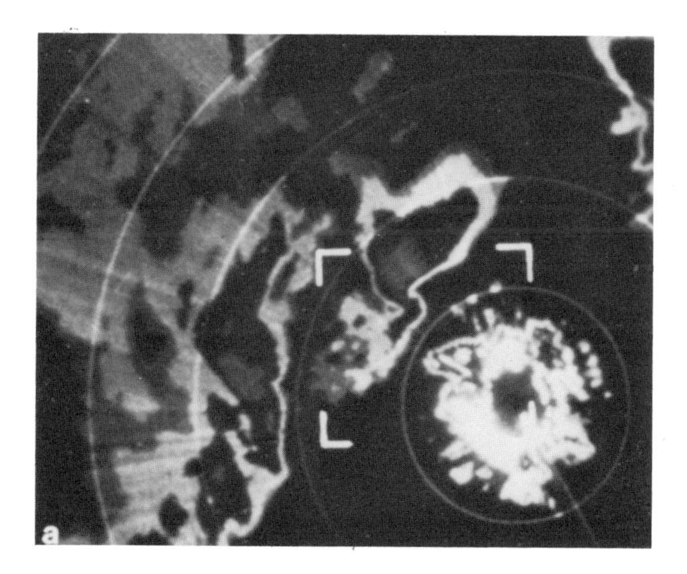

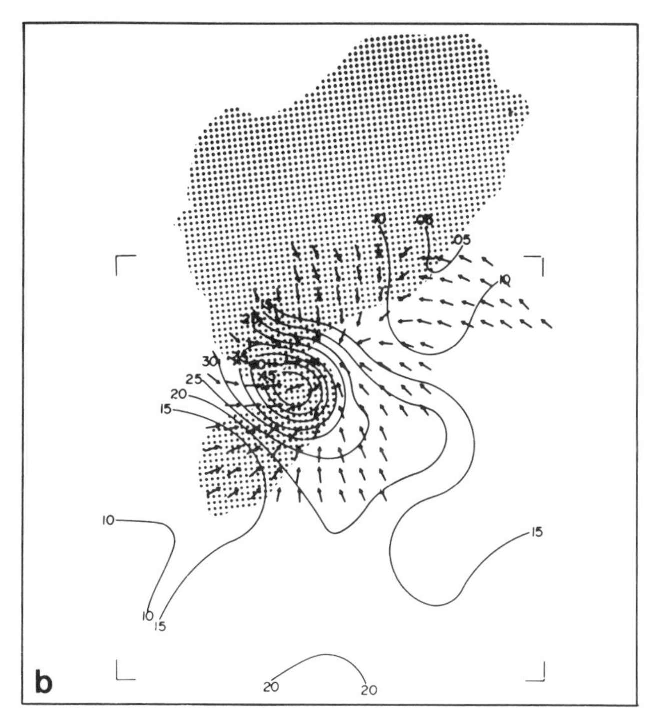

Figure 1.7. (a) WSR-57 radar display of tornado-bearing thunderstorm. (b) Surface patterns of windspeed (solid lines) and direction (arrows) as the storm crosses the NSSL mesonetwork, 29 April 1970.

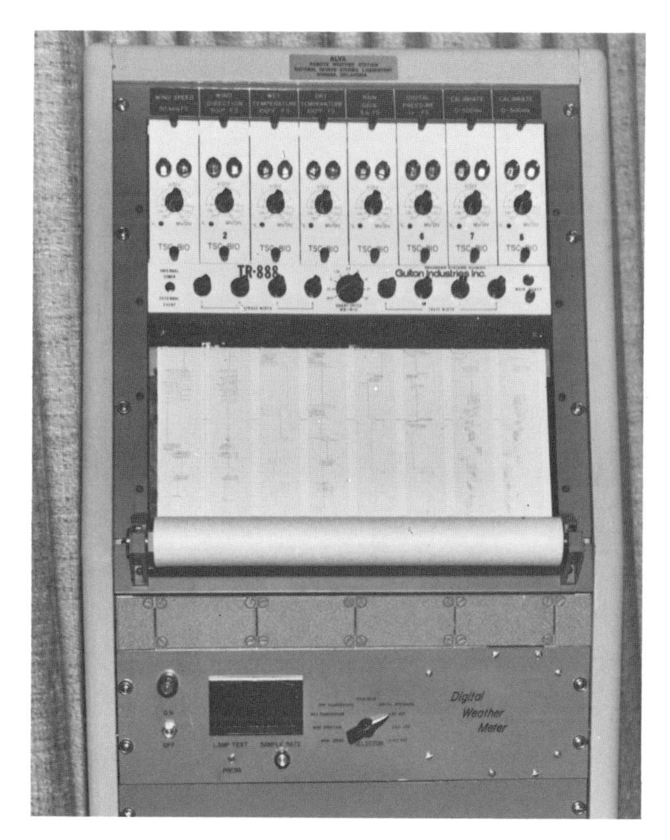

Figure 1.8. Linear, eight-channel strip-chart recorder used in NSSL's subsynoptic network, 1972–73.

Initially, the instruments are carefully cleaned, and mechanical linkages are adjusted according to manufacturers' specifications. A few days before field installation all instruments are placed in a calibration room, and adjustments on the clocks and linkage are fine-tuned to ensure agreement among all instruments. In many networks the damping fluid in the microbarographs is removed to decrease response time. (Without damping fluid both the microbarograph and the hygrothermograph respond to sudden changes within 20 sec, which is adequate for most analyses but is marginal for tornadoes.) To record detail at unattended stations over several days, the instruments may be modified to accept continuous-roll charts fed onto the clock drum from a supply spindle (Fig. 1.11). In this mode the extrathin chart paper accumulates on the clock drum, geared to rotate at four times the normal speed of one revolution per day. Buildup of paper causes a small cumulative error in the time reference as the effective diameter of the drum increases. This problem can be solved by the addition of an electrical timing circuit that deflects the pen hourly to create a time reference mark. (At NSSL the recording pens on barograph, hygrothermograph, and rain gage were deflected by a 12-v solenoid activated by a pulse from an accurate clock mechanism. In this way all analog records included the same hourly time marks.) Instruments in the field are visited once or twice a week, and comparative observations made with calibrated chronometer, precision aneroid barometer, and aspirated psychrometer are archived.

Wind equipment must sense both speed and direction. The propeller type has integrated direction sensing; alter-

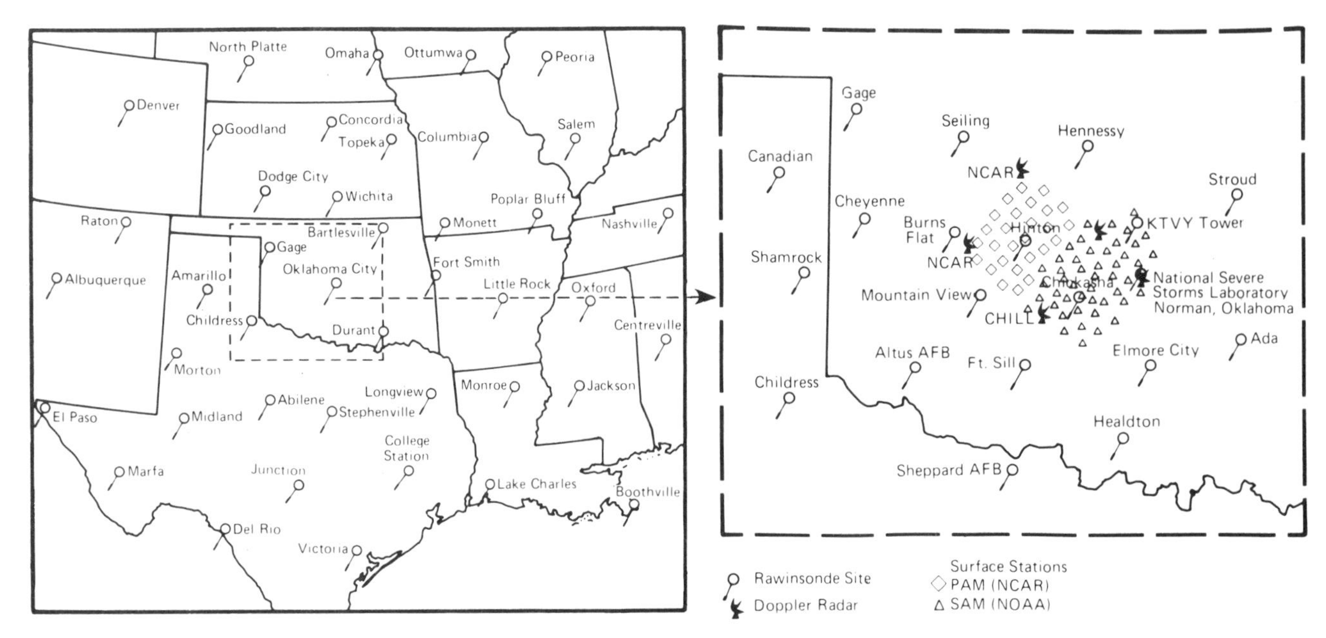

Figure 1.9. Project SESAME observational networks, 1979. Left: regional scale domain; right: storm-scale domain.

Figure 1.10. A typical site in the NSSL mesonetwork.

natively, rotating cups can sense speed, and a separate vane is used for direction sensing. Propeller devices easily adapt for digital conversion and data storage, since the speed and direction outputs are continuous. The cups can also drive a DC generator and produce an output linearly proportional to windspeed. However, early versions of direction-sensing vanes accompanying cup anemometers produced a coded output from 16 multiple-point contacts located in the sensor and connected to 8 event-pen locations in the analog recorder. This system is very dependable and has been used for many years by the Weather Bureau and several research laboratories, but it does not give sufficient resolution in wind direction. Potentiometers were added for output continuously proportional to wind direction. With a 3.6-v power source the output is conditioned easily for direct readout in degrees.

The exact placement of stations in various network configurations depends on program objectives and type of terrain. In early years exposure quality tended to be sacrificed for even spacing of stations. Now meteorologists provide a general design for network configuration, but exact selection of sites is governed mostly by the quality of exposure, considering nearness of buildings, type of vegetation, and uniformity of terrain. Problems with trees, buildings, and concrete surfaces are now dealt with only in the context of urban networks designed to gather data within a densely populated area.

After about 1973 station placement was closely related to the need to complement and verify Doppler radar data on severe thunderstorms. Also renewed research was stimulated by concerns arising from aviation accidents during thunderstorm conditions and especially during operations at busy terminals.

In 1973, NSSL completed a digital recording and telemetry system for seven levels of instrumentation on a 450-m television tower 10 km north of Oklahoma City. Various temperature, humidity, and wind sensors were tested at this facility, including devices developed by industry for envi-

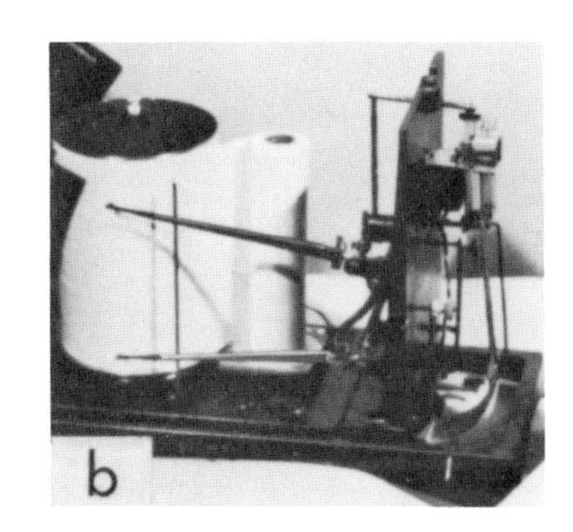

Figure 1.11. Modified (a) rain gage, (b) hygrothermograph, (c) microbarograph used in the NSSL mesonetwork from 1965 to 1978. Paper-transport mechanisms have been added to the basic instruments to facilitate continuous recording for up to a week without overwriting.

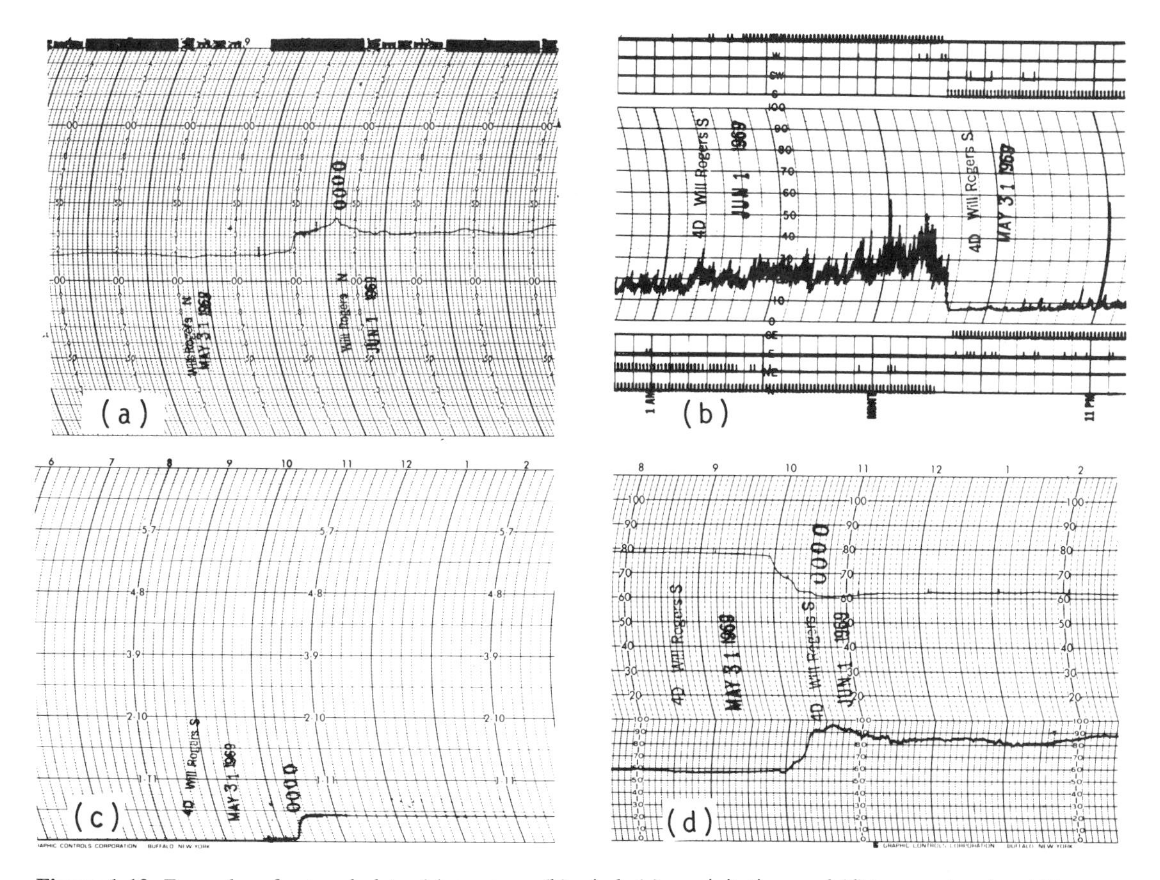

Figure 1.12. Examples of network data: (a) pressure, (b) wind, (c) precipitation, and (d) temperature-humidity.

Table 1.2. Instrumentation Used on NSSL's 450-m Meteorological Tower Facility

Parameter	Instrument
Horizontal windspeed and wind direction	Bendix aerovane model 120 (3-blade propeller)
Vertical velocity	R. M. Young, model 27100 anemometer (4-blade polystyrene propeller)
Dry-bulb temperature	Yellow Springs linearized thermistor (with aspirated radiation shield)
Wet-bulb temperature	Same as dry-bulb thermistor, but modified with moistened muslin wick attached to probe and water reservoir
Pressure	Belfort model 6080 aneroid barometer
Rainfall	Belfort model 5780 recording rain gage
Radiation	Eppley pyranometer

ronmental monitoring. Other tests included acoustical and optical radars (Danielson, 1975), sonic anemometers (Sahashi, 1972),and tetroons (Angell et al., 1973). The tower instrumentation was stabilized in 1976 with devices to measure temperature (dry bulb and wet bulb), horizontal windspeed and direction, and vertical air motion (at intervals of 1 to 10 min). Surface rainfall, pressure, and solar radiation were recorded at the base of the tower. These sensors (Table 1.2) became the reference standard for the NSSL mesonetwork, and by 1974 several surface stations in Oklahoma were similarly equipped with aspirated temperature and wetbulb sensors.

6. Modern Digital Processing and Recording Techniques

By 1975 several low-cost devices had become available commercially to provide telemetry of analog data from remote sites to a central facility. To reduce telephone costs, these systems were activated by calling the remote unit, which answered automatically, scanned channels of information, and transmitted the data by encoded frequency tone, for display by teletypewriter or analog recorder at the central facility.

The early devices did not have sufficient stability to reproduce data accurately and realiably for research needs. The experience of NCAR staff and university scientists during large field experiments between 1970 and 1975 led to development of an automatic digital system with desirable features of economy, accuracy, and reliability. Thus Brock

Figure 1.13. Digital recording station in NCAR's Portable Automated Mesonetwork (PAM), September 1980.

(1974) demonstrated at NCAR the control, processing, display, and recording of meteorological data in real time. NCAR's Portable Automated Mesonet (PAM) system was placed in operation in 1976 (Brock and Govind, 1977). Once sites were selected, the whole system was deployed in just a few days by a few persons. The system consisted of a trailer-mounted base station and a network of 40 remote stations for logging mesoscale data. Each remote station (Fig. 1.13)) sampled local sensors synchronously and transmitted data to the base station by means of a radio link. Through antennas mounted on a 50-m tower at the base station and on 12-m towers at remote stations, the communication link could extend approximately 80 km.

The PAM base station has a minicomputer for the automation of data logging, for real-time data quality checks, and for real-time data analysis and display. The data are immediately available in physical units, allowing experimenters a "quick look" at the data soon after they are recorded.

The PAM system has proved very useful for study of thunderstorms and several other phenomena, and it has provided a ground reference for testing new remote sensors. Usual station spacing ranges from 1 to 10 km, and the basic sampling rate of the PAM system is one sample per minute.

A somewhat similar automated network, but without remote call-up capabilities for continuous monitoring, was developed at NSSL and deployed for the first time (jointly with NCAR's PAM system) during SESAME in the spring of 1979 (see Fig. 1.9). It was designed to provide accurate measurements of temperature, pressure, wind, humidity, and precipitation having high temporal resolution and requiring a minimum of human intervention. The data are recorded on cassette recorders, which are collected weekly and processed for a quick look with a desk computer. The tabulated record of a squall-line passage over a station near Criner, Okla., is shown in Fig. 1.14a, with some strip chart records (Fig. 1.14b) of the same event for comparison.

It may be noted that continuous relay of data in the PAM system facilitates archiving of the data in a synoptic format, i.e., in a chronological sequence. In the NSSL system, on the other hand, the reformatting of multiple cassette tapes into series from many stations presented a formidable problem for computer operators and programmers, which was solved at NSSL about two years after initial installation of the electronic mesonetwork.

By 1981 semiautomatic and automatic archiving and retrieval of meteorological data sensed at surface stations had become a standard practice, and several systems were available commercially. A system developed under contract by the Bureau of Reclamation (then the Water Power and Resources Service), U.S. Department of the Interior, uses a communication satellite to relay data from distributed stations on the Earth's surface to a central control station and data archive.

7. Objective Analysis of Spatially Distributed Time Series Observations

a. General Description

Meteorological data are typically acquired at an irregular array of station locations. Meteorological analysis has two purposes: (1) to represent the continuous fields of variables irregularly observed, usually by mapping a set of isopleths such as isotherms or isobars, and (2) to provide a set of numerical values of the observed variables at a regular array of points for convenience in further computations.

Analysis can be done manually or by computer calculation. With a repertoire of physical and descriptive models of atmospheric processes and structures, a skillful analyst can produce maps of the distribution of variables at a given observation time that are more meaningful than present-day computer products. But the human analyst may err in application of the models (which of necessity are simplifications of the real atmosphere), may succumb to the tedium of manual analysis, and is slow to evaluate the massive data base required to depict the state of the atmosphere. For these reasons computer computation schemes have been and are being designed to aid the analyst in diagnosing atmospheric processes. Such a scheme is called "objective" analysis, although many of the decisions necessary to implement it are no less arbitrary than those required by the presumably "subjective" manual technique.

The widely used objective map-analysis method treated here has been developed from an earlier technique (Barnes, 1964), which was based on a suggestion by Sasaki (1960). The scheme uses weighted averages of observations on an arbitrary surface to determine approximate values of the observed variable on that surface at arbitrarily chosen points, usually the grid points of a square mesh. The weight assigned an observation depends on its distance from each grid point considered in turn, and this weight can be preselected to resolve the distribution of the observed variable down to the smallest details resolvable in a simple scalar analysis of a single variable, details which are limited in part by the spatial density of observations. The weighting process implies a filtering of the information contained in the observations, i.e., a smoothing of the data, which substantially alters the apparent structure of the variable's distribution. Thus data produced by the objective analysis do not necessarily agree with the observations. However, the derived fields can be made to converge to the observed values by the addition of a filtered correction field to the initial analysis, if this is deemed to be appropriate to the purpose of the analysis. The attained convergence is a function of a selectable parameter in the weight function and of the scale lengths of variations in the pattern, i.e., of the characteristic distance between extrema. One can also use the analysis method as a band-pass filter to retrieve only those features in a particular selected range of scale sizes (Doswell, 1977; Maddox, 1980).

A unique feature of this objective procedure is that observations earlier or later than the time of the analysis can be incorporated for systematic disclosure of details in the analyzed fields associated with disturbances that are traveling without dramatic change of character. Because severe thunderstorms tend to be long-lived (lasting several hours), analysis of continuous and short-interval data by time-to-space conversion of point observations has been used for many years (Fujita, 1963). In this manual analysis "off-time" observations of, for example, temperature at a given station are displaced according to a representative translation velocity of the storm system being analyzed. Each displaced temperature observation is subjectively evaluated to determine temperature distribution. Typically, some of the observations are not well accommodated, and the analyst smooths the distribution according to a conceptual model of a thunderstorm's relevant features. In such a subjective procedure the influence assigned to off-time observations is often a function of data density. When positioned among other observations, an off-time observation may be ignored or averaged with the other observations. However, when

ON	12.5000	8.7940	1.4800	5.0010	4.5520	8.0000	-8.0080	-17.3710	9.0240
OFF	10.0070	8.0050	1.6360	5.1470	4.0000	7.8570	-7.7200	-13.0930	7.8950
ON	27.52	22.89	27.90	22.90					
OFF	26.23	24.16	24.50	22.00					
ON	975.7	972.2							
OFF	974.0	968.0							

LOGREC	JUL	GMT TIIM	CORONA	C TDB	C TWB	MB PRESS	MM RG	U-M/S	V-M/S	MB PMIN	M/S SMAX	DIR	M/S SPD	RH
500	153	228	-0.0294	24.1	21.6	973.1	0.3	-0.9	2.2	973.1	3.6	158.2	2.4	79
501	153	229	-0.0294	24.1	21.6	973.7	0.3	-0.9	1.8	973.7	4.0	153.4	2.0	79
502	153	230	-0.0490	24.1	21.6	973.1	0.3	-0.9	2.2	973.1	3.6	158.2	2.4	79
503	153	231	-0.0490	24.3	21.6	973.7	0.3	-0.9	1.3	973.7	3.1	146.3	1.6	78
504	153	232	-0.0490	24.3	21.6	973.7	0.3	-0.4	1.3	973.7	2.7	161.6	1.4	78
505	153	233	-0.0686	24.3	21.6	973.1	0.3	-0.4	0.4	973.1	1.8	135.0	0.6	78
506	153	234	-0.0882	24.3	21.6	973.7	0.3	0.0	0.4	973.7	1.3	180.0	0.4	78
507	153	235	-0.0882	24.1	21.6	973.1	0.3	0.0	0.4	973.1	1.8	180.0	0.4	79
508	153	236	-0.0882	24.3	21.6	973.7	0.3	0.0	0.4	973.7	1.8	180.0	0.4	78
509	153	237	-0.1078	24.3	21.6	973.7	0.3	0.0	0.0	973.7	0.4	0.0	0.0	78
510	153	238	-0.1078	24.3	21.6	973.7	0.3	0.0	0.0	973.7	0.9	0.0	0.0	78
511	153	239	-0.0882	24.3	21.6	973.7	0.3	0.4	-0.4	973.7	1.3	315.0	0.6	78
512	153	240	-0.1078	24.1	21.6	973.7	0.3	0.4	-1.3	973.7	2.7	341.6	1.4	79
513	153	241	-0.1275	24.1	21.6	973.7	0.3	0.4	-2.2	973.7	3.6	348.7	2.3	79
514	153	242	-0.1078	24.1	21.8	973.7	0.3	0.4	-3.1	973.7	4.0	351.9	3.2	81
515	153	243	-0.0490	23.9	21.8	973.7	0.3	0.9	-3.6	973.7	5.4	346.0	3.7	82
516	153	244	0.0098	23.9	21.8	973.7	0.3	0.9	-2.7	973.7	4.0	341.6	2.8	82
517	153	245	-0.0294	23.7	21.4	973.7	0.3	0.4	-4.0	973.7	6.7	353.7	4.0	81
518	153	246	-0.0294	23.1	20.4	973.7	0.3	0.9	-4.0	973.7	6.3	347.5	4.1	77
519	153	247	-0.0686	22.7	19.8	973.7	0.3	0.9	-4.0	973.7	5.8	347.5	4.1	76
520	153	248	-0.0882	22.5	19.6	973.7	0.3	0.0	-2.7	973.7	4.0	360.0	2.7	76
521	153	249	-0.0686	22.4	19.6	973.7	0.3	0.0	-1.8	973.7	3.1	360.0	1.8	77
522	153	250	-0.0490	22.4	19.4	973.7	0.3	0.0	-1.3	973.7	3.1	360.0	1.3	76
523	153	251	-0.0098	22.4	19.4	974.3	0.3	0.4	-1.3	974.3	3.1	341.6	1.4	76
524	153	252	0.0882	22.2	19.4	973.7	0.3	0.0	-1.3	973.7	3.6	360.0	1.3	77
525	153	253	0.1471	22.2	19.4	974.3	0.3	0.0	-1.3	974.3	2.2	360.0	1.3	77
526	153	254	0.2059	22.4	19.4	974.3	0.3	0.0	-1.3	974.3	2.7	360.0	1.3	76
527	153	255	0.2451	22.2	19.6	974.3	0.3	0.0	-1.3	974.3	2.7	360.0	1.3	79
528	153	256	0.3039	22.2	19.4	974.3	0.3	0.4	-2.2	974.3	4.0	348.7	2.3	77
529	153	257	0.3039	22.2	19.6	974.3	0.3	0.0	-2.2	974.3	4.9	360.0	2.2	79
530	153	258	0.3039	22.4	19.6	974.3	0.3	0.0	-2.7	974.3	4.5	360.0	2.7	77
531	153	259	0.2059	22.4	19.8	974.3	0.3	0.4	-1.8	974.3	3.1	346.0	1.8	79
532	153	300	0.0294	22.4	19.8	974.3	0.3	0.0	0.0	974.3	2.2	0.0	0.0	79
533	153	301	-0.0294	22.5	20.0	974.3	0.3	0.0	0.0	974.3	1.3	0.0	0.0	79
534	153	302	-0.0294	22.4	20.0	975.0	0.3	3.1	1.3	975.0	7.2	246.8	3.4	80
535	153	303	-0.0294	22.0	19.6	975.0	0.3	7.6	0.9	975.0	13.4	263.3	7.7	80
536	153	304	-0.0098	21.6	19.2	975.0	0.3	-3.1	-0.9	975.0	6.3	285.9	-3.3	80
537	153	305	-0.0490	21.4	19.0	975.0	0.3	10.7	-2.2	974.3	20.1	281.8	11.0	80
538	153	306	-0.0294	21.8	19.4	975.0	0.3	10.7	-7.6	975.0	17.4	305.3	13.2	80
539	153	307	-0.0490	21.4	19.4	975.6	0.3	8.9	-8.1	975.6	15.2	312.0	12.0	83
540	153	308	-0.0294	20.8	18.8	975.6	0.3	7.2	-7.2	975.6	15.7	315.0	10.1	83
541	153	309	-0.0490	19.8	18.4	975.6	1.3	5.4	-6.7	975.6	14.3	321.3	8.6	87
542	153	310	-0.0294	19.2	18.0	975.6	1.3	4.9	-6.7	975.6	11.2	323.7	8.3	89
543	153	311	0.0098	19.0	18.0	975.6	1.3	4.9	-6.3	975.6	11.2	321.8	8.0	91
544	153	312	0.0294	18.8	17.8	975.6	1.3	5.4	-6.7	975.6	11.2	321.3	8.6	91
545	153	313	-0.0294	18.8	17.6	975.0	1.3	6.7	-8.1	975.0	13.9	320.2	10.5	89
546	153	314	-0.0294	18.6	17.6	975.0	2.3	6.3	-5.4	975.0	11.2	310.6	8.2	90
547	153	315	0.0098	18.6	17.6	975.0	2.3	5.4	-5.8	975.0	15.2	317.3	7.9	90
548	153	316	-0.0882	18.4	17.6	975.0	2.3	4.5	-5.4	975.0	10.3	320.2	7.0	92
549	153	317	-0.0098	18.4	17.6	975.0	2.3	3.6	-5.4	975.0	9.4	326.3	6.5	92
550	153	318	0.0686	18.4	17.6	975.0	3.3	2.2	-3.6	975.0	7.2	328.0	4.2	92
551	153	319	0.0490	18.2	17.5	975.0	3.3	4.5	-7.2	975.0	13.9	328.0	8.4	92
552	153	320	0.0490	18.0	17.3	974.3	3.3	5.4	-6.3	974.3	12.1	319.4	8.2	92
553	153	321	0.0686	18.0	17.3	975.0	4.3	3.6	-5.4	975.0	11.6	326.3	6.5	92
554	153	322	0.0686	17.8	17.5	975.0	4.3	4.0	-5.8	975.0	10.3	325.3	7.1	96
555	153	323	0.0098	17.8	17.3	975.0	5.3	4.0	-6.3	975.0	10.3	327.3	7.4	94
556	153	324	-0.0686	17.8	17.3	975.0	5.3	4.5	-4.5	975.0	8.1	315.0	6.3	94
557	153	325	-0.0686	17.8	17.3	975.0	5.3	3.6	-3.6	975.0	7.2	315.0	5.1	94
558	153	326	-0.0490	17.8	17.3	975.0	6.4	4.9	-4.0	975.0	8.9	309.3	6.4	94
559	153	327	-0.0490	17.8	17.5	975.0	6.4	4.9	-3.1	975.0	9.4	302.5	5.8	96
560	153	328	-0.0686	17.8	17.5	975.0	7.4	6.7	-2.7	975.0	10.3	291.8	7.2	96
561	153	329	-0.0294	17.8	17.3	975.0	8.4	6.7	-2.7	975.0	9.8	291.8	7.2	94
562	153	330	-0.0686	17.8	17.3	975.0	8.4	5.4	-1.8	975.0	8.5	288.4	5.7	94
563	153	331	-0.0490	17.8	17.3	975.6	9.4	5.4	-0.9	975.6	7.6	279.5	5.4	94
564	153	332	0.0098	17.6	17.3	975.6	10.4	5.4	-1.3	975.6	7.6	284.0	5.5	96
565	153	333	-0.0490	17.8	17.5	976.2	11.4	5.4	-0.4	976.2	7.6	274.8	5.4	96
566	153	334	-0.0294	17.8	17.5	976.2	12.5	6.3	0.0	976.2	8.1	270.0	6.3	96
567	153	335	-0.0490	17.8	17.5	976.2	13.5	4.9	-0.4	976.2	7.6	275.2	4.9	96
568	153	336	-0.0294	17.8	17.5	976.2	14.5	4.0	-0.4	976.2	5.8	276.3	4.0	96
569	153	337	-0.0098	17.8	17.5	976.8	14.5	3.6	0.0	976.2	4.9	270.0	3.6	96
570	153	338	-0.0686	17.8	17.5	976.2	15.5	4.0	0.4	976.2	5.8	263.7	4.0	96
571	153	339	-0.0490	17.8	17.5	976.2	16.5	3.6	0.0	976.2	5.4	270.0	3.6	96
572	153	340	-0.0098	17.6	17.5	975.6	16.5	3.1	-0.4	975.6	5.4	278.1	3.2	98
573	153	341	-0.0098	17.6	17.5	975.6	17.5	3.6	-0.9	975.6	4.5	284.0	3.7	98
574	153	342	-0.0294	17.6	17.5	975.6	18.6	3.1	-0.9	975.6	5.4	285.9	3.3	98
575	153	343	0.0294	17.6	17.5	975.0	18.6	3.6	-1.8	975.0	5.8	296.6	4.0	98
576	153	344	-0.0882	17.6	17.5	975.0	18.6	4.5	-3.1	975.0	6.7	305.0	5.5	98
577	153	345	-0.1078	17.6	17.5	975.0	18.6	3.6	-3.6	975.0	6.7	315.0	5.1	98

Figure 1.14a. 0300 GMT, 2 June 1981 digital record for a small portion of the period shown in Fig. 1.14b. Successive columns indicate record number, Julian date, Greenwich time in hours and minutes, an indication of corona current, dry- and wet-bulb temperatures, pressure in millibars, accumulated rainfall in millimeters, minute averages of u and v wind components, minimum pressure during the preceding minute, maximum of 60 wind measurements during the preceding minute, average wind direction, average speed, and relative humidity deduced from dry and wet bulb.

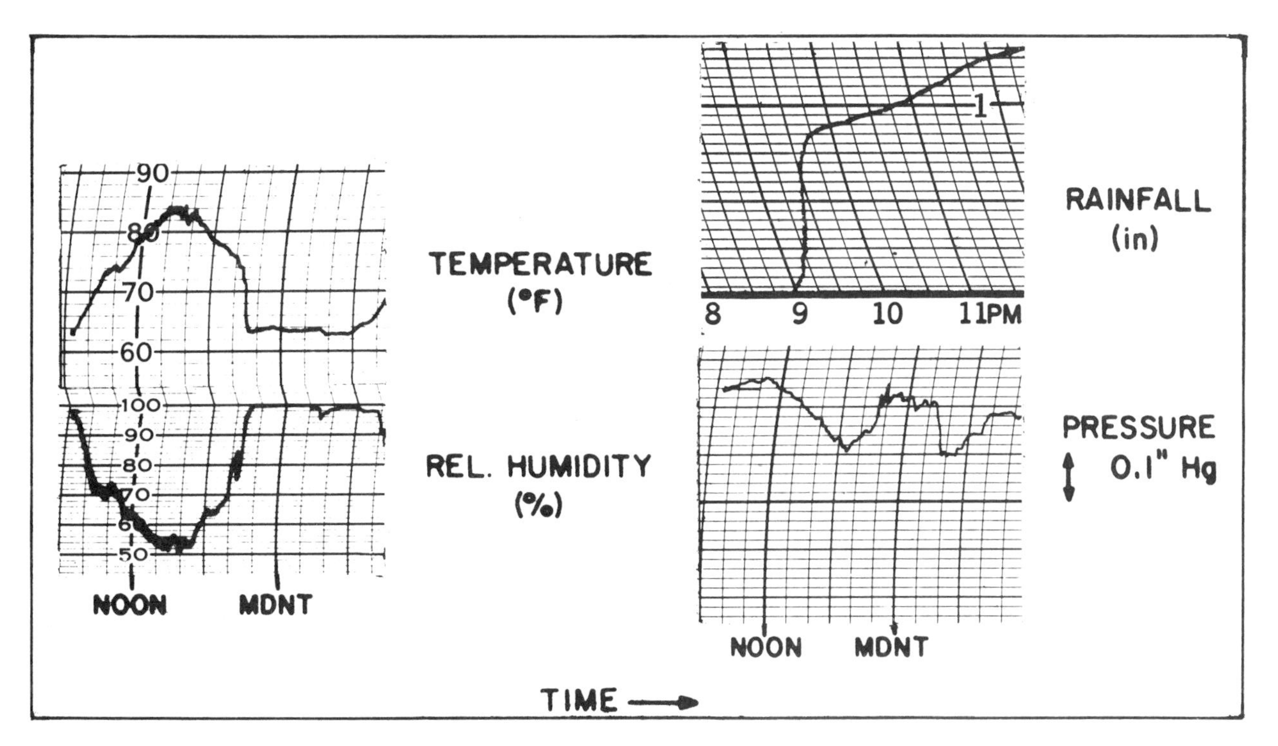

Figure 1.14b. Strip-chart records showing some parameters of a squall-line event near Criner, Okla., at about 2100 CST, 1 June 1981.

spatially far removed from other data, the off-time observation is frequently assigned a high influence simply because it does not appear unreasonable and there are no other data to consider.

In the objective technique presented here, time series observations are treated in much the same fashion as in manual analysis. First, observations are placed relative to a moving disturbance; then weight is computed for each observation according to its distance in space from the grid point being considered and in time from the nominal time of the analysis. If several observations are equidistant from the grid point, those closer to map time have greater influence.

Stations in NSSL's rawinsonde networks (Barnes et al., 1971) have been relatively closely spaced. Even so, rather large data gaps in time and space exist around individual storms. To complicate the matter, balloons are released from individual stations at different times, balloons ascend at different rates and thus reach given altitudes (or pressure surfaces) at different times, and local wind variations cause the balloons to drift from their release points along relatively different paths, thus changing both the absolute location of observations and the relative spacing between them. Analysis techniques should accommodate these effects.

Below we first indicate filtering properties implicit in the analysis scheme and then illustrate practical application, including the manner in which off-time observations are incorporated. We also illustrate use of the method with both dense surface observations and sparse upper-air data.

b. Mathematical Properties

1. Filtering Properties of the Interpolation Scheme: Consider an atmospheric variable distributed as $f(x,y)$. We can transform (filter, assign weight to) this function according to distance from an arbitrary point (x, y):

$$g(x,y) \equiv \int_0^{2\pi} \int_0^{\infty} f(x + r\cos\theta,\, y + r\sin\theta)\, w(r,k)\, r\, dr\, d\theta, \qquad (1.1)$$

where r is the distance between the grid point and the observation point (in Fig. 1.15), and the normalized weight function is Gaussian:

$$w(r,k) = \frac{1}{4\pi k}[exp(-r^2/4k)]. \qquad (1.2)$$

The parameter k determines the relative weights associated with the value of the function at various distances from (x,y); i.e., it determines w's filtering properties. It can be shown that the relationship between the value f and weighted value g at point (x,y) is of the form

$$g(x,y) = D(a,k)\, f(x,y), \qquad (1.3)$$

where $D(a,k)$ is a response function dependent on the characteristic wavelengths or scale sizes contained in f, a is a wave number $(2\pi/\lambda)$, and λ is wavelength. Holloway (1958) shows, for example, that for $f(x,y) = A\sin(ax)$,

$$D(a,k) = \exp(-a^2k). \qquad (1.4)$$

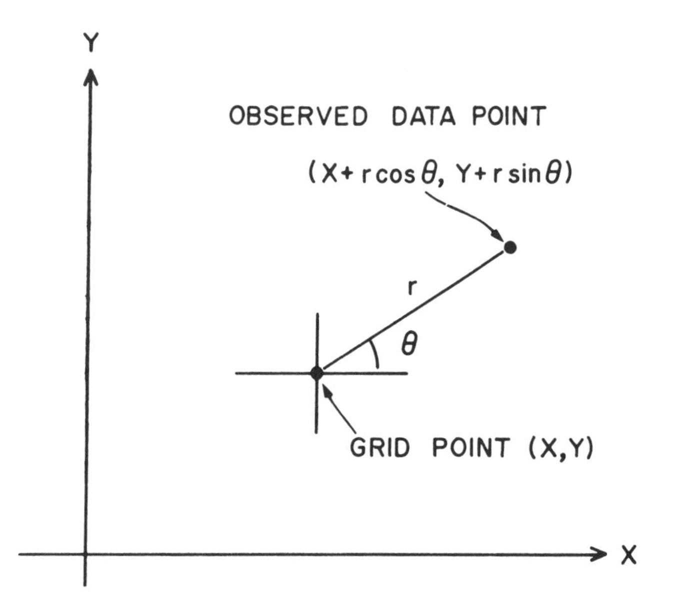

Figure 1.15. Coordinate system used in objective analysis expressed by Eq. 1.1. (x,y) is conveniently chosen as a grid point of a square mesh; $(x + r \cos \theta, y + r \sin \theta)$ represents one point where information is observed.

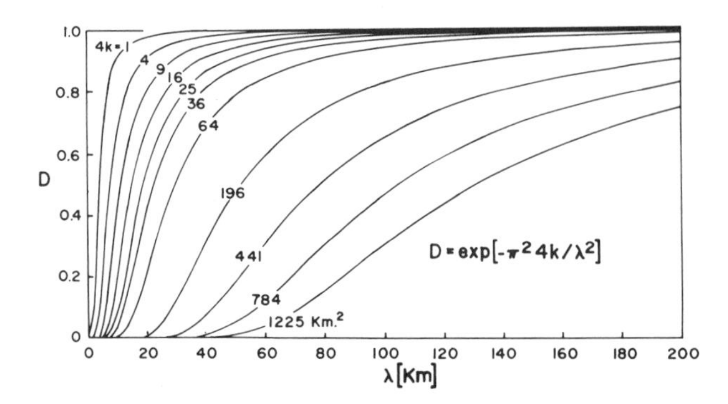

Figure 1.16. Relationship of response function D (Eq. 1.4) to wavelength λ for various choices of parameter k. These responses correspond to one pass through the data with filter (Eq. 1.2).

Note that the effect of the filtering does not alter the phase of f but acts only to damp its amplitude. Figure 1.16 indicates the effect of $D(a,k)$ as a function of wavelength λ and for various choices of parameter k. In general, response is nearly zero for very short wavelengths and approaches unity for very long wavelengths. For band-pass filtering such as that done by Maddox (1980), conceptually, two analyses of the same data set are made using differing values for k, and then one of the resultant analyses is subtracted from the other. Since both analyses retain long waves with nearly equal amplitudes and neither retains short waves, the residual of the manipulation is only those features whose wavelengths lie in a midscale range.

In practice, of course, one does not know the continuous function $f(x,y)$ but has only observations of it at an irregular array of points. In a practical application of the scheme the integrals are replaced by summations of observed values multiplied by the weighting function, as shown in Eq. 1.14. The resulting grid-point values represent an incomplete specification of the natural distribution with systematic alteration of the harmonic content implied by the observations.

2. Retrieving Amplitude Lost Through Filtering: After filtering, it is sometimes desirable to match the interpolated function g to the original observations within arbitrarily small differences. An adjustment field can be obtained through application of the filter w to residual differences between f and g, where the value of g at observations points is computed by linear interpolation from the grid-point values to the points at which f is observed. Parameter k can be either kept constant (as in Barnes, 1964) or reduced from its initially chosen value. In the former case iterations of the process described above are required to reach the desired convergence between g and f. In the latter case the desired convergence can be attained in only one application of the process with a judicious choice of k which can be determined from the following considerations.

We rewrite Eq. 1.3 as

$$g_0(x,y) = D_0 f(x,y). \tag{1.5}$$

Subscript zero denotes the first-pass result with weight function $\eta_0 = \exp(-r^2/4k_0)$. (The normalizing coefficient in Eq. 1.2 drops out in the actual computations since it appears in both numerator and denominator; see Eq. 1.14.) The second pass yields smoothed values of the residual differences between f and g, which are added to the first-pass field:

$$g_1(x,y) = g_0(x,y) + [f(x,y) - g_0(x,y)]D_1, \tag{1.6}$$

where D_1 is the response resulting from the application of

$$\eta_1 = \exp(-r^2/4k_1);\ k_1 = \gamma k_0 \text{ for } 0 < \gamma < 1. \tag{1.7}$$

Thus

$$D_1 = \exp(-a^2k_1) = \exp(-a^2\gamma k_0) = D^\gamma{}_0, \tag{1.8}$$

and we have, from Eqs. 1.5 and 1.6,

$$g_1(x,y) = f(x,y)\, D_0\, [1 + D_0^{\gamma-1} - D_0^\gamma]. \tag{1.9}$$

The resulting response function,

$$D = D_0(1 + D_0^{\gamma-1} - D_0^\gamma), \tag{1.10}$$

is now also a function of γ. Figure 1.17 shows the relationship of D to D_0 and γ. The improved amplitude response is demonstrated by the following. Suppose, for a given wavelength feature in the observed field, the response to the initial filtering pass yields only 0.3 of the true amplitude (the actual response depends not only on our choice of k_0 but

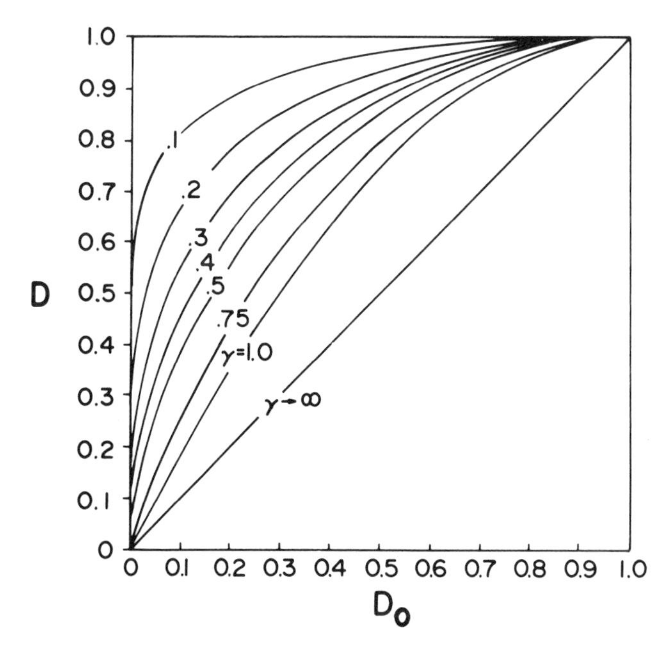

Figure 1.17. Response D after one correction pass as a function of initial response D_0 and parameter γ (see Eq. 1.10).

also on the distribution of observations in the vicinity of this feature). Then for $\gamma = 0.2$, say, the response after the second pass (first correction pass, e.g., Eq. 1.6) would be 0.85. Theoretically one could choose γ very small and achieve nearly perfect amplitude response to such a feature, but experimental results (Barnes, 1973) indicate that γ should not be chosen less than about 0.2.

3. Time Series Data: Generally, weather disturbances move from one place to another, while their overall structure varies somewhat regularly. When only a few simultaneous observations are available for reconstructing the spatial distribution of a given meteorological variable, it becomes necessary to use the atmosphere's persistence to enhance the available information. However, the manner in which temporal information is entered into analysis is critical. If observations are averaged in time before being analyzed spatially, significant features may be strongly damped. On the other hand, time-to-space conversion yields results with enhanced details, a consequence of an effective increase in observation density.

Consider now the functional representation of a moving disturbance:

$$g(x,y,t) = \int_{-\infty}^{\infty} \int_{0}^{2\pi} \int_{0}^{\infty} f[x + r\cos\theta - c(t + t'),\ (y + r\sin\theta)] \times w(r,t',k,\nu)\, r\, dr\, d\theta\, dt', \quad (1.11)$$

where x increases in the direction of motion, c is its phase speed, t is an arbitrary reference time, and t' is time difference from reference time, either positive or negative. The weight function used to interpolate and filter discrete observations on this distribution must reflect the characteristic time scale of spatial development, which depends on the nature of the phenomenon and its size. With this in mind a suitable normalized weight function can be defined as

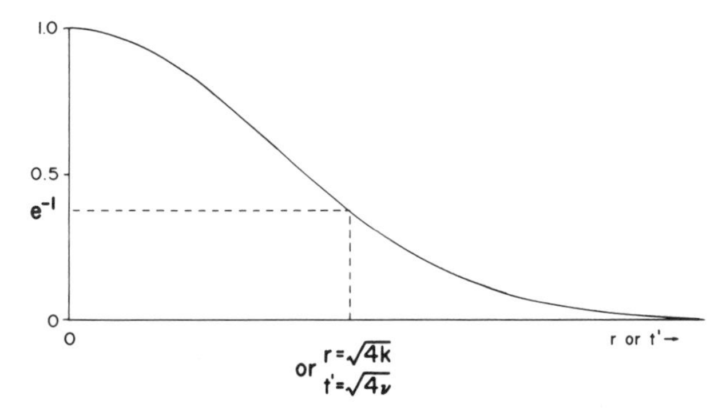

Figure 1.18. Exponential weight function for space and time analysis (Eq. 1.12 with the coefficient suppressed). Abscissa is either the radial distance from observation to grid point (see Fig. 1.15) or the time difference from reference map time.

$$\begin{aligned} w(r,t',k,\nu) &= w(r,k)\, w(t',\nu) \\ &= \frac{1}{4\pi k} \exp(-r^2/4k) \frac{1}{2\pi^{\frac{1}{2}}\nu^{\frac{1}{2}}} \\ &\quad \exp(-t'^2/4\nu) \\ &= \frac{1}{8\pi^{\frac{3}{2}}k\nu^{\frac{1}{2}}} \exp(-r^2/4k - t'^2/4\nu), \quad (1.12) \end{aligned}$$

where ν is to be related to the disturbance's characteristic decay time, as k (in Eq. 1.2) was related to its spatial scale. Parameter ν has this significance: large ν is chosen when slowly changing disturbances are to be analyzed (all data apply with nearly equal weight, regardless of when observed); small ν is properly assigned when the phenomenon is changing rapidly relative to the time interval between observations. Figure 1.18 indicates the weight function shape as it relates to k and ν.

In an analysis scheme involving observations at discrete points in space and time, the implications of t' can be explained as follows. First, t' determines the position of each observation relative to the moving disturbance, such that $r' = ct'$, where c is an appropriate translation velocity, and r' is the position vector from the station to the displaced observation (Fig. 1.19). Thus r' and r'', the position vector from a grid point to the station, add to give r for the off-time observation that will determine the weight it carries because of its spatial position (compare Figs. 1.19 and 1.15). Second, t' determines the weight (as Fig. 1.18) assigned to the observation according to its age relative to the chosen map time.

The analytical form of the response function of this time-

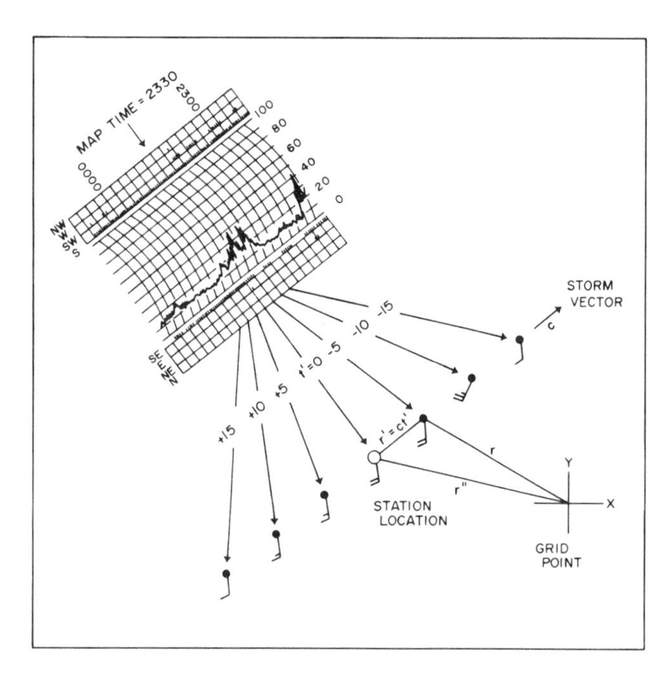

Figure 1.19. Schematic of time-to-space analysis technique illustrated for surface wind observations at 5-min intervals. Station has fixed location, r'', relative to the example grid point at (x,y). Map time observation is positioned at the station and influences interpolated value at grid point in proportion to r''^2. The time series observations are displaced along the storm's translation vector, c, by an amount $r' = ct'$ and influence grid point value in proportion to r^2 and t'^2.

space analysis scheme cannot be derived as was Eq. 1.14. Even for a simple translating wave such as

$$f(x,y,t) = A \sin [a(x - ct')], \qquad (1.13)$$

if either A *or* c varies with time or space, Eq. 1.11 becomes mathematically intractable. If both A and c are constant, however, it can be shown that the analysis yields an undampened response with proper phase for waves whose length is sufficiently long (roughly four times the average station separation) that they can be resolved by the basic station array. Smaller-scale features are depicted less accurately, but empirical results indicate that the analysis scheme retrieves details as small as a theoretical limit of twice the station separation. Features smaller than this are always aliased into longer wavelength features; i.e., the irregularities manifested by the short waves are represented at a longer wavelength by the analysis. These results relate to pure data, i.e., error-free data representing a simple wave of the form of Eq. 1.13, but they indicate the power of this analysis tool.

Of course, real atmospheric disturbances do change in amplitude and phase speed. The analysis scheme automatically incorporates amplitude changes from one reference time to another since, as reference time changes, different observations "march" through the grid. Small changes in phase velocity from one map to another, or across a given map, do not influence the results very much, since they appear as small displacement "errors" for the off-time observations. These errors become fictitious, very short wavelength perturbations that are not passed through the filtering process. If phase velocity changes significantly (a judgment), there is no reason not to incorporate this change into the analyses as a function of reference time. More complicated is a case in which different parts of the system move with different velocities. Although this too can be incorporated into the scheme, assigning values of c as a function of space, it requires a careful preliminary consideration of just what it is the system is doing and how one might best incorporate the information to yield a useful analysis. In the examples to follow, only temporal change in phase velocity has been incorporated. Spatial variation in c is not considered.

c. Examples of Analysis Based on Real Observations

1. Surface Observations: The objective time-to-space analysis procedure described above was applied for the first time to data obtained by the National Severe Storms Laboratory from severe-thunderstorm events in central Oklahoma during 29–30 April 1970 (Barnes, 1978a, b; Barnes and Nelson, 1978). In 1970, NSSL operated a 44-station network of surface recording stations, including 9 upper-air sounding sites. Figure 1.20 shows the locations of these stations relative to NSSL and the 3.175-km mesh over which the surface data were analyzed. The increased density of observations owing to the time-to-space technique is obvious in Fig. 1.21, in which 7 wind observations from each of the 44 stations are plotted in the manner of Fig. 1.19. The analysis of these data resulted in the streamline and isotach pattern presented as Fig. 1.22. Note that the smallest distance between relative extrema (half wavelength) in the isotach pattern (dashed) is about the same as the station separation, which indicates again the technique's capability to resolve wavelengths near the theoretical limit of twice the station separation. A supercell thunderstorm was traversing the northwestern corner of the network at a speed of 21.4 m s^{-1} from 233°. The sink in the streamline pattern marks the location of the updraft's roots in the surface layer of air and also the mesocyclone center. The storm's cold air outflow boundary lies along the confluence zone south and southwest of the sink.

2. Upper-Air Observations: Because of the normally large distances between National Weather Service upper-air stations and a principal concern with large-scale atmospheric systems, routine synoptic analyses of radiosonde data treat the data as though simultaneously observed vertically above the station. As mentioned above, however, it is imperative that the actual positions of the observations in time and space be considered for depicting features on the scale of an individual thunderstorm or thunderstorm sys-

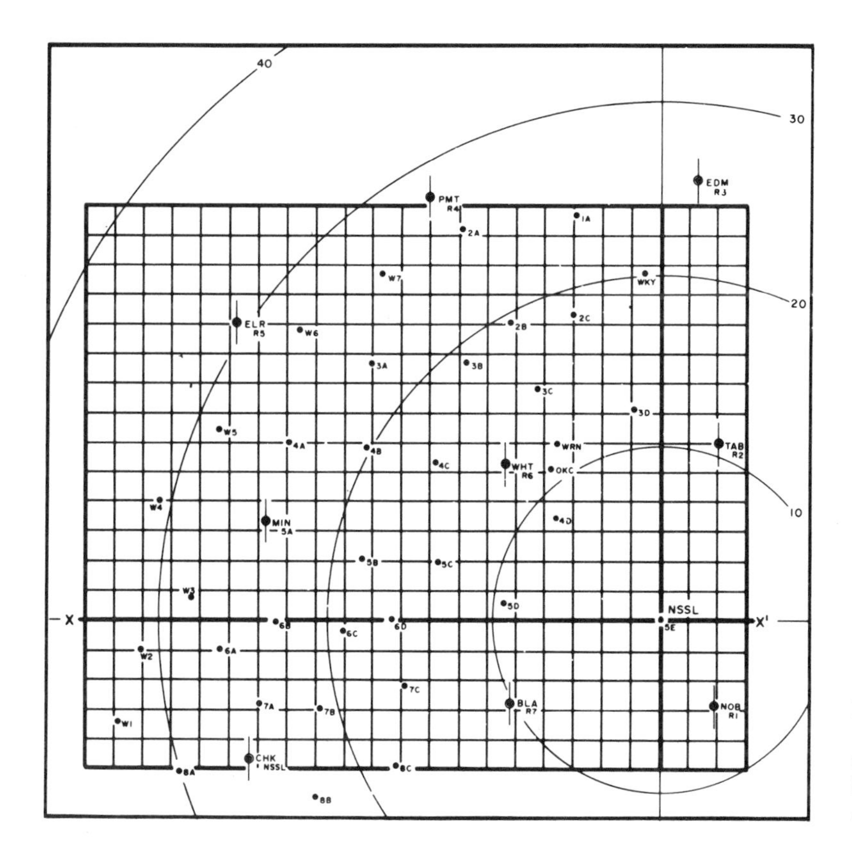

Figure 1.20. The 44-station NSSL mesonetwork for 1970. Dots are surface station locations; circles denote rawinsonde sites. Grid has mesh size of 3.175 km. Average distance between surface stations is about 10 km.

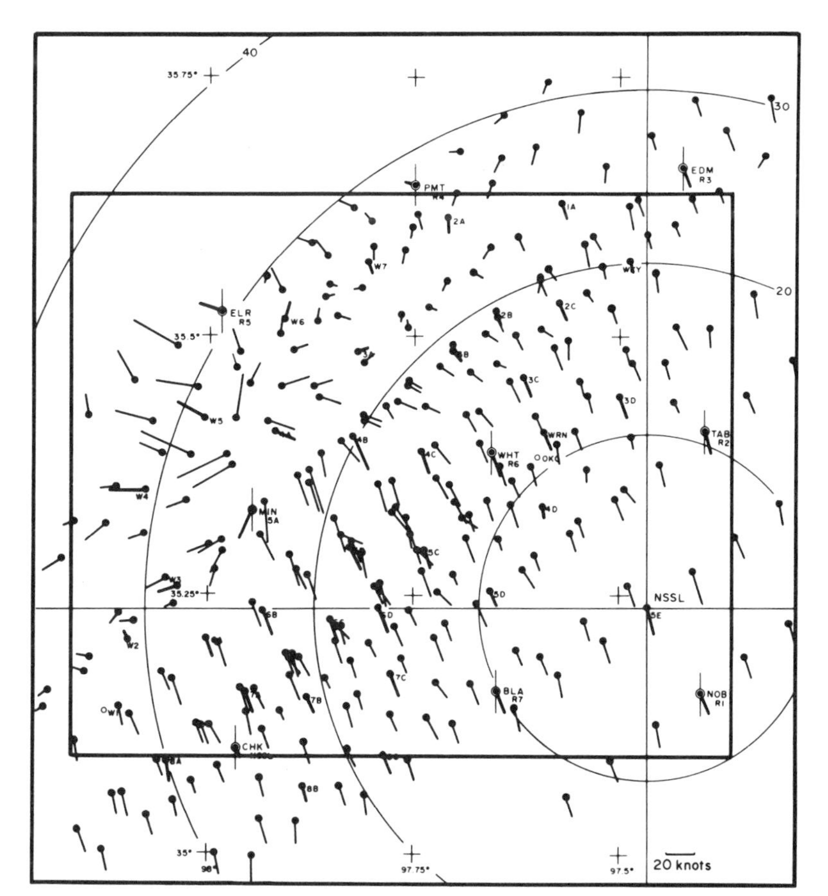

Figure 1.21. Distribution of surface wind observations averaged over 5-min intervals for ± 15 min of map time 0030 CST, 30 April 1979. Vectors project in the direction from which the wind is blowing. Off-time observations are plotted as in Fig. 1.19. A large, severe thunderstorm was present at this time in the northwest corner of the network, moving at 21 m s^{-1} from 233°.

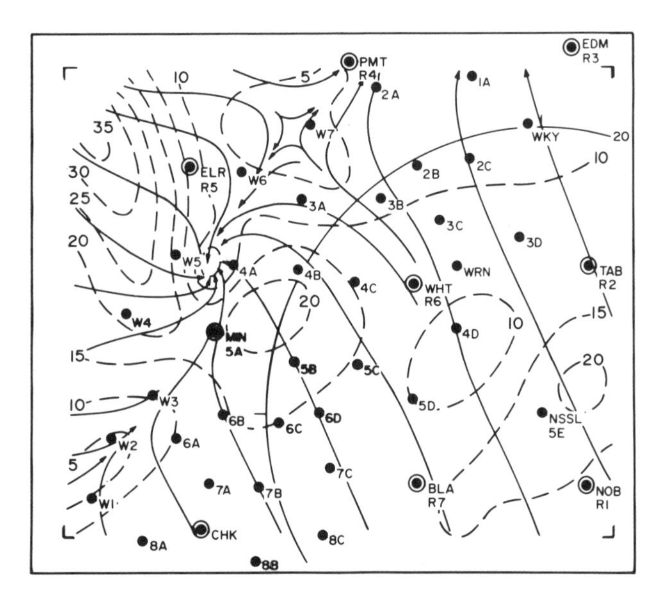

Figure 1.22. Surface wind analysis for 0030 CST, 30 April 1970, based on observations shown in Fig. 1.21. Streamlines are solid curved arrows; isotachs are dashed and labeled in knots. Parameters in Eq. 1.14 were chosen to be $k = 16$ km^2, $\nu = 32$ min^2; $\gamma = 0.5$ (Eq. 1.7).

tem. Fankhauser (1969) successfully developed a partly objective technique for accomplishing this in an analysis of a squall line. Data were first plotted on time-height sections for each station and then linearly interpolated to common times. The results were then transferred to actual balloon positions on maps for manual analysis of the spatial distributions.

As with the surface-data analysis described above, the upper-air analysis technique places observations in space relative to a moving storm, taking into account not only the actual time of the observation but also the balloon drift. Then observations are interpolated to grid points by applying the normalized, discrete formula:

$$g(r,\theta,t) = \frac{\Sigma\eta(r,t',k,\nu)\, f(r,\theta,t+t')}{\Sigma\eta(r,t',k,\nu)}, \qquad (1.14)$$

where $\eta(r,t',k,\nu) = \exp(-r^2/4k - t'^2/4\nu)$, summed over all observations, or at least over those that lie within a suitably large radius of the grid point. Such a radius leaves the influence of data beyond that radius as a small fraction of the total, e.g., less than 1%.

The upper-air analyses for the 29–30 April 1970 case study are displayed over a 25 × 17 grid with mesh size of 6.35 km. The center of the grid is placed at Wheatland, Okla. (WHT in Fig. 1.23), which is the central rawinsonde site of the nine-station network (circles). The long axis of the grid is placed parallel to the direction of storm travel to encompass off-time observations, which have been displaced along that direction. One may visualize the observational domain as a rectangular strip extending in a southwest–northeast direction and covering the spatial equivalent of the 8-h period during which soundings were obtained. Of course, not all of that information is pertinent to a particular storm. Therefore, a cutoff time (distance) was incorporated in the analysis, and the remaining observations were given weight according to their age.

The illustrated pattern (Fig. 1.23) is the distribution of wet-bulb potential temperature θ_w at 1,500 m above sea level, associated with a complex of two large thunderstorms whose radar echo outline at 2240 CST is indicated by the heavy solid line. These storms are moving from 227° at 24.5 m s^{-1}. The large cross locates the NSSL radar and also provides north orientation to the figure (the radar's storm echo is not representable near NSSL owing to ground clutter). The artificial shading between dashed isopleths of θ_w highlights the analyzed field at 1°C intervals. Observations are plotted at balloon positions (solid dots) relative to the major updrafts whose locations (based on independent data not presented here) are in the vicinity of WHT and at the southwest echo tip. Winds relative to Earth are indicated by the convention that one full barb equals 10 m s^{-1}. Interpolated values are printed to tenths of degrees Celsius, and the bottom of each rightmost digit locates a grid point.

Although the analyzed θ_w pattern within the radar echo boundary cannot be taken literally because observations therein are influenced by convective motions (i.e., they induce some degree of temporal and spatial aliasing), the general distribution of θ_w is reasonable. Two tongues of moist, warm air ($\theta_w > 22$°C) feed the updraft portions of the storms, while drier, cooler air is found north and west of the complex. The 15°C minimum for θ_w noted here at 1,500 m is indicative of downdraft air descended from above, and it is consistent with a 16°C minimum analyzed at the surface just west of WHT.

The importance of the off-time observations in providing the details of this distribution is shown in Fig. 1.24. Here the influences of observations differing by more than ±15 min from 2240 CST are diminished to less than $1/e$ (see Fig. 1.18). The information provided by the 15°C observation (at 2145 CST) is lost, and the high θ_w air associated with the updraft near WHT appears to extend well beyond the storm's northern boundary. Also in Fig. 1.24 are two notable examples of analysis response to closely spaced observations obtained at different times. The values 20.3°C and 18.0°C near ELR were observed 4 and 17 min from map time. Near TAB the values 22.9°C and 20.7°C were observed 5 and 26 min from map time. Obviously the analyzed values near those two stations are largely dominated by the observations closer to map time.

d. Concluding Remarks on Objective Analysis

There are many objective schemes for deciding what weight to give observations and what scales to represent in an analysis. The statistical properties of the observations themselves can be used for this purpose (Buell, 1960; Gan-

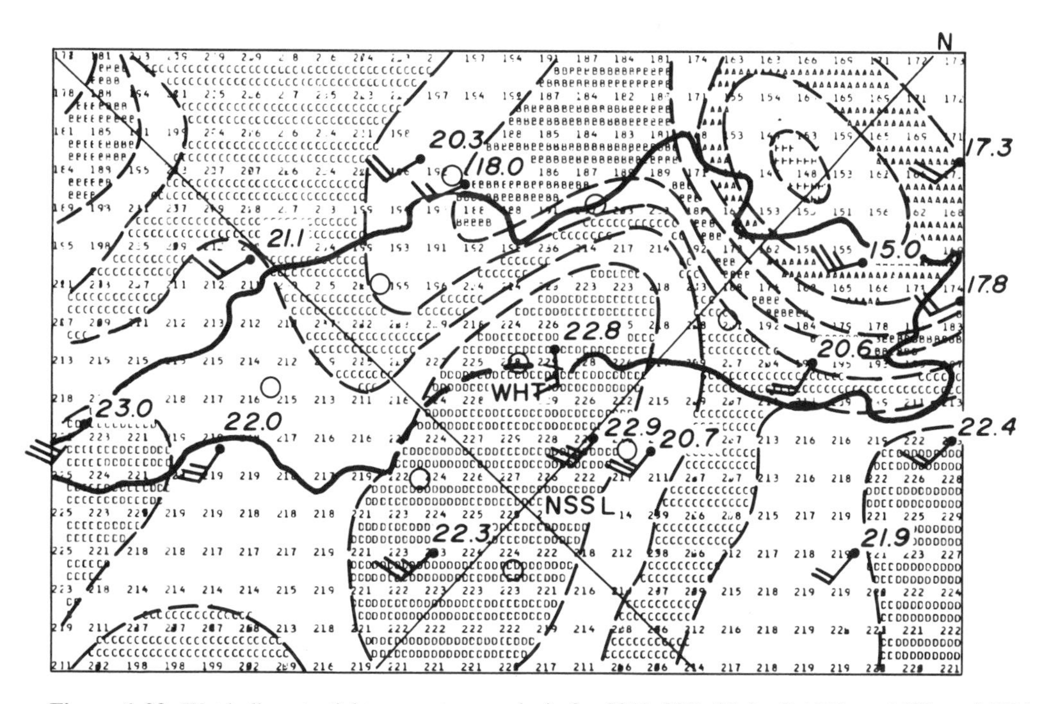

Figure 1.23. Wet-bulb potential-temperature analysis for 2240 CST, 29 April 1970, at 1500 m (MSL) with $k = 110.25$ km², $\nu = 900$ min², $\gamma = 0.34$. Interpolated values are in tenths of degrees Celsius; dashed lines are isotherms of θ_w at 1°C intervals. Open circles are rawinsonde stations identified in Fig. 1.20; dots show positions of balloons relative to the storm complex whose radar echo outline is indicated by the heavy line. One full barb represents a windspeed of 10 m s⁻¹.

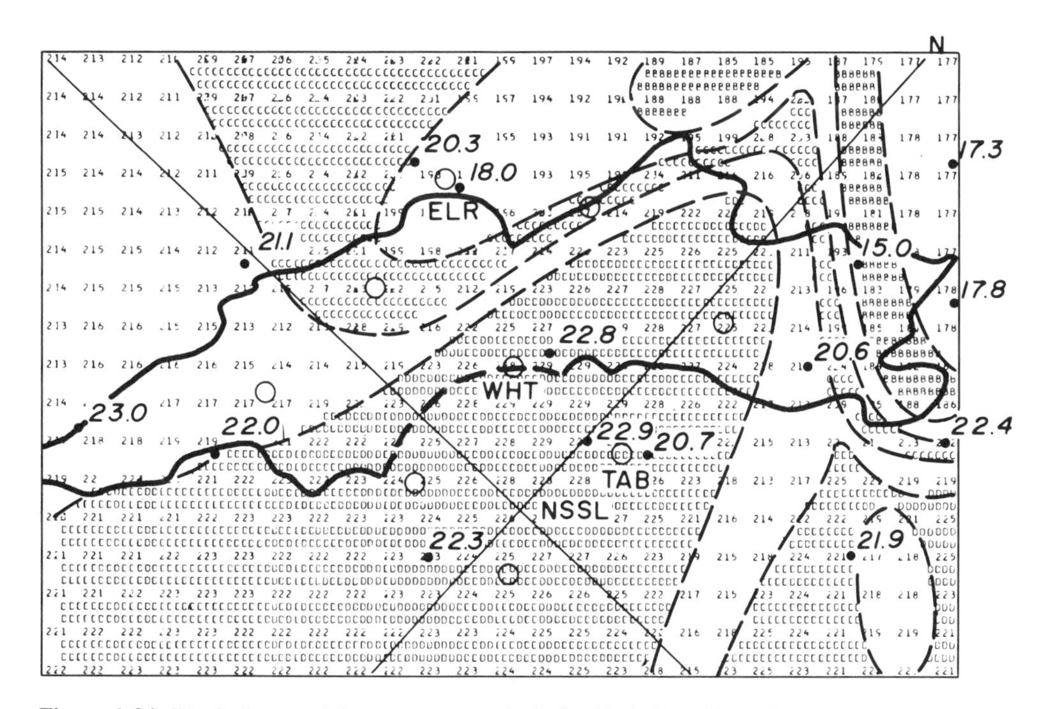

Figure 1.24. Wet-bulb potential-temperature analysis for 2240 CST, 29 April 1970. The values are the same as those in Fig. 1.23, except that observations more than ±15 min from 2240 CST have less influence ($\nu = 56.25$ min²).

din, 1963; Eddy, 1964), but they too require the arbitrary selection of a weight function that models the actual correlation between observations. Other analytic forms of weight functions can be employed, each having its own advantages and disadvantages (e.g., Cressman, 1959). We do not have space to compare these aspects here. Suffice it to say that there are no magic formulas for analysis of meteorological observations; there are only tools to help in the task of finding relevant correlations and distributions in observations that may lead to a better understanding of the physics driving the phenomenon being investigated, as discussed further in Chap. 13 of this volume, or to a better forecast based on a representative depiction of the atmosphere's initial state, as treated further, for example, by Bengtsson et al. (1981) and Barnes (1985).

With upper-air observations we do not generally have the luxury of redundant information that we often have with surface observations. It is, therefore, even more important to assess the nature of information provided by the soundings before submitting them to what may be considered a final analysis. Did the sounding sample updrafts or downdrafts that may cause aliasing in the resulting distributions? Are there unexplainable variations or biases in the vertical profiles of conservative quantities that suggest some problem in the sounding equipment or data reduction? Where, relative to a storm system, were the observations obtained?

These questions are pertinent not only to the analysis of special data from mesoscale networks but also to the analysis of larger-scale synoptic data. Unfortunately, in an operational mode one does not usually have time (or facilities) for viewing individual profiles of soundings forming the data base, although this is rapidly changing as developments in the computer industry have increased the performance and lowered the cost of the small computers that are ideal for this function. In large computer facilities such as those at national forecasting centers, cursory checks for consistency are made routinely and automatically, but they are not always sufficient to reveal problem data. It is always appropriate to view the data that produced the analysis against the background of that analysis in order to handle apparently spurious observations or to detect pathological responses of the analysis scheme to unusual data. To perform this function requires computer-displayed results that a human being can interpret and interact to adjust when necessary. The algorithms required to accomplish this interaction will eventually be developed. We believe that the ultimate "objective" analysis scheme will combine the speed, consistency, and impartiality of the computer with the capabilities of the human analyst in pattern recognition and selection of tests to apply to the data. The thinking human being is still the most important part of any analysis scheme.

2. Tornado Interception with Mobile Teams

Robert P. Davies-Jones

1. Introduction

Tornadoes, because of their infrequency, small size, and destructivity, are not easily observed and measured. Eyewitnesses are seldom trained meteorologists and consequently pay scant attention to attendant cloud structures or to the sequence of events preceding the tornado. They also do not move with the storm, so their observation time is limited. Tornadoes are too small to be resolved by radars and mesonetworks, although these sensors provide important data on the parent storms. Anemometers directly in the path of the storm are frequently damaged, and microbarographs respond too sluggishly to the rapidly passing vortex. Almost all data on the tornado scale have been gathered from visual and photographic records and damage surveys. Lack of observational data and hence of clear physical knowledge of the phenomenon hindered the construction of sound tornado theories and models (Lilly, 1965; Morton, 1966), and led Lilly (1975) to name the tornado as "perhaps the last frontier in tropospheric meteorology—the only intense and easily identifiable phenomenon whose internal structure and dynamics remain highly speculative." Although chance eyewitnesses have occasionally taken revealing tornado films, much knowledge has also been gained from visual observations made by trained meteorologists in pursuit of tornadoes.

Ward (1961) and Donaldson and Lamkin (1963) were among the first to document successful ground intercepts (i.e., close-range approaches), and Bates (1962, 1963) had similar successes from the air. These investigators went into the field irregularly and did not obtain high-quality photography, although they made skillful observations. A private citizen, David Hoadley, began intercepting and photographing tornadoes in 1956. Rossow's (1970) and Golden's (1974) airborne field experiments revealed many interesting properties of waterspouts.

The Tornado Intercept Project (TIP), a cooperative effort between the National Severe Storms Laboratory (NSSL) and the University of Oklahoma (OU), has been conducted since 1972 (Golden and Morgan, 1972; Bluestein, 1980; Lee et al., 1981). TIP is based in Norman, Okla. Its formation represented an unprecedented commitment to severe-storm intercepts and an attempt to combine close-range visual and other observations with data from fixed-base sensors such as radar. Tornado "chasing" quickly became a favorite pastime for a few meteorology students fascinated by the sublime appearance and coherent structure of rotational storms. The skills developed and the observations made by many a self-funded impromptu intercept team, relying heavily on its own visual observations for guidance, have greatly supplemented the knowledge gained on official intercepts (e.g., Moller et al., 1974; Smith, 1974; Burgess, 1976).

During the 1980s organized intercept programs were started at other bases, most notably at Texas Tech University (Rasmussen et al., 1982; Marshall and Rasmussen, 1982; Jensen et al., 1983; Marshall, 1984), at NCAR (Wilson, 1986) and PROFS (Brewster, 1986; Brady et al., 1986), in Boulder, Colorado, and at Colorado State University (as part of a graduate course on severe storms). Also, there is a sizable community of private storm chasers across the country with their own newsletter (Hoadley, 1978–86; Marshall, 1986–).

This chapter discusses TIP's goals, the intercept techniques that have been developed, the chances for success, and the knowledge gained so far.

2. TIP Goals

The original goal of approaching within close range (1–5 km) of tornadoes with conventional automobiles has been achieved many times since 1972. The continuing goals of the Tornado Intercept Project are as follows:

1. To obtain accurate times and locations of severe weather and to document changes in tornado size, shape, and tilt for use in Doppler radar analyses and other studies. Precision times and locations allow the tornado to be placed accurately relative to the radar echo and permit relations between the tornado and parent storm to be investigated.
2. To obtain high-quality films of tornado debris clouds so that tornado winds can be measured photogrammetrically.
3. To observe, videotape, and photograph the evolution of storms with the ultimate intent of constructing descrip-

tive models of tornadic storms based on visual and radar appearances.

4. To obtain visual, video, and photographic records of tornadoes and their parent clouds, of changes in cloud structure, and of sequences of events spanning tornadoes. Such qualitative observations provide an important real-world starting point for tornado models.

5. To measure the wind and thermodynamic fields near tornadoes and in mesocyclones.

6. To determine electric fields and lightning characteristics in the mesocyclone region and near tornadoes.

7. To document lightning events so that interrelationships of electrical activity with storm dynamics and structure can be determined and ground-truth data are provided for aircraft overflying the storms.

8. To launch balloons carrying a rawinsonde and an electric field meter into tornadic storms to determine the electric field and meteorological variables in specific regions, such as the updraft, weak echo region, and anvil.

9. To collect rawinsonde data in key regions of the mesoscale environment and storm proximity from moving Intercept vehicles.

3. Personnel and Equipment

Each TIP team consists of a team leader and three or four other members (one or two others in the TOTO team—see below). The team leader has the final word on all in-vehicle decisions and is responsible for his team's safety and performance. He also chooses the intercept route, on the basis of the availability of roads, visual storm observations, and information received from base. Team members are assigned driving, documentary, photographic, data-collection, and navigational duties. Emphasis is placed on accurate entries of time, location, and records of photography and observations in a log or on tape. A large van with windows on all sides is the best type of vehicle because it offers good visibility, ample space, and easy accessibility to equipment. Items carried into the field include maps, tape recorders, an insulated chest containing dry ice (for hail collection if the opportunity arises), 16-mm movie cameras, video cameras, 35-mm slide cameras, intervalometers, other photographic accessories and supplies, and portable instruments to monitor meteorological variables.

In the early years of the project microphones were deployed in the paths of tornadoes in unsuccessful attempts to record tornado sounds for scientific analysis. The best-equipped vehicle is the NSSL (formerly University of Mississippi/NSSL) mobile laboratory, which has steerable video cameras mounted on the roof and measures and records electric field, field changes from lightning, and other lightning characteristics (as described in Chap. 8), as well as meteorological variables. Currently it also serves as a mobile balloon telemetry receiving station for LORAN-C rawinsondes and electric field meters. A LORAN-C navigation receiver permits continuous recording of the van's position.

Each van is equipped with FM radio and a radio telephone. FM communications with base (NSSL) are routed through a repeater located at the 440-m level on a tall television tower 40 km north of NSSL. Beyond the FM system's range (115 km from the repeater), the radio telephone is used when an operator is within range and the channels are not busy; otherwise a public telephone must be found whenever updated information is needed. Direct, short-range, intervehicle communications, provided by a second FM radio channel, enable teams to exchange information during intercepts of storms that lie beyond the repeater's range.

In 1987 NSSL began working with amateur radio storm-spotter groups to relay verbal and packet radio messages to and from base over long distances.

Since 1981, TOTO (TOtable Tornado Observatory), a 180-kg instrument package designed for deployment in the path of tornadoes (Bedard and Ramzy, 1983), has been taken out on a pickup truck (Bluestein, 1983a, b). TOTO records windspeed, wind direction, pressure, temperature, and electrical corona at 1-s intervals. Using a winch and ramps, a two-man crew can unload and deploy TOTO in less than a minute.

First attempts were made in 1986 to deploy 10 small recording-instrument packages in a line in front of storms (Brock et al., 1987). The object of this experiment is to obtain high-resolution surface pressure and temperature data in mesocyclones and, like TOTO, to obtain data in a tornado.

Other teams with specialized assignments also go out into the field. The University of Mississippi has a project with NSSL to release balloons into severe storms to measure electric fields. The National Center for Atmospheric Research (NCAR) and NSSL have cooperated in the scientific collection of hailstones from supercell storms.

In some years as many as six scientific teams have been in the field simultaneously with NSSL and the University of Oklahoma fielding up to four photographic teams (two long range and two restricted to the area of best radar coverage). Additional coverage of storms has been provided by amateur teams with whom NSSL has traded current weather information for visual observations and photography.

The "nowcaster" is the NSSL-based person who communicates with the team leaders. After a 0-to-9-h forecast has been prepared by NSSL forecasters by 1030 CST, the nowcaster monitors the weather through the rest of the day when the potential for severe-storm development exists. His aids are the NAFAX weather maps (both analyses and prognoses), the standard teletype data reports, half-hourly satellite photographs, visual observations from the intercept teams, and NSSL radar and rawinsonde observations. By keeping abreast of the latest developments, the nowcaster seeks understanding of the current state and tendency of the atmosphere. He relays the latest nowcast and radar information to the field team members, obtains their opinions and current visual observations, and directs the vehicles to the most strategic locations. The nowcaster is also the link between the intercept teams and other participants in NSSL's Spring Observational Program. Field reports of tornadoes are relayed to the Doppler radar meteorologists (and also to the local National Weather Service Office for public alerts), while information concerning echo position, motions, characteristics, and signatures of mesocyclones and tornado vortices (see Vol. 2, Chap. X) flows out to the field.

4. TIP Strategy

To maximize intercept chances, teams enter the field well before severe weather develops and often even before the formation of echoes on NSSL's radarscopes. Immediately after the 1030 CST forecast (or even before if conditions warrant) the initial status of each team is resolved. The choices are these: (1) dismiss the teams, (2) place them on standby at NSSL, (3) dispatch them to a designated standby location in the field, or (4) vector them toward a target storm. Since the initial decision may prove to be a vital one, much depends on the forecaster's ability to assess the probability of severe storms during the day and the prime time and location for development. On potential or actual severe-storm days the status of each team is updated frequently until the mission is called off either because of darkness or because there is no hope of suitable target storms developing within range.

After storms have developed, the intercept strategy is chosen on the basis of vehicle positions in relation to storm locations and movements and expected developments. Unless a decision is made to forgo existing storms in favor of stronger ones anticipated in a different region, a target storm is chosen for each vehicle on the basis of the storm's accessibility and tornadic potential. As a field team approaches a storm, it maneuvers around the precipitation core and takes up a tracking position on the storm's right rear flank. Driving through the core is avoided whenever possible because bad driving conditions force the vehicle to slow down, large hail and strong winds may be encountered, and the team may drive "blindly" out of the precipitation core into the path of a tornado. (Incidentally, lightning, flash floods, and traffic accidents are probably greater hazards to field personnel than the tornado.)

As target distance decreases, the intercept crew's own visual observations become increasingly important, and their reliance on relayed radar information diminishes, which is fortunate since communication channels are often unavailable. The crews have become experts in interpreting cloud features, assessing the tornadic potential of a storm, picking out the part of a storm most likely to spawn a tornado, and recognizing visual precursors of tornadoes. (In fact, private intercepts have often been made by individuals acting solely on visual observations.) Once a tornado or suspicious cloud feature has been spotted (Fig. 2.1), the camera crews position themselves ahead and to the right of the extrapolated

Figure 2.1. An intercept team in pursuit of a storm.

track. This vantage point generally offers the best visibility because the tornado is silhouetted against a light background, and there is less likelihood of intervening precipitation (Fig. 2.2). Teams that have pursued tornadoes from behind have experienced poor visual contrast and fallen debris obstructing the road. The pickup truck drives into the projected path of the tornado to deploy TOTO, and other vehicles maneuver into optimal positions to carry out their particular missions. Individual intercepts are described in Brown (1976), Moller et al. (1974), Smith (1974), Gannon (1973), Doviak (1981), Taylor (1982), and Hauptman (1984).

Even though most of Oklahoma is covered by a network of section roads 1 mi apart and oriented N–S and E–W, these roads are avoided except as a last resort. Because they are unpaved, they become treacherous when wet. Also they often end at creeks and rivers. Thus the intercept vehicles stay on major highways practically all the time.

One dilemma faced by the intercept teams is deciding when to leave one storm for another. Sometimes a storm

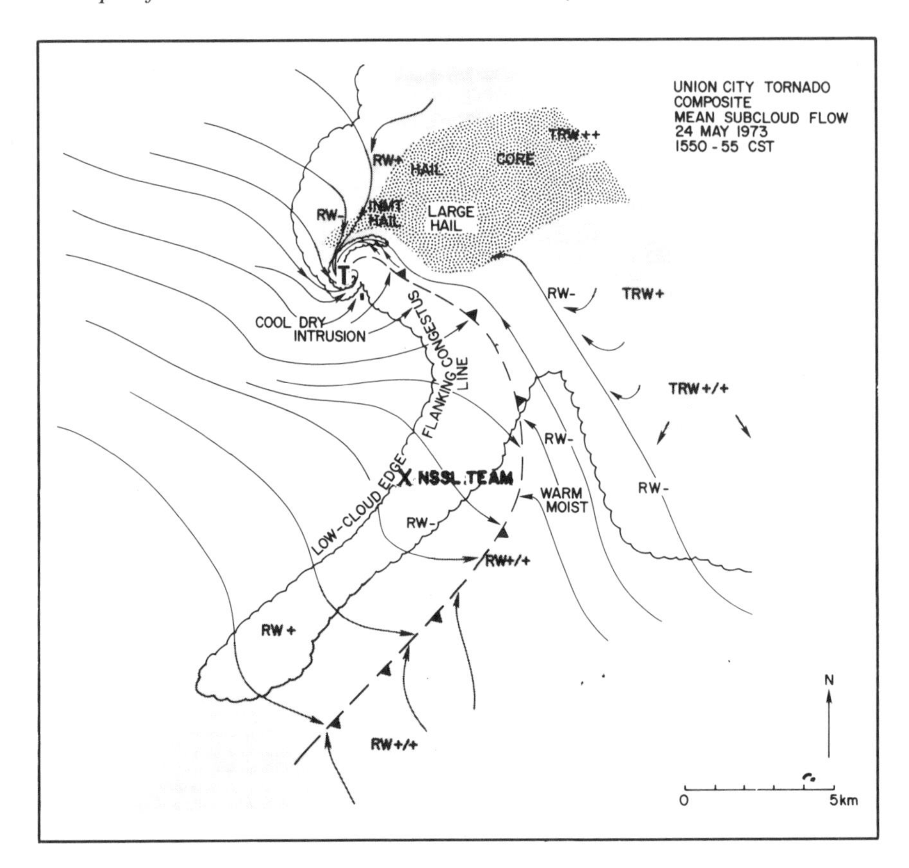

Figure 2.2. The position of the NSSL intercept team relative to tornado (T), rain, and hail at Union City, Okla., on 24 May 1973. The tornado was moving ESE at this time (Brown, 1976).

with a deteriorating visual appearance will produce a tornado after an intercept team has abandoned it.

Up to this writing, emphasis has been placed on acquiring visual, photographic, electrical and conventional meteorological data. Other experiments could be attempted, provided the equipment is portable and setup time is minimal (1 or 2 min). Longer preparation times are impractical because most tornadoes are short-lived, and stops must be short so that the vehicle can reach and maintain an optimal position for observing tornadoes as they travel.

Possible measuring systems for future field use include short-wavelength Doppler radar, Doppler lidar (laser equivalent of radar), an instrumented remote-control model airplane, and instrumented "toy-class" rockets. These devices may be carried on an airborne platform. In fact, Colgate (1982) has fired light rockets at tornadoes from the air and successfully penetrated one. However, no data were obtained because of instrument malfunction. An airborne lidar has been used successfully to collect velocity data on waterspouts (Schwiesow, 1981; Schwiesow et al., 1981; Vol. 2, Chap. 10, Sec. 9). Although less mobile than aircraft, ground vehicles have certain advantages: they are more economical, are safer in strong winds, and can stop at will.

5. The Chance of Interception

Because the intercept teams guide themselves visually over the last few kilometers of storm intercepts, their chances of success depend critically on local terrain and vegetation. The ideal "intercept country" is relatively flat with few trees so that distant cloud features at low elevation angles are visible. The region should also be sparsely populated, have few large lakes and rivers, and have a dense road network so that the teams have good mobility. The local storm environment is also important; haze, low overcast, or extensive rain areas obscure the target storm and make intercepts very difficult. Fortunately, in Oklahoma (and other Great Plains states), the storms are often clearly visible (especially in late spring and summer). Eastern Oklahoma has many hills, lakes, and trees to impede the chasers, but the farmlands of western Oklahoma are almost ideal. Intercepts are easier in late spring than in early spring because of improved visibility and slower-moving storms.

The field teams are sent out on roughly 30% of the days during the 2–3-mo period (normally April to June) of NSSL's Spring Observational Program. A field team that is permitted to range 240–320 km from base typically has a

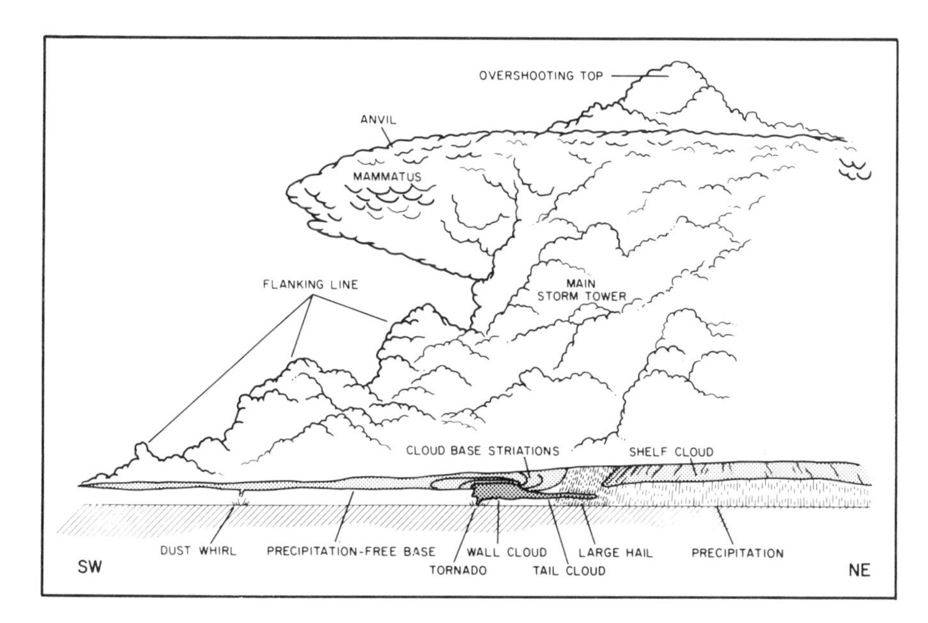

Figure 2.3. A composite view of a typical tornado-producing cumulonimbus as seen from a southeasterly direction. The horizontal scale is compressed. Not all the features shown can be seen simultaneously from a single location (diagram by C. Doswell).

successful "chase" (i.e., observes one or more tornadoes) about once in seven outings. The high percentage of fruitless missions reflects the uncertainties inherent in tornado forecasting. (These missions are fruitless only in that tornadoes are not observed; some are nevertheless scientifically valuable. Severe storms have been intercepted on 42% of field days.) Since tornadoes are rare and often unpredicted, teams are sent out even when the chances for tornado formation are considered slight. Almost all successful intercepts have been made in the western half of Oklahoma or in the Texas Panhandle.

Because of the time consumed in making a morning forecast, deciding on initial strategy, and traveling to the area where severe storms are anticipated, the teams are often not in position to make intercepts before 1300 CST. Fortunately, only 7% of Oklahoma tornadoes occur during the first half of the day (0700–1300 CST). After 1900 CST lighting rapidly becomes too low for photography. Thus we define as pursuable those tornadoes that occur between 1300 and 1900 CST (50% of the total) within 240 km of Norman. Central Oklahoma experiences in an average 2-mo severe-storm season about 18 pursuable tornadoes and 7 pursuable-tornado days. During an eight-year period (1972–79) 23% of pursuable tornadoes were sighted by field teams (both official and private).

6. Project Accomplishments

After two years it became apparent that tornadoes could be observed with sufficient frequency to make pursuit worthwhile. Counting both official and private intercepts, more than 130 Great Plains tornadoes (76 officially confirmed; only 30 of the 76 fall in the "pursuable" category described in Sec. 5) were observed between 1972 and 1979 (these statistics have not been updated recently). In most cases close-range observations of tornadic storms began well before actual tornado formation. The results and conclusions of TIP's extensive observational program and other intercept activities are these:

1. Windspeeds of four tornadoes have been measured photogrammetrically from high-quality films. Maximum measured windspeeds in these tornadoes range from 60 to 90 ms^{-1} (Lee et al., 1981; Campbell et al., 1983).
2. Between 1981 and 1986, TOTO was deployed in nine mesocyclones, within one minor tornado and within 1.5 km of two significant tornadoes (measured from the center lines of the damage tracks). The maximum recorded wind gust was 36 $m\ s^{-1}$. Maximum pressure variation (over 7- to 13-min intervals) was 3 mb, and rates of pressure change of 3 $mb\ min^{-1}$ were observed at times (Bluestein 1983a, b; Burgess et al., 1985). Nothing resembling the extremely intense 1962 Newton, Kansas, mesocyclone (see Vol. 3, Chap. 10, Fig. 10.8) has been sampled by TOTO so far, and we suspect that mesolows as strong as the Newton case are extremely rare.
3. Conceptual storm models, based extensively on visual observations, have been compiled (Lemon and Doswell, 1979; Moller, 1978; Bluestein and Parks, 1983; Doswell, 1985). Different types of flanking cloud lines have been identified by their visual appearance (Bluestein, 1986).
4. Time-lapse photography has revealed that entire convective towers rotate (Davies-Jones et al., 1976; Bluestein and Parks, 1983; Bluestein, 1984a). A few rotate anticyclonically.
5. Rarely is any cloud-to-ground lightning seen near (closer than 2–3 km) or within tornadoes (Davies-Jones and Golden, 1975).
6. Most tornadoes form close to but outside precipitation

Figure 2.4. A tornado near the edge of the cloud at Quail, Tex., 16 May 1977.

Figure 2.5. A wall cloud (discrete lowering of the cloud base).

areas. A few do form in precipitation, however, and many end their lives in rain. Large hailstones often fall near the tornado, generally on its left forward side (Fig. 2.2).

7. The tornado usually forms in a region of the storm where the cloud base has always been free of precipitation and lightning. Thus there is no observational evidence to support theories that require cloud-to-ground lightning (Vonnegut, 1960) or a burst of precipitation (Rossman, 1960; Danielsen, 1975; Eskridge and Das, 1976) to initiate tornadoes. (See Vol. 2, Chap. 10, for a discussion of tornado theories.)

8. The tornado generally develops from convective towers on the storm's right rear flank (Fig. 2.3). By the end of its life the tornado may be very near the edge of its parent cloud (Fig. 2.4).

9. Strong tornadoes form from wall clouds (Fig. 2.5). Thus wall clouds identify potentially very dangerous parts of the storm, and storm spotters are now trained to recognize and report wall clouds before tornado formation (Moller, 1978; Moller and Boots, 1983). Eyelike features sometimes seen in wall clouds may indicate strong subsidence at the center of an occluding mesocyclone (Bluestein,

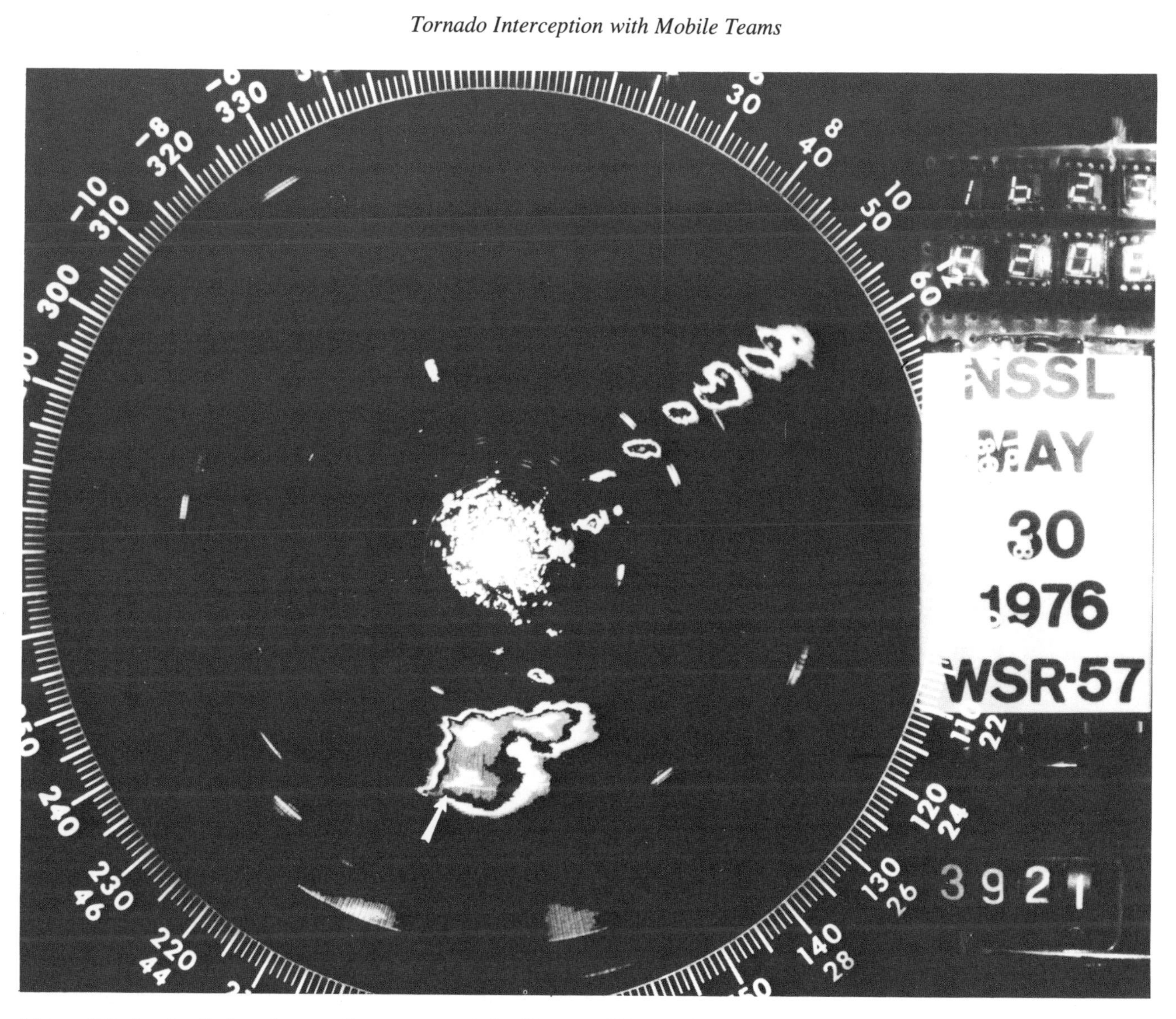

Figure 2.6. A radar display of a tornadic storm recorded at Norman, Okla., on 30 May 1976. The range to the edge of the scope is 200 km. The position of the tornado is indicated by the arrow.

1985). Twin rotating wall clouds observed revolving around each other (Bluestein, 1984b) are evidence for "vortex splitting" (see Vol. 2, Chap. 10, Sec. 16) on the mesocyclone scale.

10. Gradual evaporation of cloud is often observed in the wall cloud, first to the rear of the tornado and then slowly propagating around its right side forming a relatively cloud-free area known as the "clear slot" (Lemon and Doswell, 1979). This observation apparently indicates the presence of a developing unsaturated rear-flank downdraft. Intense upward motions seen ahead and to the left of the tornado indicate that the tornado is located near an updraft-downdraft interface.

11. Mesocyclones often regenerate periodically along the storm's "pseudo-cold front," the old mesocyclone occluding and a new one on its right succeeding it (also apparent on Doppler radar—see Vol. 2, Chap. 10, and Burgess et al., 1982). In this way a storm can spawn a succession of tornadoes over a several-hour period (Rasmussen et al., 1982; Taylor, 1982; Jensen et al., 1983).

12. Isolated storms or storms with inflows that are unaffected by neighboring storms are the best tornado producers (Fig. 2.6), presumably because they are not competing with neighbors for available warm, moist inflow air and are not ingesting harmful amounts of rain-cooled outflow air from other storms.

13. Sometimes surface weather features pinpoint small areas of maximum tornado threat. Tegtmeier (1974) showed

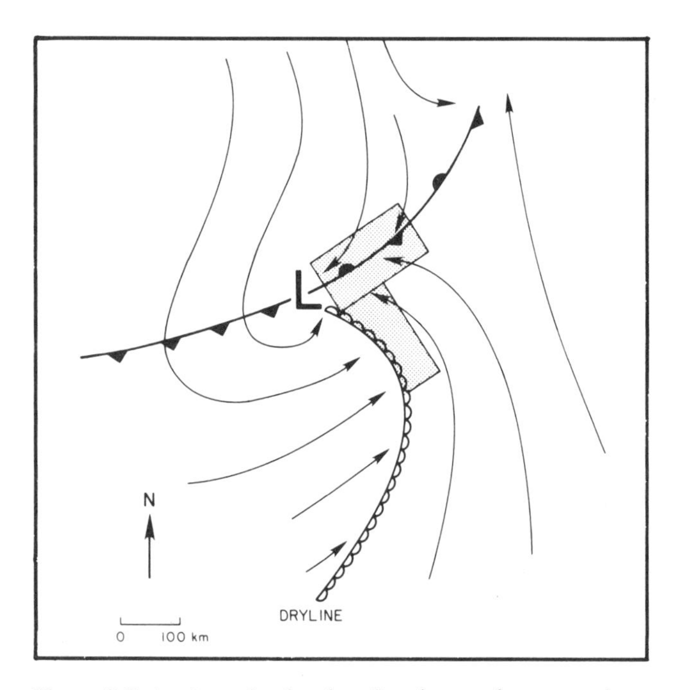

Figure 2.7. A schematic of surface-flow features for a case where the small-scale low and a dryline bulge are present simultaneously. Stippling indicates maximum-threat areas.

that favored locations for tornadoes are the northeast sides of developing small-scale lows and bulges in the dryline (Fig. 2.7).

14. Anticyclonic tornadoes definitely exist (Fig. 2.8) (Burgess, 1976; Jensen et al., 1983). Their typical location within the storm is discussed in Vol. 2, Chap. 10.

15. TIP storm observations and photography have been incorporated in a training film (NOAA, 1977) and slide collection (NOAA, 1980) for tornado spotters (Moller, 1978).

16. A new type of rotational severe storm that fails to fit the classic model, produces little precipitation at the ground, and looks benign on conventional radar has been identified (Davies-Jones et al., 1976; Burgess and Davies-Jones, 1979; Bluestein and Parks, 1983; Bluestein, 1984a).

17. TIP has provided vital information to Doppler radar meteorologists involved in both operational projects and the analysis of research data. For example, well-documented TIP tornado observations verified 11 of 34 experimental tornado warnings issued solely on the basis of Doppler radar observations during 1977 (Burgess et al., 1978).

18. Quantitative measurements of electric fields and of electric-field changes and optical transients associated with lightning have been made in the inflow region of tornadic storms while the vehicle is moving as well as stationary.

19. Lightning that lowers positive charge to ground has been shown to emanate from the following areas of severe

Figure 2.8. An anticyclonic tornado near Alva, Okla., 6 June 1975. Note the condensation funnel at the top center (photo courtesy of J. Leonard and E. Sims).

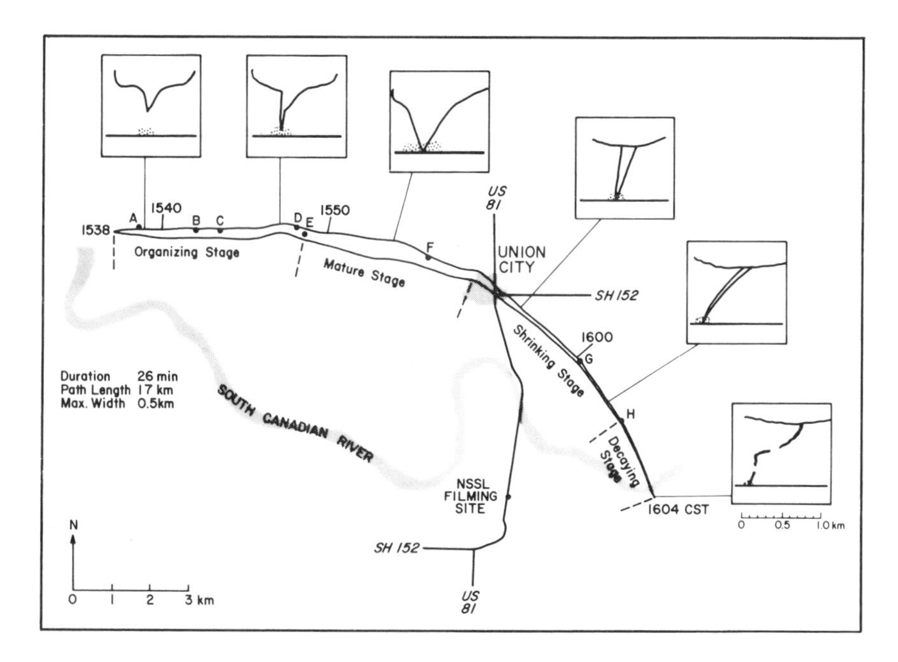

Figure 2.9. The damage path of the 24 May 1973 tornado at Union City, Okla., with sketches of the funnel at different life-cycle stages (Brown, 1976).

storms: high on the main convective tower, on rare occasions from beneath the wall cloud, and from well out in the downshear anvil (Rust et al., 1981). Positive flashes had not been documented previously in severe storms.

20. A portable radiosonde unit and a theodolite have been used to obtain soundings near drylines and in and near severe storms, including one inside a tornadic storm (Bluestein, et al., 1987). Also, balloons instrumented to measure the electric field, temperature, pressure, and humidity have been successfully flown into anvils, where fields of 94 kV m^{-1} have been found 65 km downwind of the storm core (Marshall et al., 1984), and into the inflow region of severe storms. Results from initial attempts to fly balloons from a mobile base have been encouraging and have demonstrated the feasibility of the concept.

21. The mobile laboratory provided important verification data for evaluating lightning strike locating systems (Mach et al., 1986).

An early successful intercept involved the Union City, Okla., tornado of 24 May 1973 (Brown, 1976). In this instance films that were photogrammetrically analyzable for tornadic windspeeds were obtained, and the closeness of this tornado to NSSL (~40 km) made possible high-resolution Doppler radar data collection and detailed damage survey. Tracking debris in the film yielded windspeeds to 80 m s^{-1} at 200-m radius, 90-m elevation during the tornado's mature stage, and 65 m s^{-1} at 25-m radius, 50- to 80-m elevation during its shrinking stage. The intercept teams' documentation of the tornado's position in time and space and its tilt with height helped establish the significance of Doppler radar's tornadic vortex signature, first discovered in the Union City storm. The complete life cycle of the Union City tornado was observed and filmed (Fig. 2.9; Moller et al., 1974; Brown, 1976), and a subsequent damage survey revealed that changes in tornado structure were accompanied by corresponding change in damage intensity and debris configuration (Vol. 2, Chap. 10, Sec. 3).

7. Further Discussion

Obtaining films of debris has proved more difficult than first anticipated. Nevertheless, high-resolution, analyzable 16-mm films of roughly a dozen tornadoes have been obtained. Several other tornadoes have also been filmed, but photogrammetric analyses of these films could not be performed because the tornado was too distant (or too near on one occasion!), the contrast was too poor, or the tornado traveled over terrain devoid of potential debris material. The films obtained by intercept crews are more valuable scientifically than those taken by chance eyewitnesses because the high-resolution film used by TIP permits small features to be seen and tracked.

Sending meteorologists into the field is beneficial in many other ways. Intercept teams can record accurate times—difficult to obtain post facto—of tornadoes and hailfall, and they can collect hailstones for scientific analysis (Knight and Knight, 1974, 1976; Ziegler et al., 1983). The rise rate of convective towers measured photogrammetrically can be compared with vertical velocities measured by Doppler radar. Occasionally they may see severe storms that are so unadorned and have such elementary structure that their basic internal dynamics can be deduced from their visual appearance alone (e.g., Davies-Jones et al., 1976; Bluestein and Parks, 1983). The ever-changing visual outlines and motions of the clouds they see and record on film contain valuable information about severe-storm structure and the dynamic processes that foreshadow tornadoes. Many fruitless tornado theories would never have been conceived if the instigators had witnessed tornadogenesis firsthand. The field crews may also make fortuitous discoveries of great research value, such as the first proved anticyclonic tornado (Burgess, 1976).

In the future, scientists will continue to attempt new techniques for measuring tornadoes and mesocyclones and, perhaps, will even try to modify them. The art of tornado interception will be important logistically to the success of these field experiments.

3. Observations from Instrumented Aircraft

John McCarthy and Donald L. Veal

1. Introduction

In basic flight training we are told in hundreds of ways, "Thunderstorms and airplanes do not mix!" Yet our knowledge of thunderstorms and airplanes has grown in a nearly symbiotic relationship.

Although a history of the relationship between thunderstorms and aviation is given in Vol. 1, Chap. 7, a brief review is nevertheless useful. As the use of airplanes for transportation became established in the 1930s and for weapon delivery systems during World War II, more close encounters with thunderstorms occurred. With good reason pilots learned to fear them, avoiding them at almost any cost. Partly because of this fear, as World War II came to a close, little was known about thunderstorm interiors.

As civil transportation and military needs expanded, aviators realized that complete avoidance of thunderstorms was impractical, and thunderstorm research efforts were intensified. Analyses of data from the first comprehensive field efforts were reported in Byers and Braham (1949). The exercise on which the study was based included the first research-oriented aircraft penetrations of thunderstorms. Results represented a substantial advance in understanding of thunderstorms; much of our knowledge concerning the cellular nature of thunderstorms came from the Thunderstorm Project and was a foundation of pilot training in aviation hazards.

As various aircraft types were developed and route coverage was increased, particularly in the 1950s and early 1960s, much practical experience was gained in unavoidable thunderstorms. The vast increase in flight numbers and range by piston-engine aircraft such as the C-121 Superconstellation, the DC-6, and the DC-7 and jet aircraft such as the Boeing 707, the Convair 880, and the DC-8 increased the emphasis on understanding thunderstorms and thunderstorm hazards. The use of radar to probe remote thunderstorms considerably increased our understanding of the structure of thunderstorms and made possible additional studies of thunderstorms using aircraft for direct probes.

In recent years understanding regarding risk of thunder-

Figure 3.1. The NCAR Research Aviation Facility fleet during the mid-1980s showing Electra (N-308-D), Queen Air (N306D), King Air (N-312-D), and Sabreliner (N-307-D) (NCAR photograph).

Figure 3.2. The Beech King Air cloud physics airplane developed by the University of Wyoming for the U.S. Bureau of Reclamation. This aircraft has specialized instrumentation to measure cloud-icing and for other microphysical research.

storms has been elaborated with respect to their downdrafts and outflows, recently called microbursts and macrobursts (Fujita, 1985), depending on their horizontal extent. These events, even though not severe in terms of the public weather service definition (e.g., winds greater than 50 mph), can cause crashes of jet transport aircraft that penetrate them below 1,000 ft above the ground, during takeoff or landing. Flight data recorders on commercial aircraft involved in accidents with turbulence and wind shear as contributing factors have provided substantial insights regarding aircraft and pilot performance during such episodes.

In 1958 the U.S. Weather Bureau established the National Severe Storms Project in Kansas City, Mo. The operational arm of this project remained there as the National Severe Storms Forecast Center when the research arm moved to Norman, Okla., in 1964 to become the National Severe Storms Laboratory (NSSL). Since then numerous thunderstorms have been penetrated by aircraft, usually of military type, bearing special instruments and operating in the NSSL program. Studies of the near-thunderstorm environment have also been conducted with instrumented aircraft that are not suitable for flight within thunderstorms.

Several major aircraft facilities have been developed and have devoted part of their effort toward understanding the thunderstorm. The Research Flight Facility (RFF) of the National Oceanic and Atmospheric Administration (NOAA) was established in 1962, with initial and primary emphasis on hurricane operations and research. In 1964 the Research Aviation Facility (RAF) of the National Center for Atmospheric Research (NCAR) began flight operations in support of NCAR and university research. Important flight research programs have also been undertaken by the Illinois State Water Survey; Colorado State University; the Universities of Chicago, Nevada, North Dakota, Washington, and Wyoming; and the National Aeronautical Establishment of Canada, alone and in collaboration with other groups such as NCAR and NSSL. Figures 3.1–3.3 show several research aircraft for storm penetrations and for flight in the near-thunderstorm environment. Table 3.1 lists meteorological research aircraft generally available in the United States today.

2. Comparison of Airborne and Ground-based Thunderstorm Probes

Why should aircraft be used in thunderstorm research? Would it not be more sensible to use a system for remote probing such as ground-based radar? Since radar is the most often used thunderstorm probe, a comparison of the two approaches is worthwhile.

An aircraft is usually used as a direct in situ probe; radar is indirect (airborne radar systems, of course, complicate our comparison). The obvious advantage of ground-based radar is its relatively nonhazardous position compared with that of a penetrating aircraft. Furthermore, radar is able to sample a much larger volume of the atmosphere in a much shorter period of time.

Let us examine sample volume, sample time, and representativeness of sample. For simplicity we treat a small thunderstorm as a cubical volume, 10 km on a side. First

Table 3.1. Thunderstorm Research Platforms

Airplane	Facility	Profile
F-4-C Phantom	USAF	Thunderstorm penetration
F-101 Voodoo	Colorado State University	Thunderstorm penetration
F-101 Voodoo	NASA	Thunderstorm penetration
T-28	South Dakota School of Mines	Thunderstorm penetration (cloud physics)
Beech King Air	University of Wyoming	Thunderstorm environment (cumulus congestus)
Sabreliner	NCAR	Midlevel environment
Beech King Air	NCAR	Cloud microphysics
Electra	NCAR	Subsynoptic environment
P-3 Orion	NOAA	Thunderstorm-hurricane environment
Cessna Citation	University of North Dakota	Thunderstorm environment
U-2/ER-2	NASA	Thunderstorm environment at high altitudes and air quality in stratosphere

Figure 3.3. The NOAA WP-3-D research aircraft, used in a variety of research areas, but particularly noted for hurricane missions.

consider the radar looking at our storm volume of 10^3 km^3 in approximately 1-min sample time. The radar samples an individual volume that is dependent on the illumination area A_i:

$$A_i = \frac{\pi R^2 \theta^2}{4}, \qquad (3.1)$$

where R is the range to the target and θ is the beam width in radians. The sample volume depends also on the pulse length h, typically 300m. The individual sample volume is $A_i(h/2)$; for a 1° beam and an average distance to the storm of 75 km, this is approximately 0.2 km^3. If our storm is sampled uniformly, in 1 min we can obtain 5×10^3 total radar samples distributed throughout the storm.

In comparison, let us assume that our typical airborne sensor samples i cm^2 while flying through the storm at 100 m s^{-1} (a typical jet penetration speed); the horizontal penetration time would be 100 sec. The total penetration sampling volume would be 10^6 cm^3, or 10^{-9} km^3. If the airplane instrumentation system were able to sense and record data at 8 Hz, then each individual sample would refer to 12.5×10^2 cm^3, and during the pass there would be 8×10^2 individual samples.

Clearly the radar sees a much larger volume than does the aircraft. However, the aircraft has much greater resolution along the flight path: the aircraft senses a minimum volume of 12.5×10^2 cm^3; the radar cannot resolve a volume smaller than 2×10^{14} cm^3. But is the aircraft observation representative of the storm? The ratio of the total aircraft sample volume to the total radar sample and storm volumes is 10^{-12}, but the airplane data have much higher resolution. As with most other measurement systems, there are trade-offs: in airborne instrumentation we gain resolution at the expense of volume and representativeness of the sample. Also, ground-based and airborne instruments sense different weather parameters, and the airborne system serves relatively specialized needs. Nevertheless, the airborne in situ platform allows us to obtain measurements that cannot be taken by radar; for cloud microstructure, high-resolution velocity and turbulence, and atmospheric chemistry there is no replacement for the airborne platform (see Sec. 3a below). Clearly the proper use of an airborne system involves its combination with other measurement systems, such as conventional and Doppler radar, rawinsonde, and surface recording networks (see Sec. 5).

The unique visual perspective from aircraft is especially complementary to radar observations. The airborne observer is sometimes in a good position to direct the operation of data acquisition systems overall and later to propose

Figure 3.4. A tornado-bearing thunderstorm as viewed from the University of Wyoming Queen Air for a University of Oklahoma study of an Oklahoma dryline.

meaningful approaches to data synthesis. Also, a number of thunderstorm features (see, e.g., Fig. 3.4) are best detected by real-time visual observations; gust fronts, tornadoes, and tornado cyclones occurring in the subcloud region of the planetary boundary layer are easily seen from a low-flying airplane. Research mission objectives can be varied on the basis of these real-time visual observations.

3. Measurements by Airborne Sensors

a. State of the Atmosphere

The synthesis of state-of-the-atmosphere parameters is basic to many important calculations in thunderstorm research. Thus measurements of these parameters are an essential part of the data obtained from the aircraft platform. Figure 3.5 shows a complement of state sensors mounted on an aircraft, and Table 3.2 lists research measurements and associated sensors for the NCAR King Air aircraft.

Figure 3.5. Instruments on the NCAR Queen Air (decommissioned in 1986): A: Reverse flow probe for in-cloud temperature. B: Johnson-Williams Liquid Water Content Meter. C: PMS two-dimensional optical imaging probe (courtesy NCAR).

Table 3.2. NCAR King Air Instrumentation

Variable measured	Instrument type	Range	Accuracy	Resolution
Aircraft latitude	Inertial navigation system	±90°	≤1 nmi/flight hr	0.0014°
Aircraft longitude	Inertial navigation system	±180°	≤1 nmi/flight hr	0.0014°
Aircraft lat./long.	LORAN-C	±90°/±180°	0.19 km	0.0002°
Aircraft position	DME	0 to 240 km	Geometry- and reception-dependent	0.0002°
Static pressure	Variable capacitance	300–1,035 mb	1 mb	0.07 mb
Total air temperature reverse flow	Platinum resistance	−60 to +40°C	±0.5°C	0.006°C
Total air temperature fast response	Platinum resistance	−60 to +40°C	±0.5°C	0.006°C
Dew-point temperature	Thermoelectric	−50 to +50°C	±0.5°C	0.006°C
Absolute humidity	Lyman-alpha Hygrometer	−45 to +30°C	±5%	0.2%
Geometric altitude	Radio altimeter	0–762 m	5%	0.1 m
Geometric altitude	Radio altimeter	0–21,000 m	±9.7 m	0.1 m
Cloud liquid-water content	Hot-wire	0–5 g m^{-3}	—	0.005 g m^{-3}
Aerosol spectrum	Laser spectrometer	0.12–3.12 μm	—	0.025 to 0.375 μm
Cloud-droplet spectrum	Laser spectrometer	0.5–45 μm	—	Selectable
Cloud-droplet spectrum	Laser spectrometer	10–620 μm	—	10 μm
Cloud-droplet spectrum	Laser spectrometer	20–280 μm	—	20 μm
Hydrometeor spectrum	Laser spectrometer	300–4,500 μm	—	300 μm
Cloud-particle spectrum 2-dimension	Laser spectrometer	50–1,600 μm	—	50 μm
Hydrometeor spectrum	Laser spectrometer	200–4,500 μm	—	200 μm
Wind vector (horizontal)	Gust probe—INS	±0.1 ms	—	0.012 m s^{-1}
Wind vector (vertical)	Gust probe—INS	±0.1 ms	—	0.012 m s^{-1}
Wind direction	Gust probe—INS	—		—

1. Temperature: The principal problem in measuring free-air temperature from an aircraft is placement of the measuring instrument on the airplane. The instrument is subject to dynamic heating, radiation, conduction and viscous effects, cloud and precipitation elements, and side effects of deicing. A number of techniques have been attempted for observing free-air temperature, the most common involving a platinum resistance element mounted on an aircraft so as to be subjected to the airstream. This approach yields the total temperature, which is the actual temperature of the sensor and which yields the air temperature after corrections. These consider dynamic heating, or the rise in temperature in the airstream at the thermometer when the air is brought momentarily to rest, and water or ice on the element. Effects of water and ice are particularly difficult to treat; hence the reverse-flow probe of Rodi and Spyers-Duran (1972) has been devised, shown as A in Fig. 3.5. This probe shields the element from direct exposure to the airstream; the airflow is reversed with separation of the ice and water particles within the airstream, and the temperature-measuring element mounted in the probe is not wetted. Spyers-Duran and Baumgardner (1983) examined in-flight response of temperature probes.

The physics of in-flight temperature sensing is complicated by the sensors' rapid motion through the atmosphere. The compressional warming of the probe is termed dynamic heating. The dynamic heating factor can vary from 1 to 50°C at airspeeds from 85 kn to the speed of sound, so a correction clearly must be applied. A typical expression suitable for dynamic heating correction is

$$T_s = T_f + \alpha(1.3189 \times 10^{-4}\,\mathrm{TAS}^2), \qquad (3.2)$$

where T_s is the sensor temperature in degrees Celsius, T_f is the free-air temperature in degrees Celsius, α is the recovery factor, and TAS is the true airspeed in knots.

The recovery factor reduces to 1 when all kinetic energy of the airstream is converted to thermal energy; the probe is then called a total-temperature probe. However, varying amounts of ventilation do occur, and values of α range from 0.5 to >0.9; α, presumably determinable by theoretical considerations, is usually determined by an empirical "speed run." The airplane is flown through potentially isothermal air at varying airspeeds, and a plot of 1.3189×10^{-4} TAS^2 versus $T_s - T_f$ is made; α is the slope of the line of plotted points. Since *TAS* depends on T_f, an iterative procedure is actually required here, as indicated in the discussion following Eq. 3.6 below.

When a temperature sensor is flown through a cloud, the heating process cannot be assumed to remain free of wet-bulb effects. Whether this is a problem has been the subject of much discussion. Telford and Warner (1962) and McCarthy (1974) believe it is not; Lenschow and Pennell (1974) believe it is.

Finally, calibration of any temperature probe installed on an airplane involves not only the recovery factor but also the baseline (*DC*) offset, if one exists in the data system. Flight past a tower at different temperatures and airspeed or immersion of the probe in an ice bath while on the ground establishes the appropriate offsets.

2. Humidity: Humidity has been measured from aircraft in several ways: (1) use of wet-bulb and dry-bulb thermometers; (2) direct measurement of dew point by deposition of dew or frost on a nonsorptive, inactive surface whose controlled temperature is the dew point when the dew or frost remains the same thickness; (3) measurement of the absorption cross section of water vapor in the Lyman-alpha spectral region with a spectral hydrometer; or (4) measurement of the microwave refractive index with a microwave resonance cavity.

Early airborne measurements of humidity used a pair of simple wet- and dry-bulb mercury thermometers mounted outside or through the skin of the aircraft. In the 1940s mercury thermometers mounted in special housings were the standard humidity instruments for the United States. For slow aircraft, or when response time was not a problem, this type of instrumentation is moderately satisfactory. The wetted-wick measurement must be used in the psychrometric equation, which relates various moisture parameters (List, 1958). An alternate approach is to attach a wetted wick to platinum resistance wire or to the thermocouples and expose it to the air either directly or mounted in a housing. For example, copperconstantan thermocouples were mounted in a reverse-flow housing and used to determine the wet-bulb depression, with a dynamic heating correction applied. This system must also be corrected for dynamic heating with the appropriate wick pressure used in the psychrometric equation. If the liquid water is not separated from the air before the sensor is read, the dry-bulb measurement has no significance.

Mullin and Wolver (1964) describe a hygrometer that measures directly by sensing dew or frost on a thermoelectrically cooled surface. The Cambridge Systems dewpoint hygrometer is now probably the reliable and most widely used instrument. The response time of this instrument is usually about 1 sec; however, it may be much longer at low ambient dew points, which tax the cooling capacity of the instrument. An inlet that deflects rain and cloud droplets from the air sample has been developed by the Naval Research Laboratory (Ruskin and Scott, 1974). If the cooled surface becomes dirty or if hygroscopic material collects on the surface, the control point must be adjusted to yield the proper dew point. Periodic cleaning of the surface usually prevents this problem.

Ruskin (1967) and Buck (1973) describe the Lyman-alpha hygrometer, which responds in a few milliseconds to absolute humidity (g m^{-3}), which is convertible to other humidity units when pressure and temperature are known. This instrument has problems of drift of the source, detector performance, and window deterioration under moist conditions. These problems cause calibration instability, but if they can be overcome, the system should be ideal for airborne use; it is especially applicable in aircraft that operate at high rates of speed, such as turbojets.

3. Pressure: In principle, pressure is the simplest atmospheric parameter to measure from an airborne platform. The most important problem is proper location of the static-pressure source on the aircraft. Pitot-static probes have been designed in which the static ports are located symmetrically about the sideship, or yaw axis. When a pitot-static probe is used, wind-tunnel experiments can be conducted and empirical corrections established as a function of the angles of attack and sideslip. To minimize the reduction in static pressure resulting from increasing angles of attack and sideslip, the static-pressure source is often located on the empennage of the aircraft. For accurate measurements the static-pressure defect or error must be experimentally evaluated; such evaluation can be accomplished by installing an accurate pressure device in the aircraft and flying at different airspeeds past a station (tower) at an altitude of known pressure. Depending on the aircraft and the static-source location, corrections ranging from zero to 10 or more mb may be necessary. Once the static-source errors are resolved, the determination of flight-level pressure depends upon only the aircraft pressure sensor. Pressure sensors are available that are stable over a wide range of environmental conditions, operate on various principles, and yield resolutions of fractions of millibars. A straightforward process converts the pressure values to pressure altitude using the ICAO (International Civil Aeronautical Organization, a United Nations affiliate) standard atmosphere, or other pressure-height relationship.

b. State of the Aircraft

1. Indicated Airspeed: Indicated airspeed (IAS) for an aircraft is determined from the difference between the pitot and the static pressure at the aircraft flight level. Pitot pressure is not particularly difficult to observe, especially for small angles of attack and sideslip. A pitot tube that aligns itself with the relative airflow can be used to minimize errors that arise in evaluation of large angles of attack. An additional source of error occurs in precipitation or in the presence of supercooled water. By heating the pitot probe, both effects can be minimized. The main error in IAS or pitot–static-pressure difference is static-pressure defect or static-pressure error.

The conservation of energy in an airstream may be expressed as

$$C_p dT + d\left(\frac{V}{2}\right)^2 = 0, \qquad (3.3)$$

where C_p is heat capacity of air at constant pressure, T is temperature, and V is true airspeed.

By using the Poisson relation between temperature and pressure for adiabatic processes, it can be shown that

$$V = \{2C_p T_f\,[(\frac{P + \Delta P}{P})^{R/C_p} - 1]\}^{\frac{1}{2}}, \qquad (3.4)$$

where T_f is the free-air temperature, P is the total pressure (static pressure), ΔP is the difference between pitot pressure and static pressure, and R is the gas constant of air.

Indicated airspeed may be defined as follows:

$$\text{IAS} = \{2C_p T_a[(\frac{P_a + \Delta P}{P_a})^{R/C_p} - 1]\}^{\frac{1}{2}}, \qquad (3.5)$$

where T_a and P_a are temperature and pressure of the ICAO standard atmosphere at sea level.

2. True Airspeed: For cases in which ΔP is much less than P, the indicated and true airspeed (TAS) are related by

$$\text{TAS} = \text{IAS}(\rho_0/\rho)^{\frac{1}{2}}, \qquad (3.6)$$

where ρ_0 is the density of the air at sea level of the standard atmosphere and ρ is the density of the air through which the aircraft is flying. Because elevation of the density terms in Eq. 3.6 involves knowledge of the temperature, it is necessary to use an iterative procedure to solve Eqs. 3.4, 3.5, and 3.6. If accuracy in TAS is required, considerable care must be taken in selecting, locating, and calibrating the pressure and temperature sensors.

3. Aircraft Position: A general treatise on avionics navigation has been presented by Kayton and Fried (1970); the following summary focuses on aspects of major interest in meteorological research. Aircraft location in space and time is, of course, critical to most airborne meteorological research; the accuracy required of this positioning depends on the nature of the study and may particularly depend on whether the study requires integration of a variety of data from other systems (i.e., satellite, ground-based radar, rawinsondes, and other research aircraft). Various positioning aids have been used over the years, including standard VHF navigation (VOR-DME)[1], Doppler and inertial navigation, global and regional low-frequency systems such as OMEGA and LORAN, and, most recently, satellite-based systems such as the GPS (Global Positioning System).

The simplest technique for position keeping involves continuous recording of two lines of position from two VOR stations; this approach requires the aircraft always to be within radio reception range of both stations. It usually gives an aircraft track within 2 km of the true track, though accuracy is a function of range from the VOR station and calibration of the VOR receivers. Alternatively, one line of position (LOP) and the distance along the LOP can be continuously recorded from one VOR station. The VOR-DME system locates the aircraft track with about the same accuracy as the dual VOR system. If two (preferably three) or more VOR stations are within radio range, multiple DME fixes can be obtained for better track accuracies (± 0.5 km of true track). However, the need for reception of more than one VOR may be quite limiting since this en route navigation system was laid out to provide navigation signals to a given area without much overlap, except in congested areas near metropolitan airports. Recent developments in very low frequency (VLF) navigation systems make aircraft position keeping reasonably inexpensive over a much wider range of altitudes and areas.

By integrating the ground speed and drift angle defined by a Doppler navigation system, the position of the aircraft as a function of time can be estimated. Because of errors in ground speed and drift angle during turns of the aircraft, and because these errors are cumulative, this technique is not suitable for long-term position keeping. It is, however,

[1] VOR (Very high frequency Omni Range) is the national airspace principal navigation aid, which permits the determination of the aircraft's magnetic bearing to or from the ground VOR station. The system is line of sight. DME (Distance Measuring Equipment) allows determination of the slant distance between the DME site on the ground and the aircraft in flight.

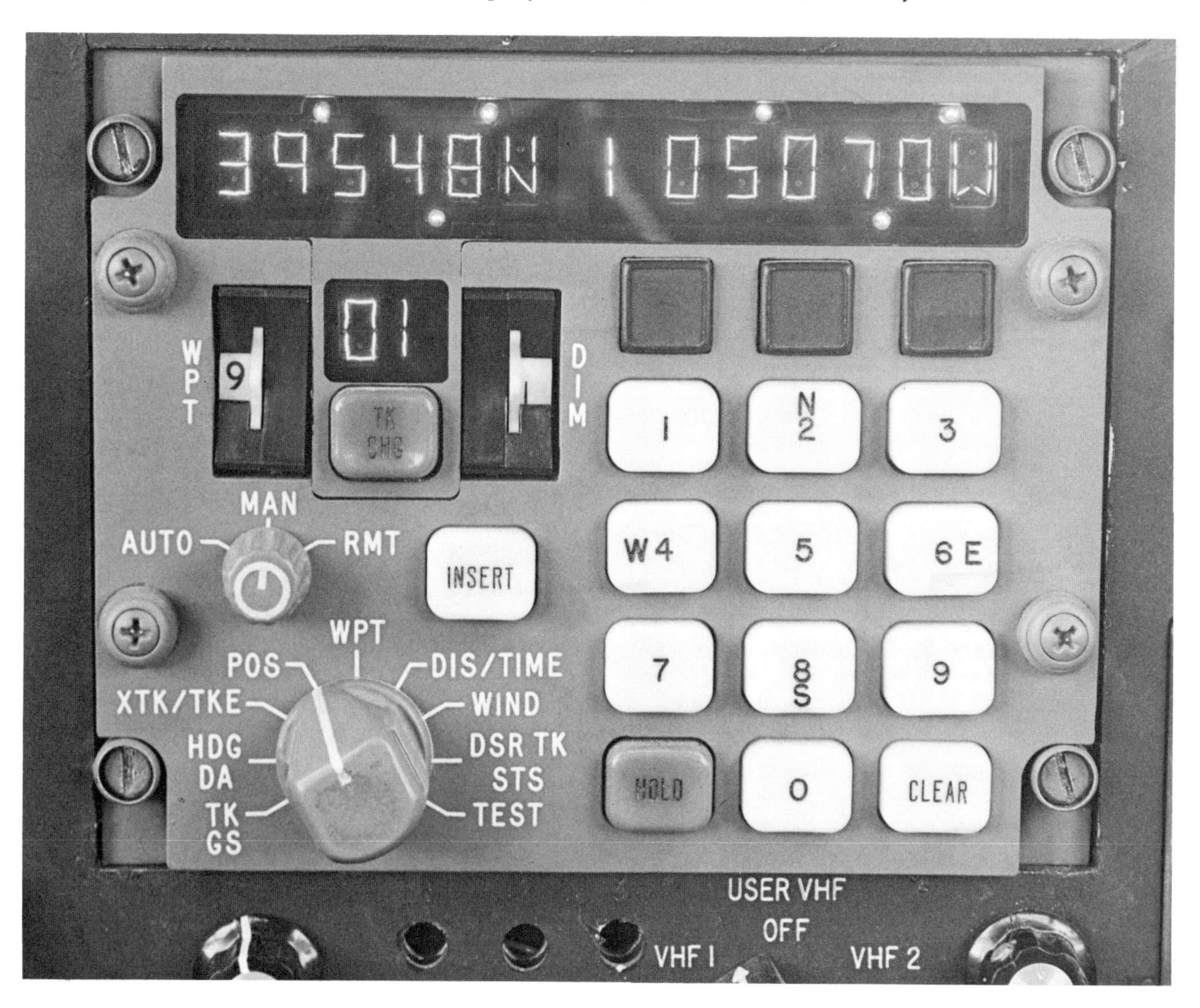

Figure 3.6. The display unit of the inertial navigation system (INS) on the NCAR Queen Air. The device, a Litton LTN-51, is an important component in the wind-measuring system and in the determination of aircraft position (courtesy NCAR).

good for short-term dead reckoning, especially with frequent ground or other references.

An inertial navigation system (INS) comprises a gyro-stabilized platform and a computer. Accelerometers sense velocity changes along all three Cartesian axes and hence define the quantities required to determine aircraft position at all times. The INS locates the aircraft during flight from a known position without reference to any signals from stations on the earth's surface. Although the INS is the most accurate system for continuous tracking, position errors do occur, and they accumulate with time. For greatest position accuracy external position references (i.e., visual reference points) should be provided at various locations along the flight path. The "front-end" display of the NCAR inertial navigation system is shown in Fig. 3.6.

The low-frequency navigation aids, OMEGA and LORAN, offer moderate position accuracy at rather low cost, and they have come into more widespread use as the global implementation of ground transmitters has been completed. However, for precise positioning with respect to thunderstorm-scale phenomena, these systems are generally considered less than acceptable. Estimates of the accuracy of low-frequency navigation systems can be found in Nicholls (1982).

The latest-generation system, GPS, when it becomes available late in the 1980s, will revolutionize position finding, with accuracies reported to be better than 100 m, worldwide.

Real-time onboard display of the aircraft position can be accomplished by driving an x-y plotter or CRT display with any of the position-keeping techniques. An example of a reconstructed flight track using VOR-DME and airborne Doppler navigation radar is given in Fig. 3.7. Aircraft-position information is particularly critical when an overall synthesis of data systems (e.g., aircraft and radar) is required. Too little emphasis has been placed on aircraft position in many research endeavors, and efforts should be made to upgrade this measurement.

c. Measurement of Air Velocities

Measurements of air motion are usually divided into three categories: (1) large-scale vertical motion, (2) large-scale horizontal motion, and (3) small-scale motion or turbulence. Large-scale vertical motion can be determined when the aircraft is in equilibrium with the vertical air motion; then either the aircraft itself can be used as a sensor or various sensors that detect the motion of the aircraft with respect to the Earth as well as the relative motion of the air can be attached to the aircraft.

Measurements of horizontal air motion require sensors such as an INS on the aircraft to measure the position or velocity of the aircraft with respect to the Earth, with the INS capability coupled to airspeed and heading information. In general, the velocity of the air with respect to the ground is the sum of the velocity of the aircraft with respect to the Earth and the velocity of the air with respect to the aircraft.

The most elementary technique for observing large-scale vertical air motion uses the aircraft as the primary sensor and requires the pilot to maintain a constant airspeed and power setting, thus allowing the aircraft to move up and down with the vertical-air-motion field. This type of flying is typical; aircraft usually maintain an assigned altitude with power adjustments to keep the airspeed within preselected limits.

While this simple technique has been replaced by much more sophisticated concepts, it can be more finely tuned if the vertical motion of the aircraft, at operational gross weight, is calibrated at various airspeeds and power settings in a quiet atmosphere. The response time of an aircraft to a change in the vertical motion of the air is given by equations of motion of the aircraft, which can be found in standard texts. Lenschow (1976) has shown, using simplified airplane equations of motion, that responses of the North American T-28 and the Beech Queen Air are fast enough to distinguish a "top-hat" profile of thunderstorm updraft velocities from a "smooth" profile. The advantages of using the aircraft itself as a sensor are simplicity, low cost, and smoothing of the vertical-motion data. The disadvantage is the lack of accuracy over a broad range of conditions, which is a result of the difficulty in maintaining aircraft altitude or airspeed within close tolerances.

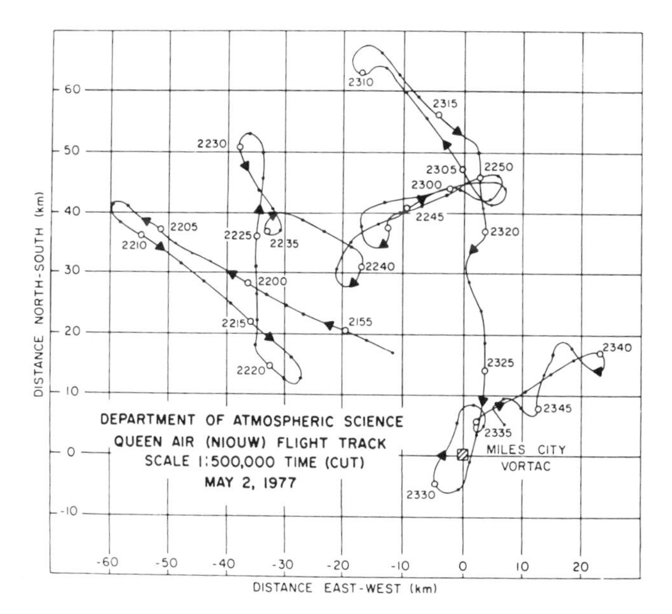

Figure 3.7. The flight track of the Wyoming Queen Air, reconstructed using VOR-DME and airborne Doppler navigation radar. The aircraft position is with respect to Miles City, Mont., VORTAC navigation aid; the research supported the High Plains Experiment.

Greater detail in the vertical-motion field can be obtained with sensors installed on the aircraft to quantify the motion of the aircraft with respect to the Earth and that of the air with respect to the aircraft. The velocity of the air with respect to the aircraft is usually measured by sensors on a boom in front of the aircraft, and the velocity of the aircraft with respect to the Earth is typically determined from accelerometers or radar altimeter measurements. Inertial navigation systems have been used extensively to provide the necessary vertical accelerations, which are integrated to determine the vertical velocity of the aircraft with respect to the Earth. Associated problems relate to the angular accuracies of the platform and the angular drift rate of the gyros.

Turbulence or small-scale vertical velocities are determined with a gust probe, which consists of fast-response pitch and yaw vanes and a pitot tube. The vanes can be either rotating or constrained. The rotating vane turns freely in the wind gust, and the angular departure of the vane from level is related to accelerations. In the constrained vane strain sensors are calibrated to accelerations. Gust probes sometimes use pressure sensors in place of vanes. The gust probe is always mounted on a forward boom to obtain measurements in the undisturbed flow ahead of the aircraft. If the inertial platform is physically attached to a rigid boom, the velocity of the vanes with respect to the platform is found by a simple geometric calculation that takes rotation into account. If the inertial platform is well removed from the boom and/or the boom has oscillations independent of the platform, then the boom is referred to as a "floating"

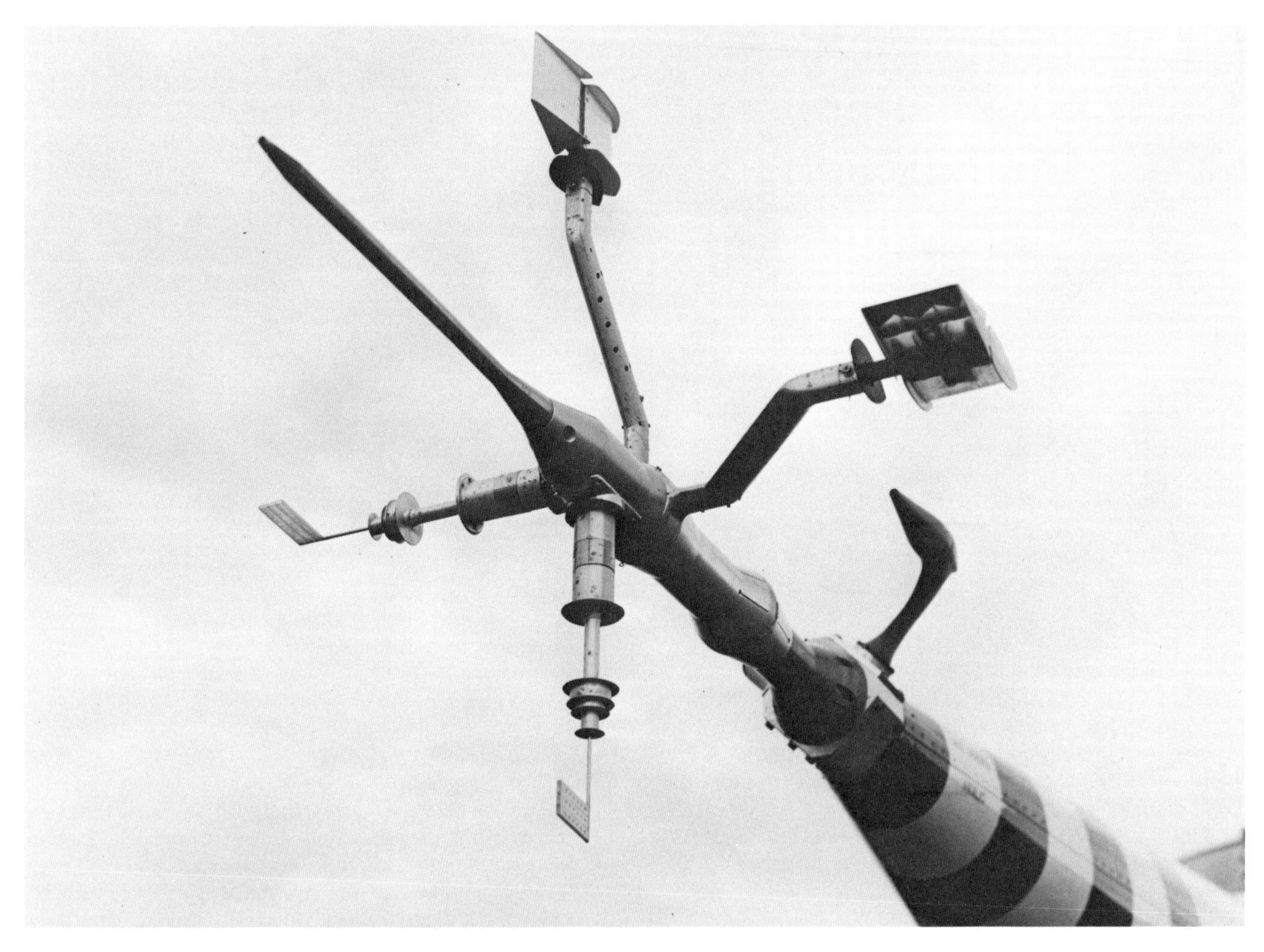

Figure 3.8. The gust-probe assembly of the NCAR Electra aircraft, including the fast-response pitot tube, fixed and rotating vanes for momentum sensing, and a fast-response temperature sensor (courtesy NCAR).

boom and must have vertical and lateral accelerometers attached to calculate the velocity of the vanes with respect to the platform. Figure 3.8 shows the NCAR gust probe mounted on a nose boom on the NCAR Electra.

Determination of horizontal air motion from an airborne platform requires knowledge of the track, heading, ground speed, and true airspeed of the aircraft. Inertial, low-frequency, and GPS navigation systems measure the ground speed and drift angle (the angle between true course or ground track and true heading of the aircraft); when these are combined with the aircraft heading and true airspeed, horizontal wind can be determined. Without a yaw vane the aircraft is the sensor that limits horizontal resolution of the wind field to the product of aircraft response time and horizontal speed. Since the reference system measures drift angle, errors in the compass heading affect determinations of wind direction rather than magnitude. If a yaw vane is used to account for relative air motion about the vertical axis of the sensor, resolution of the horizontal-wind field is determined by the Doppler system response, which is approximately 1 km.

Until recently the high resolution air-motion sensing required an elaborate gust probe vane system (e.g., Fig. 3.8). Brown et al. (1983) have demonstrated a much simpler radome gust probe system that utilizes a differential pressure technique and provides quality high-rate air-motion data without the use of an externally mounted vane system.

Finally, several hybrid systems show promise. A hybrid system is one in which an independent measurement of the position of the aircraft is used to update the long-term drift inherent in state-of-the-art inertial navigation systems. Independent position measurements include pressure altitude, DME-VOR, multiple DME, VLF, and even multiple inertial navigation systems on the same aircraft.

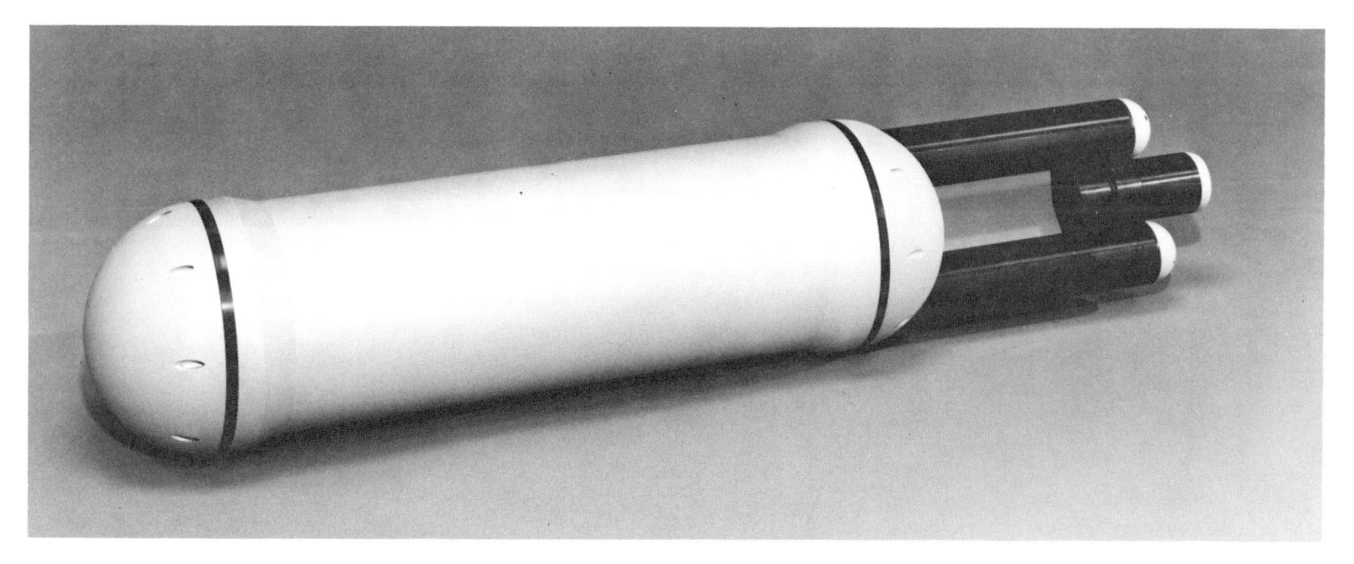

Figure 3.9. The PMS Forward Scattering Spectrometer Probe (FSSP) (courtesy PMS, Inc., Robert Knollenberg; see also Knollenberg, 1976).

4. Microphysical Measurement Techniques

The basic microphysical parameters required for understanding precipitation processes within severe storms are liquid-water content and the sizes and concentrations of cloud droplets, raindrops, and ice particles. Measurements of cloud condensation nuclei, ice nuclei, and electrical parameters are also desirable but are not addressed here.

Measurement of hydrometeors within the severe-storm environment is difficult for a variety of reasons:

1. Difficulty in obtaining specimens over many orders of sizes and concentrations, 10^{-6} to 10^{-1} m in diameter, and 10^2 to 10^9 per cubic meter, respectively.
2. Problems with instrument calibration stability.
3. Difficulty in obtaining suitable sensors and aircraft sturdy enough to withstand turbulence and hail.

To the mid-1980s the armored T-28, operated by the South Dakota School of Mines and Technology, has been the best system for making microphysical measurements within severe storms.

a. Cloud-Droplet Size and Concentration

Observations of the cloud-droplet spectrum have been made using impactor sampling, optical devices, and electrostatic disdrometers (see Veal et al., 1978). Impaction sampling collects cloud droplets on a depressible surface; droplet craters are measured, and a known relation between droplet and crater size is applied. Cloud droplets can also be protected in oil and photomicrographed or replicated in a material such as Formvar plastic. The necessity for continuous or at least frequent sampling, exposure to an undisturbed airstream that may be difficult to identify, difficulty in determining collection efficiencies, and difficulty in automating for minimum operator attention all reduce the value of impaction devices.

At least three devices have been used for optical sensing of cloud-droplet size and concentration: the Particle Measuring System spectrometer (Knollenberg, 1976), the Blau nephelometer (Blau et al., 1970), and the devices developed by Soviet scientists and described by Laktionov et al. (1972). These devices count individual cloud droplets by detection pulses of scattered light and determine their size by electronic analysis of pulse height.

The most important use of these devices is to observe droplet concentration and spectrum shape. The behavior of the tails of spectra (concentrations of larger droplets) is often of particular importance. Ease of data analysis and good spatial resolution of spectra are also highly desirable. The optical devices and the electrostatic disdrometer continually monitor the spectrum with about 2-mm-size discrimination, and the main part of the spectrum is in readily analyzable form. Optical and disdrometer techniques are superior to any impaction technique with respect to sample continuity and ease of data analysis.

Figure 3.9 shows the Forward Scattering Spectrometer Probe (FSSP), manufactured by Particle Measuring Systems, Inc., and Fig. 3.10 illustrates its optical system. FSSP in-cloud flight examples are given in Fig. 3.11.

Critical to the success of the various optical sensing techniques has been the difficult job of calibration, primarily in wind-tunnel tests. Baumgardner (1983), Baumgardner and Dye (1983), Baumgardner et al. (1985), and King et al. (1978) have led this standardization.

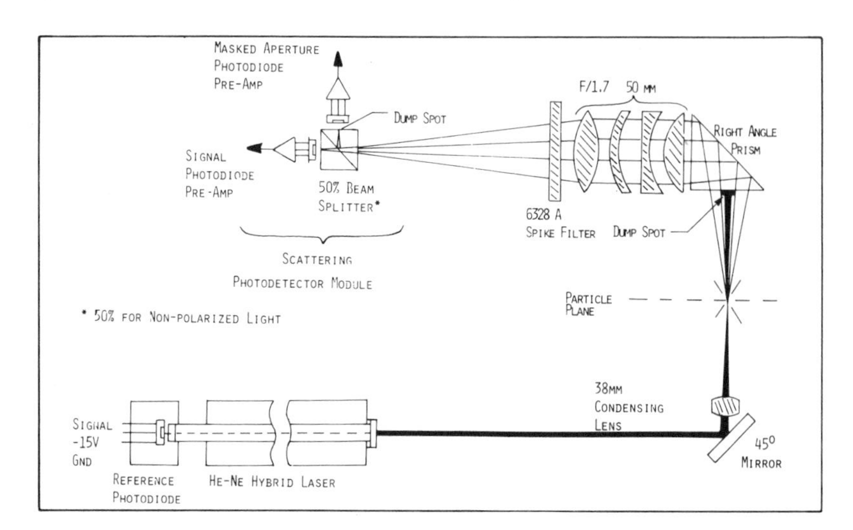

Figure 3.10. The optics diagram for the PMS FSSP probe (courtesy PMS, Inc.).

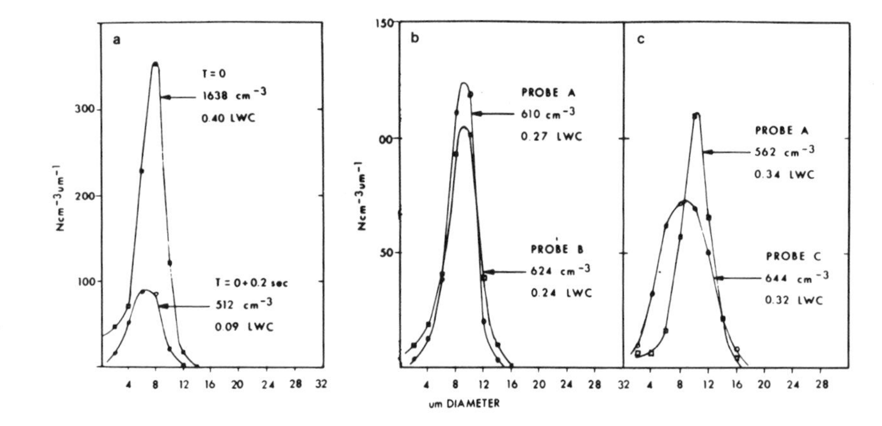

Figure 3.11. Examples of FSSP cloud-droplet spectra. An exciting aspect of the Knollenberg systems is the essentially real-time availability of such spectra (courtesy PMS, Inc.).

b. Liquid-Water Content

Liquid-water content can be calculated by summation of drop-size spectrum (observed by the previously discussed devices) or by direct bulk measurement techniques, using devices such as evaporators (Ruskin, 1967; Kyle, 1975), optical flowmeters (Brown, 1973), transmissometers (aufm Kampe and Weickmann, 1952), microwave radiometers, paper tapes (Warner and Newnham, 1952), synthetic filaments (Sasyo, 1968), and hot-wire total water indicators.

The most frequently used device is the Johnson-Williams (J-W) hot-wire instrument (B in Fig. 3.5). The device is very convenient to use, and the data obtained are readily analyzable, but the unit has a limited response to droplets with diameters larger than ~3 mm. An improved version alleged to have faster response time, response to larger droplets, and greater stability has been developed by Merceret and Schricker (1975). The claimed characteristics, however, have not been well demonstrated. Because of poor response to larger droplets and because calibration and stability are difficult to maintain, the hot-wire devices may yield misleading information. Total water evaporators have better sample volume, response time and sensitivity to all forms of water. Window determination and source and detector drift are potential problems with some Lyman-alpha sensors.

c. Raindrop Size and Concentration

Raindrop-sized hydrometeors can be measured by foil impaction, optical devices, and momentum sensing. When hydrometeors hit lead foil backed by mesh, an impression is made. The size of the impression may be correlated with the size of the hydrometeor. The lower limit of detection is typically ~250 μm, and for 750 μm it is sometimes possible to distinguish water from ice on the basis of image shape (Knight et al., 1977). Impaction devices usually have adequate sample volume and can be either continuous or

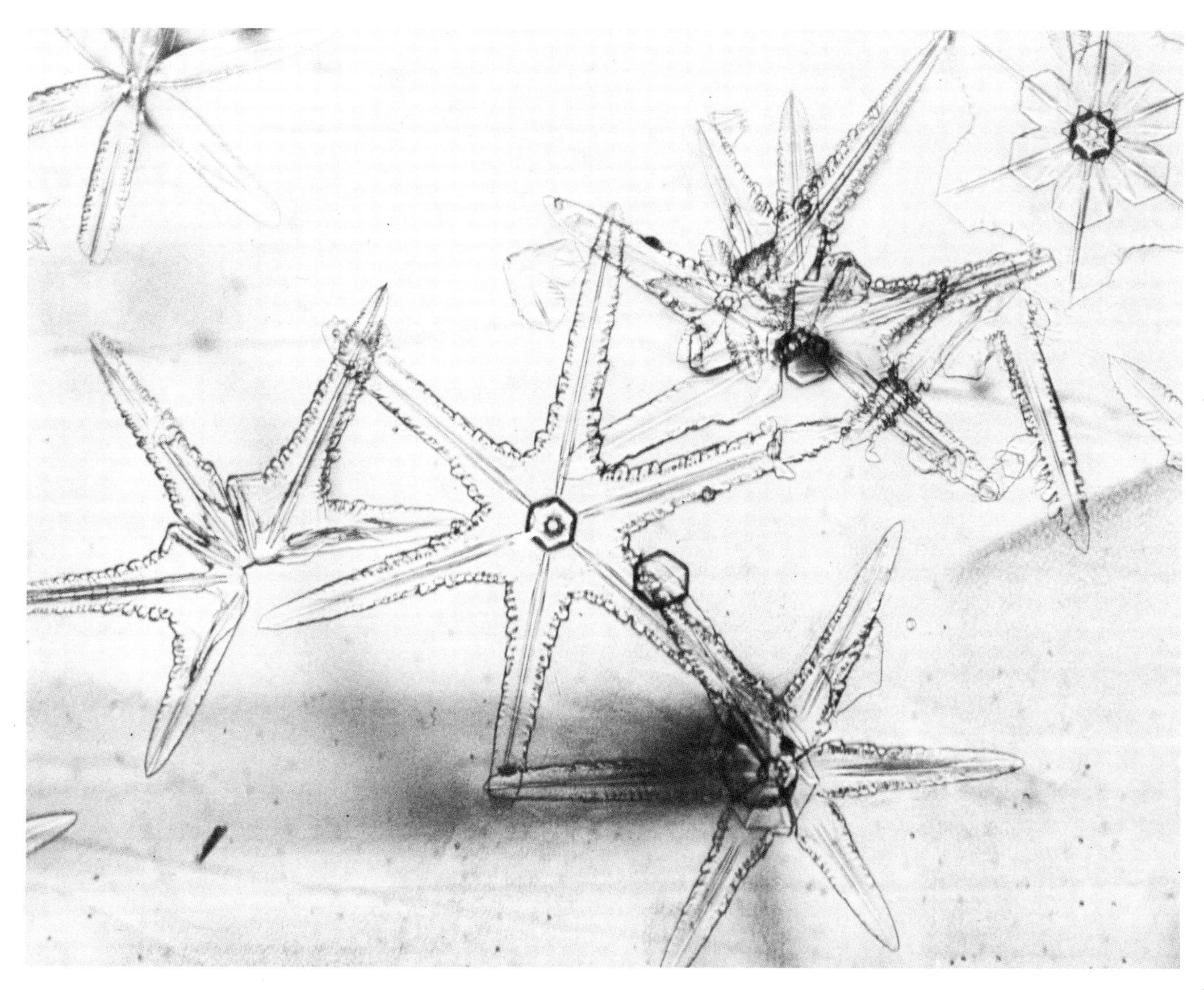

Figure 3.12. Ice crystals captured by a single-shot oil-coated slide (courtesy University of Wyoming).

single shot. Data reduction is difficult to automate and is therefore time-consuming.

Data from optical devices are available without extensive hand analysis. A device developed by Knollenberg (1970) uses a linear photodetector array onto which passing drops are imaged as shadows by a collimated light beam. The number of shadowed elements indicates the drop size, and the concentration is determined by counting the number of times one or more elements of the array are shadowed. A histogram of the drop spectrum can be viewed in real time, and the information can be recorded on magnetic tape for easy data analysis. The cloud-droplet size distributions shown in Fig. 3.11 are similar to those obtained by the precipitation-sized drop sampler described here. One disadvantage of the Knollenberg instrument is that it does not discriminate between liquid water and ice particles. In addition, when the image of a particle overlaps the end of the array, the event is disregarded. Because such events are thrown out, the sample volume is reduced by greater amounts for larger particles, although we desire larger-sample volumes for larger particles, which are rarer. An alternate approach is to employ two-dimensional optical imaging whereby the sample volume increases with particle size.

d. Ice-Particle Size and Concentration

Weickmann (1947) practiced direct capture of ice crystals on oil-coated slides. This approach has also been used by scientists at the University of Wyoming for single-shot collections (Fig. 3.12). In this technique ice crystals are collected on a slide coated with mineral oil to ensure adher-

Figure 3.13. The two-dimensional optical imaging probe on the laboratory bench (shown mounted on the NCAR Queen Air in Fig. 3.5) (courtesy PMS, Inc.).

ence, after which the slide is immersed in a silicone oil (Dow Corning 330 fluid) whose temperature is maintained below 0°C. Since mineral oil is insoluble in silicone oil, the crystals remain on the slide even under mechanical agitation; the oil prevents evaporation. The original crystals can then be viewed at any desired magnification. The advantages of collection over replication are the absence of artifacts, simplicity of operation because of temperature controls, improved resolution, and shorter drying time.

The most difficult problem to overcome with either collection or replication techniques is the fracturing of ice crystals on the collection surface. Decelerators have been designed (Mossop et al., 1967; Mossop and Wishart, 1970; Hobbs et al., 1973; Davis and Veal, 1974) to reduce impaction velocity, thereby allowing collection or replication of even fragile crystals. Hallet et al. (1972) used a continuous replicator in which a stagnant air layer was maintained between the sampling aperture and the collecting surface to slow down incoming crystals. Schreck et al. (1974) developed a system in which crystals entered a tube placed outside the aircraft. Air flowed into a container of silicone oil kept at dry-ice temperature. Because only a small amount of air was allowed to pass through the tube, deceleration of the crystals was accomplished. The collected crystals could then be examined in a cold room at any desired magnification.

Devices that detect ice particles by optical techniques include the Mee (named after its inventor), the University of Washington counters, Cannon's camera (and others), the holocamera, and the PMS two-dimensional spectrometer.

The Mee (Sheets and Odencrantz, 1974) and the University of Washington ice-crystal counters (Turner and Radke, 1973) use the property of an ice crystal to rotate the plane of polarization of polarized light as the basis for their operation. Both devices make possible real-time display of ice-crystal concentration and provide for rapid data reduction and analysis. No size information is available from either instrument. Because of significant size-dependent efficiency corrections (Turner et al., 1975), which may also depend on crystal shape, independent shape and size information is desirable to obtain more reliable crystal concentrations.

The two-dimensional optical imaging probe developed by Particle Measuring Systems, Inc. (Knollenberg, 1976), is C in Fig. 3.5 and is also shown in Fig. 3.13. This device incorporates an optical array similar to that in the liquid hydrometeor device described above. Particles shadow an array, and by frequently recording the status (shadowed or illuminated) of each element in the array as the particles pass, a two-dimensional picture can be reconstructed. The aircraft unit uses a 32-element array, with a typical resolution of about 25 μm per element. The maximum resolution per element is a function of aircraft velocity. Various configurations of the probes are available for detection of cloud sizes of precipitation-size particles. Figure 3.14 is a block diagram of the optical system. The concentration of particles as well as the images can be displayed in real time in the aircraft. Examples of images obtained from rain and dendritic ice crystals are shown in Fig. 3.15.

The obvious advantages to obtaining direct photos of cloud elements are in situ sampling and more detailed information on the shape of the particles. The principal difficulties are achieving sufficient sample volumes with the required magnification and resolution and the necessity to "stop the motion" of the particles being photographed. Elliot (1947), McCullough and Perkins (1951), Lavoie et al. (1970), and Cannon (1974) describe various systems for photographing cloud elements. Analysis of the data collected using any of these systems is extremely demanding in time and manpower. Perhaps, however, computerized analysis could enhance the utility of direct photography of cloud elements.

Holographic imaging has also been used to observe cloud elements. Its main advantages are large sampling volumes and good rendition of spatial distributions. Analysis of holographs is performed by illumination with a laser light fol-

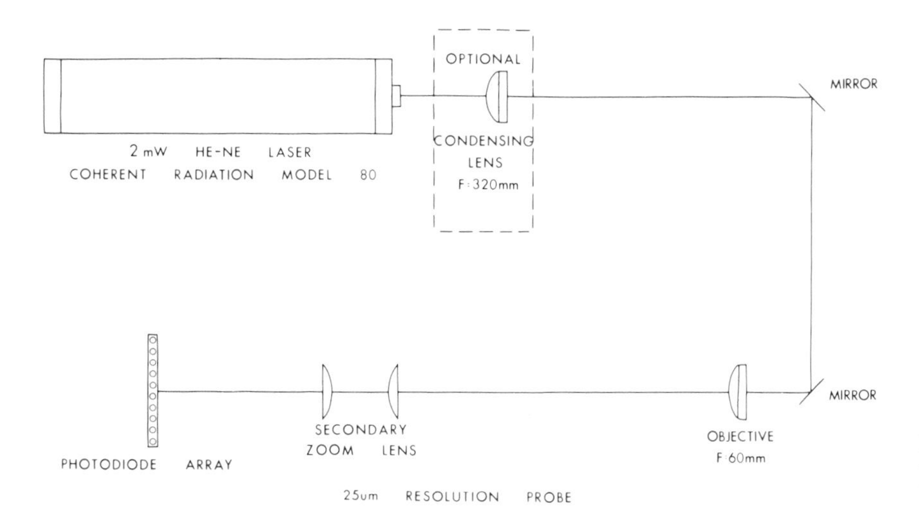

Figure 3.14. The optics diagram for the two-dimensional PMS probe (see Knollenberg, 1976; courtesy PMS, Inc.).

lowed by scanning the reconstructed three-dimensional image field with a microscope or camera arrangement (see Wilmot et al., 1974).

5. Airborne Radar

Airborne radar is invaluable for studying severe storms with an aircraft since it can provide the aircraft crew with immediate knowledge of storm intensity on the intended flight path. High-quality commercial radar systems with many antenna sizes and scope displays are available. The frequency and power of the system selected are functions of the intended use and the size of the aircraft on which the unit is to be installed. An appropriate system allows weak-echo regions to be entered and allows close maneuvering near severe storms with reasonable safety. Several airborne radar displays are illustrated in Vol. 1, Chap. 7, and weather radar science and engineering are detailed in Chap. 10 of this volume.

Recently airborne radar development has been extended to include Doppler capabilities. A 3-cm (x-band) Doppler system is being flown on the NOAA P-3 Orion aircraft and has provided detailed information regarding thunderstorm structure. Hildebrand and Mueller (1985) and Mueller and Hildebrand (1985), using the P-3, have shown that dual Doppler radar analyses of thunderstorm wind and reflectivity structure can be obtained from a single aircraft. These techniques represent potential future significance, and development of additional airborne Dopper radars is under way at both NCAR and NASA.

6. Airborne Communications and Data-recording Systems

Data systems employed in atmospheric research range from knee-pad notes to computer-directed devices. Choice of system for a particular mission depends on need for real-time decisions during the mission and on the resolution required in the recorded data.

Because of weight or space limitations on research aircraft, on-board data recording may not be desirable. A telemetry link to a ground station can replace certain recording systems on the aircraft. The critical factor in a telemetry link is its range-altitude capability.

If the aircraft crew are to optimize observations of storms and interact with other personnel, descriptive parameters must be available in near real time. The best solution at present appears to be the use of a real-time system that provides for three important time scales, or learn-time cycles, of data use.

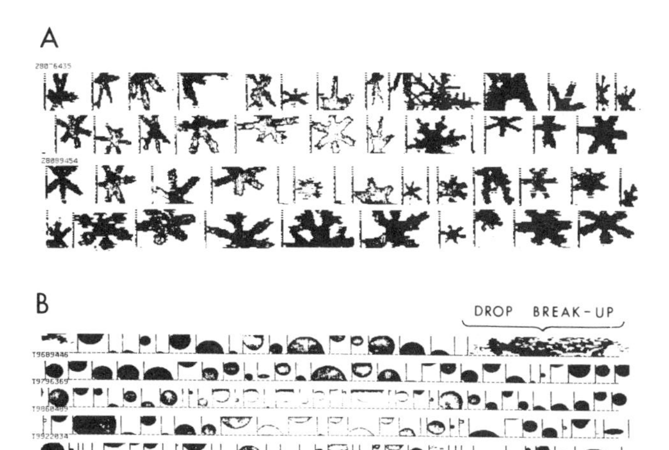

Figure 3.15. Examples of PMS two-dimensional probe imaging. A: Snow sampling for identification and sizing of ice-crystal types. B: Examples of rain sampling. The aircraft airspeed was 80 m s^{-1} (courtesy University of Wyoming).

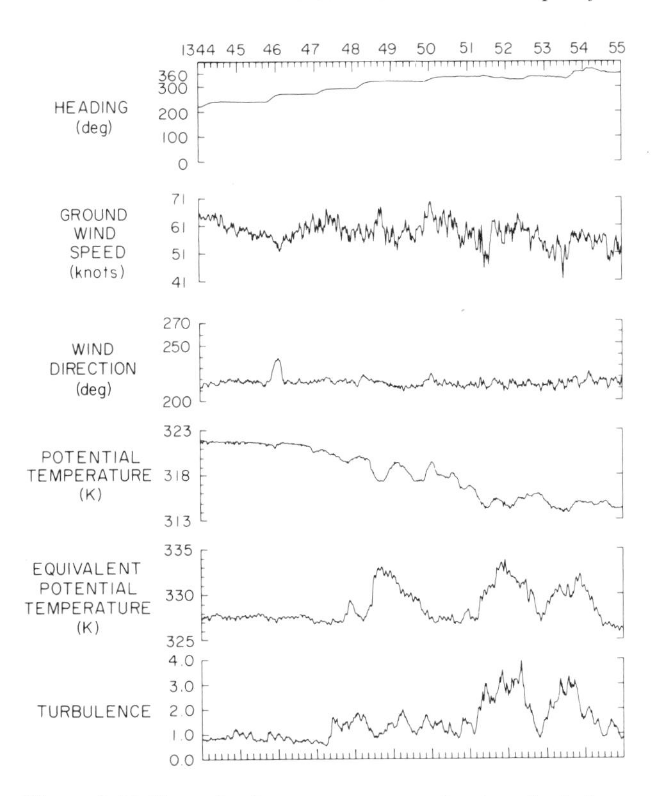

Figure 3.16. Example of computer-processed meteorological variables as a function of time (Bensch and McCarthy, 1978).

At the shortest time scale, the real-time cycle, the aircraft crew can use the measured parameters to aid the conduct of the research flight in real time and relay information to ground sites to aid in overall operation of the experiment. With the aid of the computer on board the aircraft, the necessary conversions and corrections for various parameters can be made, and variables such as potential temperature, specific humidity, and wind can be determined and displayed.

The next scale in a research project is the single-mission or one-day cycle. With a properly configured data system hard copies of selected portions of the data can be obtained very soon after each flight, enabling the scientists to reconstruct and review mission highlights and compare their data with other observations.

The third, or seasonal, cycle is the review of accumulated daily data and many types of detailed analysis that may have been suggested from the real-time or single-mission learn-time cycles. An example of an in-depth postexperiment data-examination capability is the Research Data Support System (RDSS) at NCAR's Atmospheric Technology Division. The recorded data should be in the fundamental form to facilitate new computations or analysis schemes.

Airborne computer-directed data systems are usually designed to be operated in one of three modes selectable by simply entering a new program into the on-board computer. These modes are data acquisition, self-check or preflight-postflight program, and data processing.

In the first mode the sensor outputs are signal-conditioned, transferred to the computer through a multiplexer, and then formatted and recorded. The computer also calculates various derived parameters for real-time display. Selected fault detection and identification during data collection can also be provided. Fault indications are desirable if the instantaneous values or rates of change of variables fall outside a predetermined range. The computer can also monitor the status of the tape recorder and perform a "read-after-write" check on the data recorder. If any malfunctions are detected, a signal or code that identifies the fault can be displayed in the aircraft.

The second mode of operation is very useful as a preflight-postflight or troubleshooting tool. The data system can be designed with sensor simulation circuitry that can be controlled by the computer to produce any desired output over the range of the sensor being simulated.

Operation in the third mode takes place on the ground after the mission. Line printers and teletypes are interfaced to the aircraft system, and processed data are obtained on hard copy. The ability to view data with only limited processing during field projects is often very useful. With computer-compatible formations, extensive processing of the data on other computer facilities is straightforward. Figure 3.16 is a typical example of the final derived parameter output from an airborne data system.

Internal and air-to-ground communications among the crew are usually a vital component of the data system and overall research mission but are too often neglected. Communications requirements for research often exceed the installed capacity in aircraft used principally for transportation. For a flying laboratory to be effective, the scientific crewmembers and pilot must be able to communicate easily with each other. In addition, the system should allow each crew member to isolate his conversation from that of others. This is required when the pilot communicates with FAA controllers while other crew members carry on other discussions.

Flying laboratories with computer-directed data-entry systems and an array of modern sensors are especially valuable when versatile communications capabilities allow close coordination with other aircraft, ground mobile systems, ground radar systems, and FAA controllers.

The sophisticated airborne data system used by NCAR has been reported by Walther (1985). It provides output in all three modes described above.

7. Special Uses of Aircraft in Thunderstorm Research

a. Chaff as an Air-Motion Sensor

Since World War II radar reflective chaff has been used to jam radar installations as an antiradar surveillance tech-

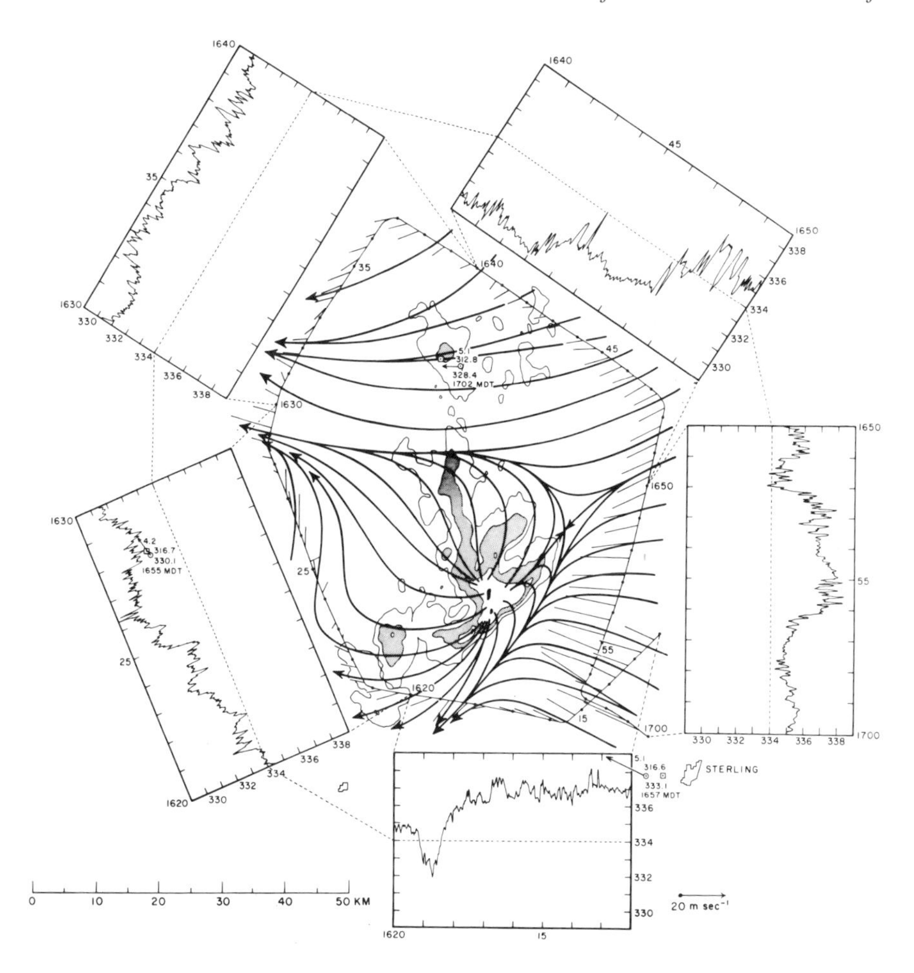

Figure 3.17. NCAR de Havilland Buffalo aircraft data combined with other aircraft and radar data. The Buffalo track is at 3 km, positioned relative to radar echo at 1655 LST. Wind vectors (scale at lower right) and the streamline pattern represent flow relative to the storm. Equivalent potential temperatures based on variables measured along the track are distributed on surrounding panels as a function of time and space. Sharp gradients near 1628 and 1647 are interpreted as boundaries between inflowing and outflowing branches of circulation (Foote and Fankhauser, 1973).

nique. It was soon discovered that chaff released from an airplane could be used as radar tracer to determine air motion. Jessup (1972) of NSSL made extensive use of point chaff packets to look at wind flow in thunderstorm environments. Marwitz (1973), of the University of Wyoming, used chaff to trace vertical motions within the echo-free vaults of thunderstorms over the National Hail Research Experiment research area.

McCarthy et al. (1974) used chaff to fill a large volume of echo-free air in the near-thunderstorm environment. A unit that includes five chaff packets is dropped from an airplane; individual chaff pockets are released every 500 m during descent. The result is a column of chaff exending from near the airplane flight level to approximately 2,500 m below flight level. By a series of releases a large volume can be filled, and a Doppler radar system can measure the instantaneous winds in an otherwise nonreflective volume.

b. Lightning-Suppression Experiments

Heinz Kasemir, of NOAA, has developed a technique of lightning suppression that uses chaff distributed from aircraft. The idea behind the technique is simple: the corona discharge property of a chaff fiber causes a silent conduction of current and neutralizes separated charge centers. Results of several experiments have been promising but not definitive, as noted in Vol. 2, Chap. 16.

8. Synthesis of Airborne Data

Although we are not describing research achievements of thunderstorm flying in detail, we do offer several examples in which instrumented aircraft data have been combined with other direct and remote sensors to yield valuable insight on thunderstorms. This combining of sensor data is an essential step in production of whole pictures.

The example in Fig. 3.17 is taken from research at NCAR, as part of the National Hail Research Experiment. A severe hailstorm was observed by surface-based radar, airborne instruments, and other surface sensors. NCAR de Havilland Buffalo aircraft data are super-imposed on other data. Radar intensities, airborne-measured equivalent potential temperature, and storm-relative airflow streamlines

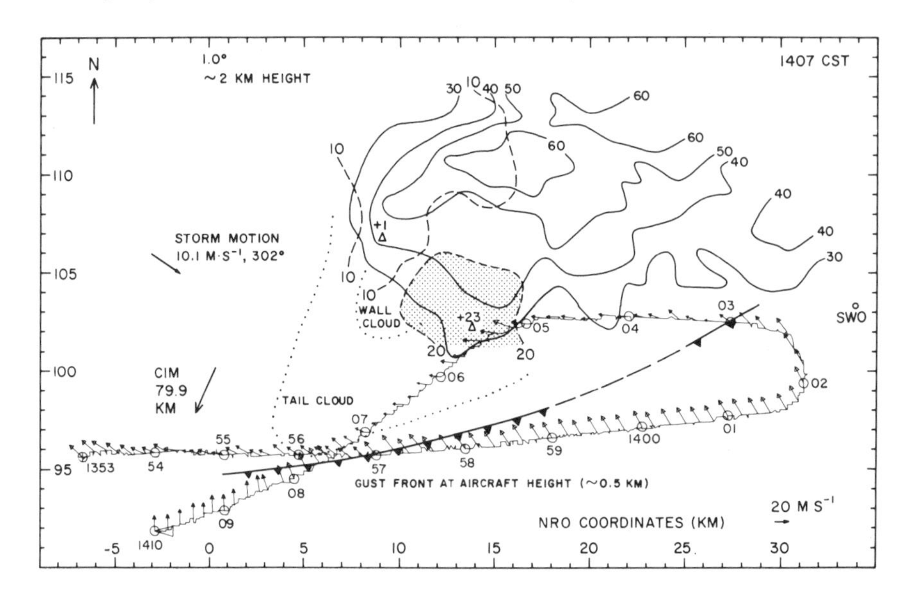

Figure 3.18. A synthesis of NCAR Queen Air data and NSSL Doppler radar data: Cimarron Doppler radar, 1° elevation angle, reflectivity (solid contours, dBZ), and radial velocities (corrected for storm motion) at 1407 CST. Maximum and minimum values of radial velocity are indicated by Δ symbols, and numerical values are shown. Area of radial velocities greater than 20 m s^{-1} is shaded. NCAR research aircraft storm-relative track and wind vectors are shown for 1353–1410. The position of the inflow cloud was determined from aircraft photography. SWO indicates the location of the Stillwater Municipal Airport (Bensch and McCarthy, 1978).

are shown. Such combined data present elements of storm structure at a level of detail that supports new inferences on the processes governing storm development.

Figure 3.18 is another example of synthesis, this time using NSSL Doppler radar data with NCAR Queen Air aircraft. The position of the thunderstorm gust front has been deduced from aircraft-measured winds and superimposed on NSSL Doppler wind and intensity measurements.

9. Conclusion

We have described equipment utilized on research aircraft and have indicated why such deployments are made in the context of thunderstorm research. While aircraft measurements are important standing alone, their greatest strength lies in the synthesis of various sensors and analysis techniques.

4. Photogrammetry of Thunderstorms

Ronald L. Holle

1. Introduction

Clouds are visible tracers of atmospheric motion; for this reason pictures of clouds are among the most revealing measures available to the meteorologist. When the pictures are combined with information from other sensors, the evolution of the condensation process can be followed, the influence of internal and external atmospheric motions on the cloud matter can be seen, and the ultimate disappearance of the cloud can be viewed. Photographs can capture the tracers of a wide variety of processes in an immediately comprehensible form. Although many atmospheric phenomena on larger and smaller scales are not directly visible, thunderstorm evolution is ideally suited to photographic description. The thunderstorm, in many cases, is small enough that the whole of one side is visible from one location.

A working definition of photogrammetry appears in the American Society of Photogrammetry's monthly journal, *Photogrammetric Engineering and Remote Sensing:* "Photogrammetry is the art, science, and technology of obtaining reliable information about physical objects and the environment through processes of recording, measuring, and interpreting photographic images and patterns of electromagnetic radiant energy and other phenomena."

Photogrammetric analysis can proceed to every transient feature in thunderstorms, but the process can be tedious. In many situations qualitative information can be found easily from a photo without formal analysis. As in all other scientific research, the motivation for study must be clearly in mind before analysis is begun. In most cases the motivation is supplied by examination of the same clouds using meteorological data from other sources, and little may be gained by detailed examination of aspects of a cloud not readily associated with other features of interest.

In view of the potential benefits from photographic analysis in conjunction with other data sources, every thunderstorm study should have a photography program. If the procedures of photographic analysis are known in advance, the photography program can be designed for the desired accuracy and the time available for analysis.

Photogrammetry has been a substantial aid to thunderstorm research, but its use may be limited by (1) impressions that photoanalysis is always difficult, (2) inability to see inside large clouds or to photograph them at night or during days with poor visibility (photographs are complemented by radar and other remote sensing techniques that give information on interior condensates), and (3) the abundance of data in photographs. Less detailed data sources provide general information about a cloud rather easily; careful planning is essential to effective analysis of the large quantities of data inherent in photographs.

This survey of cloud photogrammetry treats pictures in the visible portion of the spectrum, as well as the use of various cameras and techniques.

2. Basic Equations for Photogrammetric Analysis

The angular field of view of a camera is

$$\tan \frac{\alpha}{2} = \frac{d}{2f}, \qquad (4.1)$$

where α is the full horizontal or vertical angle of the lens, d is the horizontal or vertical dimension of the original film frame, and f is the focal length of the lens (Wolf, 1974, p. 63). The dimensions of d and f are in the same units, such as millimeters.

If the horizon is in the picture and the camera focal length is known, heights and widths of every cloud element can be calculated with appropriate equations dependent on camera tilt. For photos from an untilted camera, with the horizon in the center of the picture, McNeil (1954, problem 43) gives

$$\tan \theta = \frac{x}{f} \qquad (4.2)$$

and

$$\tan \sigma = \frac{y \cos \theta}{f}, \qquad (4.3)$$

where θ is the horizontal angle to the right or left of the picture's center, σ is the vertical angle from cloud feature to horizon, and x and y, the coordinates of the measured cloud point, are the horizontal and vertical distances from vertical and horizontal lines that divide the picture into halves. In practical application, when the photograph is projected, x and y may be measured in millimeters, and f is multiplied by the amount of magnification M. The value of M is the ratio of the projected image size to the original image size.

When the picture is tilted, the equations are more complex. Then, to find angles to the cloud feature, one must find the camera tilt. First, find the exact center (the principal point P) of the entire photograph by locating the intersection of diagonals that connect the picture corners. Second, measure the perpendicular from point P to the horizon, at point A. Then substitute into $\tan t = PA/f$ to find the camera tilt angle t above or below horizontal.

Exact equations for a cloud point on a picture tilted at angle t are

$$\tan \theta = \frac{x}{f \cos t - y \sin t} \tag{4.4}$$

and

$$\tan \sigma = \frac{\cos \theta (y \cos t + f \sin t)}{f \cos t - y \sin t}. \tag{4.5}$$

Tilt is 0° for a camera looking horizontally, 90° directly upward, etc. Note that when $t = 0°$, Eqs. 4.4 and 4.5 reduce to Eqs. 4.2 and 4.3. A derivation of a similar pair of equations is given by Wolf (1974, p. 366). McNeil (1954, problem 43) gives almost the same equations for a different definition of t.

Heights of cloud features can be calculated once the angles θ and σ are found from one of the two previous pairs of equations. To find the height, use

$$h = r \tan \sigma, \tag{4.6}$$

where r is the range to the cloud feature. Methods of finding r are given in Secs. 3 and 4 below for several types of pictures. The importance of determining range as accurately as possible is seen clearly from Eq. 4.6. For example, for a change in r of 1 km, h changes by 360 m when the cloud-top vertical angle is 20°. Effects of earth curvature and atmospheric refraction should be included by using $C = 0.0676\, r^2$, where the correction C to the cloud height is in meters and range r is in kilometers. The value is always additive to the cloud height. C is substantial at large distances, but may be omitted for larger clouds at ranges less than 30 km, where C is less than 60 m.

The cloud width is found by determining the horizontal angle θ to the right and left sides of the cloud feature. To find cloud width, take the difference γ between these two angles and use

$$w = r \tan \gamma. \tag{4.7}$$

Width is as sensitive as height to changes of range. No correction C is applied here because the atmospheric variations that cause refraction are systematic only in the vertical direction. The basic equations, Eqs. 4.1 through 4.7, are not difficult to use, although the preparation of an accurate diagram for measuring x and y can be time-consuming. Considerably more complex expressions may be necessary for special applications or corrections to unique situations. Stereoscopic analysis, tornado photography, and downward-looking pictures usually use different mathematical expressions.

3. Photogrammetry of Ground-based Pictures

In the following sections the broad categories of thunderstorm photography and analysis proceed from simplest to most difficult. The level of difficulty is defined by the time period, personnel, and equipment involved in data collection and the expertise, equipment, and sophistication required for analysis. In any case, the product collected substantially determines the method to be followed in full use of a picture.

a. Hand-held-Camera Pictures

The simplest photographic data collection involves still photography, as opposed to motion pictures.

Field personnel can take pictures with very little time and cost, and the prospect of a photo of the exact cloud of interest justifies the effort. A polarizing filter is often helpful for increased contrast between clouds and sky. The picture must be labeled with time and location, and the horizon must be visible. An example of a short-lived, rapidly moving cloud feature photographed during a field project is shown in Fig. 4.1.

Analysis of the picture may include qualitative documentation of cloud organization, orientation, timing, shear, and stage of life cycle. However, one may calculate the cloud's azimuth if the angle of view of the camera is known from Eq. 4.1 and angles to distant landmarks in the picture are measured. The analysis potential of hand-held-camera photos is substantially improved if some supporting evidence gives distance to the cloud feature. For example, Holle and Maier (1980) took a series of pictures of a tornado. Its location was found by examining the patterns of rain-gage reports and radar echoes in the vicinity and then combining this information with the photographs' views of the relative positions of tornado and rainfall. The standard radar scan near 0.5° elevation angle is frequently appropriate for determining r to lower levels in clouds. Vertically scanning radars, however, usually provide information on top locations, which may differ significantly from the lower-level scan position of the cloud. Given range from these methods, or from pictures taken at another location, Eqs. 4.2 to 4.7 can be used to measure sizes of visible cloud features at picture time. Hand-held-camera photographs from the ground have been taken and analyzed in virtually every thunderstorm study and can help in the selection of case-study days on the basis of especially illustrative photographs such as Fig. 4.1.

Figure 4.1. A hand-held-camera photograph taken from the surface at 1540 EDT on 18 June 1973 during the Florida Area Cumulus Experiment (FACE). At this time an outflow was moving across the mesonetwork at 15 m s^{-1} (photo from Michael Maier, Titusville, Fla.).

b. Single Time-Lapse-Camera Pictures

The time-lapse movie camera is the next-simplest camera system to operate in the field. When an unattended camera is operated all day, it can be located next to a radar site or observation station. The camera should be pointed toward a mountain range that can be expected to produce afternoon thunderstorms, or along a lake shore or ocean beach, where large clouds may be produced by the lake or sea breeze convergence zone, or toward a target area for an experiment, as shown in Fig. 4.2. A frame from a time-lapse film of thunderstorms over mountains is shown in Fig. 4.3. The camera can be placed perpendicular to the wind, so that clouds move across the view, or toward an advancing cloud system. One inexpensive system amenable to detailed analysis is the 16-mm camera (Fig. 4.2) with color film and a wide-angle lens, taking pictures at intervals ranging from one frame per second to one frame per minute, depending on cloud motions. Automatic exposure controls can be used to cover different cloud lighting conditions. Videotape systems have also become quite popular for such applications.

An accurate record of the motion of condensed water vapor can be obtained by use of Eqs. 4.1 through 4.7 if several variables are known. The necessary information is similar to that needed for hand-held cameras: the range to visible cloud features, angles to distant landmarks on the horizon (such as the trees in Fig. 4.2), lens focal length, and time recorded directly on the film to determine the exact frame rate. The stop-motion projector or a variable-speed video playback system is the only special equipment needed to find precise heights and widths of clouds visible on a time-lapse film or tape. A sequence of values can then be incremented to give horizontal and vertical growth rates of clouds. In addition, the following qualitative data can be estimated from time-lapse motion pictures: differential cloud motion at various altitudes and locations, shear, or-

Figure 4.2. A time-lapse 16-mm camera system operating at the University of Miami Doppler radar site at Pahokee, Fla. The camera is using color film and is pointing toward an experimental seeding area and related surface mesonetwork and tri-Doppler radar coverage region during the 1975 Florida Area Cumulus Experiment.

ganization, extent of glaciation, presence of rain, and the time history of each parameter. In one of many applications Bhumralkar (1973) studied these qualitative factors from time-lapse film and related them to a model of flow over a heated island.

Readily available optics now permit attachment of all-sky lenses to movie cameras so that the entire sky from horizon to horizon can be seen in every direction. Cloud evolution, direction, and speed, and percentage of cloud cover can be documented easily from all-sky photos (Holle et al., 1979; Lerfald and Erickson, 1979). However, the heights and widths of clouds near the horizon at a distance cannot be obtained, since such clouds cover a very small area of the picture, just as they cover a very small area of the actual sky. Horizontal-facing time-lapse cameras are needed to study clouds along the horizon.

From the data-collection methods described above, useful pictures can be obtained from the field on a part-time basis. The following kinds of photographs require a nearly full-time photographer and/or increasingly sophisticated analysis equipment.

c. Panoramic Photographs

Panoramic photographs of the sky at fixed intervals (e.g., hourly) provide benchmarks of cloud evolution in every direction from a single location. Ideally, a wide-angle camera with a large format and large vertical angle (such as the Hasselblad Super-Wide) is used to take six to nine photographs with some overlap. A sunshade and a polarizing filter or other filter is usually desirable. Figure 4.4 shows a series of these panoramas taken on Grand Bahama Island during a field project. Thunderstorms grew over the center of the long E–W island and became well organized just north of the station. The three successive mosaics show clearly how clouds changed over the 3-h period, which was part of a day being studied for modeling the flow over a heated island. Directions are not in perfect alignment from one time to the next because individual pictures were cropped with respect to cloud entities and edges; very accurate azimuths were available in the individual photos.

A photographer must be at the field site to take pictures at the appointed time; his time may be divided with other observing duties at a radar, upper-air, or communications site. Since the panoramic photos do not give fine time resolution, in a thunderstorm study they are best used in combination with a time-lapse camera. The time-lapse camera shows a narrower region of convection, and the panoramas document developments on the sides of the time-lapse camera's pictures that may be of great importance in understanding the narrower view.

As with photographic methods described previously, the analysis with Eqs. 4.1 to 4.7 requires only the usual parameters: horizon, lens focal length, and distance to cloud feature. Analysis is usually made on an enlarged print. Typically, hundreds of panoramas are taken during an extensive field project, so, for economy, they are taken in black and white. Contact prints are frequently examined by personnel from other subprograms of the field project; more often than not it seems that interests in cloud parameters are not those anticipated in the original analysis program. Thousands of panoramas were taken in the Barbados area (Holle, 1968) for cloud population statistics, in FACE (Florida Area Cumulus Experiment), and in GATE, the GARP (Global Atmospheric Research Program) Atlantic Tropical Experiment.

d. Multiple Time-Lapse-Camera Photographs

Multiple time-lapse-camera photographs of thunderstorms can be taken from either one or several locations with cameras usually operated unattended. If a specific cloud is to be kept in view, the cameras may need to be moved if the field of view (Eq. 4.1) is small. If one photographer must oversee two or more sites or three or more cameras, about half the day may be needed for setup, take-down, time checks, and other duties. Several time-lapse cameras looking in different directions from one site can provide nearly all the information that panoramic picture sets can give. However, one local storm may obscure the views of all cameras. If

Figure 4.3. A time-lapse photograph taken on 16 July 1975 from Socorro, N. Mex., with 16-mm color film. The thunderstorm was over the Magdalena Mountains, where the Langmuir Laboratory stands (photo from Charles Moore, New Mexico Institute of Mining and Technology, Socorro).

storms are isolated but numerous, it may be better to place one camera in each of several locations. This situation prevails in the tropics and subtropics; during FACE 1975 there were as many as five time-lapse cameras at four locations (Staff, Cumulus Group, 1976), each positioned at a Doppler radar or operations center and all looking into a common region. In Socorro, N.Mex. (Fig. 4.3), where thunderstorms frequently occur over the same mountain, a network of three 16-mm cameras has been permanently installed, equally spaced around the mountain (Aloway et al., 1970). In both Florida and New Mexico additional cameras with larger film sizes have been established at some of the regular sites. Since 16-mm film provides relatively small negatives for enlargements, 35-mm cameras with timers and large magazines have been operated intermittently throughout the day when an especially good cloud view is apparent.

e. Stereoscopic Photographs

Stereoscopic photography has the advantage of producing its own distances to clouds and affords a qualitative view of relative distances without any analysis. There are proportional disadvantages in that more careful data collection and analysis are needed than for ordinary ground-based photography. For example, the lens focal lengths must be known very accurately, and angles to distant landmarks must be measured precisely. Over a summer the operation of a stereoscopic system during a field program requires a half-time

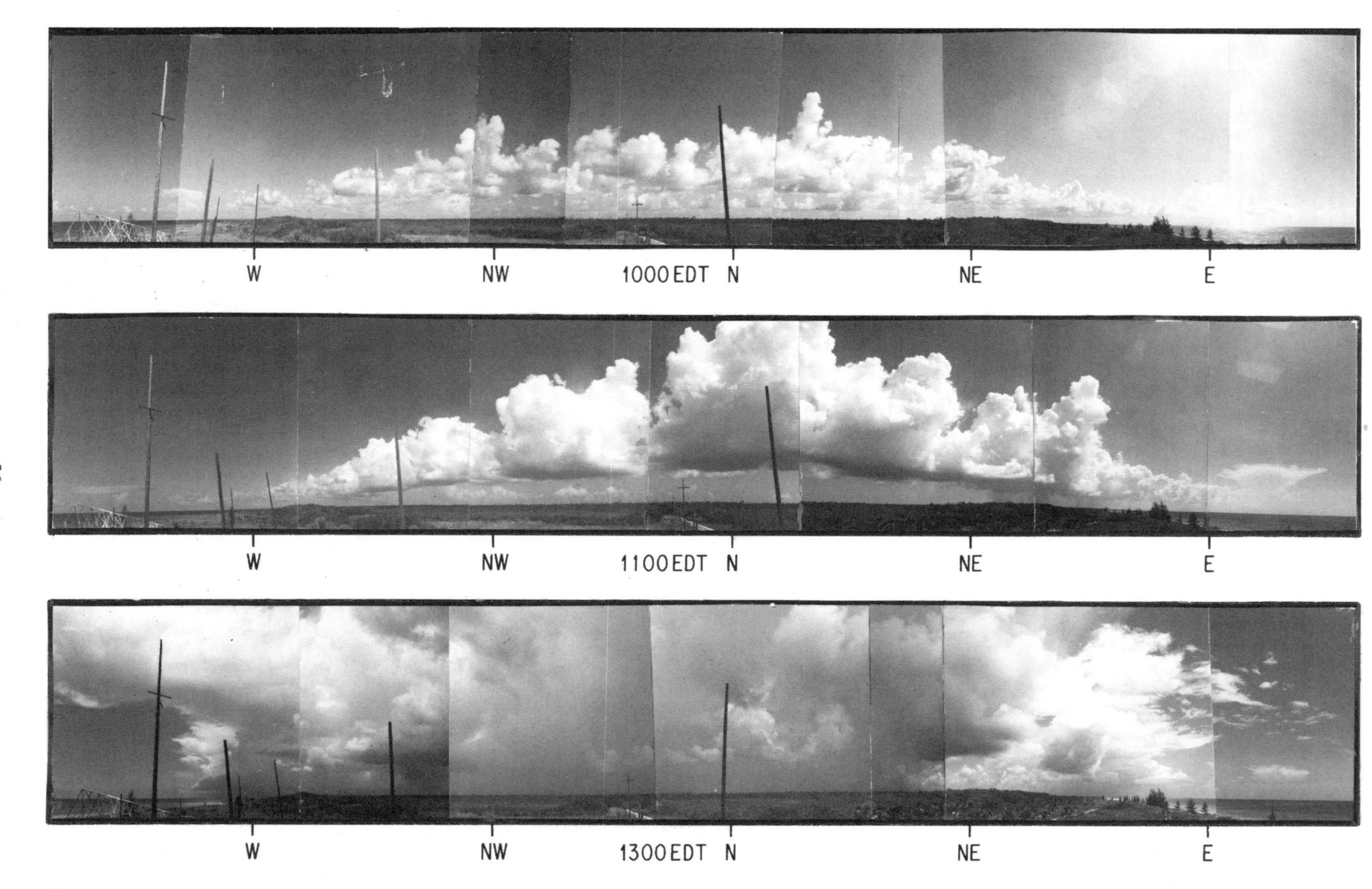

Figure 4.4. Mosaics of panoramic pictures of thunderstorms taken on the south coast of Grand Bahama Island on 20 August 1970 at 1000, 1100, and 1300 EDT. Mosaics from other days in this project are presented by Bhumralkar (1973).

Figure 4.5. A stereoscopic pair of pictures showing hailstorms over Stettler, Alberta, on 15 July 1968 at 1929 MST. This view looks east from two 16-mm color cameras located at Penhold, Alberta. The towers were growing explosively and produced hail of 1.5-cm diameter (right tower) and 1-cm hail (left). The pair is properly oriented for viewing with a pocket stereoscope (photos from Charles Warner; the photographs were taken as part of the Alberta Hail Project).

photographer. In a typical field operation an electronic signal fires two cameras simultaneously. A baseline of several kilometers provides useful parallax between pictures of the same thunderstorms located tens of kilometers away. This baseline is a compromise between range resolution, needing a long baseline, and a view similar enough for common cloud elements to appear in both pictures, needing a short baseline. The optimum situation has the cloud subtending about 10° across the baseline. Closeness of two cameras increases the prospect that a local storm or cloud obscures the view of more distant clouds that must be visible on both cameras for a stereoscopic effect. This problem is considerably reduced in drier areas, where fewer cumuli block the view and where storms grow over a fixed geographic location, such as a mountain range, where most stereoscopic systems have been placed (e.g., Simmon, 1978). An example of a stereoscopic pair is shown in Fig. 4.5.

Different methods and equipment have been used for taking and analyzing stereoscopic pairs. For example, Shaw and Marshall (1972) employed two rotating-lens cameras with fields of view of 55° vertical by 140° horizontal. Warner et al. (1973) used two 16-mm time-lapse cameras and a specially designed analysis system for studying the Alberta clouds in Fig. 4.5. Such figures can be viewed either with an inexpensive pocket stereoscope or with the cross-eye method described by Fraser (1968). The latter technique requires no special equipment. Both viewing methods provide relative distance information but do not measure any parameter directly.

Detailed measurements from stereoscopic pairs have been made with such a wide variety of devices and concepts that few generalizations can be made. One interesting approach is to use stereoscopic plotters commonly employed for mapping terrestrial features (roads, mountains, etc.) from stereoscopic aerial photography. These complex and costly machines normally measure phenomena that are stationary in time, as described by the American Society of Photogrammetry (1980). Adaptation of standard mapping procedures to stereoscopic cloud photos is not simple, but it holds promise for projects with large numbers of stereoscopic pairs to be analyzed.

f. Photos of Tornadoes and Severe Weather

Motion pictures, videotapes, and still photos from handheld cameras have given valuable information on tornadoes, waterspouts, hail, severe windstorms, and squall lines. Such pictures cannot be well planned in advance, so the analyst often encounters difficulties that result from incomplete data collection. A major problem in analyzing such pictures is introduced because the horizon and distance to storm may be missing. In this situation the sites of the photos and the tornado path must be marked and surveyed to determine the angles to power poles and other landmarks in the picture. In a research project the pursuit of such pictures requires one or more dedicated photographers in an automobile, van, boat, or aircraft; contact may be maintained with other field personnel at, for example, a radar site. Such a tracking and photography project for gathering tornado pictures has been described by Golden and Morgan (1972). An example of an analyzed tornado photo is shown in Fig. 4.6. Fujita (1960, 1975; see also Vol. 1, Chap. 3) and Agee et al. (1977), among others, have analyzed photographs of tornadoes from the public.

g. Summary

Considerable qualitative insight can be gained from still and time-lapse pictures with little more information than logs on time, location, and equipment. A good cloud observer can use photographs to prove a point of contention or to clearly disprove a possibility that other data have raised. The latter is a frequent case in which pictures provide negative information—for example, the expected cloud configuration is clearly absent. Limited quantitative information can be gained from one camera if the range is only roughly

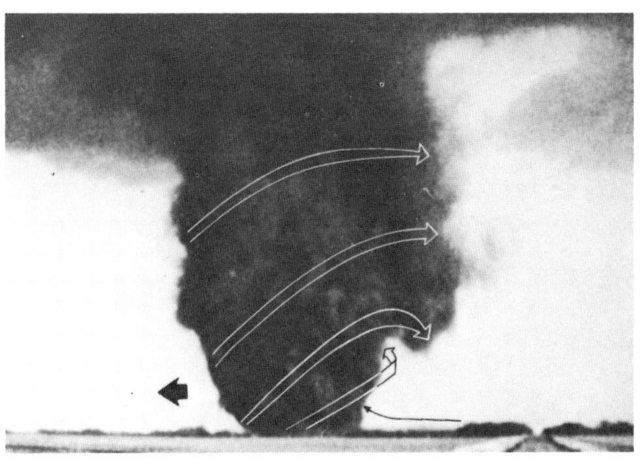

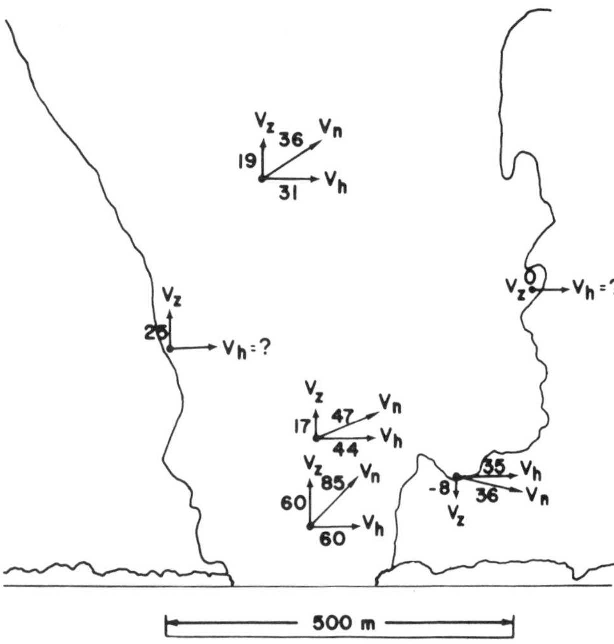

Figure 4.6. The Great Bend, Kans., tornado of 30 August 1974. Above: The vortex from a 16-mm motion-picture frame taken about 0005 GMT; debris trajectories are based on photogrammetry of the movie loops covering 0005 to 0010 GMT. Below: The most accurate vectors from a larger set of calculations, made during this period, which lead to trajectories above (Golden and Purcell, 1977).

known, but if range is very accurate, single-picture analysis is easy and rather precise. Finally, sophisticated cameras and analytic procedures performed on stereoscopic pairs and hand-held-camera pictures of rare targets of opportunity can provide highly detailed time histories. Ground photographs often provide the only way to see low-level thunderstorm features such as rainshafts, squalls, tornadoes, and cloud bases.

4. Photogrammetry of Aircraft-based Pictures

a. Special considerations

Pictures from aircraft may show important aspects not visible from the ground. However, since an aircraft changes position continually and undergoes altitude changes while it is moving, analysis of airborne photography begins at a more difficult level than that for ground photos. The additional parameters of aircraft altitude, location, time, and horizon must be considered.

The track of the aircraft must be known if clouds are to be located. Where only qualitative illustration of the cloud is needed, notes taken during flight may be sufficient, either in reference to geographic locations or in relation to specific clouds when they are easily identified features of the cloud field. However, when frequent pictures are taken, as with time-lapse-camera operation throughout a flight, an automatic data-collection system must record aircraft position. Plots of the sometimes complex flight patterns can then be produced, and these form a basic component of the photogrammetry.

Time must also be determinable for aircraft pictures. For qualitative analysis, a photographer taking still pictures may record time to the nearest minute. If a time-lapse camera is operating, however, a time mark must appear on the film itself. It is quite risky to assume that one picture is taken every 5 sec through an entire flight, for example, without being able to verify it. In typical flights the aircraft may be moving at 100 m s^{-1} so that very small time errors are associated with large location errors.

Horizons are more complicated features in aircraft pictures than in ground-based photographs. The apparent horizon is the line on the picture where sky and land meet, while the true horizon is at the intersection of the photo plane and the true horizontal plane containing the camera lens. From an aircraft the observer's apparent horizon always dips below true horizontal in a manner dependent on aircraft height. This difference, the dip angle, is more than 2° at 6-km altitude. The dip angle results mainly from earth curvature, but there is also some contribution from atmospheric refraction. In addition, the apparent horizon is frequently not visible from an aircraft when there are thunderstorms because of obscuring clouds, haze, etc. Consequently, calibration of the camera relative to the aircraft in flight is necessary during flight segments with clear views of the horizon.

b. Hand-held-Camera Pictures

Photographs from hand-held cameras frequently show the qualitative nature of clouds seen during a flight. Even though an aircraft may also have time-lapse cameras, a set of 50 to 100 35-mm slides, for example, from an afternoon's flight is a valuable index to the day's events, while providing larger negatives than a movie frame for making prints. In addition, the time-lapse film may not have the most reveal-

Figure 4.7. A hand-held-camera photo taken on 4 July 1976 at 1722 MDT, looking west, in south-central Alberta. The pedestal cloud on the left is ~1,700 m above ground level, near aircraft flight level, where updrafts of 2–3 m s^{-1} were measured. Later the adjacent area grew from a higher base near 2,700 m above ground (photo from Charles Warner; the photograph was taken as part of the Alberta Hail Project).

ing view of a feature because of its fixed times of exposure (see Fig. 4.7). The following information may be recorded for each picture: time, aircraft location, direction of flight path, direction of camera, altitude, lens focal length, camera, photographer, film roll and picture numbers, and descriptions of the thunderstorm's features.

When the apparent horizon cannot be found, only the relative heights and dimensions of cloud portions can be determined, together with approximate location, organization, time history (from a series of photos), shear, and visible ice-water contrasts. In many situations these data are sufficient (Fankhauser, 1983), or are all that can be recovered (Warner, 1977). If the apparent horizon can be found, then the basic equations, Eqs. 4.1–4.7 are applicable to analysis.

c. Stereoscopic Views

Stereoscopic pairs can be obtained in flight by a photographer looking from a side window. Proper stereoscopic parallax is gained in two steps. First, the distance to the cloud of interest is estimated. Then the pictures are spaced along the flight path at an interval appropriate to the aircraft speed. An interval between 1 and 3 sec will often be appropriate for pictures taken from a jet aircraft (Scorer, 1972). Fraser (1968) describes in more detail how to calculate the optimal time between pictures in a stereoscopic pair taken from aircraft. A useful equation for the time interval between pictures is

$$T = 300 \frac{D}{V}, \qquad (4.8)$$

where T is the time in seconds, D is the distance to the closest cloud in miles, and V is the speed of the plane in miles per hour. Not every cloud in the view of a stereoscopic pair can be made to fuse in that pair alone. For closer or much more distant clouds extra pictures may be needed.

Instruments for stereoscopic viewing include inexpensive pocket stereoscopes, sophisticated stereoscopes, or a small hand mirror; the cross-eye method is also employed for stereoscopic viewing (Fraser, 1968). An example of the pocket stereoscope method is shown for ground-based pictures in Fig. 4.5. If a large number of stereoscopic pairs is required, it is preferable to operate an automatic time-lapse camera

Figure 4.8. Right, left, and nose-camera views of thunderstorms taken 29 July 1975 at 1951:20 GMT. The pictures were taken during FACE from the NOAA DC-6 aircraft. Side-camera photos were taken on 35-mm black-and-white film, and the nose camera used 16-mm color film. Lines in the front-facing photo are located in the image plane of the lens, but finding the picture center by crossed diagonals is the better method (see Eq. 4.3).

looking out the side of the aircraft. During analysis of side-camera film the distance to visible clouds will be roughly known. When depth perception is needed, an equation such as Eq. 4.8 can be used to choose the proper photos to print for stereoscopic viewing.

d. Time-Lapse Photographs

Airborne time-lapse cameras are the most sophisticated systems for taking thunderstorm pictures in the air. The cameras may look out the aircraft sides, front, or downward. A complete set of horizontal-facing pictures is shown in Fig. 4.8. The motion of the clouds across the screen, when the film is viewed at normal projection speeds, reduces the need for stereoscopic pairs in many, but not all, situations. Once the entire camera system is operational at the beginning of a flight, the cameras themselves may not need frequent checks in the air.

If only one time-lapse camera is available for an aircraft during a field project, the choice of direction depends on the flight pattern. When flight tracks are complex, having, for instance, a turn every 5 min, the nose camera is usually preferred. It provides a visual record of clouds penetrated along the aircraft track, shows distant clouds ahead of the plane, and produces pictures that lend themselves to extensive analysis. The most common nose-camera frame rate is one frame every 2 or 3 sec, with slower rates at higher altitudes. This rate allows a sufficient number of frames to show an upcoming cloud in several frames and then to indicate how many seconds the aircraft was in cloud during

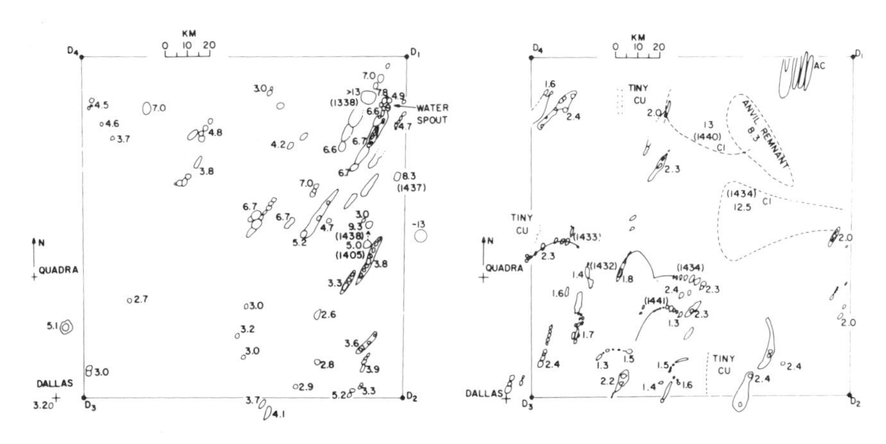

Figure 4.9. Cloud maps made from aircraft time-lapse photographs during GATE on 18 September 1974 between 1300 and 1445 GMT. The pictures were taken with side and nose cameras aboard four of five aircraft flying the box circuit $D_1D_2D_3D_4$. The left panel shows active cloudtops with heights of 2.5 to more than 13 km; the right panel shows tops less than 2.5 km. Numbers are altitudes in kilometers and times of measurement. Thin solid lines represent arcs of clouds, and dashed lines indicate anvil shapes. GATE ships *Quadra* and *Dallas* are to the southwest (Warner et al., 1979).

penetration; the rate is not unreasonably tedious on long, straight flights. Color 16-mm film has usually been used in nose cameras, although its quality is not as high as that of 35-mm film as Fig. 4.8 shows. Forward-looking cameras have one special complication: when repeated penetrations of supercooled clouds are made, the window in front of the lens must be heated to remove ice accumulation.

On long, straight flight legs, a complete record of clouds visible on one side of the aircraft track can be gained from a side camera looking exactly perpendicular to the aircraft track. Large areas are seen continuously, and detailed mapping can be accomplished (Malkus and Riehl, 1964). Side cameras are typically operated at a rate between 1 and 5 sec per frame. Although the side-looking wide-angle lens covers a large area, enough pictures must be taken to locate clouds by triangulation or stereoscopic methods. Side cameras have been operated with both 16-mm and 35-mm film in both color and black-and-white formats.

Calculations of cloud height and width (Warner, 1981), or detailed mapping, become rather difficult from the air, but the greater data potential from aircraft compensates for the complexity. For example, an aircraft flying at 6 km in the area of a thunderstorm is above most of the lower obscuring clouds. The larger towers and the cloud complexes, which are not often visible from every surface site, penetrate flight level (see Fig. 4.8). These clouds constitute the major dynamic elements of the thunderstorm and are readily identified.

e. Mapping of Clouds

Cloud maps can be developed from aircraft photos with the aid of lines drawn to connect the flight track to specific clouds when they are perpendicular to the side cameras and when they are at the azimuths corresponding to their entering or leaving the side cameras' view, as defined by Eq. 4.1. Triangulation of these azimuths on a large flight-track map is an easy way to locate clouds that are seen but not penetrated. When a cloud is penetrated, the distance to clouds on the flight track is found from nose-camera film by multiplying aircraft speed by the number of seconds to penetration. Or the distance to a cloud may be found in using a forward camera by noting the time interval needed for the cloud to double in angular dimension on the screen during aircraft approach to the cloud. The distance covered during this time is half the range to the cloud from the first view (Ronne, 1959). Another procedure is to use radar echoes to locate rain from the subject clouds, although precipitation and visible cloud boundaries are not exactly the same. The accuracy of aircraft navigation can be increased when geographical features are seen on radar. Warner (1978) described a method of mapping with nose camera film during straight and level flight.

Another method of mapping unpenetrated clouds during level flight is the crossing-angle procedure developed by Ronne (1959) and expanded by Whitney and McClain (1967). This method uses side cameras to give range r to a cloud from

$$r = \frac{VFR}{2 \tan \frac{\alpha}{2}}, \qquad (4.9)$$

where V is the ground speed of the aircraft along the portion of the track of importance, F is the number of frames that a cloud requires to cross the field of view of a side camera, R is the interval between frames in seconds, and α is the total horizontal angle of the lens from Eq. 4.1. From Eq. 4.9 useful tables can be derived. The crossing-angle procedure and other mapping methods mentioned here have been used to produce the cloud maps in Fig. 4.9. It should be noted that some time-lapse projectors do not show the entire horizontal film frame.

f. Measuring Widths and Heights of Clouds

Widths and relative heights are not difficult to measure from the air. One qualitative rule sometimes overlooked is that any cloud above the true horizon is above the aircraft flight level, and clouds below that level appear below the true

Figure 4.10. The view from a downward-pointed camera aboard B-57-F aircraft flying at 60,000 feet over western Oklahoma, 11 June 1967, 1445 CST. This is one-half of the full print and extends from the nadir (center of lower margin) to a horizon normal to the line of flight. Each full negative of the original film measures 5 by 11 in.

horizon. However, the higher the aircraft, the greater the dip angle (see Sec. 4a), and this idea becomes less useful. Given distance to cloud from one of the techniques in the preceding paragraphs and the lens focal length, Eqs. 4.1–4.7 are appropriate. The only caveat is that tilt in Eqs. 4.4 and 4.5 must be known in order to use the equations exactly. Absolute heights and widths are calculable if the apparent horizon is visible, or if the camera has been calibrated to locate the true horizontal plane and aircraft roll was recorded. The earth's curvature and atmospheric refraction correction mentioned for Eq. 4.6 also must be applied for aircraft film. A graphic method described by Cantilo and Woodley (1970) includes the use of grids for the 35-mm side camera film shown in Fig. 4.8. A key step for photogrammetry is the calibration of the apparent horizon in clear-sky conditions at the desired altitude. This is an important consideration; once the horizon is located and cloud position is found from triangulation, absolute height can be measured in cloudy areas where the apparent horizon is totally obscured. Then absolute heights can be found rather quickly from the grids. Techniques have been devised (Fujita, 1974; Warner, 1981) that use basic photogrammetric equations, with an accurate determination of the camera's orientation relative to horizontal, to allow cloud-height measurements in regions where no reference is available to any absolute height except that of the aircraft itself.

g. Pictures from Downward-looking Cameras

Downward-looking time-lapse pictures of thunderstorms from aircraft are valuable for detailed mapping of cloud distributions at a given time, but not for surveillance of rapidly changing clouds. An aircraft with a downward-looking camera must fly rather high (more than 10 km) to see a wide-enough area. With thunderstorms, cirrus clouds often cover some important features during storms' later stages, as satellite views from much higher levels also show. The aircraft may collect photos over a large region, so that each location is seen only a few times during a flight, or it may circle a small area for better time resolution and less spatial coverage of cloud developments (Plank, 1974). The typical film is black and white and rather large, ranging up to a 5-by-11-in negative. Plank (1969) describes a project, with both vertical and oblique pictures, in which the vertical pictures provided cloud diameters and the heights of cloud bases from their shadows on the ground. Oblique pictures were taken with the horizon across the top of the photograph for finding cloud heights. Other downward-looking cameras have been used to take elongated pictures, sweeping from horizon to horizon, giving detail in a small area under the aircraft and less detail over larger areas toward the horizons (Fig. 4.10). Vertical or oblique photos can be difficult to analyze; use of special grids developed for specific cameras or larger stereoscopic plotting machines may prove economical only if there are many analyses.

5. Concluding Remarks

It is frequently anticipated that analysis of cloud pictures can be automated to eliminate time-consuming procedures. Although computers reduce the routine of long series of calculations, often only a few thunderstorm pictures taken with each type of camera from the same location are analyzed in the same way. Even more important, substantial automation is not feasible in many situations because of the nature of the clouds themselves. One must ask these questions: Where are the sides of a cloud? When have two elements become one? Can the same feature be seen from both ground and air? The meteorologist must be involved in decisions in light of experience, understanding of the storm from other data sources, and the level of detail that is required.

Photographic data collected in visible wavelengths from a camera situated near or below the tropopause have been discussed in this review. New methods of photography of terrestrial features from high-altitude aircraft and satellites are continually being developed. High-altitude balloons (near 30 km) have occasionally been used to gather downward-looking and oblique photos of thunderstorms with better time resolution than aircraft and better space resolution than satellites (Vonnegut, 1970). Although satellite views have somewhat supplanted high-altitude vertical photography for cloud mapping, new coordinated studies that match detailed vertical cloud views with high-resolution satellite pictures may be appropriate.

There has been improvement in sensors for penetrating haze, poor visibility, and darkness with instrumentation outside the visible wavelength band. Such sensors and related techniques are given in the *Manual of Remote Sensing* of the American Society of Photogrammetry (1975). As new instrumentation and methods are developed, it is likely that renewed impetus will be provided to relate newly probed phenomena to more familiar visual records.

Conventional photo archives often represent some of the most interesting and cost-effective data in a thunderstorm study. Although myriads of new cameras and recording methods have been developed, Ronne (1959) still holds true: "The lowliest box Brownie can measure off an angle or record a fact with an accuracy far greater than the eye or memory of the most skilled observer." [1]

[1] Joanne Simpson (NASA, Greenbelt, Md.), Michael Garstang (University of Virginia), and William Woodley (Boulder, Colo.) are gratefully recognized for providing the opportunity, encouragement, and time to pursue photogrammetry of cumuli for more than a decade. The late Claude Ronne, who for many years was associated with Woods Hole Oceanographic Institution, described many of his ideas on cloud mapping from aircraft, as well as other basic concepts of photogrammetry. Charles Warner (United Kingdom) reviewed this manuscript on the basis of his wide photogrammetric background. Charles Moore (New Mexico Institute of Mining and Technology) and Joseph Golden (NOAA, NWS, Silver Spring, Md.) kindly provided illustrations of their photographic data. I also thank Michael Maier (Titusville, Fla.), John Brown (Weather Research Program, NOAA/ERL, Boulder, Colo.), and Richard Decker (NOAA, Miami, Fla.) for comments on the manuscript.

5. Storm Acoustics

Roy T. Arnold

1. Introduction

Study of storm sounds is as old as science itself (Aristotle, 384–22 B.C.; Lucretius, 98–55 B.C.) and as new as the most recent technological advance. In modern probing of convective storms this approach has been underexamined, and literature devoted to storm acoustics is relatively sparse. More atmospheric scientists should be aware of the merits and limitations of acoustical studies; this chapter is intended to serve as an introduction for those unfamiliar with the subject.

Research in storm acoustics generally involves both active (e.g., acoustic sounding; Brown and Hall, 1978) and passive (e.g., study of thunder) techniques; this chapter is restricted to the passive methods. It is restricted also to audible frequencies, except for thunder studies, which include subaudible frequencies. Pressure waves at frequencies below the audible range (<20 Hz) are infrasounds. Infrasounds below 1 Hz from thunderstorms, (discussed in Chap. 6) are ultralow-frequencywaves that propagate with very little attenuation and probably have quite different origin from that of sounds and infrasounds described here.

Sounds most frequently associated with severe storms are generated primarily by lightning discharges (thunder) and by interactions of both precipitation and wind with terrain. Once produced, sounds pass through a complex atmosphere, and other mechanisms generate audible signals. For example, there have been reports of continuous rumbling (Arnold and Bass, 1977) apparently emanating from rather isolated storms and varied descriptions of unusual sounds from tornadoes and funnel clouds aloft (Flora, 1954). Verification and documentation of uncommon sounds are difficult, and records suitable for scientific analysis are practically nonexistent. Consequently this chapter is limited to descriptions of experimental and theoretical studies of thunder, analysis of sounds generated by severe tornadoes, and a brief comment on noise generated by wind.

2. Propagation Effects

Sound in the atmosphere is subject to the following alterations that affect the propagation direction and the amplitude of a pressure wave:

1. Refraction caused by both temperature gradient and wind field.
2. Spreading losses related to geometry of the sound source.
3. Absorption in which sound energy is converted into random molecular motions.
4. Reflections that occur at boundaries between air parcels having different physical characteristics, e.g., different density.
5. Reflection and absorption at the earth's surface.
6. Finite amplitude in the vicinity of the source that results in dissipation far in excess of spreading losses and a lengthening of the original pressure waveform.

Many important physical characteristics of the atmosphere near a severe convective storm are unknown, and it is difficult to treat propagation effects quantitatively. Some insight can be gained into just how important the effects are by using mean values of temperature, pressure, humidity, and windfield velocities in propagation analyses. It is also useful to consider some propagation effects applied to a sound source (such as a lightning channel segment) located a few kilometers above the ground: initially sound is radiated outward from the source in directions determined by the shape of the source; e.g., spherical and point sources radiate energy equally in all directions. As waves travel from source to recording station, refraction occurs because the velocity of sound is temperature-dependent and because sound direction is altered by winds. Sound refracted upward in the mean vertical-temperature field accounts for some observed effects, e.g., the lack of sound from a nearby cloud-to-ground lightning flash. Variations in the wind field and turbulence produced by wind changes might account for other observations, e.g., seeing a cloud-to-ground lightning flash in one direction and hearing the thunder from another.

Sound velocity in a gas can be expressed as

$$c = \sqrt{\gamma r T}, \tag{5.1}$$

where c is the speed of sound, γ is the ratio of the specific heat of the gas at constant pressure to that at constant volume, r is a constant whose value depends on the particular gas involved, and T is the absolute temperature in kelvins (Kinsler and Frey, 1962). In most instances of importance in storm acoustics, changes in sound velocity in the atmosphere are much more strongly influenced by variations in temperature than by variations in composition and pressure. For air at 1,013-mb pressure we may write

$$c = c_0 \sqrt{T/273}, \tag{5.2}$$

where c_0, the speed of sound in air at 0°C, is 331.6 m s^{-1}. Equation 5.2 can be applied to any region of the troposphere with good accuracy.

The importance of temperature-related refraction can be easily illustrated. We rewrite Eq. 5.2 as

$$c = c_0 \left(\frac{T_s - ay}{273}\right)^{1/2}, \tag{5.3}$$

where T_s is surface temperature, a is lapse rate, and y is altitude above the surface. Binomial expansion of the radical in Eq. 5.3 shows that the speed of sound is approximately a linear function of temperature over the vertical extent of the troposphere, if one assumes a constant lapse rate no greater than 6.5 K km^{-1}. Consequently, the vertical velocity gradient, g_T, is approximately constant and is given by

$$g_T = \frac{dc}{dy} = \frac{-ac_0}{2} (T_s/273)^{1/2} \tag{5.4}$$

It is clear from Eq. 5.4 that in this model of the troposphere sound velocity decreases with altitude and rays will refract upward. To demonstrate the importance of the effect, we consider an initially horizontal ray at ground level radiated from a cloud-to-ground lightning flash. Near the ground a ray will follow (approximately) a circular arc of radius $R(R = c_0/g_T)$ if the velocity gradient is constant. For a ray at ground level that has propagated horizontally 1 km, the ray path will be approximately 6 m above the ground with a direction of travel about 0.65° above the horizontal.

To illustrate refraction by wind-velocity gradients, we select a plane boundary of given thickness between two layers of air moving in opposite directions with identical speeds. The wind-velocity gradient in a direction perpendicular to the boundary is assumed constant. A sound-ray incident normally on the boundary will follow a curvilinear path until it reaches the second boundary and then continue straight-line propagation with a sound-velocity component in the prevailing wind direction. A 100-m layer between two air masses moving at speeds of ± 15 m s^{-1} would produce an angular deflection of approximately 7.5° for a normally incident ray. Therefore, one can see that wind effects are often quite pronounced but because of lack of knowledge of wind fields are difficult to evaluate. The relative importance of temperature and wind-field refraction effects in a storm environment has not been determined.

Reflection and scattering of sound by the earth's surface is an important modifier of sound signals received near the surface. Sound arriving at a receiver is composed of a direct wave from the sound source and a reflected wave from the surface. Interference between direct and reflected waves is frequency-dependent because phase shift upon reflection is frequency-dependent. The phase shift upon reflection, the magnitude of which is a function of both the angle of incidence and the surface acoustic impedance, is π at very low frequencies and falls off slightly with frequency increase. A phase change of approximately π occurs at nearly all frequencies for grazing incidence ($\lesssim 5°$), in which case almost complete cancellation occurs. Consequently, at distances beyond 10 km on a relatively flat terrain almost no direct sound can be heard from the visible portion of a cloud-to-ground lightning flash. Similarly, tornado noises emanating from very near the ground will be greatly diminished by interference only a few kilometers away. A detailed review of sound propagation near the earth's surface is presented by Piercy et al. (1977).

Sound waves are attenuated with distance because sound energy is lost to molecular motions in the atmosphere. Absorption of higher acoustic-wave frequencies is most rapid and is determined by temperature, pressure, and molecular composition of the atmosphere. Reliable calculations of the absorption coefficient for sounds propagating in the atmosphere can be incorporated into storm acoustics (Bass and Losey, 1975; Arnold et al., 1976). Total absorption of sound in air as a function of frequency and relative humidity is presented in Fig. 5.1; note the magnitude of attenuation.

Sound traveling outward from a source is also subject to spreading losses; i.e., the intensity of sound emanating from either a point or spherical source diminishes proportional to the inverse square of the propagation distance.

Some other propagation effects also modify sound waves in the atmosphere. Reflection of sound at boundaries between air masses with different temperature and/or wind fields undoubtedly contributes to echoes and variations in sound amplitude such as rumbles of thunder. Large-amplitude effects such as wave steepening and increased absorption are important when the wave overpressure is large (>1 mb). In addition, ducting of sound in a layered atmosphere, diffraction, and interference from multiple sources are often effects to be considered in some problems.

3. Thunder

Stored electrostatic energy is the energy source of both audible and infrasonic thunder (see Sec. 3a). When electrostatic energy is released from a thundercloud, perhaps 99% is available to audible thunder, which is a mechanical disturbance produced by great overpressure within the segments of a lightning channel (Krider and Guo, 1983). There is first a shock wave propagating outward somewhat faster than the speed of sound (Zel'dovich and Raizer, 1967).

With both absorption and spreading losses, the mechanical wave passes from a strong-shock into a weak-shock regime and finally evolves into a finite-amplitude acoustic wave from which thunder follows. Other mechanical disturbances can be generated by a lightning discharge. For example, neutralization of charged droplets in a large volume can produce an observable pressure pulse as the region undergoes partial collapse (Wilson, 1920).

Thunder and infrasonic pressure variations associated with lightning have been studied by numerous investigators. Uman (1969) surveyed the thunder literature, and Few (1982) presents a good review of the electrical processes that give rise to thunder. Few (1982, 1974b, 1970, 1969), Bass (1980), Balachandran (1983), Bohannon et al. (1977), Dessler (1973), and Holmes et al. (1971) list pertinent research papers.

a. Sounds of Thunder

1. Audible Thunder: Claps and rumbles are the characteristic thunder sounds. Thunderclaps are abrupt changes in loud sounds that occur when the observer is located where sound from major channels and branches is at a maximum. Occasionally when one is very near a vertical channel, absorption of high-frequency components of sound is diminished, and a loud crack is heard. Rumbles occur because of arrival of many erratic sound signals from the entire extent of channels and branches. Major channel elements are approximately 50 or more m long, and the minor segments are a few meters or less in length (Hill, 1968). Channel segments represent individual cylindrical sound sources that radiate energy perpendicular to the channel axis, and the sound one hears originates in thousands of sources along the lightning channel; thus thunder following a lightning flash is perceived as a superposition of sounds from many segments of a very tortuous lightning channel (Few, 1974b). An approximate quasilinear theory of thunder generation by tortuous lightning channels has been developed by Ribner and Roy (1982), which is in effect the mapping of the shape of lightning into the sound of thunder. Since pressure variations produced by lightning are very complex, careful recording and data reduction techniques are required before analytical benefits can be derived. More detailed qualitative descriptions of what we hear as thunder can be found in Few (1974b, 1975) and Uman (1969).

Loudness, a qualitative aspect of audible thunder, merits a brief description. Excluding propagation effects, what we perceive as loudness is not only a function of sound wave intensity but also a function of frequency content of the wave, sound duration, and spectral response of the ear. Consequently, comparison of thunder with other noises common in our experience is of little value unless we consider each variable.

Frequency content of typical thunder sounds is shown in Fig. 5.2a. Note that the spectrum peaks at about 100 Hz, and most of the acoustic energy is within a frequency band

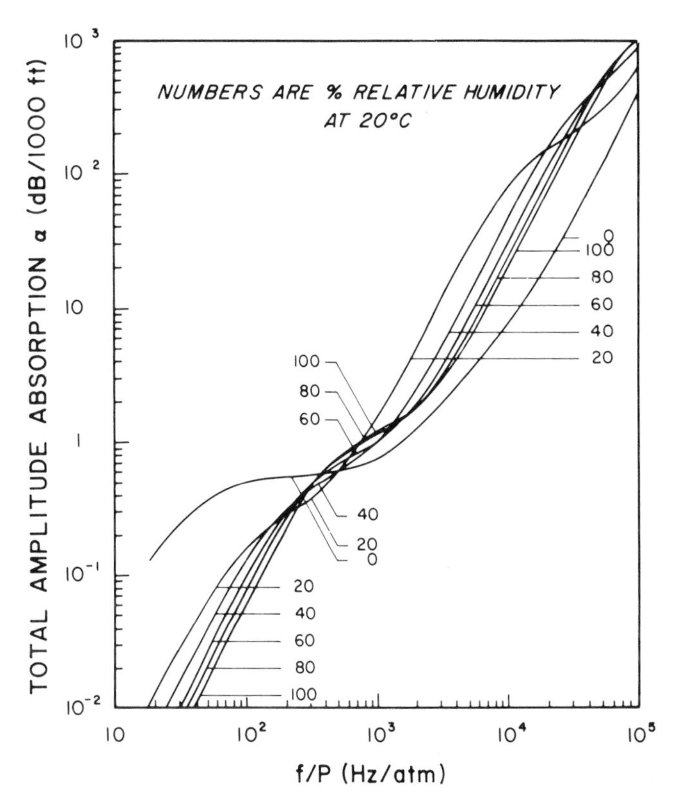

Figure 5.1. Total absorption of sound in air as a function of frequency (after Evans et al., 1972).

approximately 200 Hz wide. If we correct the power spectrum in Fig. 5.2a for the frequency variations of the ear's response, the peak maximum is shifted to 250 Hz, and the frequency interval of significant power contributions is 300 Hz. The corrected total acoustic intensity is 2.7 ergs cm^{-2} s^{-1}, which corresponds to a loudness of 94 dBA.[1] In overpressure this acoustic intensity is 1.3×10^{-6} atm (representative of thunder heard 5 to 6 km from a flash). It is not uncommon to have overpressures of 10^{-3} to 10^{-4} atm for nearby lightning (tens of meters) (Uman, 1969). This means that the peak loudness might be as high as 123 dBA (3 dB above the threshold of pain), but more commonly one might expect values between 90 and 110 dBA.

Until the duration of sound exceeds a few oscillation periods of the wave, sounds are perceived as clicks. If duration is longer than about 0.5 sec, loudness is independent of duration except for a slight diminution of loudness if the sound persists for several seconds. For duration shorter than 0.5 sec, the time required to reach maximum loudness is frequency-dependent, higher frequencies reaching maxima

[1] To provide measurements of noise as perceived by the ear, a weighted, relative, logarithmic sound-pressure scale is used that corrects direct noise levels in dB for the acoustic response of the human ear. This corrected scale is called dBA (Kinsler nd Frey, 1962).

first. For typical thunder (as presented in Fig. 5.2), the duration of the peak intensity is 0.5–1 sec; it has little or no effect on loudness. However, very near the lightning channel the duration can be substantially less than 0.5 sec. There are two effects: what we hear is shifted to higher frequencies, and the perceived loudness is diminished. To calculate a value of loudness would require an exact knowledge of the sound-wave form. For further considerations of sound perception and loudness see Kinsler and Frey (1962).

2. Infrasonic Thunder: Pressure waves at both audible and subaudible frequencies are associated with lightning. No reports of significant energy content in thunder at frequencies above ~600 Hz have been published as of this writing, but several workers (e.g., Uman, 1969) reported pressure variations at frequencies well below the lower limit of the audible range (<20 Hz), and other interest in infrasonic thunder blossomed (Bohannon et al., 1977; Dessler, 1973; Holmes et al., 1971; Colgate and McKee, 1969). Although our perception of thunder is based mainly on audible pressure changes, we often sense that significant infrasonic energy is present; we literally "feel" thunder even when audible sound is not excessive. Most subaudible sounds associated with a lightning flash are greater than 0.1 Hz and less intense than the much lower frequency sounds produced by the convective storm as a whole (see Chap. 6). Evidence suggests that at least part of the infrasonic wave observed after a lightning flash does not originate in the rapidly expanding shock wave but is generated before the flash itself (Bohannon et al., 1977; Balachandran, 1979, 1983).

b. Measurements and Theory

Description of experimental and theoretical work is divided here on the basis of measurement techniques required to obtain useful data.

1. Thunder power spectra: One might think that recording thunder is an almost trivial undertaking, but in any measurement of thunder a very difficult experimental problem is posed by noise from mechanical interaction between wind and microphone and from pressure fluctuations (turbulence) in the wind itself. One would appropriately use a microphone and recorder having uniform frequency response between 0.1 and 600 Hz. Enclosure resonances and interference between direct and reflected waves can be minimized if the microphone is placed on the ground far from other major reflecting surfaces such as buildings. Rain damage to the microphone's sensitive element is easily eliminated by use of a rain shield, but both microphone and rain shield are directly exposed to surface winds, and interaction between them and the wind produces significant background noise. Most commonly noises associated with the terrain (e.g., grass and bushes) do not pose problems as serious as wind-microphone interactions unless windspeeds are quite high (>20 kn), or there is unusually thick vegetation.

To minimize interaction between wind and microphone, two techniques are commonly used. A windscreen similar to those used on voice mikes provides an 8-to-15-dB noise reduction in the frequency interval of concern. A second technique is to mount the microphone as near as possible to the ground, where wind is at a minimum. A microphone mounted on a flat, hard surface that is flush with the ground is preferred for a single-microphone recording station; flush mounting reduces significantly the effect of interference between the direct and ground-reflected waves. Thunder is most easily heard in a storm's cool outflow, but in this region the surface winds are often strong. Good data can often be gathered in the region of surface inflow, where winds are milder, except in some severe storms.

Turbulence advected by the wind field generates the dominant wind noise once the wind-microphone interaction has been dealt with satisfactorily. For low-to-moderate windspeeds the noise is principally below 3 Hz and cannot be removed by electronic filtering techniques without loss of the low-frequency thunder data. One possible way to reduce this noise is to add signals from an array of microphones and employ spatial averaging. If the wind noise is random at each point in the array, then the wind noise will tend to cancel, and the thunder signal will add constructively; i.e., there will be an increase in the signal-to-noise ratio. Whether or not this approach is taken, wind noise is a problem that must be solved if very-low-frequency data are to be obtained.

Figure 5.2 shows power spectra of two separate thunder events recorded for a convective storm; it is evident that wind-associated noises can be appreciable.

Experimental recordings of the thunder power spectrum are described by Holmes et al. (1971), Uman (1969), Few (1968), and Few et al. (1967). Typically a four-channel FM magnetic tape recorder is used: a voice channel for supplementary data, a channel for time code, a channel for electric-field change and a channel for the sound. Putting thunderstorm data in digital form facilitates theoretical study. This can be done in different ways; one particularly useful scheme is to present the analog data to a real-time spectrum analyzer, which digitizes the thunder power spectrum directly.

The starting point for theoretical calculation of the power spectrum for thunder is the sound source, i.e., the lightning channel. Many have attempted to approximate the evolution of the shock wave into a sound wave by treating the lightning channel as a cylindrical acoustic wave source (Bass, 1980; Few et al., 1970; Troutman, 1969; Remillard, 1969; Colgate and McKee, 1969; Bhartendu, 1968, 1969; Jones, 1968; Jones et al., 1968). Because the lightning channel is tortuous (i.e., the channel is broken up into many short, straight segments; see Hill, 1968), the cylindrical calculation is invalid for sound recorded farther than a few segment lengths from the channel. Few (1969) considered the effect of a channel tortuosity on the evolution of thunder and concluded that the channel is more properly considered a series of short cylindrical segments that at distances greater than

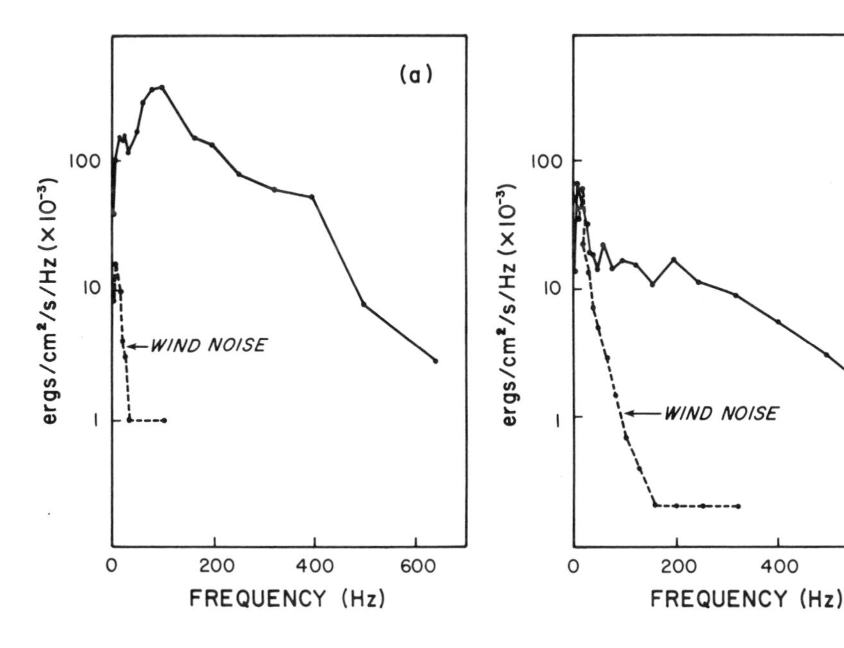

Figure 5.2. (a) A typical analog thunder spectrum with a peak frequency of 100 and a wind-noise spectrum. (b) A thunder spectrum with a low-frequency peak owing to wind noise. The wind-noise analysis is based on the pressure data just preceding the thunder (after Holmes et al., 1971).

a segment length appear more like a string of spherical wave sources (string-of-pearls model) than a cylinder. Later Few et al. (1970) illustrated the effect of lightning channel tortuosity photographically. The string-of-pearls model accounts for the varying sounds of thunder, i.e., claps and rumbles, and is graphically demonstrated in the theoretical treatment of Ribner and Roy (1982).

Few (1969) calculated the acoustic power spectrum for a short-line source of sound, choosing the total mechanical energy available per unit length to be 10^6 J m^{-1}. Qualitative arguments infer the actual thunder power spectrum, and thunder data are in general agreement with qualitative predictions of the string-of-pearls model. However, a detailed comparison of individual thunder spectra with calculated spectra is needed if the model is to be analyzed critically. Calculations based on the model reveal that the peak frequency in the acoustic-power spectrum varies inversely as the square root of the energy per unit length along the channel. Brode (1956) suggests that the pressure wave will have a length about 2.6 times the radial extent of the volume that can be created if all the energy from the discharge is used doing *PdV* work against the ambient atmospheric pressure. That is, the frequency of the power spectrum maximum is given by

$$f_m = (0.63)c(P_0/E_1)^{1/2} , \qquad (5.5)$$

where c is the local speed of sound, P_0 is the atmospheric pressure, and E_1 is the energy per unit length of the channel available to thunder. Holmes et al. (1971) observed low-frequency peaks (5 Hz and lower) apparently inconsistent with the mean energy input into the lightning channel as measured independently by electrical means. A conclusion is that is some instances thunder energy is contributed by a process other than channel expansion, a matter not easily resolved on the basis of power spectra alone. The more recent work of Bohannon et al. (1977) and Balachandran (1979, 1983) suggests that the low-frequency peaks in these measurements might have been produced by electrostatically produced acoustic waves.

Theoretical calculations of the thunder power spectrum are far from complete. Evolution of high overpressure in a lightning channel into a shock wave is not understood. In any theoretical treatment of the evolution of a shock wave into an acoustic wave, one must first choose a suitable initial pressure profile for the wave, which is a formidable task in itself. Calculation of the pressure as the wave evolves in time is then involved but straightforward, although propagation effects for a real atmosphere are difficult to include quantitatively (Few, 1969; Bass and Losey, 1975; Bass, 1980). Fourier analysis of the waveform (Fig. 5.3b) provides an acoustic-power spectrum of the pressure wave (Fig. 5.3a). But how to choose an initial pressure pulse will be subject to question for some time. The waveform in Fig. 5.3b is comparable with the general wave shape observed for laboratory sparks (Klinkowstein, 1974; Uman et al., 1970; Dawson et al., 1968; Wright and Mendendorp, 1967), but laboratory sparks are different in many ways from a lightning discharge. Extreme overpressure present in a lightning discharge might generate a pressure profile that is independent of the detailed structure of the lightning channel. In this case thunder power spectra would contribute little detail about the actual electrical discharge process; i.e., the evolution of a high channel overpressure into audible sounds would be more strongly influenced by propagation effects than by the detailed atomic and molecular processes of the lightning channel.

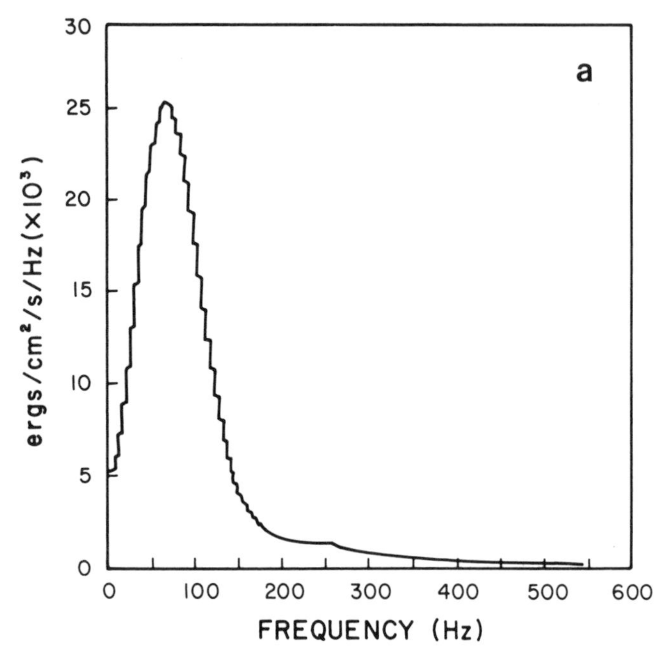

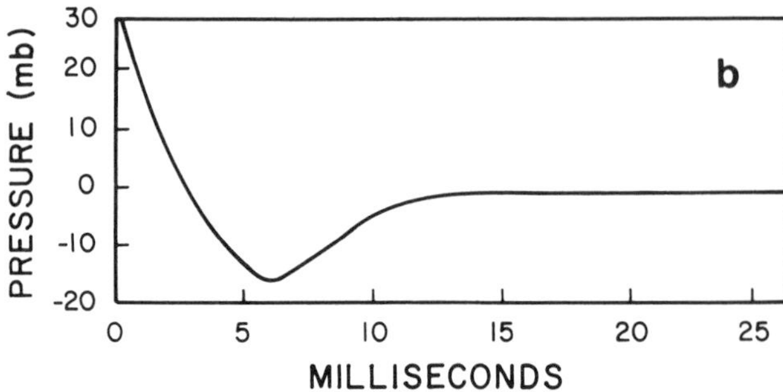

Figure 5.3. (a) Calculated acoustic-power spectrum for thunder 17 m from a short-line source with an energy per unit length of 10^6 J m^{-1} available to thunder. (b) The pressure pulse (sound wave) used to calculate (a) (after Few, 1969).

2. Lightning-Channel Reconstruction: The output of an array of microphones can be used to reconstruct the geometry of lightning channels (Few, 1970), and studies have been conducted on channel reconstruction for both audible and infrasonic thunder (Bohannon et al., 1977; Few and Teer, 1974; Few, 1974b; Teer and Few, 1974; Nakano, 1973; Teer, 1973). Further investigation is needed, but Few and Teer (1974) concluded that the shape of a lightning channel can be determined accurately, especially if one properly includes the effects of wind and temperature. An example of channel-reconstruction results compared with a lightning channel photograph is shown in Fig. 5.4 (Few and Teer, 1974). The reconstruction technique proves to be an important tool in studying spatial and temporal evolution of lightning, which adds to our knowledge of how electrical charge changes in space and time in convective storms.

In one reconstruction technique an array of microphones used in a ray-tracing technique may include three microphones deployed at the vertices of an equilateral triangle 100 m on a side (Teer, 1973). Sound waves generated as shock waves propagate outward from a lightning channel and pass over the microphone array. The waves can be treated as plane waves if the source is distant compared with the microphone spacing. A quantitative measure of similarity between two different signals shifted in time relative to one another is provided by cross correlation. The peak cross correlation of thunder signals defines relative time of arrival of wavefronts at each microphone, and if the speed of sound is known, the wave vector for an incident wave can be determined. The time lapse between the lightning flash and the sounds of thunder can be used to calculate the range of the source from the array, thus locating the wave source. Space between microphones is no more than about 100 m since wave fronts become incoherent if spaces are greater. Maximum spacing is also limited by atmospheric turbulence and by the lengths of discrete sources in major lightning channels (Few, 1974a).

A second reconstruction technique employed by Few (1974a) has been aptly termed thunder ranging. Microphones are placed at the corners of a right triangle with sides of approximately 1–2 km. The time of arrival of a thunderclap at each detector is measured by comparing the time of the thunder record with the time of origin as recorded in terms of electric-field change. Range from the lightning channel segment is computed from

$$R_i = ct_i, \tag{5.6}$$

where R_i is the distance of the ith detector from the source, t_i is the time for the sound of thunder to travel from the source to the ith detector, and c is the mean speed of sound between the source and detector. Separation of the microphones can be increased since coherence of the acoustic waves is not important, and as a consequence, accuracy in locating the sound source is greatly improved. Source location is at the intersection of arcs with radii R_i around each microphone. Accuracy of the calculated source position is a function of the relative positions of source and detectors; the technique is not immune to atmospheric propagation errors but is an exciting addition to thunder measurements.

Electrical data provide evidence that there is significant horizontal development of lightning channels inside storm clouds (Ogawa and Brook, 1969; Pierce, 1955; Workman et al., 1942), but the scientific community was slow to perceive the real importance of nonvertical development until the work of Teer and Few (1974). They used the ray-tracing technique with microphones deployed at the vertices of an

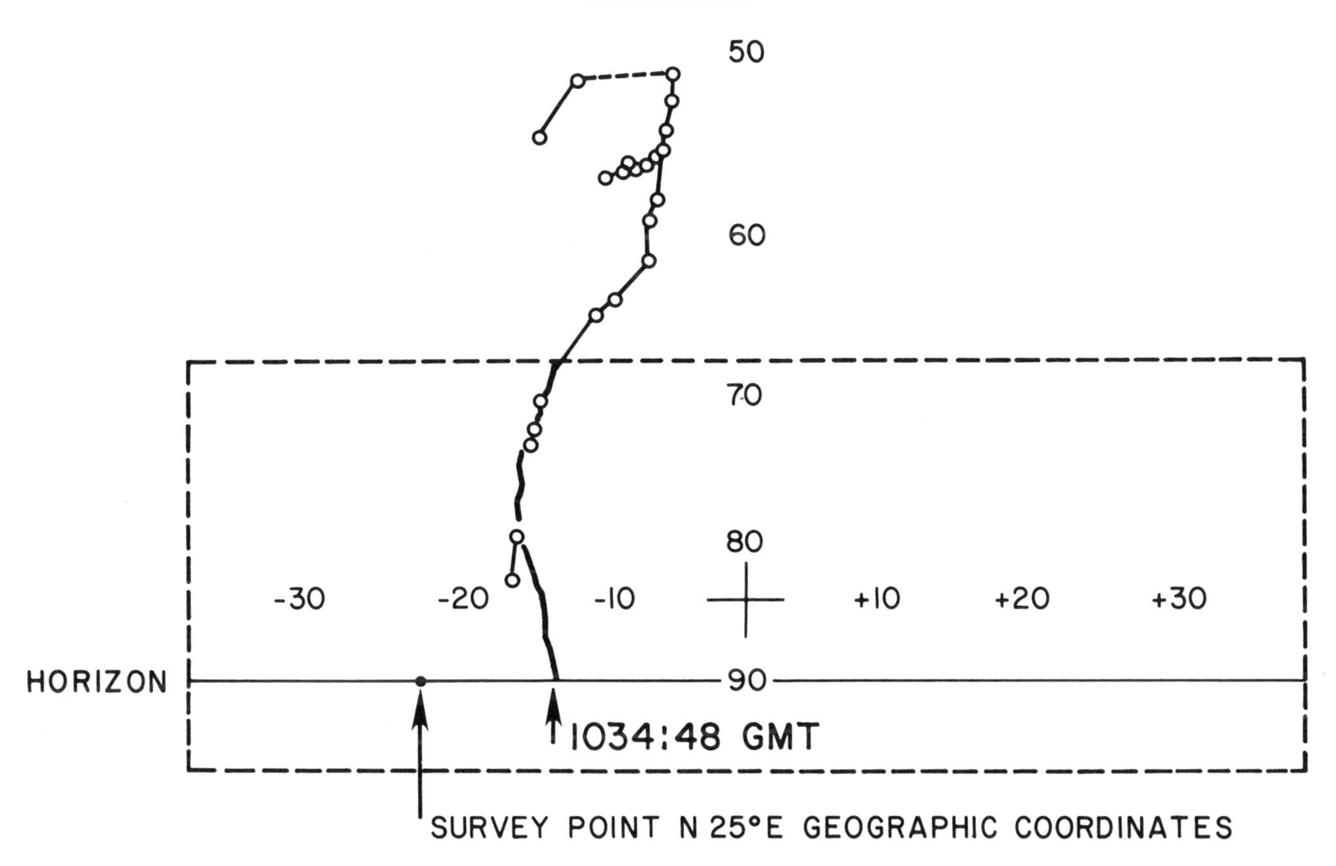

Figure 5.4. Overlay of an acoustically reconstructed lightning channel on the photographed channel. The field of view of the camera system is indicated by the dashed-line rectangle, the look direction of the camera is given by cross hairs in the center, and angles of azimuth from the look direction and from the zenith are indicated in degrees. The open circles are the reconstructed source points for the thunder; to produce this reconstruction, a model atmosphere based on rawinsonde measurements was used (after Few and Teer, 1974).

equilateral triangle with a baseline distance of 100 m, and from their reconstructions they concluded that the intracloud structure of both cloud-to-ground and intracloud lightning is near the $-10°C$ isotherm of the thundercloud, a point of view generally held today.

How, where, and when charge separation occurs within a cell is still a puzzle that must be pieced together. However, reconstruction (Few, 1974a) and other techniques such as that reported by Taylor (1978) have begun to provide information about the life history of electrical charge within a storm. For example, Few concluded that in one storm he studied the electrification process occurred near the freezing level and in a region of low radar reflectivity, and the work of Taylor (1978) seems to confirm Few's conclusions.

3. Infrasonic Thunder: Dessler (1973) has considered the possibility of infrasonic thunder that may arise from a mechanism quite different from rapid expansion of a lightning channel. Before the discharge by lightning, equilibrium between electrostatic forces and fluid dynamic forces defines the charged volume. Following charge neutralization by the discharge, a new equilibrium is reached only by a fluid adjustment (in this case a collapse) in response to the reduction of electrostatic force, and a low-frequency acoustic pulse is generated. Wilson (1920) first considered this process, and others (Colgate and McKee, 1969; Holmes et al., 1971; Colgate, 1967) have discussed it as a possible explanation of infrasonic signals sometimes observed in thunder. Bohannon et al. (1977) used ray-tracing techniques on a storm on 12 August 1975 near Socorro, N.Mex., and observed that infrasonic emissions are indeed associated with the lightning process. On the basis of the absence of infrasonic signals from the vertical portion of cloud-to-ground flashes, they concluded that the waves do not arise from channel heating and subsequent expansion. Dessler's predictions have been verified in part by Bohannon et al. (1977) and Balachandran (1979; 1983).

4. Tornado Acoustics

In contrast to thunder, tornado acoustics, a relatively new subject of study, has received very little analytical attention. Experimental (Arnold et al., 1976) and theoretical (Abdullah, 1966) treatments of acoustic emissions from tornadoes indicate a lack of data. For the most part, eyewitness descriptions cite loud-noise sources of the times (e.g., a freight train or a jet plane; see Abdullah, 1966; Flora, 1954), although tornadoes are often seen and not heard. Qualitative descriptions of tornadic sounds have provided little insight into the nature of the sound source.

a. Tornadic Sounds

Major interruptions in fluid flow by ground obstacles as the base of a funnel moves along the earth's surface are most likely what produce the principal tornado sounds in the near field. Eddying of the flow generates sound, and acoustic power output is determined by the flow velocity (Powell, 1964; Lighthill, 1952, 1954). Other sounds might arise from resonant behavior of the vortex itself (Abdullah, 1966; Sozou, 1968). There is little doubt that sounds are produced aloft; quite often noises are reported from funnels aloft. Abdullah made a serious attempt to predict fundamental frequencies of tornado sounds, but there has been no quantitative calculation of acoustic-power emissions. Interest in the genesis of tornadic sounds has not been high because of the complexity and paucity of experimental data; whether activity increases in the future will be influenced by the supply of sound recordings to the scientific community.

Existing data for tornadic sounds are derived from recordings made by the public.[2] Each recording has utility limited primarily by poor frequency response of the recording microphone (Arnold et al., 1976), and information below 100 Hz is nonexistent. Figure 5.5 presents the estimated relative acoustic-power spectrum for a very damaging nocturnal tornado that went through Guin, Ala., on 3 April 1974; in spite of large errors reported, it is evident that virtually all of the energy was below 500 Hz. Abdullah's (1966) hypothesis that the air mass involved in the tornado circulation executes free vibrations provides a simple formula for the fundamental frequency in the special case of radially symmetric vibrations:

$$\nu_{max} = \frac{207}{a}, \qquad (5.7)$$

where a is the vortex radius expressed in meters. For the Guin tornado, whose damage path was about 200 m wide, the maximum frequency of a free oscillation would be near 1 Hz. Analysis of the power spectrum shown in Fig. 5.5 offers little insight into the accuracy of Abdullah's prediction because of the low fidelity of the recording.

Instrumentation required for making a high-fidelity tornado sound recording is basically the same as that for recording thunder, but it is good to employ a microphone with a broad frequency response since the nature of tornadic sounds is unknown. Arnold and Bass (1977), working in conjunction with the National Severe Storms Laboratory, used storm-tracking techniques similar to those reported by Golden and Morgan (1972) (see Chap. 2). A sound-recording package with a uniform frequency response between 10 Hz and 10 KHz is used. Since the recording system might be damaged or lost, microphone fidelity has been sacrificed for economy. The nearest the sound package has been placed to a tornado track is about 2 km, and background wind noise on that occasion masked any audible tornado sounds that might have been otherwise detected.

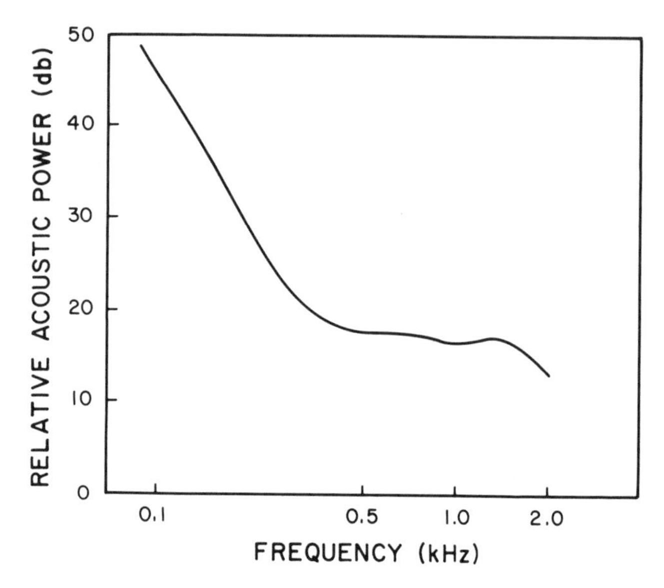

Figure 5.5. Tornado relative power spectrum for the Guin, Ala., tornado. Corrections have been made for microphone response (after Arnold et al., 1976).

Dozens of tornadoes have been seen and photographed by the storm-chase teams, but only in rare instances have any sounds been heard. As more tornado-producing storms are encountered, a high-quality recording will eventually be obtained, and the value of recorded tornadic sounds as a data base from which we study the tornado can be assessed.

Propagation effects are also important for interpretation of acoustic emissions from tornadoes. What is observed at a recording station not only will be a function of the intervening atmosphere but also will depend on the terrain, since much of the sound originaters very near the ground. Further, refraction of sound by the tornado core might well influence the direction of sound travel (Georges, 1971).

b. Temporal Evolution of Tornadic Sounds

The most promising analysis of available tornado sound tapes is the study of how the sounds evolve in time (Arnold et al., 1976). As a tornado moves relative to a recording

[2] Sound tracks have been made by various persons for the following tornadoes: Carrol County, Ind., 20 March 1976; Omaha, Nebr., 6 May 1975; Tulsa, Okla., 8 June 1974; Xenia, Ohio, 3 April 1974; Guin, Ala., 3 April 1974; Clay Center, Kans., 25 September 1973; Alto, Ind., 11 April 1965; and Winfield, Kans., 1964. Copies of these recordings are with the author.

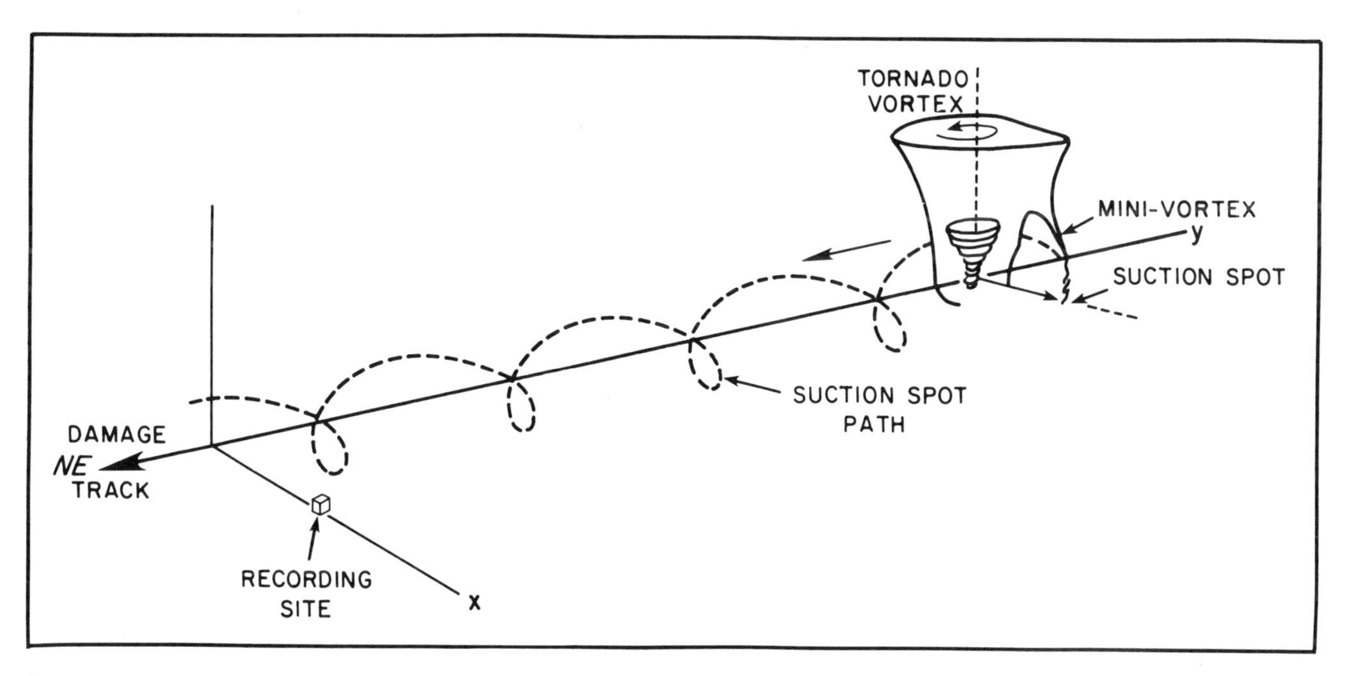

Figure 5.6. Geometry assumed for the track of the Guin, Ala., tornado (3 April 1974) and a possible minivortex. Variations in elevation (−10 to +7 meters) were neglected (after Arnold et al., 1976).

station, both frequency content and sound intensity change, because distance changes when the tornado describes a complex path and because the vortex evolves. The most important sources of sound to be found on a tornado recording are these: (1) wind-surface noises, (2) noises from within the tornado vortex, and (3) noises associated with the parent storm. Arnold et al. (1976) consider only the first of these sound sources in the analysis of evolution of tornadic sounds. Figure 5.6 shows the geometry assumed for the track of the Guin tornado, and Fig. 5.7 shows the measured temporal variation in the relative acoustic power at 350 Hz. In the rather speculative model sounds are assumed to be generated at the bases of both a small vortex moving about a larger vortex and the larger vortex itself. It is also assumed that the source of noise is the interaction of the vortices with terrain. In calculating the relative power received, absorption by terrain and atmosphere was an important correction. Figure 5.8 shows a comparison of observations and calculations. Agreement is far from perfect, and the regular variations are suspect because they are twice the observational interval; i.e., they may represent analysis error rather than a true physical phenomenon. However, if the loudest sounds were assumed to coincide with the point of nearest approach by the main vortex, then the shift in phase of the regularly spaced intensity peaks coincident with the tornado's passage supports the contention that the variations are real. Consequently, it is reasonable from sound study alone to postulate that there was multiple vortex structure. A more detailed analysis can be made when the assumed vortex tracks can be compared with the actual tornado-damage survey.

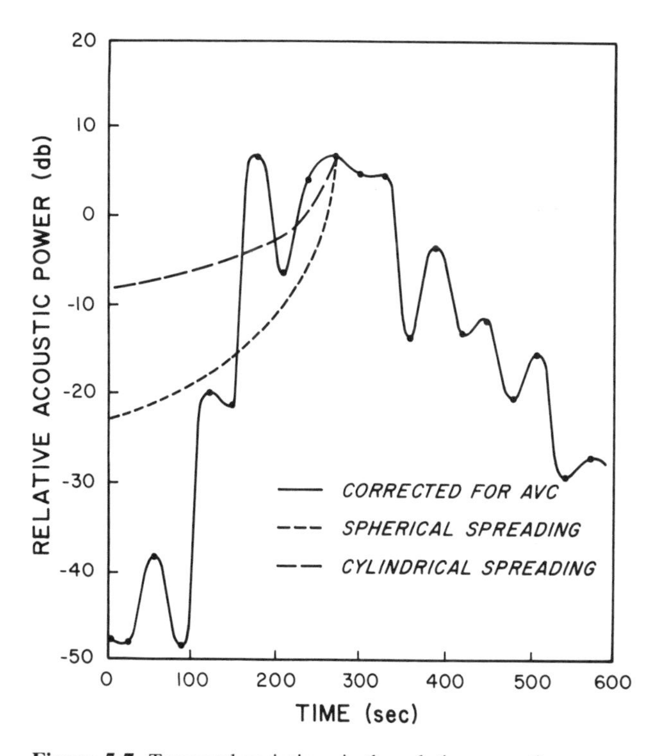

Figure 5.7. Temporal variations in the relative acoustic power at 350 Hz for the Guin tornado compared with power variations expected for spherical and cylindrical spreading. The tornado vortex was assumed to track at a constant speed of 96 km h^{-1} in computing spreading losses. The relative power for the tornado has been corrected for atmospheric absorption, assuming 90% humidity at 20°C (after Arnold et al., 1976).

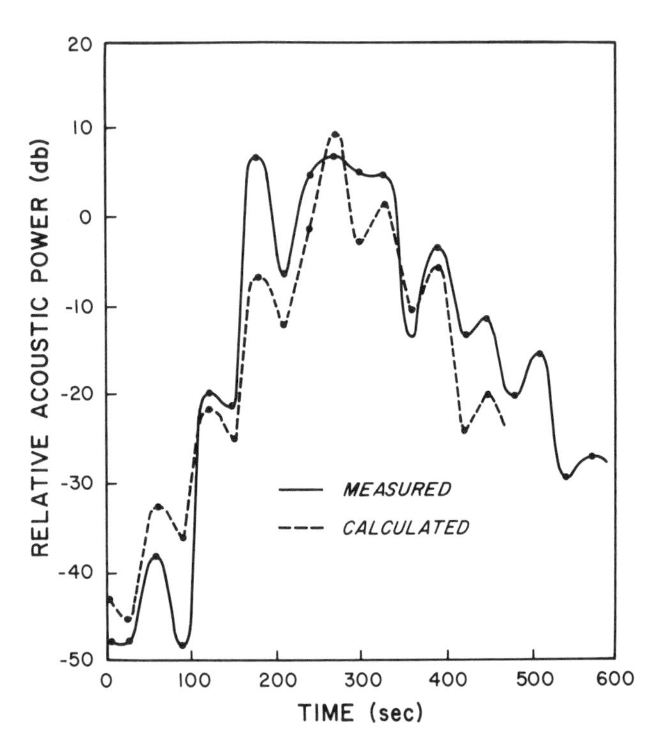

Figure 5.8. Temporal variations in the relative acoustic power at 350 Hz for the Guin tornado compared with the variation expected from two point ground sources, i.e., from the large and small vortices (after Arnold et al., 1976).

5. Wind Noise

Wind fluctuations can be as large an influence in sound refraction as a temperature gradient. Also, turbulence itself generates pressure waves, which produce a response in sound-recording equipment. The intensity of sound produced by turbulence in the free atmosphere depends strongly on the wind velocity. Although howls and whistles associated with high winds are commonly observed, they probably arise from surface features and not in the free atmosphere.

6. Conclusion

The energy per unit length released in a lightning flash, determined from peaks in thunder power, and the total channel length from reconstruction provide a direct means of measuring the mechanical energy produced by a lightning discharge. Observations of power peaks at infrasonic frequencies have led to an explanation of low-frequency thunder based on electrostatic collapse of a charged volume rather than on channel expansion. Reconstruction analysis of lightning channels from thunder data shows great promise in studies of space-time evolution of lightning in convective storms.

Thunder studies, however, cannot alone provide a complete picture of what are inherently electrical processes within storms. It is important that both active and passive electromagnetic mapping techniques be studied simultaneously with sound measurements. Future experiments might use the lightning flash as source of intense sound with which to probe the atmosphere in and near a severe storm.

Recorded sounds from a tornado are in short supply, but there is encouraging evidence from the one sound analysis reported in the literature that useful information on the tornado itself can be derived from detailed sound study.[3]

[3]Active storm research at the University of Mississippi would not be possible without support from the Office of Naval Research, the National Science Foundation, the Nuclear Regulatory Commission, and the National Severe Storms Laboratory. To each I express a sincere thank you. A special note of thanks to my colleague Henry Bass for his constructive critique.

6. Infrasound from Thunderstorms

T. M. Georges

1. Introduction

Everyone is familiar with the sounds for which thunderstorms are named; thunder is probably the loudest natural sound we are regularly exposed to. Chapter 5 reviews some of the physics of thunder and other acoustic emissions from storms, but the attention of that chapter is confined mainly to audible storm sounds and entirely to waves whose frequencies exceed about 1 Hz. This chapter is concerned with storms that emit acoustic waves at infrasonic frequencies, specifically at frequencies less than 1 Hz.

If we "listen" to these ultraflow frequencies with suitable instruments, we find acoustic waves whose intensities are comparable with those of natural audible sounds. But we also find an interesting property not shared with the audible spectrum. Sounds like thunder are detectable no more than a few tens of kilometers from their sources, because of strong atmospheric attenuation; infrasound below about 1 Hz can travel hundreds or thousands of kilometers relatively unattenuated. The difference occurs because the classic coefficient of acoustic absorption is proportional to the square of the wave frequency (Reed, 1972; Procunier and Sharp, 1971). Thus infrasound sensors can monitor events not just on a local scale but on a global scale. This chapter explores the remote-sensing potential of such a capability, with particular application to severe storms.

2. Historical Perspective

As is often the case in scientific discoveries, natural infrasound was initially regarded as interference on instruments designed to observe something else—in this case acoustic-gravity waves from distant nuclear explosions. But as better ways to monitor nuclear tests evolved, attention turned to the sources of the natural background "noise." Empirical associations with various known geophysical events soon uncovered a whole "zoo" of infrasonic wave types. Attempts to classify the different species were published by Cook and Young (1962) and by Georges and Young (1972). Bibliographies on both natural and man-made infrasound and related subjects have been presented by Thomas et al. (1971) and the Swedish Defence Materiel Administration (1985).

One source turned out to be severe thunderstorms, some over 1,000 km from the sensors (Chrzanowski et al., 1960; Goerke and Woodward, 1966; Bowman and Bedard, 1971; McDonald, 1974). The waves often had pressure amplitudes of a few microbars and periods of tens of seconds and often lasted for hours. They were distinguished from local pressure fluctuations because they traveled at essentially the speed of sound. The association with storms was established by comparing the times and directions of arrival of a certain kind of infrasound event with severe-storm reports and with national weather radar charts.

Another, quite different manifestation of severe-storm infrasound was independently discovered in 1967. During a survey of wavelike ionospheric motions revealed by HF-CW (high-frequency, continuous-wave) Doppler soundings at Boulder, Colo., certain nearly monochromatic oscillations of approximately 3-min periods and durations of several hours emerged as a distinct type of event. Georges (1968a, b) correlated observations of these ionospheric waves with the occurrence of severe thunderstorms below the ionospheric observation point. He also explained the waves' frequency content in terms of a narrow-band "window" for acoustic waves reaching the ionosphere from the troposphere. These observations have been since confirmed by Detert (1969), Baker and Davies (1969), Davies and Jones (1971, 1972a, b), Prasad et al. (1975), Smith and Hung (1975), and Hung et al. (1975).

Davies and Jones (1972a) observed that frequency spectra of the ionospheric waves consistently reveal a "fine structure" with distinct peaks near 3.5- and 4.5-min periods. Chimonas and Peltier (1974) recognized that these peaks occur at the frequencies of modal resonances revealed by theoretical analyses of acoustic wave-guide propagation in the atmosphere's temperature structure (e.g., Francis, 1973). Jones and Georges (1976) constructed a mathematical model of an acoustic window to the ionosphere that explains the major features of the observed spectra. Moo and Pierce (1972) have proposed a competing source mechanism that does not invoke atmospheric filtering but is based on buoyancy oscillations and nonlinear frequency doubling inside storms.

Georges (1973) cataloged the phenomenologies of these two kinds of severe-storm waves, i.e., the waves in air pressure at the surface and wavelike ionospheric motions, and found that they share many features, leading him to suggest that they may be just two different manifestations of the same emission process. He also reported some examples of simultaneous observations of the two different kinds of waves from the same storm system. Possible use of the emissions for storm warning has been examined by Georges and Greene (1975), and Georges (1976) has critically examined several proposed emission mechanisms.

3. Instrumentation and Some Representative Observations

a. Surface Pressure

Modern microbarographs can record pressure changes of less than 10^{-8} atmospheres (Matheson, 1964; Cook and Bedard, 1971; Hubbard and Bedard, 1969; Bedard, 1971). For comparison, the threshold of hearing, 0 dB, is 2.2×10^{-10} atm. The infrasound from severe storms has pressure amplitudes that are very small in meteorological terms but are comparable with those of ordinary speech (about 60 dB, or 0.22 μb). But because its frequency is much lower (0.1 to 0.004 Hz, or 10-to-250-s wave period), it must compete with much higher levels of nonacoustic atmospheric pressure fluctuations, or "noise."

Because the power spectrum of atmospheric pressure fluctuations exhibits something like an inverse-square frequency dependence (Kimball and Lemon, 1970; Meecham, 1971), band-pass filtering is normally used to focus on the particular decade or so of interest; otherwise long-period pressure fluctuations tend to dominate. Signal-to-noise ratios are usually less than unity, so that cross correlation of pressure fluctuations obtained at spatially separated microbarographs is required to separate acoustic waves from the noise produced at the sensors by wind-advected turbulence. Optimum separation is determined by the spatial coherence of the acoustic waves; for severe-storm infrasound, a sensor spacing of about 10 km has been found to be the best compromise between angular resolution and spatial decorrelation. Sampling the atmosphere through long pipes (300 m or more) with resistive ports spaced approximately every meter provides additional spatial filtering to smooth out small-scale pressure fluctuations (Daniels, 1959; Priestley, 1966; Burridge, 1971; Grover, 1971; McDonald et al., 1971). This technique reduces the effects of pressure fluctuations whose temporal scales fall in the frequency band of interest but whose spatial scales are much smaller than the wavelengths of acoustic waves of the same frequency (at 0.1 Hz, the acoustic wavelength is more than 3 km).

Arrays of microbarographs lend themselves to analog and digital array-processing techniques, which transform an array of pressure-time data into spatial and temporal spectra (Brown, 1963; Young and Hoyle, 1975; Smart and Flinn, 1971). This permits automatic searching and filtering in frequency-wave-number space and greatly simplifies the task of identifying particular kinds of waves from particular directions.

One way to display infrasonic waves from severe storms is shown in Fig. 6.1, which is the superposition of four analog recordings of pressure versus time made at Boulder, Colo., with a quadrilateral array of microbarographs whose average spacing is about 8 km. The four pressure traces have been shifted a few seconds in time with respect to each other; the shifts correspond to the acoustic delay across the sensor array for a wave arriving from an azimuth of 128° at a speed of 340 m s^{-1}. The frequency passband used to make this record has 20-dB response points at wave periods of about 2 and 100 sec. These particular fluctuations were associated with a severe-storm system about 800 km away, near Oklahoma City. Emissions from this storm were recorded at seven different observatories, the most distant at College, Alaska. The Boulder observatory recorded the emissions for more than 5 h. This was an exceptionally strong and long-lived event; the estimated maximum acoustic power radiated by this storm was 5×10^7 W (watts). However, a typical severe-storm cell might release latent heat at an average rate of about 10^{13} W (Georges, 1976), nearly a million times the observed acoustic power.

Wind noise now imposes the most severe limitation on the sensitivity of microbarographs at infrasonic frequencies. A statistical study of the sensitivity of the NOAA microbarograph array near Boulder, Colo., showed that the wind-noise level exceeded 0.5 μb (about the average level of severe-storm waves) about 50% of the time (the sensors themselves can detect 0.01 μb). Boulder is a very "quiet" location, and the array is equipped with 300-m noise suppression pipes. Average wind-noise levels at two other locations were two to three times as great (Georges and Greene, 1975).

Because the average sensitivity of such arrays is wind-noise-limited, advances in infrasound measurement technology should focus on the development of noise-reducing probes (Smith and Bauer, 1970; Elliott, 1972) and on assessing the effects of meteorology, terrain, and vegetation on the noise level at a given location.

b. Ionospheric Waves

The ionospheric effects of severe-storm infrasound were discovered on records of HF-CW Doppler sounders, a technique invented by Watts and Davies (1959) to monitor the sudden ionization induced by solar flares in the ionospheric *E* and *F* layers. A transmitting station emits a continuous, unmodulated, high-frequency radio wave that reflects from the ionosphere and is detected at a receiving station tens to thousands of kilometers away. The transmitted frequency is so stable that fluctuations of the received frequency indicate only the Doppler shifts imposed by motions of the reflecting ionospheric layer. Acoustic waves that pass through the

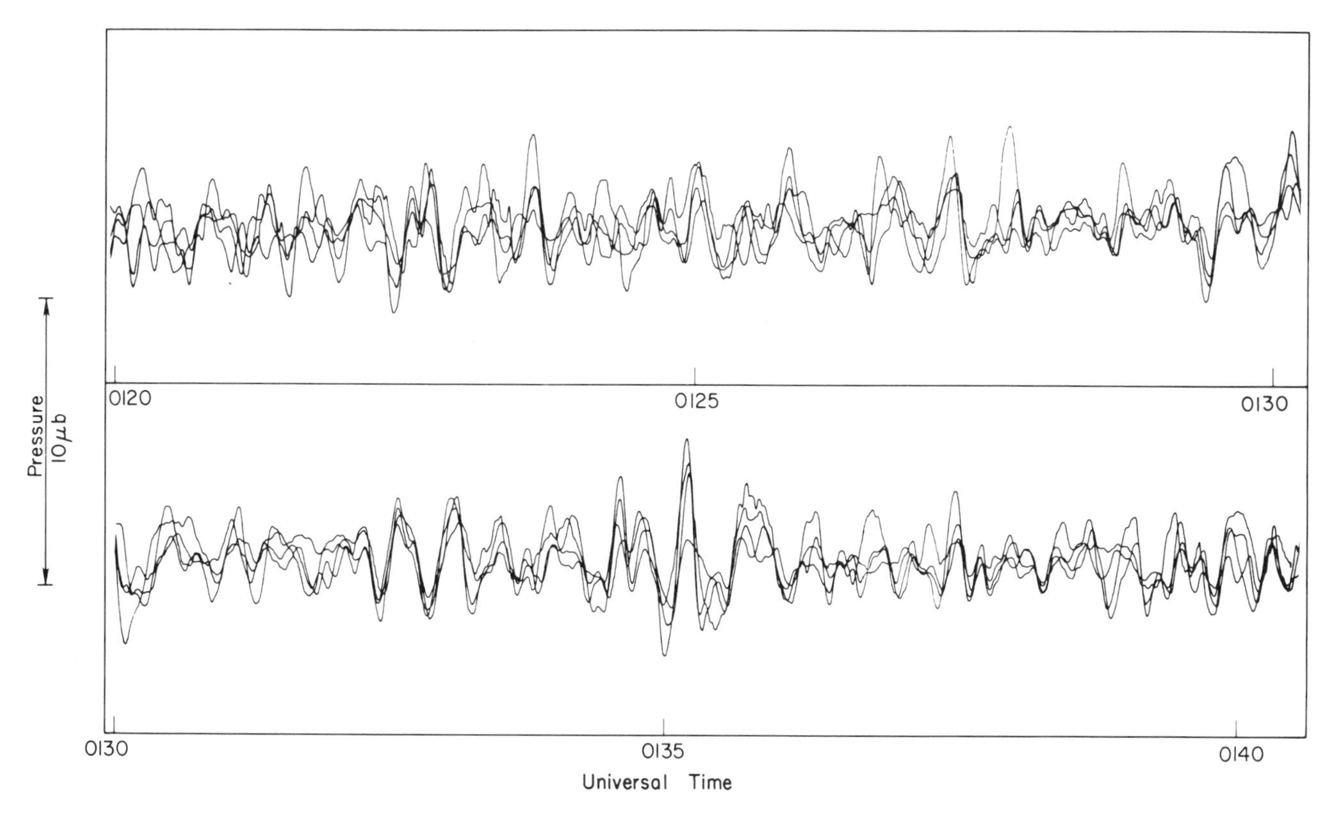

Figure 6.1. Four superimposed microbarograph traces recorded at Boulder, Colo., during a severe-weather outbreak (including tornadoes) in Oklahoma on 5 June 1973. The four traces have been shifted in time for a maximum cross correlation; the shifts correspond to delays appropriate to a plane acoustic wave passing over the sensor array at 340 m s^{-1} from 128° azimuth.

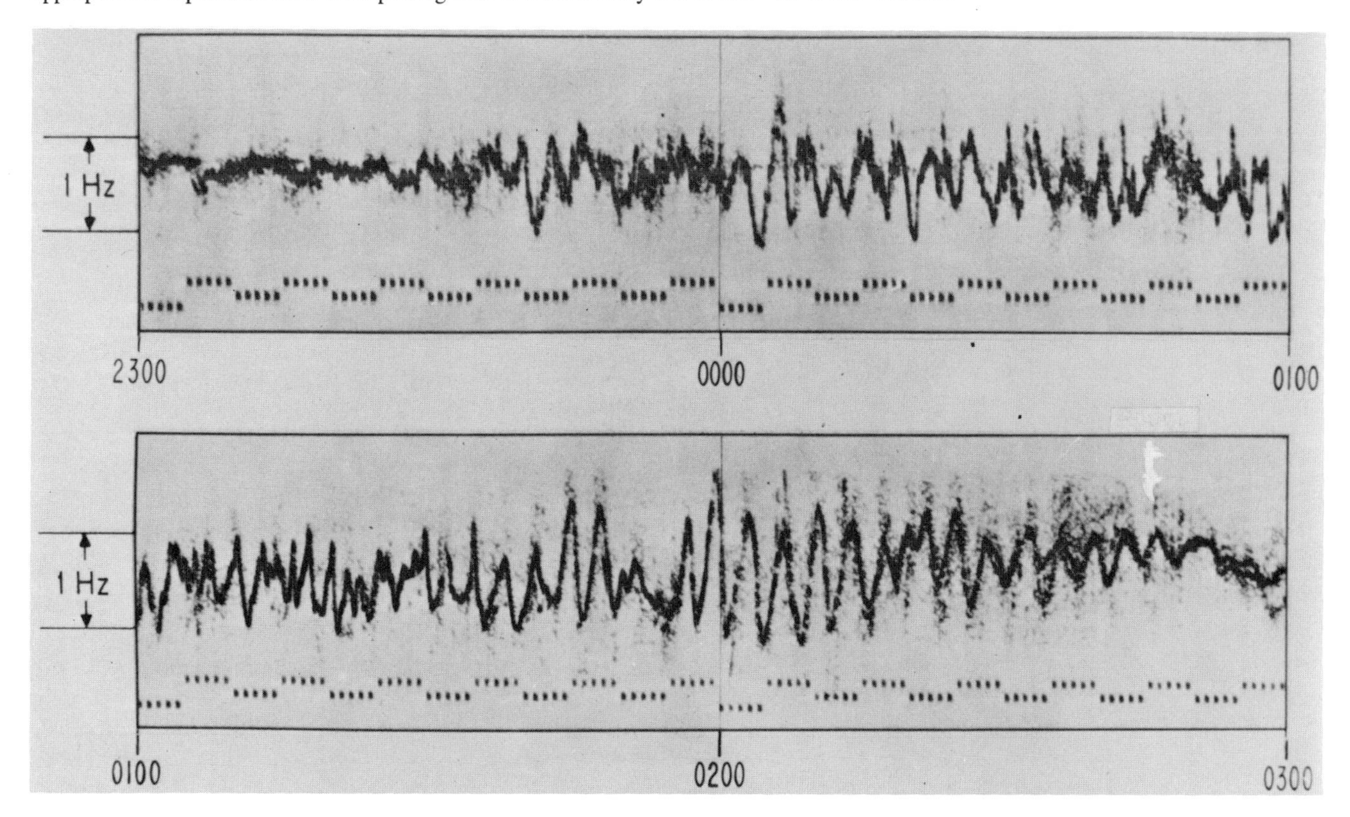

Figure 6.2. Frequency fluctuations of a 3.3-MHz radio wave reflected from the ionosphere during a 4-h period on 27–28 June 1969. The undulations whose period is about 4 min (2300 UT to 0300 UT) are unusually periodic and are attributed to the upward passage of severe-storm acoustic waves through the reflecting ionospheric layer (after Davies and Jones, 1972a).

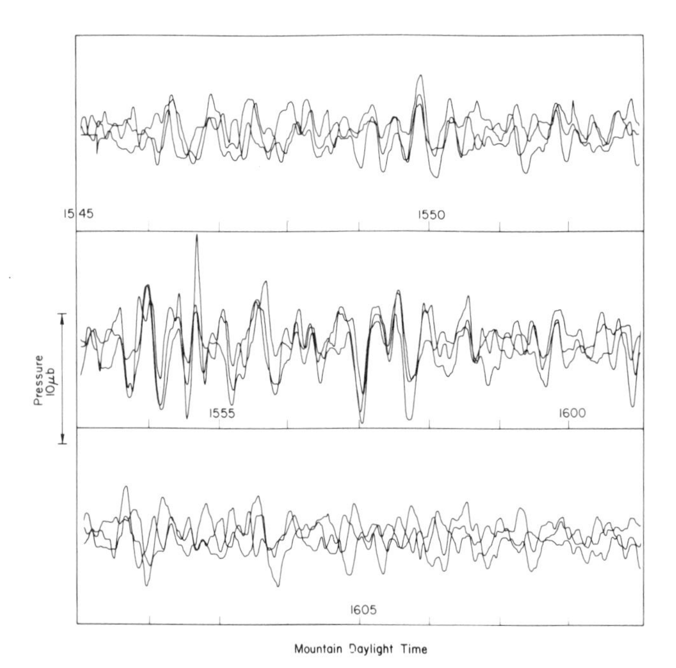

Figure 6.3. Three superimposed microbarograph traces recorded at Boulder, Colo., on 4 June 1976 during the passage of a tornado about 27 km NE of the observatory. The traces have been shifted in time to correspond to a plane acoustic wave from 71° azimuth at 360 m s^{-1}, which coincides with the direction of the tornadic storm.

ionosphere move the reflecting layer up and down, and the resulting periodic frequency fluctuations show up on the frequency-time records the sounders produce (Georges, 1973). An example of such a record, obtained during the passage of a severe-storm wave, is shown in Fig. 6.2. A simple theory relates the Doppler fluctuations to acoustic-wave amplitude (Georges, 1968b, 1973).

Arrays of such sounders can probe the ionosphere at spatially separated points and, using different sounding frequencies, can probe different heights in the ionosphere (Georges, 1967, 1968a). Such spaced soundings permit the measurement of acoustic-wave speeds and directions of propagation. A disadvantage of the method is that the height where the radio waves are reflected is usually not known accurately and must be independently measured if vertical wave velocities are desired. Another disadvantage is that horizontal ionization gradients cause nonvertical sounding paths that introduce error in the estimate of the location being probed.

4. A Case Illustration: Infrasound from a Nearby Tornadic Storm

On 4 June 1976, a medium-sized tornado passed 27 km northeast of an array of microbarographs near Boulder, Colo. Immediately following the time the tornado was on the ground, the microbarographs recorded, from the direction of the parent storm, the passage of acoustic waves of 20- to 30-s period and up to 7-μb amplitude. Figure 6.3 shows three superimposed, but time-shifted, Boulder microbarograph traces for the entire event. No ionospheric soundings were made at the time.

Although infrasound is commonly observed from more distant severe storms, this is the first occurrence of which we are aware of a well-documented tornado passing so near a microbarograph array without the winds and turbulence accompanying the thunderstorm activity causing excessively high acoustic noise levels at the instruments. This fortuitous event allowed us to study in more detail the spatial and temporal relations among visual and photographic tornado records, weather-radar echo development, and the infrasonic waves, especially since the infrasound observatory was so close that propagation effects could not seriously distort the directions of arrival. The location and size of the tornado are well known because it was photographed from two locations, and its damage track was located following the storm. The visible funnel was at least 375 m in diameter at ground level, and its intensity was estimated from a damage survey at between F0 and F1 on Fujita's scale (see Vol. 1, pp. 42–43).

Figure 6.4 is a sequence of radar-scope photographs of the storm system at approximately 5-min intervals preceding and during the event. The frame labeled 1530 MDT shows the location and time of first sighting of the tornado (T) and the location of the Boulder microbarograph array (B). The precipitation cell associated with the tornado shows an extraordinary growth rate compared with other storm echoes in the sequence.

The direction of infrasound arrival was measured at about 5-min intervals between 1545 and 1605 MDT. The wave arrival direction changed smoothly from about 60° at 1545 MDT to about 95° at 1605. The bearing of the tornado's ground damage track from the observatory was between 60° and 65° azimuth, and the tornado was visible from about 1530 to 1540 MDT. Therefore, the emissions apparently began near the end of the visible tornado's life; they appeared at first to come from near the tornado's location, but then the apparent source moved clockwise. Perhaps the infrasound is related to the appearance of the appendage to the tornadic cell that appeared on radar at about 1540 MDT and then grew to join the larger cells to the southwest. This feature of the emissions is common: they are apparently connected with the tornadic storm, but not with the location and time of the tornado itself.

5. A Possible Source Mechanism

Although hundreds of instances of infrasonic emissions from severe storms have been identified, the physical mechanism of the acoustic radiation remains unexplained. Several superficially reasonable mechanisms have been suggested, and these have been critically examined by Georges

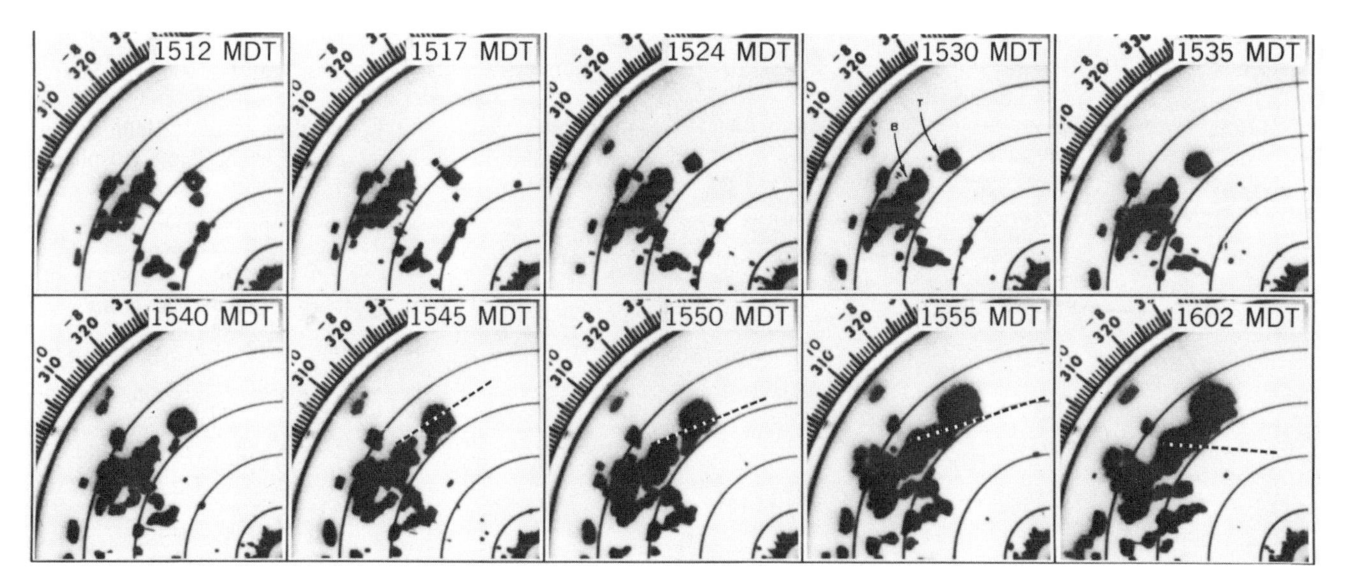

Figure 6.4. Weather radar echoes at approximately 5-min intervals from the National Weather Service radar at Limon, Colo., during the development of a tornadic storm that passed near Boulder on 4 June 1976. The dashed lines in the last four frames show the variation of the direction of infrasound arrival. The range circles are spaced 25 nmi apart.

(1976). Among the candidates are these: (1) simple acoustic sources (monopole, dipole) related to the expansion of heated storm air and to the storm-scale circulation pattern; (2) the random acoustic noise from "turbulent" motions inside storms; (3) low-frequency thunder, i.e., the acoustic radiation from lightning discharges or from electrostatic relaxation inside storms; (4) vortex sound, for example, the radiation from several kinds of dynamic instability in vortices, or the interaction of multiple-vortex systems; and (5) thermomechanical oscillations driven by the interaction between latent-heat release and a storm's updraft.

Deciding among these candidates means formulating for each a quantitative model that makes predictions, for example, of radiated power and frequency spectrum, that can be compared with observations. Furthermore, the model selected should be compatible with known storm processes. These calculations and comparisons are discussed in detail by Georges (1976).

A major observational clue to the source mechanism is a marked tendency for the emissions to come from tornadic storm systems. However, attempts to correlate the infrasound emission time and direction of arrival with specific tornadoes have generally been unsuccessful. In fact, some emissions appear to begin 1 h or more before the first observations of tornadoes (Georges and Greene, 1975). Thus the evidence points to some process characteristic of tornadic storms, but not the tornado itself, and possibly a process that precedes tornado formation.

Doppler-radar observations inside storms have revealed just such a process: the consistent formation of concentrations of vertical vorticity a few kilometers in diameter inside storms that, as much as 1 h later, spawn tornadoes (Kraus, 1973; Brown et al., 1973; Lemon et al., 1978). Such a vortex is shown in Fig. 6.5 for a nonsevere Colorado thunderstorm. Radar resolution is insufficient to reveal any subvortex structure. Agee et al. (1976) have suggested that such mesoscale vortices are members of a hierarchy of vortices of decreasing diameter that characterize some tornadic storms. The hierarchy extends through (1) the mesoscale tornado cyclone just mentioned, which probably coincides with radar hook echoes and may be exhibited visually in the wall or pedestal clouds that often accompany tornadoes; (2) a mini-tornado cyclone, perhaps several hundred meters in diameter, corresponding to the parent of multiple tornadoes; (3) tornadoes themselves, up to a few tens of meters in diameter; and (4) the subtornado-scale suction vortices, up to a few meters in diameter, proposed by Fujita (1970, 1971) and Agee et al. (1975). The members of this hierarchy could be linked by a dynamic instability that causes large vortices to decompose into a few satellite vortices instead of being dissipated by turbulence. Many such instabilities are known in the theory of rotating flows (Lord Kelvin, 1910; Davy et al., 1968).

Ward (1972) investigated the conditions under which multiple vortices form from larger ones in a laboratory model simulating tornado flow. He found that certain relations between the inflow angle and the updraft speed in his apparatus favored the formation of multivortex flows, in which several vortices rotate about a common center. Furthermore, changing a parameter such as the inflow angle could cause single vortices to decay into multivortex systems. These simulations suggest that, under the conditions that exist in the vicinity of some severe thunderstorms, transitions through successive stages of the vortex hierarchy described by Agee et al. (1976) may be common.

Multivortex systems suggest a simple model from which the acoustic radiation is readily calculated. Consider, for example, the radiation from a corotating vortex pair, as calculated by Powell (1964). The radiated acoustic power is

$$\Pi = \frac{16}{15\pi}\frac{\rho}{c^5}\left(\frac{\Gamma}{4\pi r}\right)^8 l^2, \qquad (6.1)$$

where Γ is the vortex strength (circulation), $2r$ is the distance separating the pair, l is their axial length, and ρ and c are the atmospheric density and sound speed. The quantity $\Gamma/4\pi r$ is the tangential speed u at which the vortices spin about their common axis, so that the rotation frequency is $u/2\pi r$. Powell shows that the pair radiate acoustically like a rotating acoustic quadrupole.[1] The radiated frequency is thus twice the rotation frequency, or $u/\pi r$. For example, if $u = 43$ m s^{-1} and $l = 3$ km, then $\Pi = 10^7$, and the vortex separation required for radiation at a 30-s wave period is about 820 m. The required source power estimated by Georges (1973) for infrasonic waves of about 0.6-μb amplitude observed at a distance of 500 km is approximately 2.3×10^7 W. More realistic models could be constructed, but not without guidance that can be supplied only by more detailed measurements of storm-vortex structure and breakdown dynamics.

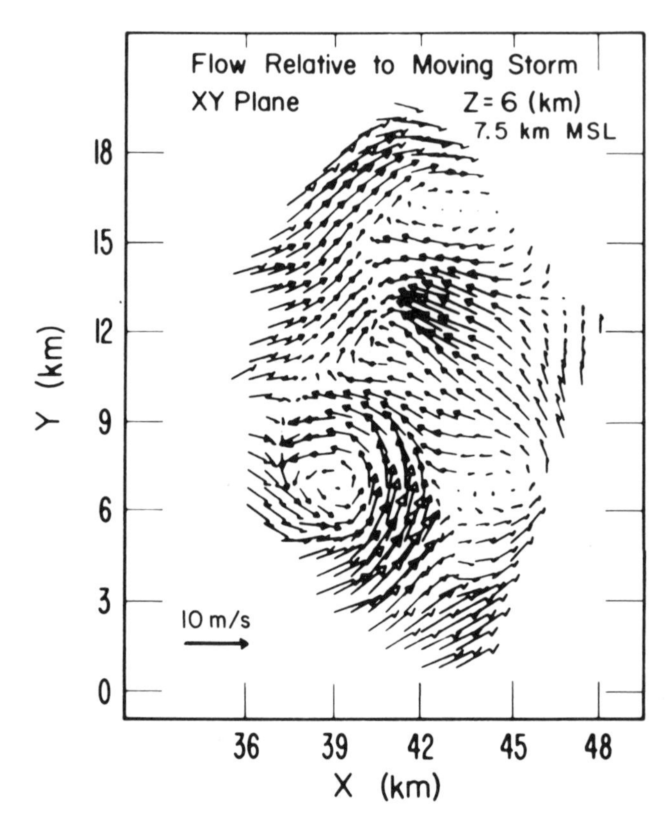

Figure 6.5. A map of horizontal air velocity vectors 6 km above ground inside a Colorado thunderstorm. The velocity field was mapped with the WPL dual Doppler radar (Kropfli and Miller, 1976). The size of each vector is proportional to airspeed, with mean removed, at that location, with the highest speeds reaching 10 m s^{-1}. A strong vortex is visible at the lower left, and two weaker ones at the top and at the lower right can also be seen. The radar resolution does not allow the subvortex structure to be seen with any detail.

6. Is It Useful for Storm Warning?

Georges and Greene (1975) examined the storm-warning potential of the infrasound observed at ground level by comparing the observability of the waves at three observatories (spaced about 550 km apart) with weather-radar data and severe-storm reports. They evaluated their results on the basis of four "indices of usefulness":

1. False-alarm rate, which tells how often infrasound from other sources is mistaken for that from storms. They devised a sorting procedure that reduces the false-alarm rate to 15–20%, and even lower rates seem achievable.
2. Detection rate, which tells what fraction of all severe storms in a specified region is detected. Here the major problems are defining severe storms objectively and verifying their occurrence. They estimated a 65% detection rate for tornadic storms, a 31% detection rate for tornadoes themselves, and a 33% detection rate for storms with radar tops above 50,000 ft. (15,000 m).
3. Timeliness, which tells how much advance warning the emissions might give of dangerous storm effects. Only tornadoes were considered from this viewpoint, because of the difficulties of specifying onset time for other storm features. It was discovered that many infrasound emissions begin 1 h or more before tornado sightings, with some emissions preceding tornado onset by many hours (Fig. 6.6). However, the relatively slow speed of acoustic waves (about 1,000 km h^{-1}) imposes a fundamental limitation on the density of a practical observing network. Furthermore, some emissions apparently come after tornado formation.
4. Location accuracy, which tells how well the emissions could be used to locate and track storms. Of 177 events that were classed as storm emissions during one season, fewer than expected (about 18%) were recorded at more than one observatory, and fewer than 3% were recorded at all three stations. Therefore, source location by triangulation was

[1]A monopole would represent local fluctuations of mass or energy; a dipole would represent a local flux of mass or energy, for example, an oscillating fluid parcel; a quadrupole would represent a cloverleaf-shaped distribution of fluid density, for example, with alternating positive and negative lobes.

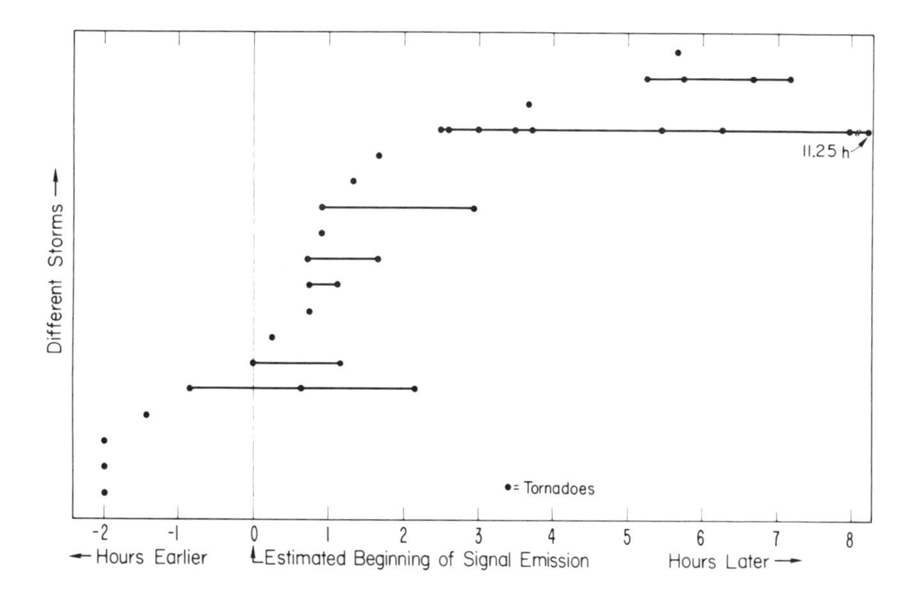

Figure 6.6. Tornado time relative to signal emission time for 18 tornadic storms for which the infrasound onset time was well defined. Infrasound travel time has been removed from these estimates. The horizontal lines connect different tornadoes (dots) occurring within the same storm system.

not usually possible. Source identification was usually made by comparing the apparent source direction measured at one observatory with national weather radar charts. When triangulation was possible, the indicated intersections usually fell close to radar-indicated severe-storm systems, but positive identification and tracking of a particular storm was seldom possible. Atmospheric refraction of the acoustic waves in azimuth apparently permits no better than about 100-km accuracy in source location at a nominal range of 700 km.

If ways to reduce background noise can be found, severe-storm detection rates should improve, and it may be practical to consider deploying a prototype network of microbarograph observatories in something like a 1,000-km grid to supplement the present storm-warning system. Compared, for example, with a radar station, the costs and complexity of a microbarograph observatory are modest indeed, and the automation of all its signal-processing tasks seems straightforward. Integration of such a capability into the existing storm-sensing and warning system may be a cost-effective way to improve the reliability and timeliness of tornado warnings.

The warning potential of the ionospheric waves seems more remote than that of the ground-level waves, mainly because the ionospheric effects appear to be more localized (generally within a few hundred kilometers of the source storm), and because directional information is more difficult to extract (Davies and Jones, 1972a).

7. Conclusion

Although the acoustic energy that a storm radiates represents only a minuscule part of a storm's energy budget, the emissions are interesting and possibly useful from several viewpoints:

1. They may be a usable precursor of tornadoes.
2. As a remote indicator of vorticity concentration inside storms, they may supplement meteorologists' array of storm research sensors.
3. They have helped us formulate a model of acoustic propagation into the upper atmosphere.
4. And finally, they have provided a recurring example of the natural aerodynamic generation of strong infrasound. A complete explanation of the storm emission process may also help explain other natural infrasound, such as the emissions traced to several windy, mountainous regions of the Earth (Larson et al., 1971).[2]

[2]I thank Gary Greene for processing the infrasound records of the case illustrations and Joe Golden for providing the tornado observations he collected.

7. Sferics and Other Electrical Techniques for Storm Investigations

Edward T. Pierce

1. Introduction

An atmospheric is the electromagnetic signal radiated by a complete lightning discharge or by any subsidiary phase of the flash. The abbreviated form "sferic" is often used. The plural "sferics," however, has a different and more specialized meaning, and it is this meaning that is the main subject of this chapter. We define sferics as referring to techniques using atmospherics for the detection of lightning occurrence and, in more sophisticated applications, for the location of the flash position.

The use of sferics methods ranges from use within largely social contexts through more technical applications to basic research. An example of the social use of sferics techniques is their potential utility in warning populated areas of the approach of severe storms. A typical technical application is the identification in forestry of areas of potential lightning-caused fires. Many sferics studies of storms less than 200 km away in particular have the research objective of better defining the electrical characteristics of the storm; an example is the mapping of the lightning channels within a thundercloud.

Sferics techniques can be conveniently distinguished in three categories according to distance: long distance (> 1,000 km), intermediate (200–1,000 km), and close (< 200 km).

2. Long-Distance Techniques

a. Introduction

Sferics methods have been used to locate storm centers at great distances (> 1,000 km) for two main reasons. First, they are adjunct input to meteorological forecasting, supplying immediate data for regions where surface observations are not readily available. Coverage by satellite meteorology has greatly diminished the need for such sferics information in routine forecasting. Nevertheless, some past successes have been substantial, notably the hourly location for many years of any eastward-moving active frontal system in the North Atlantic by the British sferics network. The second motivation for long-distance sferics studies, less immediate and in some respects more basic, is to establish the climatology of thunderstorm occurrence ideally on a worldwide basis. Such information can be applied to fundamental studies of the global circuit in atmospheric electricity (Vol. 2, Chap. 13).

There are possibilities for the global monitoring of thunderstorms by sferics systems carried in satellites. These systems would operate at frequencies in the VHF (30–300 MHz) band or higher so that the source signals would pass through the ionosphere. However, for various reasons the resolution that can be achieved from a satellite is limited, and for the accurate sferics location of distant lightning, ground-based observations are now preferred. This implies reliance on lightning signals in the LF (300–30 kHz), VLF (30–3 kHz), and ELF (< 3 kHz) bands, and consequent restriction of sferics location to that obtained from the lightning return stroke. It is only at frequencies below 300 kHz that propagation is relatively stable and well understood and propagation losses are sufficiently low for the source properties to be recognizable and detectable at great distances, and to be therefore useful in sferics studies.

Propagation below 300 kHz is channeled in the waveguide formed by the earth and the lower ionospheric D region (Wait, 1970). A general description of the propagation has already been given (Vol. 2, Chap. 13); some further details of particular significance in sferics methods are added here.

Waveguide propagation is frequency-selective in its attenuation properties. There are two windows of minimum attenuation: below 300 Hz and between about 10 and 30 kHz. Figure 7.1 illustrates the influence of these windows on the received-sferics spectrum. The source spectrum at d = 100 km is typical of a return stroke in which the initial violent current surge is succeeded by an intermediate current. The maximum amplitude is at 5 kHz. After propagation, the spectrum at 3,000 km has split into two components corresponding to the two windows of minimum attenuation at ELF and VLF. Note the shift in the maximum amplitude from 5 kHz to 10 kHz with propagation.

The behavior shown in Fig. 7.1 results from attenuation in two dominant waveguide modes in which the magnetic field is transverse to the direction of propagation (TM modes). Other modes have only subsidiary effects. Each

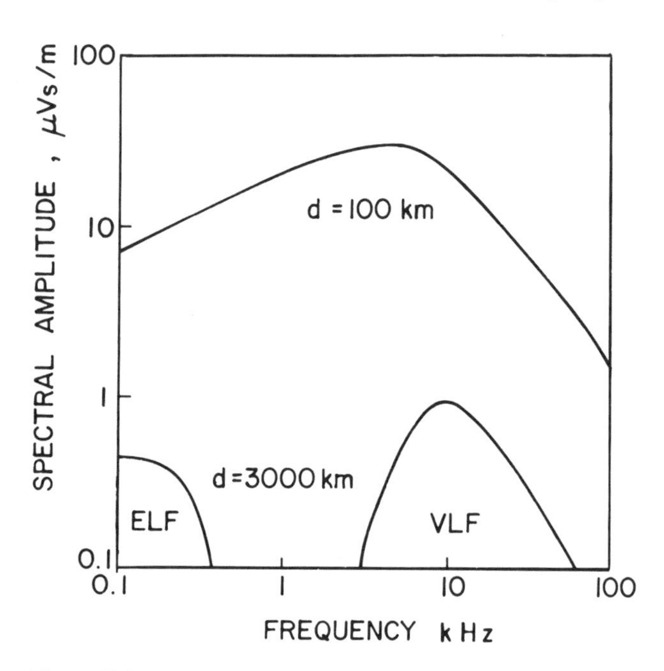

Figure 7.1. Return stroke sferic spectra at two distances.

mode, besides having its own frequency-selective attenuation characteristics, is also dispersive; i.e., propagation speed varies with frequency. This variation is complicated, but the ELF component lags increasingly behind the VLF signal as the propagation distance increases.

b. Single-Station Methods

A common approach to all single-station techniques for locating a lightning source is to establish separately the azimuth of the incoming sferic and the distance, d, of the discharge. The range is deduced from some property of the incoming sferic that is distance-dependent.

1. Crossed Loops: All long-distance, single-station sferics techniques rely on a crossed-loop antenna system to yield the azimuth information. In such a system two identical vertical loop antennas are oriented N-S and E-W. Consider an incoming atmospheric with its electrical field in the plane of propagation and the magnetic field transverse to this plane. Such a vertically polarized atmospheric incident at an azimuth angle Φ with respect to north will induce signals in the loops proportional respectively to cos Φ and sin Φ. This information can be displayed as an azimuth on a cathode-ray tube by applying the loop outputs after suitable amplification to the respective plates of the tube. The information can also be digitized and the value of Φ indicated accordingly. In its simplest form the loop-antenna system includes an ambiguity of 180° in direction of arrival, which can be removed by adding an auxiliary vertical whip antenna and using the output suitably phased to eliminate the appropriate diametrical half.

Polarization, the orientation of the vibration vector of the electric field, can introduce errors in the operation of a crossed-loop system. An incoming atmospheric will produce a clear-cut response in the system only if the signal does not include horizontally polarized components; when such components are present, the response on a cathode-ray-tube display, for example, is elliptical and not linear. Horizontal polarization results from lightning channel orientation and propagation effects. During propagation there is a conversion of polarization state by ionospheric reflections so that even an initially vertically polarized signal will develop horizontally polarized components. Because of the propagation the later stages of an atmospheric of distant origin are particularly liable to include abnormally polarized components. For this reason, gating techniques in which the azimuth is effectively determined from the earlier parts of the waveform are of some use. However, with an optimum choice of frequency of operation, the errors owing to polarization and channel orientation are rarely serious. Both experience and theory indicate that, within the VLF band, operation at or near a frequency of 10 kHz is a satisfactory choice for minimizing errors in direction finding on distant atmospherics, if the bandwidth is relatively wide.

Errors in crossed-loop work are also introduced at the site (Horner, 1964). Even topographical irregularities that are small compared with the wavelengths involved introduce distortion of the wave front and consequent abnormal signals into the loop system. The site should be not only level but also as electrically homogeneous as possible. Long lengths of cable, buried or exposed, are very likely to produce errors, because the cables represent preferred paths for the currents induced by the impressed sferic, and the resultant reradiation is a source of unwanted abnormal signals in the loop system. As a final source of errors with a crossed-loop system, there is the possibility, with tuned loop receivers, of the disturbance or ringing response produced by an incoming sferic not having decayed completely before another sferic arrives. The duration of the disturbance is inversely proportional to the bandwidth, so that with a typical system of 1 or 2 kHz bandwidth the ringing is on the order of 1 ms. This time is not usually sufficiently long for interference between successive bearings to be a problem. Even if signals were accepted from all the global lightning activity (100 flashes per second), the average interval between return-stroke pulses is still several milliseconds. However, if narrow bandwidths of approximately 100 Hz or less are employed, interference between successive responses becomes significant.

2. Distance-dependent Parameters: The magnitude of an incoming sferic depends in a complicated fashion on the distance of its origin. However, the variability of the source signals is such that, even when the distance is known, magnitude is only an extremely crude indication of distance. For instance, the amplitude distribution of broadband VLF return-stroke pulses at the source extends over a range of at least 20 to 1. With this great source variability it follows that, even when a specific frequency is selected for which the attenuation with distance is considerable, the uncertainties in distance estimation by magnitude alone are immense.

A historically well-tried method of estimating the distance of an atmospheric is to interpret the waveform by use of reflection or ray theory. With this approach, and assuming the earth and the ionosphere to be good reflectors, the waveform will consist of a series of pulses whose sequence of temporal arrival is uniquely defined by the height of the ionospheric layer and the great circle distance. Unfortunately, the few waveforms suitable for analysis by the reflection mechanism are usually restricted to short ranges and/or a night environment (Kinzer, 1974). Under other conditions the pulses are severely modified by the ionospheric reflection. The ionosphere cannot ever be regarded as a surface of constant height since there are changes with pulse order and obliquity of incidence, and there is much superposition of pulses. The waveform analysis method, although of historical interest, cannot now be regarded as of practical importance.

Examination of the spectral content and dispersion of an atmospheric provides significant ways of estimating distance. As indicated in Fig. 7.1, the attenuation with distance is a function of frequency. Thus if the source spectral distribution—not its magnitude—is constant and established and the attenuation as a function of frequency and distance is known for the specific propagation conditions, then the ratio of received amplitudes at two frequencies, for example, is potentially a sensitive indicator of distance. Potentially is a key word because of the assumed constraints of source spectral shape constancy and knowledge of the propagation factors. Dispersion in an atmospheric depends on distance, but the method can be accurately applied only if the dispersion at the source is constant and the propagational dependence of the dispersion is known.

3. The Atmospherics Analyzer: The equipment to analyze atmospherics is the most elaborate yet devised for single-station sferics work. It has been extensively developed (see, for example, Harth [1972]) since the original concept by Heydt and Volland (1964). The equipment combines separate measurements of three distance-dependent parameters as a function of azimuth for all incident atmospherics exceeding a certain threshold level. The chosen parameters are these:

1. The amplitude at a specific frequency (spectral amplitude—SA). A typical choice would be 9 kHz.
2. The ratio of the amplitudes at selected frequency pairs (spectral amplitude ratio—SAR). The selection might be 5 and 9 kHz.
3. The difference in times of arrival of the signals on two adjacent frequencies (group delay time difference—GDD). Frequencies of 6 and 8 kHz are typically chosen; the parameter is expressed in milliseconds per kilohertz at the average frequency of the selected pair.

The conversion of basic data from the atmospherics analyzer into distance information has received much attention. Since the measurements are confined to a narrow range of frequencies in the lower part of the VLF band, the assumption of propagation in a single dominant waveguide mode is not reasonable. Extensive, theoretical calculations for this mode yielding the variation with distance of SA, SAR, and GDD are available (Wait, 1970). These calculations cover a variety of propagational conditions including ionospheric temporal variability and direction of propagation with respect to the geomagnetic field. Comparison of calculated values with actual data enables the analyzer to be calibrated in terms of distance. The calibration is particularly sound if a comparison can be made with independent estimates of storm location. Extrapolations at zero distance allow estimations of the average source characteristics.

The information from the atmospherics analyzer is usually presented on an oscilloscope. One deflection represents azimuth and the other the specific distance parameter being studied. Each incoming atmospheric brightens the tube so that a photographic record of the screen, taken over a few minutes, reveals clusters of dots corresponding to the respective storm centers. These clusters are usually much more elongated in the distance direction than in the azimuthal coordinate. Undoubtedly much of this elongation is due to source variability between individual atmospherics in SA, SAR, and GDD. The extent of this source variability is far less well understood than are the propagational factors. Further improvements in the precision with which the atmospherics analyzer can identify storm centers are likely once the extent of the source variability in the different parameters is established. However, the very existence of the source variability means that the analyzer can never locate an individual atmospheric with accuracy, but can only fix the position, on a statistical basis, of an aggregate of atmospherics that is a storm center.

c. Multistation Techniques

The standard method of fixing the position of an atmospheric is to use crossed-loop systems at two or more stations. Intersection of the respective bearings then yields the location. All the errors already discussed in relation to a single crossed-loop installation apply; for this reason, the redundancy furnished by more than two stations is very desirable. Typically, stations are separated by 500 km or more to obtain good fixes at distances beyond 1,000 km. The accuracies are approximately 10%. Some form of interstation timing is necessary to ensure that bearings are taken on the same atmospheric at the different stations. However, this requirement is not severe; a timing accuracy of 10 ms is usually quite adequate. Indeed, if the location information is not quickly required, the patterns of pulse reception at the different stations can often be matched, particularly when there is no local activity. In these circumstances the need for interstation timing is eliminated.

A time-of-arrival (TOA) method is by far the most accurate way of fixing the source of an individual sferic. It is also, understandably, the most elaborate and expensive. Suppose we have two stations and measure the difference in the times of arrival of the same feature—preferably the start—of an atmospheric waveform. Then a given time difference implies that the source lies on a certain hyperbola, with the stations as foci (Hughes and Gallenberger, 1974). Two time differences can be achieved with three stations arranged in two pairs, thus providing two hyperbolas, but four stations (three hyperbolas) are necessary if all ambiguity is to be avoided. TOA systems are much less subject to errors than are crossed-loop techniques. Polarization errors are effectively nonexistent; site errors are very small. However, if the potential accuracy of a TOA system is to be realized and confusion between separate atmospherics is to be avoided, interstation timing of approximately 10 μs is required. This implies the installation of accurate time standards at each station. Furthermore, the hyperbolic geometry involved makes plotting and fixing more difficult than with the straightforward directions yielded by crossed loops. Also, the TOA approach requires more stations than does a crossed-loop system. All these considerations suggest that, unless exceptional accuracy is required for special reasons, a crossed-loop multistation system is economically preferable to the TOA approach.

3. Intermediate Distances

Many of the long-distance techniques discussed operate equally well or even better at shorter distances. This is so for multistation TOA systems, as well as for some of the methods of estimating distance from a single station. Among these, the successive reflection technique is more powerful at intermediate distances (200–1,000 km). The spectral-amplitude-ratio method can also be effectively used, although the pair of frequencies must be chosen with care. A compromise is necessary between pairs of frequencies that are insufficiently sensitive to distance changes and pairs that are overattenuated.

Conventional crossed-loop techniques tuned to 10 kHz give serious polarization errors at shorter distance for two main reasons. The loop antennas sense the abnormal fields owing to long horizontal sections of channel before these have been attenuated by propagation. Also, the ground and ionospherically reflected waves containing horizontally polarized components are of comparable amplitude and arrive almost in superposition. Consequently, polarization errors of many degrees are common at close distances.

Gating techniques in which only the first part of the lightning signal is sampled are relatively ineffective at VLF, since most of the VLF signal originates at a later stage of the return stroke. However, a crossed-loop system has been perfected by Krider et al. (1976) in which a gating method is used at broad band. This system represents a very sophisticated development that is being extensively applied in forestry for locating potential lightning-caused fires.

The principle of the gating method devised by Krider et al. (1976) is easily understood. Records of the magnetic field owing to the return stroke of a close flash to the ground show that the peak value is attained within 1 to 5 μs of the start of the return stroke. Suppose we assume a zero-to-peak time of 3 μs. Then since a typical return stroke upward velocity is 5×10^7 m s^{-1}, it follows that the initial peak fields must originate in channel length of only 150 m above ground level. This channel length close to ground will almost always be approximately vertical, so that the complications of horizontal polarization are avoided if the azimuth information from the loop antennas is obtained by sampling close to the initial radiation field peak.

Broad-band antenna systems are essential, and the loop antennas involved cover a frequency range of approximately 1 kHz to 10 MHz. The combined signals from the antennas trigger a sample gate whenever they exceed a certain threshold value. At the same time the azimuthal information from the loops is tracked pending the initiation of a display gate. In the preferred mode of operation, the display is initiated by a coincidence of the sample gate and the first zero crossing of the differentiated signal, i.e., at the peak output. Display initiation holds the azimuth information at the instant of initiation; this information is then presented as a bearing with a decay constant of 1 ms on an oscilloscope.

In the application to forestry the interest is in cloud-to-ground lightning alone and therefore in return strokes. For this reason various discriminants against background noise and intracloud lightning are incorporated into the direction finder. A return stroke pulse from the first stroke of a flash rises abruptly to a peak and then decays through a series of subsidiary peaks. The discriminants are such that only pulses characteristic of this temporal and peak sequence behavior are accepted.

There have been extensive tests of the broad-band direction finder, involving comparisons with visual observations and television scanners and also, from greater distances, with storm locations revealed by weather radar. Generally the results have been excellent. Bearing errors of 1° or less are usual for ranges from 10 to 600 km. The principal exception is for discharges in mountainous terrain when errors of a few degrees are possible.

4. Close Distances

a. Lightning Mapping

Quite recently some very interesting applications of sferics techniques at VHF have been made, making it possible for the first time to map the VHF sources within a cloud. The principle of the method—a TOA technique—was first formulated by Oetzel and Pierce (1969). It is illustrated in Fig.

7.2 for a baseline very short compared with the distance to the flash. Under these circumstances the hyperbolas of a conventional TOA system degenerate into asymptotic straight lines. With the illustrated three-antenna system, the difference in times of arrival of the same impulse at antennas S_1S_2 and S_2S_3, respectively, are given by

$$\Delta t_1 = (2l/c) \cos \theta$$

and

$$\Delta t_2 = (2l/c) \sin \theta \qquad (7.1)$$

where $2l$ is the baseline and c the velocity of light. Short baselines (< 100 m) are desirable to avoid ambiguities, to ensure similarity of pulses at each end of the baseline, to preserve the approximate geometry, to achieve compactness, and so on. However, the shorter the baseline the less the accuracy. Fortunately, fast logic circuitry components now available can measure times of arrival to 1 ns. It follows that even with antennas separated by only 15 m, cos θ or sin θ is measured to 0.02, and the bearing accuracy is therefore approximately 1°.

Taylor (1978) has extended the short-baseline arrangement to measure azimuth and elevation for incoming VHF pulses. His equipment covers a bandwidth of 20 to 80 MHz and uses baselines of only 13.74 m for both the vertically separated (elevation) and horizontally spaced (azimuth) antennas (Fig. 7.3). The angle of elevation ϕ is obtained from

$$\phi = \text{arc sin} \left(\frac{c}{L_1} \mathrm{T_{el.}} \right), \qquad (7.2)$$

where c is the velocity of light, and

$$\mathrm{T_{el.}} = t_b - t_a \,, \qquad (7.3)$$

where t_a and t_b are the times of arrival of an impulse at antennas a and b. The azimuthal angle of arrival is obtained by

$$\theta = \text{arc sin} \left(\frac{c}{L_2} \frac{\mathrm{T}_{az}}{\cos\phi} \right), \qquad (7.4)$$

where

$$\mathrm{T_{az}} = t_c - t_d \,, \qquad (5)$$

and where t_c and t_d are the times of arrival of an impulse at antennas c and d. L_1 and L_2 are the separations between corresponding antenna pairs used in obtaining the elevation and azimuth angle.

Figure 7.4 is a picture of one observing site at the Ken-

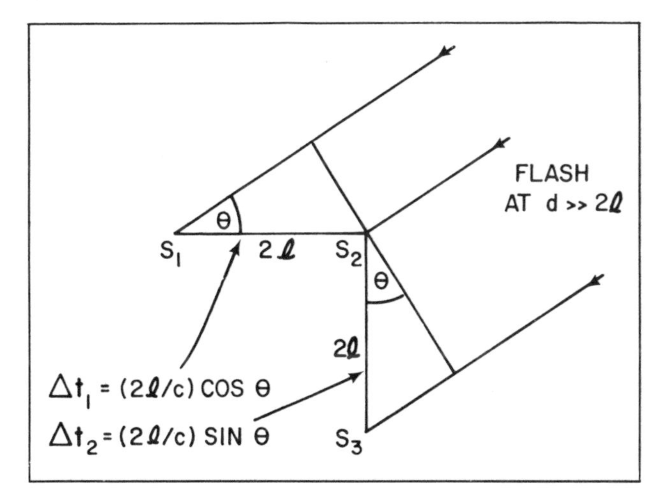

Figure 7.2. Illustrating principle of time-of-arrival VHF azimuth indicator.

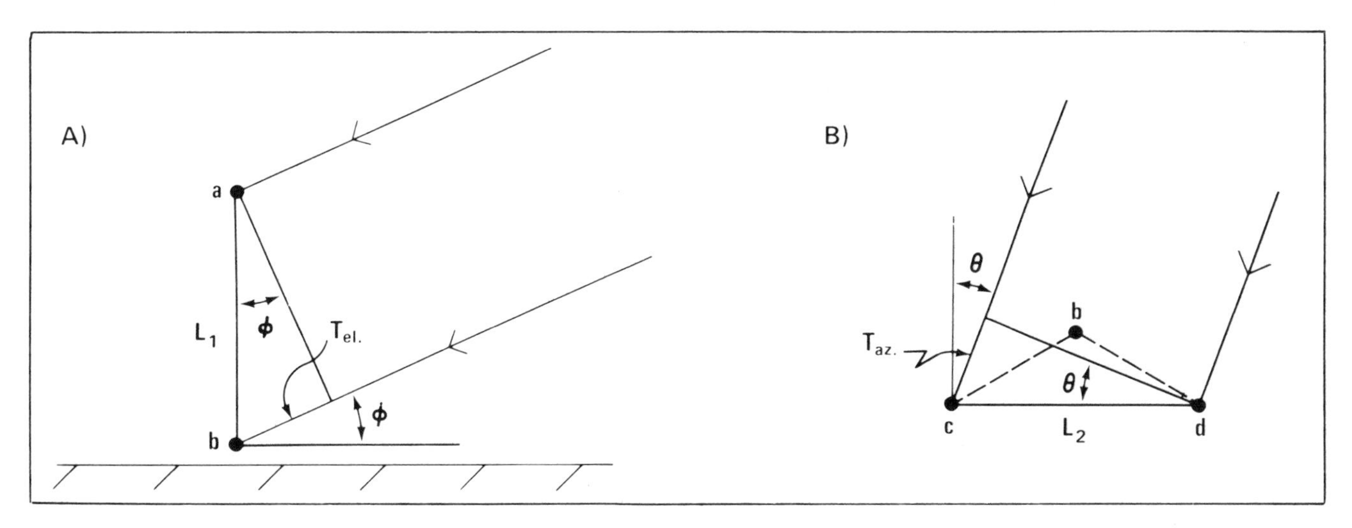

Figure 7.3. Antenna arrangements showing how the (A) elevation and (B) azimuth angles are related to the time-of-arrival differences.

Figure 7.4. Mapping-site antenna mast and equipment van.

nedy Space Center in 1976. The mast at the left supports antennas a at the top, with antenna b situated at the base. Twelve other antennas surround the mast and are electrically selected corresponding to antennas c and d to restrict impulse acceptance to a 60° azimuthal sector that can be shifted in 30° increments. The antenna on top of the van is for communications.

Although there is no problem with ambiguity in determining elevation angles with the use of only two antennas, a third antenna must be used to eliminate ambiguity in measuring azimuth. A simple solution is to use antenna b, as shown in Fig. 7.3. By requiring that the receiving impulses arrive at antenna b before arrival at either antenna c or d, ambiguity is eliminated and a limit in azimuth is imposed so that only impulses from within a preferred azimuthal sector will be accepted. This sector width is 180° minus the angle c-b-d in Fig. 7.3B, unless relative time delays are imposed in the signals received from the three corresponding antennas. Accuracies of 0.5° are obtained, and the system can accommodate up to 25,000 impulses per second. This corresponds to times somewhat shorter than the usual VHF pulse separation of 50 to 100 μs within a flash.

An example of a cloud-to-ground stroke with associated intracloud processes is shown in Fig. 7.5, a pictorial view in elevation and azimuth obtained at a range of about 14 km. The highest dot represents the source of an impulse located at a height of 7.1 km. There appear to be three branches to the right of the main vertical configuration of dots. The lowest branch, containing three dots, is at 2.7 km. Below this height there are four dots representing the main return-stroke channel.

At times pulses can occur as short as 10 μs apart. This means that with baselines exceeding 3 km the possibility of confusion arises. However, this will not always be the case, and an elaborate VHF locating system known as LDAR (Lightning Detection and Ranging System) that uses baselines of approximately 10 km has been constructed and is successfully operated at Kennedy Space Center. Five stations are employed, one central and four peripheral. The information from any four of the stations yields the location of the VHF signal using true three-dimensional hyperbolic geometry; the asymptotic approximations are not applicable since the baselines are now not short compared with the distance to the flash. Various discriminants are incorporated in the LDAR system to ensure that the measured times employed in computing a location are physically realistic and that confusion between unrelated pulses is avoided. Time differences are measured to within 50 ns. The accuracy of the system varies with the configuration of the selected stations and with the range and altitude of the source. Typically, for a range of 32 km and an altitude of 10 km, errors of 2% are obtained in each parameter. However, the errors in altitude estimation are much greater for low sources.

b. Warning Techniques

Perhaps of most public interest is the use of sferics in monitoring the approach of severe weather. Thunderstorms emit radio signals over almost the entire radio-frequency spectrum, but there are strong indications that the spectral distribution of the emitted noise varies with the severity of the storm. In particular there is now some belief that severe storms that may spawn tornadoes generate more noise at higher frequencies than do nontornadic storms. This belief is the basis for the much-publicized Weller method of detecting tornadoes (Biggs and Waite, 1970).

The Weller technique advocates monitoring the brilliancy of a TV receiver set at its lowest channel (channel 2—54 to 60 MHz). Lightning flashes produce a transient brightening of the screen, but if there is a tornado in the vicinity, the brightening is said to be much more continuous. This behavior would be consistent with an increased output of VHF noise owing to the tornado. However, even if this increased output does indeed occur, the Weller method cannot be recommended as an alternative to the organized tornado warning information disseminated by weather agencies. TV receivers, with their intrinsic variability in internal and antenna characteristics, cannot really be used as precision indicators. They are no substitute for equipment specifically designed for warning purposes.

Taylor (1973) has made some thorough studies of the utility of sferics as detectors of severe storms. After examining the outputs of receivers at frequencies from 10 kHz to well above 10 MHz, Taylor established that the channels at frequencies exceeding 1 MHz were uniformly more responsive to severe storms. In particular, the burst rate at 3 MHz was a sensitive indicator of the severity of a storm. With a given threshold level, corresponding typically to a storm at 70 km, a burst occurs when the received signals exceed the threshold level at a rate greater than 500 per second over a time of more than 0.1 s. Burst rates exceeding 20 per minute were present in 82% of tornadic storms, in about 22% of other severe storms (funnel clouds, hail, and/or strong winds), and in only 1.4% of nonsevere thunderstorms. This apparently favorable result is, however, tempered by the fact that the vast majority of storms are nonsevere. In fact, if the 20-per-minute burst rate were used as an indicator of tornadoes, the false-alarm rate would be about 70%. This result applies with an omnidirectional arrangement—addition of a directional capability reduces the false-alarm rate since much of this is due to the aggregation of several nonsevere storms simultaneously active at similar ranges.

Sometimes the severity of the storm is not a prime consideration in warning techniques, for example, for operations in which lightning is a hazard, and for which, in consequence, all thunderstorms are dangerous. Typical of such operations would be a protracted loading or unloading of explosives. Under such circumstances a simple omnidirectional threshold sferics warning device can be useful. It is best for the device to function in the "near-field" regime within its operating range. This enables the range to be better defined; the operation thus depends inversely on the cube of distance of the near-field electrostatic component, rather than inversely with distance as with the far-field radiation component. Choice of the electrostatic field regime implies selection of an extremely low frequency for the warning device. This is dictated by the near-field requirement $f < (c/2\pi R)$, where f is the frequency, R the range, and c the velocity of light. Thus a nominal range of 50 km would entail a device functioning at frequencies of less than 1 kHz.

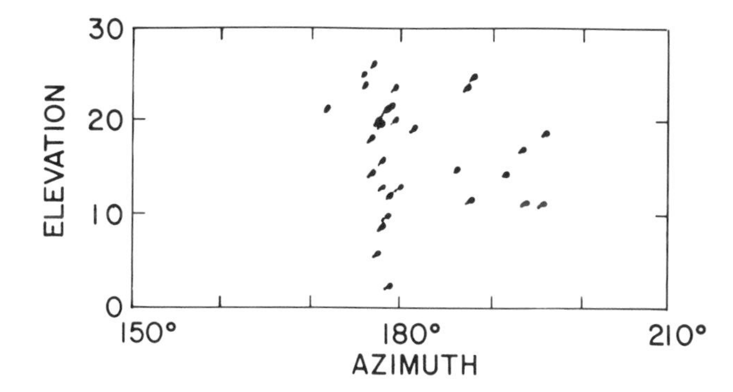

Figure 7.5. Example of impulse configuration associated with a cloud-to-ground stroke.

c. Studies of Electrical Fields and Field Changes

The preferred methods for research studies of the electrical effects of close thunderstorms do not fall under what is conventionally referred to as sferics. However, no account of experimental studies would be complete without some discussion of methods of the measurement of the electric fields associated with storms and of the field changes caused by lightning.

The standard instrument for measuring electrical field is the field mill. Many designs of field mills exist, but all operate on similar principles (Chalmers, 1967). Consider a conducting plate of area A exposed to the earth's field E. Under the influence of the field a surface charge of $\varepsilon_0 EA$ will reside on the plate, where ε_0 is permittivity of free space. If the plate is alternately exposed to and shielded from the field, the bound surface charge will vary alternately with an amplitude that is proportional to E. The alternating charge is conveniently amplified and then demodulated to produce an output signal dependent on E.

A typical field mill consists of a set of insulated plates

alternately shielded from and exposed to the ambient field by a sectored rotor. The rotor is often driven at a speed of about 2,000 rpm, and, in the usual arrangement, the exposed area $A(t)$ of the plates varies as a triangular function of time. A synchronous signal derived from the rotor is used for demodulation after the output from the sensing plates has passed through a charge amplifier.

The modulating frequency of a field mill depends on the number of sectors in the rotor and rotor speed and is typically about 100 Hz. However, the output of the mill is usually integrated over several modulation cycles to give a time constant exceeding 0.1 s. This integration, and the basic limitation that the exposed plate area does not remain constant but varies with time, renders the field mill unsuitable for studying the rapid field changes owing to lightning. These must be investigated with some kind of fixed antenna system.

Many antenna arrangements are possible, including vertical whips, ball antennas, and plates. With some of these arrangements calibration in terms of the ambient field is very straightforward, notably so for the case of a plate antenna of area A mounted flush with the earth's surface. As in the case of a field mill sensor, there is a bound charge $\varepsilon_0 E(t)A$ on the antenna and the output voltage of a charge amplifier will be proportional to $\varepsilon_0 E(t)A/C$, where C is the capacitance of the antenna system. The capacitance will be shunted by a resistor R to produce a time constant RC. Choice of a time constant appropriate to the lightning phenomenon being studied is a fundamental requisite; the time constant should be at least 10 times the duration of the event being investigated. Thus if the field variation over the entire discharge (duration $\sim$ 1s) is being examined, a 10-s time constant is reasonable. An antenna with this value is termed a slow antenna. At the other extreme a fast antenna with a time constant of about 100 μs is appropriate for studying the rapid ($<$ 10 μs) changes accompanying return strokes.

Whatever the type of antenna or field mill being used, the choice of a site is of prime importance. The area should be flat, and there should be no shielding of the antenna or mill by adjacent objects. An artificial ground plane surrounding the antenna helps to ensure correct operation and to simplify calibration. For a plate antenna, mounted flush with the earth and surrounded by a ground plane, calibration is simply effected by using an insulated screen above the antenna. Calibrating voltages, preferably in the form of square waves, are applied to the screen. The dimensions of the screen should be appreciably larger than the size of the antenna to avoid fringing effects.

8. Techniques for Measuring Electrical Parameters of Thunderstorms

W. David Rust and Donald R. MacGorman

1. Introduction

Although knowledge of atmospheric electricity has long indicated the kinds of measurements needed to advance this branch of meteorology, it is only since the advent of solid-state electronics that many required measurements can be made properly in the harsh environment of storms. We report here on some of the most common and a few of the unique instruments used in various thunderstorm research programs and confine the presentation mainly to electrical measurements that are related to other storm processes. For this reason, some new and interesting techniques developed to study the physics of lightning, e.g., triggering of lightning, are not discussed.

2. Electric Field Measurements

Perhaps the most fundamental measurement is that of the atmospheric electric field E. The devices used to measure E are termed "field mills" or "field meters." Although field mills have been used for decades, modern technology has contributed significantly to new designs and applications. Measurements of E aloft have been made with the use of rockets, balloons, and airplanes. To date all rocket measurements have essentially measured only the horizontal component of the field (Winn and Moore, 1971). Since the vertical field is often dominant and is fundamental to understanding cloud electrification processes, instruments and systems for measuring all three components are discussed here.

a. Rotating-Vane Field Mills

A rotating-vane field mill senses charge induced on different plates, or stators, alternately shielded from and exposed to the atmospheric electric field by a grounded rotor. Geometrically opposite sensor plates are connected to each other, and each pair in turn is connected to the inverting input of an amplifier. The amount of charge induced on the stators, and thus flowing in the amplifier circuit, is a function of the magnitude of the local electric field. A simplified circuit diagram and a photograph of such a field mill are shown in Figs. 8.1a and b. Since the amplifier (A_1) is connected as a charge amplifier, the sensor plates are held at virtual ground, i.e., essentially at circuit ground potential, and the signal is not a function of the rotation rate of the rotor as long as its rate exceeds a few revolutions per second. These two design parameters make the measurement of the electric field at the Earth's surface straightforward. The theoretical signal out of each amplifier is a triangle wave, but electric field fringing effects at the rotor edge as it crosses a stator cause the actual wave shape to be more sinusoidal. The two amplifier signals are 90° out of phase and are sent through solid-state analog switches (labeled DG 307 in Fig. 8.1) for combination into a single wave. The switches are driven by a reference square wave generated by a black plate, identically shaped to the rotor, that rotates through a light beam shining on a detector. The reference square wave is used to determine the rotor position in relation to the peak of the charge amplifier signal, a necessity for determining the polarity of the field. The combined output from the switches is a full-wave rectified sine wave. This wave is generally filtered at < 10 Hz, and the final field-mill output is a varying direct current (DC) voltage whose polarity and magnitude are a function of the atmospheric electric field.

The field mill is calibrated directly by placing it in a uniform electric field E made by applying known voltages V across parallel conductive plates, separated by a known distance d (then $E = V/d$). To keep calibration errors minimal, the plate dimensions should be several times greater than the diameter of the field mill. The top of the rotor should be level with one of the parallel plates.

If the field mill is mounted flush with the Earth's surface and facing upward, calibration is facilitated since there is no significant perturbation of electric field lines, and the geometrical form factor of the instrument is nearly unity. Thus the instrument does not significantly affect the local electric field. In this case the output voltage V can be directly calibrated to the atmospheric electric field E by

$$E = KV, \tag{8.1}$$

where K is the constant circuit calibration factor for a particular gain setting (most field mills have switchable gains or at least two output channels of different sensitivity).

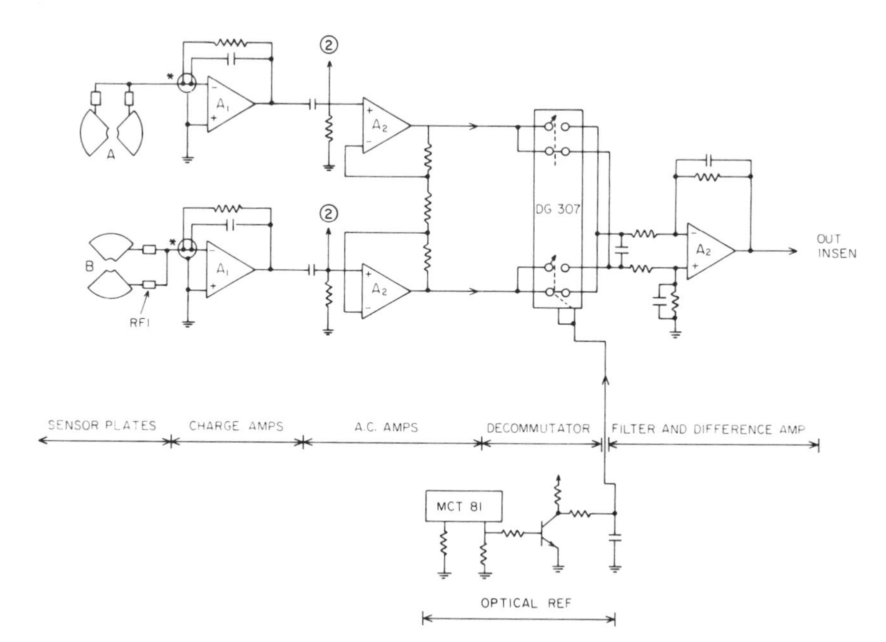

Figure 8.1a. A simplified circuit schematic of an electric field mill. A and B denote the two pairs of sensor plates, which are connected through radio frequency interference filters (RFI) to the operational amplifiers (A_1) with field effect transistor (FET) inputs. The asterisk (*) denotes the location of amplifier input connections that have high isolation from ground. Either guard-ring circuit layout or Teflon standoffs are used. The circled 2 denotes connection points to a duplicate set of electronics with a higher gain that are the sensitive channel of the mill.

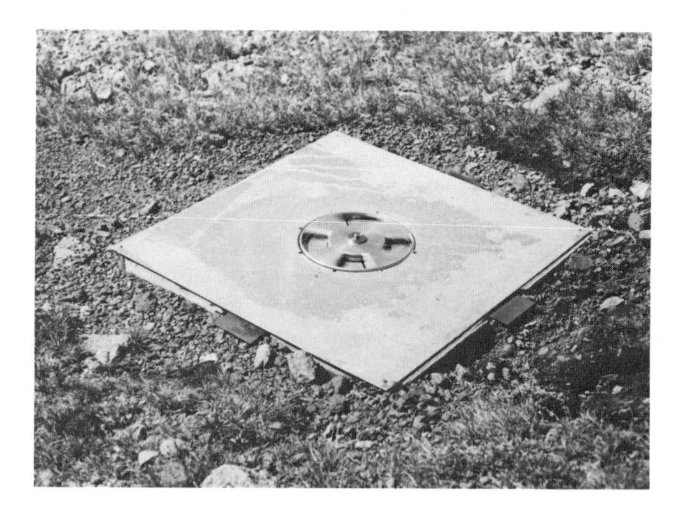

Figure 8.1b. Photograph of a field mill flush-mounted and upward-facing, at Langmuir Laboratory, in New Mexico.

Mills for ground-level recording are designed to measure from about 100 V m^{-1} (fair weather) to 20 kV m^{-1} (intense thundercloud overhead). Fields at the Earth generally do not exceed about 15 kV m^{-1} because they are limited by space charge originating in corona breakdown from vegetation and other projections aboveground. Sometimes other installation configurations besides flush-mounted are needed; in particular, the mill can be mounted a meter or so aboveground, facing downward. This arrangement is beneficial if measurements are to be made in snow or very intense rain, where covering or electronic shorting of the sensors can occur. If a configuration other than flush mounting is used, a determination of the geometrical form factor G is made with an upward-facing, flush-mounted mill operated nearby during selected periods. The instrument itself is still calibrated as before, but Eq. 8.1 becomes

$$E = KGV. \tag{8.2}$$

Major practical problems and precautions to overcome them include the following:

1. Drifting electronics can cause DC shifts with shifts of the true $E = 0$ output voltage. An electronic "zero" can be obtained by disconnecting the sensors.
2. Leakage currents can develop across the stator insulators (generally Teflon or Kel-F) from a combination of high humidity with dirt, spider webs, blades of grass, etc. Very smooth surfaces from proper machining and routine cleaning with a low-residue solution such as Freon TF generally prevent this problem.
3. Local changes in the electric field near the mill sensors can be induced by varying charges on nearby dielectric surfaces.
4. Corona ions are created under intense fields and influence the field-meter reading by the electric field they cause. The sensor plate and other metal edges must be rounded to decrease this effect.
5. The same material, preferably stainless steel, should be used to fabricate stators, rotor, and nearby housing to avoid local electric fields arising from differences in contact potential. Such local fields limit the minimum measurable field and add "noise" as a resultant DC signal at the instrument output.
6. Nearby structures and tall vegetation should be minimal to avoid shielding the local electric field or causing

corona-induced disturbances. Note that, with a surface wind, space charge emitted by corona can be blown over a field mill hundreds of meters away.

Even with field measurements at the ground and aloft within thunderstorms, the charge magnitude and its distribution in the storm cannot be determined uniquely without an infinite number of such measurements. However, much has been learned about the polarity and distribution of charge within storms.

An excellent example of the use of a set of rotating-vane field mills to make measurements in storms is on the instrumented, powered Schweitzer sailplane (Fig. 8.2) operated by the New Mexico Institute of Mining and Technology; for a sample of results obtained in isolated thunderclouds, see Gaskell et al., 1978). Five mills are used. One pair is mounted vertically and another pair horizontally on the fuselage; the fifth mill faces aft. Each mill is calibrated as described above. However, in an airplane installation, changes in field-mill output owing to form factors, aircraft charge, and contributions from all three components must be determined. A very large local field is often generated by charge accumulated on the airplane from engine exhaust and especially by impaction of cloud particles and precipitation. The field caused by aircraft charge must be removed from the data to determine the undisturbed components of atmospheric electric field in the vicinity of the airplane. The effects of airplane charging can be determined by artificially charging the airplane aloft in fair weather where the ambient field is only a few volts per meter (preferably above any inversion layer and several kilometers aboveground). Here virtually all of the field will be due to airplane charge. Airplane charging is accomplished by (1) changing from cruise to nearly full engine power, which increases charge emission, or (2) "pumping" charge overboard through the use of a corona point attached to a high-voltage power supply. The corona ions put one sign of charge into the atmosphere and leave the airplane with the opposite charge. The second technique is preferred since it is easily done, and each polarity of charge can be applied to the aircraft by interchanging the high and low sides of the power supply relative to airplane ground. Note the implicit assumption that the aircraft surface near the field mills is conducting and free of dielectrics such as nonconductive paint and antenna radomes. The presence of such dielectrics can result in charge being trapped in nonuniform distributions, which can in turn cause nonrepeatable local fields that vitiate the measurements.

The form factor of the airplane can be determined first from the vertical component of E by flights just above (e.g., about 10 m) a field mill operating at the ground where the field is large enough to give a usable signal for sensitive mills. Other flight maneuvers, such as pitch-up and pitch-down and banking turns at known angles to the vertical, allow calibration of the horizontal field-mill system by comparison with the vertical component. In addition, the airplane can fly by a tethered balloon carrying a calibrated field meter in the vicinity of a thunderstorm to obtain an independent calibration of the field mill system under larger field conditions. Kasemir (1972) describes repeated figure-eight flight paths near a large storm where horizontal field components exist.

Figure 8.2. The powered Schweitzer sailplane equipped with field mills and other instrumentation. The mills are symmetrically located on the fuselage, a pair of mills each in the vertical and horizontal (wing tip–to–wing tip) planes. A fifth mill is mounted on the tail.

Each field mill can have all three components of E plus E_Q, the field caused by airplane charge, present in its output signal; e.g., for field mill no. 1 (fm 1), the field is $E_{\mathrm{fm\,1}} = a_1E_x + b_1E_y + c_1E_z + d_1E_Q$. The set of equations includes one for each field mill; thus they can be written in matrix form. By choosing geometrically symmetrical locations on the aircraft, unwanted components of E for a particular mill are reduced though not eliminated. Numerical modeling of the aircraft shape and measurements in the laboratory of charge on a scale model are also used to help determine the optimal locations for field mills. Inside storms E_Q is often quite large. The coefficients, a–d, are unique to each mill installation and are determined by the calibration and analysis procedures. For example, the calibration procedure can be carried out by (1) calibrating each field mill in a parallel plate, (2) flying high in clear air where $E_x \simeq E_y \simeq E_z \simeq 0$ and charging the aircraft to determine the d coefficients, (3) flying low in fair weather where $E_x \simeq E_y \simeq E_Q \simeq 0$ but E_z is adequately large to determine the c's, and (4) doing banks (rolls) and pitches to determine the b's and a's, respectively. Procedures (3) and (4) can be replaced by, or, better, done

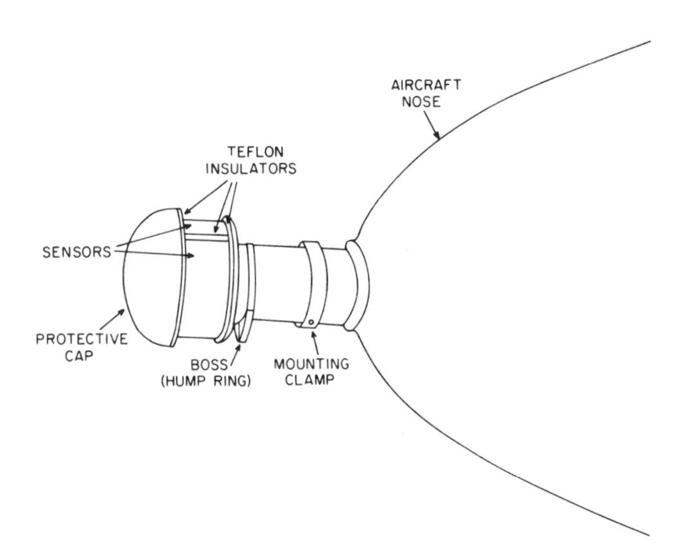

Figure 8.3a. Diagram of cylindrical field mill on the nose of an airplane. The sensors, cap, and hump ring are made of stainless steel.

Figure 8.3b. Photograph of a cylindrical field mill mounted on the nose of a T-29 (Convair 240) twin-engine aircraft. Note, left to right, aluminum tape over the fuselage for good conductivity near the mill, quick-release (Marmon) clamp, stainless steel hump ring to neutralize the effect of charge on the aircraft, the field-mill sensors (hemicylinders), and grounded end cap.

in combination with, flights by a calibrated balloon-borne field meter in the vicinity of a thundercloud, but in clear air where E_Q can be maintained near zero.

The signals from all mills can be combined electronically or processed analytically to determine the three components and charge on the airplane by simultaneous solution of the set of equations like that shown for field mill no. 1. The analytical technique is preferred because the data from each mill are recorded without processing to facilitate real-time monitoring of mill operations. Monitoring is straightforward since the output wave shapes of all mills are nearly identical (Winn, 1984).

b. Cylindrical Field Mill

A second type of field mill designed for use on airplanes is the cylindrical mill. Although its mechanical design is quite different from that of the rotating-vane type, major circuit elements are the same. It is unique in its measurement of two orthogonal components of the field with a single mill. Shown in Figs. 8.3a and b is the cylindrical mill developed by Kasemir (1972). When the mill is mounted on the nose of an airplane, the vertical and wing tip–to–wing tip components of E can be measured. A second cylindrical mill mounted on the top or underside of the fuselage measures the two horizontal components, giving a redundant measure of the wing tip–to–wing tip component plus the fore-aft component.

The sensing elements for this type of mill consist of two hemicylinders. One is grounded and insulated from the other with Teflon. The external electric field induces an alternating charge on the sensors as the mill rotates. This results in an alternating current that flows from the insulated sensor through a rotating capacitor (see Fig. 8.4) to the stationary electronic circuitry, where it is amplified, rectified, and filtered (mechanical slip rings are sometimes used to couple the rotating and fixed frames). Two sinusoidal reference signals, 90° out of phase, are produced by a signal generator driven by the shaft of the field-mill motor. These are used in the phase-sensitive rectification circuits so that the two orthogonal components of the external field can be determined. The outputs corresponding to the field components are aligned with the geometrical axis of the airplane by electronically phase-shifting the reference signals. After phase-sensitive rectification the signals are filtered to give DC outputs proportional to the electric field. Use of several gains allows field from about 10 V m^{-1} to 500 kV m^{-1} to be measured.

The field mills themselves are calibrated in the laboratory between parallel plates. Calibration of the field-mill system on the airplane is similar to that for a rotating-vane field-mill installation.

The effects of charge on the airplane are made negligible through a different technique from that for rotating-vane field mills. The airplane is charged in flight as before, but the electric field caused by charge on the aircraft is nulled out by using a smoothly rounded metal boss called a "hump ring," which is mounted on the cylindrical housing just below the sensing elements (see Fig. 8.3). The charge on the hump ring is so close to the cylindrical mill that the field from this charge can offset the field from charge on the entire airplane when the hump ring is positioned correctly. To do this, the hump ring is rotated and translated until the charging procedure causes no noticeable outputs from the field mill, even on sensitive ranges. This is a slow, tedious task since the hump ring cannot be moved during flight, but the technique is reliable. Recently a remote-controlled, motorized set of four spherical bosses, mounted symmetrically about the mill body in place of the hump ring, has been adjusted during a single calibration flight to produce the same charge nulling.

c. Balloon-borne Field Meters

Several types of field meters have been used on balloons. The term "meter" is used instead of "mill" to denote the different mechanical design. Winn et al. (1978) have described a simple "disposable" balloon-borne field meter to determine the electric field vector. The sensor (Fig. 8.5) consists of two metallic spheres held apart by a dielectric mount. One sphere is connected to the inverting input of a charge amplifier, and the other sphere is connected to circuit ground. The spheres are made to rotate about the vertical and horizontal axes. As the spheres rotate, the atmospheric electric field induces an equal and opposite charge on the spheres, and the resulting output signal is sinusoidal. To determine the position of the sensing sphere and thus the polarity of E, a reference pulse is generated with a mercury switch that rotates about the horizontal. Although these meters could be calibrated in a large parallel-plate arrangement, they have been calibrated theoretically by solving for the charge on two spheres in an electric field and using the relationship of the charge on the sphere to the voltage out of the charge amplifier. The measured field is also compared with the indication of a field mill at the ground near the launch site. In addition to sensing E, the spheres form the antenna for a 400-MHz transmitter, which is modulated by the combined (multiplexed) output of voltage-controlled oscillators (VCO's), each from a different circuit and with a different center frequency. Data are received at the ground and recorded on analog tape. Subsequently the data are replayed through discriminators that convert each sensor's frequency shifts into voltage changes. Such a technique is called an fm-fm telemetry system (fm = frequency modulation). The meter is typically suspended about 60 m below a balloon that also carries a standard rawinsonde, transmitting at 1680 MHz. The rawinsonde provides both meteorological information and a determination of field-meter location by standard automatic tracking with an AN/GMD-1 or similar device. The meter (Figs. 8.6a and b) makes electric field soundings during ascent and also upon descent after the balloon has burst and the instruments ride to Earth on two parachutes. Such balloon-borne field meters have been flown in small thunderstorms in New Mexico (Winn et al., 1981; Marshall and Winn, 1982) and more recently in severe storms in Oklahoma.

Another balloon-borne field meter, called the Balloon Electric Field Sensor (BEFS), is made from a spherical, overpressurized balloon of aluminized Mylar (Christian and Few, 1977). As shown in Fig. 8.7, four electrically isolated patches of the balloon skin are used to measure the field by means of charge amplifiers. The internally housed electronics of this instrument and its spherical shape reduce the production of corona and make calibration straightforward (a sphere enhances the undisturbed field by a factor of 3). The electronic circuits also sense instrument azimuthal orientation and atmospheric pressure. The flow of air past the nonconducting spin paddles causes the instrument to rotate and wobble about the vertical as it rises or descends. These motions allow calculation of all the components of the field from the changing charge on the sensor patches and balloon orientation. In this system the rate at which the field, and thus field changes owing to lightning, can be sensed is 8 ms and is limited by the telemetry.

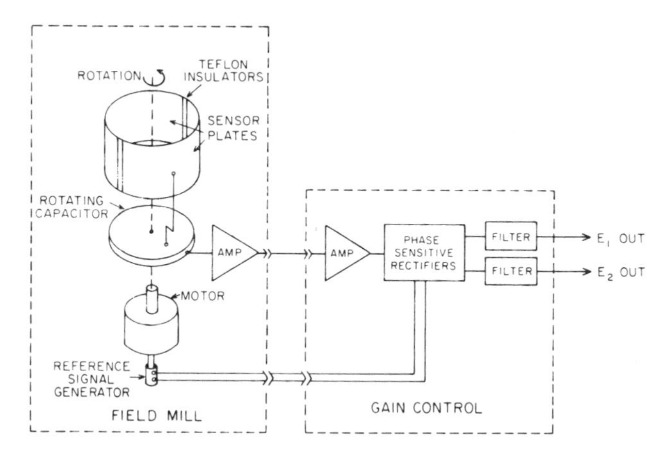

Figure 8.4. Simplified design of cylindrical field mill. A rotating capacitor is one device for coupling the signal from the rotating sensors to the electronics; slip rings or optical techniques can also be used. The other circuits are similar in design and functionally equivalent to those in the rotating-vane field mill. E_1 and E_2 represent the two orthogonal components of the electric field measured with the cylindrical mill.

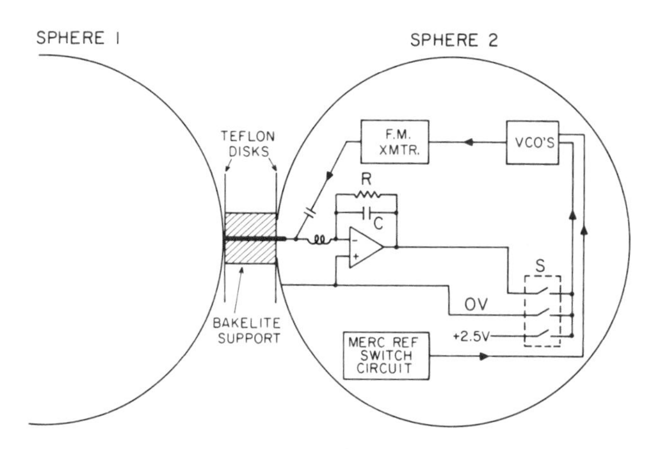

Figure 8.5. Circuit diagram of balloon-borne two-sphere field meter. The instrument telemeters electric field, sensor orientation (in the vertical plane) from the mercury reference switch, and calibration signals of 0 and 2.5 V. The switch S is controlled by timer circuits (not shown) to switch from E-field to reference voltages.

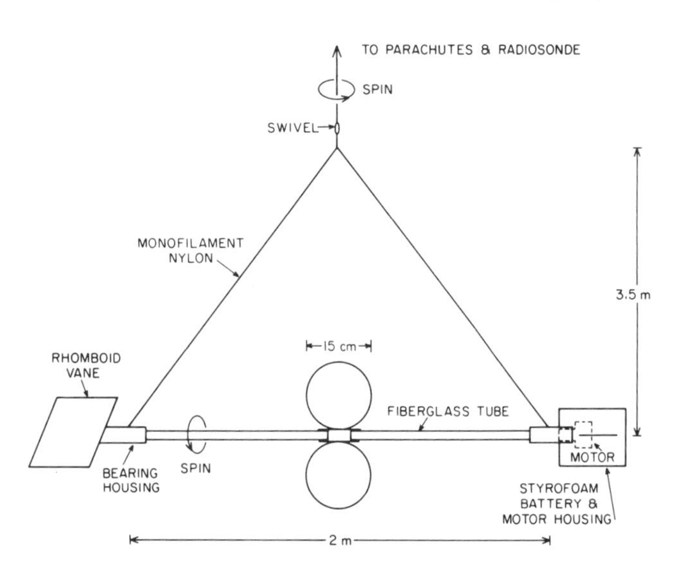

Figure 8.6a. Functional diagram of a balloon-borne, two-sphere field meter. The motor spins the meter about the horizontal at about 2 rev s^{-1}. The rhomboid-shaped vane causes the device to rotate about the vertical at a slower rate, which is a function of the balloon's vertical rate and generally is about 1/12 rev s^{-1}. Since the meter sweeps the entire horizontal component of the electric field each one-half rev, the vector **E** can be determined every 6 seconds.

Figure 8.6b. A "mobile" balloon launch in the updraft of a severe thunderstorm. The instrumentation train consists of two parachutes; a standard rawinsonde for temperature, pressure, and humidity; and the electric field meter depicted in Fig. 8.6a. The tracking and meteorological information allows determination of instrument location relative to storm structure.

Errors in determining the vertical field component are larger when the balloon ascends more slowly; the wobble angle is then smaller, and there is a corresponding reduction of the signal owing to the vertical field. The determination of horizontal components is not affected by reduced wobble. The instrument measures field components from fair weather to thunderstorm fields up to about 100 kV m^{-1}.

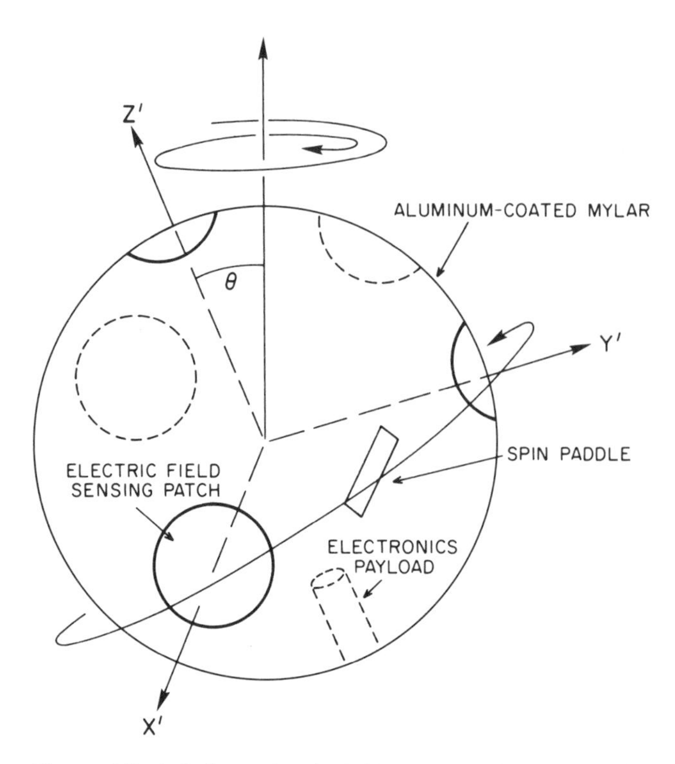

Figure 8.7. A balloon electric field sensor (BEFS). The entire balloon has a conducting skin, except for the isolation rings around the sensing patches. The spherical shape is maintained by an overpressure of helium gas.

There is evidence that at about 100 kV m^{-1} there may be corona discharge from the balloon with possible effects on the measurements.

Although this instrument is unique, its use is less appealing than use of the two-sphere field meter because (1) the balloon is relatively expensive, (2) its large size precludes easy launch except from a fixed-base balloon hangar, (3) the dependence of measured vertical field on wobble, which often seems to be too small (Weber et al., 1982), can result in large uncertainties in the vertical component of **E,** and (4) data reduction is difficult owing to the large number of variables. It has been demonstrated, however, that the instrument reliably measures thunderstorm fields; additional details can be found in Christian and Few (1977) and Weber et al. (1982).

3. Maxwell Current Measurements

A fundamental parameter of storm electrification is the Maxwell current density. It is defined by one of Maxwell's equations, $\nabla \times \mathbf{H} = \mathbf{J} + \partial\mathbf{D}/\partial t$, where **H** is the magnetic field, **D** is the electric displacement, and **J** is the electric current density. As mentioned in the appendix of Vol. 2, Chap. 13, the Maxwell current density can be estimated from the slope of the electric field change (dE/dt) when $E \simeq 0$ and no precipitation is falling at the field mill. Note that E refers only to the vertical component since it is at the Earth's surface. This has been shown theoretically by Krider and Musser (1982), who used the Maxwell $\mathbf{J}_M$, or total air-Earth current density

$$\begin{aligned} \mathbf{J}_M &= \mathbf{J} + \frac{\partial \mathbf{D}}{\partial t} \\ &= \mathbf{J}_{PD} + \mathbf{J}_{PR} + \mathbf{J}_L + \mathbf{J}_w + \frac{\partial \mathbf{D}}{\partial t} \end{aligned} \qquad (8.3)$$

Krider and Musser stated that $\mathbf{J}_{RD}$, the current density owing to point discharge, and $\mathbf{J}_w$, the convection current density, are widely distributed in the storm environment and tend to vary slowly with time. The current density owing to lightning transients is $\mathbf{J}_L$. When $E \simeq 0$ and there is no lightning (or as the field "relaxes back" after lightning flashes), both $\mathbf{J}_{PD}$ and $\mathbf{J}_L$ are zero. $\mathbf{J}_w$ is an unknown and contains currents owing to transport of charge by air motion and falling precipitation. In the absence of precipitation at the field mill only an air transport mechanism remains. Under the assumption that $\mathbf{J}_w << \mathbf{J}_M$, the Maxwell current density becomes

$$\mathbf{J}_M \simeq \frac{\partial \mathbf{D}}{\partial t} \bigg|_{E \simeq 0}, \qquad (8.4)$$

where $\frac{\partial \mathbf{D}}{\partial t}$ is the displacement current density, or

$$J_M \simeq \varepsilon \frac{\Delta E}{\Delta t} \bigg|_{E \simeq 0}. \qquad (8.5)$$

This technique has been verified by comparison with direct measurement of the Maxwell current density by Blakeslee and Krider (1984). They used a section of natural turf, isolated from ground except through an electrometer circuit, to measure current. Not only was the theoretical prediction of Eq. 8.5 demonstrated correctly, but also it was found that the average Maxwell current density changes slowly throughout the storm on time scales similar to those of storm development, thus supporting the suggestion of Krider and Musser (1982) that $\mathbf{J}_M$ is an electrical parameter that may be correlated directly with storm evolution.

4. Conductivity

Another fundamental parameter of atmospheric electricity is the electrical (ohmic) conductivity $\lambda = \mathbf{J}/\mathbf{E}$. The reason for the air's conductivity and the relationship of conductivity to other electrical parameters are described in Vol. 2, Chap. 13. Conductivity is usually measured in fair weather and is most often applied in studies of the global circuit. Clearly, knowledge of conductivity within thunderstorms would increase our understanding of cloud electrification. Before we discuss the instruments and problems of measuring conductivity, it is worth restating that we are measuring a property of the air identified with small, fast ions. Since currents flow in clear air below and above thunderstorms, it is important to measure conductivity there. However, the situation inside a thunderstorm is entirely different. Conductivity measurements made in the lower regions of thunderstorms with the instrument to be described show that cloudy air is only about one-tenth as conductive as clear air at the same altitudes. This condition is shown theoretically to be due to capture of the small ions by cloud particles (Brown et al., 1971).

Even though the ohmic (small ion) conductivity is low, there is an array of larger charged particles that moves under the influence of electrical forces and air motions. Thus the total electrical conductivity of the storm medium is not a simple concept. Conductivity measurements have yet to be made in regions of largest electric fields. Conductivity there may be both large and non-ohmic; i.e., conductivity within a thunderstorm may tend toward independence from the electric field strength. In these high-field regions, charge transport may vary in a quite nonlinear way with the applied electric field (thus $\mathbf{J} \neq \lambda\mathbf{E}$), because the field affects not only the motion of charged particles but also the population of the particles as a result of particle-charging processes.

With the above basis of understanding, we now examine some attempts to measure the ohmic (small ion) conductivity in the thunderstorm environment.

Conductivity measurements can be made through the use of a cylindrical capacitive device (sometimes called a Gerdien capacitor), through which air is pulled with a fan. The fundamentals of such a device were described by Chalmers (1967). The major problem involves accuracy in measurements when this sensitive device is immersed in the large external electric field and high humidity characteristic of thunderstorms. The large atmospheric electric fields and field changes from lightning are difficult to shield totally from such sensitive electronics, and high humidity can cause large-error currents by leakage from the central cylinder support to circuit ground. Rust and Moore (1974) designed a balloon-borne device (Fig. 8.8) using a 30-cm-diameter spherical aluminum housing with a gently rounded recessed intake orifice to prohibit, or at least reduce, the production of ions by corona breakdown processes, which would then be mistakenly measured as naturally present in the storm. The rudder helps keep the intake pointing into air uncontaminated by any corona production that might occur from other parts of the instrument. The instrument design also shields all but the largest external field changes from the sensor electrode, and large hydrometeors and splash are prevented from entering by use of a silicone gutter above the reentrant air-intake port.

A potential difference of 10 V is applied between the capacitor wall and the central cylinder, and air to be measured is pulled through with the fan. Small ions are moved away from or toward the central cylinder, depending on their po-

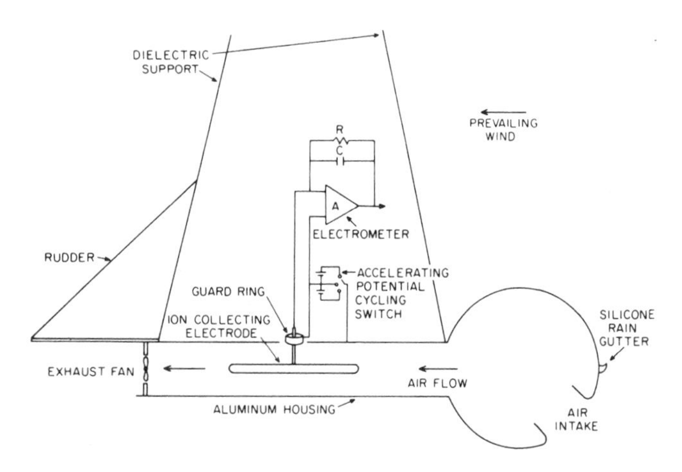

Figure 8.8. Balloon-borne conductivity apparatus. The major components are a standard electrometer amplifier having a guard ring around the input to reduce leakage currents and a motor-driven switch that supplies positive, negative, and zero accelerating potential within the cylinder. The RC time constant is 3.7 s.

larity, by the coaxially distributed electric field within the capacitor. Those that move to the central cylinder cause a current to the cylinder that is very small (about 10^{-14} A), and a sensitive electrometer-type amplifier is needed. A full measurement cycle is about 12 min—two 5-min intervals with opposite polarity of the accelerating potential to measure alternately both conductivity polarizations and 2 min with zero potential. The latter is used to determine a "dynamic" zero, i.e., a zero for the conductivity. The dynamic-zero interval is used to measure the electrometer current i caused by impaction of charged cloud particles on the central electrode. The difference in electrometer voltages when a potential is applied and during the dynamic zero, ΔV, allows the determination of conductivity λ:

$$\lambda_{\pm} = \frac{\varepsilon \Delta i_{\pm}}{C \Delta V_{\pm}}, \tag{8.6}$$

where C is the cylinder capacitance, which can be measured (6.5 ± 0.1 pF for this instrument) as well as calculated from theory. The subscript $\pm$ refers to the polarity of small ions and thus the conductivity; ε is the permitivity of the air. Two amplifier outputs with different gains are telemetered to ground, with resulting conductivity measurement range of 5×10^{-16} to 1×10^{-13} per ohm m.

5. Particle Charge, Size, and Current

To understand the role of precipitation in cloud electrification, it is necessary to measure the charge on particles within the cloud. Note that precipitation particles capture ions as they descend, and hence the magnitude and even the polarity of charge on precipitation can vary substantially between cloud base and the ground. This was hypothesized by Wilson (1929) and demonstrated probable by Rust and Moore (1974). There are methods for measuring both individual particle charges and the bulk charge flux, i.e., the current density.

Individual particle charges are most often measured with induction rings. Amplifiers connected to such a ring measure the charge induced with the passage of the particle.

Two major experimental difficulties are associated with measuring small charges in thunderstorms: (1) the large field changes caused by lightning must be shielded from the sensors, and (2) measurement devices must be built so that splashing particles and corona ion production do not contaminate the measurements. Examples of airborne measurements of particle charge with the use of balloons and airplanes follow. Although these measurements of particle charge have yielded useful information, the balloon-borne and early airplane measurements are deficient in that the number of particles with very small or no charge could not be ascertained. However, the information provided by these instruments has resulted in increased understanding of fundamental electrical parameters in storms.

a. Balloon-borne Charge/Size Instruments

Most balloon-borne instruments have utilized induction rings as described by Rust and Moore (1974) and Marshall and Winn (1982). Figure 8.9 shows the device used in the former study. It is a combination of two instruments, one for measuring individual particle charge and size and the other for measuring current density. Precipitation current density is determined by capturing all particles that fall into a Faraday funnel, recessed within a cylindrical shield to reduce signals from field changes. The funnel is connected directly to a charge amplifier with a 10-s decay time; this integrates the contribution of individual particles to produce a measure of current density produced by falling precipitation. The metal intake orifice, a highly truncated conical section with a top diameter of about 35 cm (area $\simeq$ 960 cm^2), is edged with Teflon to prevent corona breakdown, and thus possible particle charge alteration, and provides a well-defined intake area that reduces the collection of splashing drops from the inside of the shielding housing. Water drains from the funnel through an orifice well shielded to prevent any inductive effects of drops leaving in the presence of an electric field. The device is calibrated by applying known currents to the funnel; current densities from about 0.5 to 10 nA m^{-2} can be measured.

The instrument for determining size and charge on individual particles has an intake orifice of 9-cm diameter (area $\simeq$ 64 cm^2) and is designed to prohibit the capture of drops splashing from the outer housing. Beneath the intake are two induction rings each 10 cm in diameter and 4 cm high. The centers of the rings are 10 cm apart. Beneath the rings is a Faraday funnel to provide a redundant measurement of particle charge. This funnel contains stainless-steel wool to break up raindrops without splashing. Each ring is connected to a charge amplifier with a 0.4-s decay time, and the funnel is connected to a charge amplifier with a 0.1-s decay time. The outputs from all four sensors are connected to VCO's that modulate a transmitter as in the fm-fm telemetry system described for balloon-borne field meters.

As a detectable charged particle falls cleanly through the rings, a characteristic twin-peaked signal is produced. The peak amplitude is related to particle charge, and the interval between the peaks divided by the ring separation yields the fall velocity from which raindrop size can be determined. As the particle falls into the Faraday funnel, its charge is determined from the peak amplitude of the output pulse.

The induction ring instruments are calibrated by allowing water drops or metal spheres with known charge to fall through the rings and into the funnel. In the metal sphere calibration technique, the sphere is held magnetically to the underside of the top of a parallel-plate arrangement. Upon release and fall through a known vertical field, a sphere of radius r acquires a charge (Winn, 1968)

$$q = 1.83 \times 10^{-10} Er^2. \tag{8.7}$$

With two channels of output sensitivity, the Faraday funnel device has a useful range of about $\pm$ 0.15 to $\pm$ 135 pC; the equivalent raindrop diameters that can be measured are about 0.2 to 3 mm.

Individual charge amplifier outputs are calibrated through the entire system, i.e., from sensor to playback onto hard copy. As part of the calibration, the device is suspended on monofilament lines between two large parallel plates. Electric fields in excess of 100 kV m^{-1} are applied while drops of known charge fall into the instruments. In addition, the plates are shorted to produce lightninglike field changes. In both tests no significant effect on the particle charge and size sensors has been observed. The current density device also gives correct current values, upon which are superimposed small impulses from the large field changes. Electrostatic shielding is not perfect, owing to the large intake opening. Only lightning-flash rates as fast as one every few seconds will introduce errors in the measured precipitation current density; the size-charge instrument is unaffected.

The device is suspended several tens of meters below and at the side of tethered balloons with the use of nonconductive lines and dielectric booms. The current device is large and relatively heavy. However, it has been found that current densities calculated from individual particles measured in the induction ring device are only about one-sixteenth those measured directly, apparently the result of the smaller collection area of the induction ring device. Thus measurements of individual particle charge are reliably made with the induction ring device, but caution must be exercised when interpreting current density values from induction ring data.

The instrument flown by Marshall to measure precipitation charge and size is shown schematically in Fig. 8.10. The use of two 15-cm-diameter metal spheres to house the sensors, electronics, and batteries produces a device similar in exterior design to that of the balloon-borne field meter made from two spheres (Fig. 8.5). The principles of operation and calibration are as described for the induction ring device. However, in this instrument the top ring is exposed to the ambient field and is used only to detect the presence of charged particles that hit the side of the intake cylinder; the two better-shielded middle cylinders are used to measure fallspeed and charge. The bottom cylinder is attached to the sphere and helps shield the two measuring cylinders from the external field. The intake orifice is 8.3 cm in diameter (area $\simeq$ 54 cm^2). The device measures minimum charges of $\pm$10 C and maximum charges of -350 pC and $+450$ pC; the asymmetry is due to amplifier design. Compared with the instrument described previously, this device has advantages of lighter weight, smaller size, an automatic calibration pulse, and spherical shape. Disadvantages include the high minimum detectable charge of 10 pC and the absence of any direct measurement of current density.

A significant advantage in using a lighter-weight device is that it can be flown on a small, free balloon. Since such balloons move with the wind, there is no significant horizontal component in the particle fall at the instrument. The instrument is suspended beneath the balloon with a 6-m-

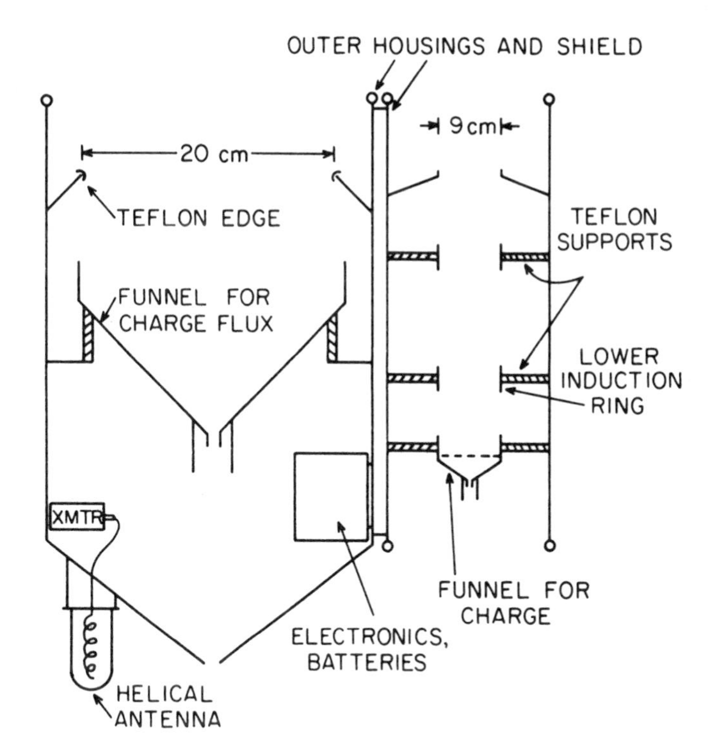

Figure 8.9. A device for measuring precipitation charge flux (current density) and size and charge of individual particles. Identical units have been simultaneously flown on balloons and used at the ground. The outer housing is thick aluminum foil wrapped around on open grid of vertical rods that form the cylindrical housing. Each sensor is connected to a charge amplifier and other electronics; these in turn drive VCO's, and the multiplexed VCO signal is telemetered to the ground.

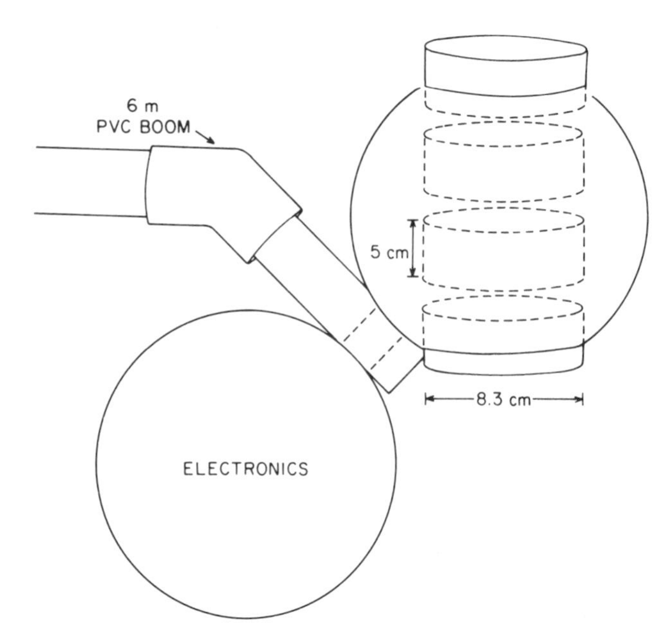

Figure 8.10. A balloon-borne device for measuring precipitation charge and size. The top ring detects particles that collide and splash, the middle two rings measure fallspeed and charge, and the bottom ring is grounded to the housing to help shield the inner rings. The sensing rings are connected through the hole in the PVC boom to the electronics in the other sphere.

long, horizontally oriented dielectric boom to keep water from dripping off the balloon into the cylinders.

b. Charge/Size Instruments on Airplanes

An instrument for use on an airplane has been developed to sense optically the particle and its size (Gaskell et al., 1978). This device can detect particles with small or no charge. The optical detection and size-determination instruments generally function by measuring the amount of light blocked as a particle passes between a light source and a detector. A linear array of 64 photodiodes, each spaced 100 μm apart, is used to sense light from a parallel beam emitter. The shadowgraph optics look across a cylinder 5 cm in diameter and 8 cm long; the sensing cylinder is encased in another larger, grounded cylinder to provide shielding. A charge amplifier with a decay time constant of 22 ms is connected to the sensor. Charged particles passing through cleanly give a symmetrical pulse with a width of about 2 ms; drops hitting the cylinder have a fast rise to peak followed by a slower decay. Spurious data from collisions and splash can thus be ignored. The instrument has been mounted under the wing of the Schweitzer sailplane operated by the New Mexico Institute of Mining and Technology. The sailplane typically penetrates storms at an airspeed of only 40 m s^{-1}. The instrument can measure

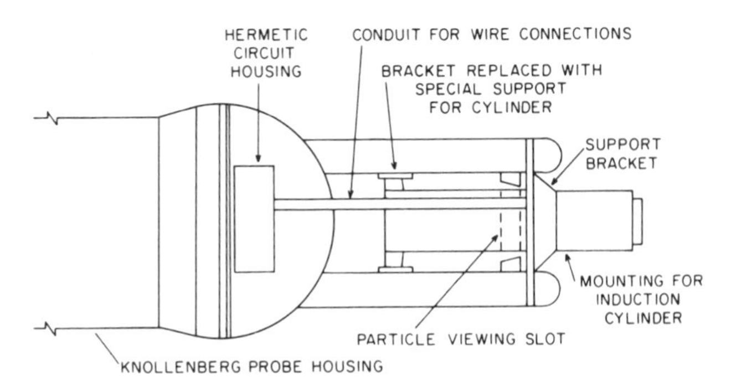

Figure 8.11. A modified Knollenberg probe to detect cloud and precipitation particle size and charge. The device is designed to be flown on an airplane.

precipitation particles with radii from about 0.25 to 2.5 mm and charge from a few picocoulombs to about 130 pC.

This methodology is being improved through the use of a modified Knollenberg probe (see Chap. 3). One modified version uses a 64-element linear diode array so that particles from < 50 μm to 3 mm in diameter can be measured, and particles 10–50 μm can be detected but not sized. The sensing cylinder is connected to a charge amplifier followed by a standard noninverting amplifier with a gain of 25. Directly out of the charge amplifier, the system has a dynamic range to measure particle charge from 150 fC to 307 pC. The sensitive channel output allows charge measurement down to about 6 fC. The cylinder providing the charge measurement contains a gap 0.8 cm across through which the sizing optics view passing particles (see Fig. 8.11). Placed in front of the sensing cylinder is another cylinder to detect splashing drops. When a particle strikes the front cylinder, the characteristic waveform is produced, and the spurious measurements can be ignored.

The instruments described notwithstanding, particle-measurement capability is still deficient because there is no instrument for routine measurement of the size and charge of small cloud particles. A significant fraction of the total storm charge may reside on such particles.

6. Electric Field Changes Associated with Lightning

The electromagnetic fields generated when a lightning flash occurs have been measured by various instruments at many different frequency bands. Some information appears in Vol. 2, Chap. 13, and in Chap. 7 of this volume; here we present fundamentals of field change and interpretative examples. Note that we change our terminology when discussing lightning. It is still the electric field E that is measured, but the term "electric field change" is used to denote the electrostatic component, i.e., the magnitude of field

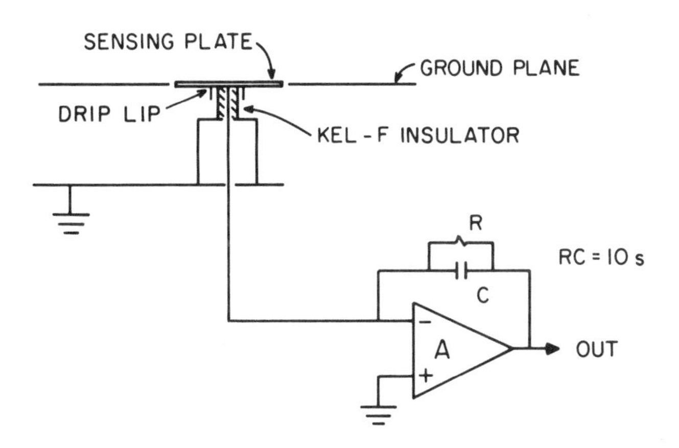

Figure 8.12a. Electric field change sensing device for recording lightning. The circuit is a charge amplifier whose gain is controlled by and inversely proportional to the magnitude of the capacitor. The parallel resistor merely controls the decay time constant. Charge amplifiers are generally made with operational amplifiers having field effect transistor (FET) inputs. The drip lip is needed to keep the insulator from being shorted out by a film of water.

Figure 8.12b. A field change instrument designed by Marx Brook and Paul Krehbiel, of the New Mexico Institute of Mining and Technology. The unit can be equipped with a rain shield, which reduces sensitivity but keeps charged rain from saturating the instrument.

changes identified with charge redistributed by all the processes of the lightning flash. The term "electric field waveform" is used to denote the radiation component, i.e., higher frequencies (lasting nanoseconds to milliseconds) owing to individual processes within a flash. When examining lightning E, we are generally not concerned with the quasi-static value of the atmospheric E before or after the flash. The electrostatic field change is the portion of the field change that is measured close to lightning (within a few tens of kilometers).

Modern sensors are conceptually quite simple. One widely used version is shown in Fig. 8.12a and b. The sensing plate is a disk connected to the inverting input of a charge amplifier. The insulator around the input connection should be Kel-F or Teflon. In addition, the cylindrical insulator should have wide grooves or depressions machined in it to increase the path length along it, thus reducing the possibility of leakage currents. The insulator must be kept clean. A circuit like that shown in Fig. 8.12a responds to an impulse transient within microseconds, but it is the slower components of the recorded field change that are electrostatic. The circuit must allow decay of the response; otherwise the instrument would saturate and remain so. The decay time, $\tau = 1/RC$, is selected to be longer than the phenomenon of interest (as in the charge-measuring devices described in the preceding section). Since lightning generally lasts less than 0.5 s, a decay time of 10 s allows faithful recording of the electrostatic field change throughout the event (too short a time constant would cause the output signal to shift quickly back toward zero, and the relatively slowly varying components of the field would not be recorded). These instruments have often been called "slow antennas" in the literature, owing to their relatively slow decay time.

In examining field changes, it is often easy to ascertain whether a flash is cloud-to-ground (CG) from the characteristic steplike changes that are indicative of each return stroke of a ground flash (see Fig. 8.13). Occasionally very close intracloud flashes exhibit similar changes, but flashes more than a few kilometers away generally do not. In addition to flash type, characteristics of CG flashes can be inferred, e.g., the number of return strokes, the presence of continuing current, and the polarity of charge "neutralized" or effectively lowered from the cloud to the ground. The field change instrument can be calibrated by applying pulses of known amplitude through a capacitor connected to the input, i.e., the sensing plate, or by placing the sensor flush

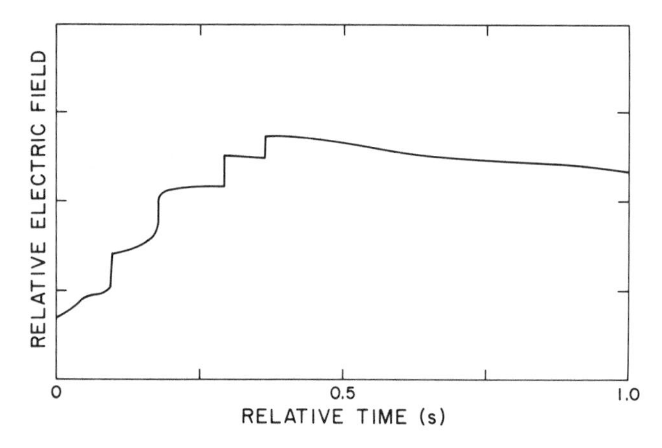

Figure 8.13. Electric field change from a cloud-to-ground lightning flash. Each step in the field change is due to the occurrence of a return stroke in the flash.

with a plate parallel to another to which a pulse is applied. This procedure calibrates the electronics, but the geometric form factor of a particular installation must be determined, as for field mills, by comparison with an identical instrument mounted in a flat ground plane at the surface of the Earth. The electric field change for each return stroke and the entire flash can then be determined from

$$E = \frac{CGV}{\varepsilon_0 A}, \tag{8.8}$$

where C is the capacitance in the feedback circuit, V is the voltage out, A is the sensing-plate area, and G is the geometrical form factor determined from the ratio of the voltage outputs recorded for the same lightning transient by the flush-mounted and the other instrument. With a network of such instruments that have been carefully calibrated, the location of individual charges neutralized during ground flashes and the dipole moments of intracloud lightning processes can be determined and compared with storm structure (Krehbiel, 1981). Such a network is discussed in Sec. 8b below.

The same instrument can be used to observe the higher-frequency components of flashes as well. This is done by merely reducing the time constant, common values being 100 μs to 1 ms. Since saturation is less likely, owing to the fast decay, these instruments can often be set to higher gains than a slow antenna for the same flash, allowing small pulses to be recorded. These instruments have been called "fast antennas," though they are also referred to (imprecisely) as "field change instruments." A time constant of 1 ms allows the major aspects of a return-stroke waveform to be recorded faithfully, which is necessary if the measurements are to be used quantitatively. Uses of these electric field waveforms have included examination of intracloud processes and of physical characteristics of return strokes. The latter includes risetime, with typical values of a few microseconds for lightning over land (the risetime slows as a wave propagates over an imperfect conductor such as the Earth). Over salt water, rise times less than 1 μs have been measured. In addition, return-stroke measurements permit modeling of peak currents in the lightning channels to ground. Both risetime and peak currents are relevant to design of means for protection of ground and airborne systems, which utilize sensitive microelectronics such as computers. An excellent detailed summary of electric and magnetic field waveform measurements and calculations of this type is found in Uman and Krider (1982).

Extremely low frequency (ELF) radiation is closely associated with CG flashes. For flashes more than a few tens of kilometers away, ELF apparently provides an unambiguous determination of the polarity of charge lowered and millisecond accuracy in determination of the time of each return stroke. Use of ELF systems is resurging because they are needed to test and evaluate systems that detect and locate CG flashes automatically.

An ELF system can be constructed quite simply (Fig. 8.14). A characteristic frequency band pass is 400–1,000 Hz. With sensitivity that can be adjusted over a range of 1,000, the instrument illustrated can detect CG flashes to distances of about 600 km. It is proving invaluable for comparisons of data from several systems and in providing confirmation of positive CG flashes (see Vol. 2, Chap. 13, App.).

7. Optical Measurements of Lightning

A fundamental optical measurement is that of the wave shape produced by various light-emitting processes in a flash, e.g., the stepped leader as it moves from the cloud to the ground, and the wave front that travels up the channel during the return stroke. Sensors are either photomultiplier tubes or solid-state devices. The latter have the advantage of small size, low voltage and power requirements, and comparatively low cost. An example of the simplicity of a solid-state device is illustrated in the modification of a sound super-8 movie camera by Vonnegut and Passarelli (1978), who placed an inexpensive silicon photocell behind a plastic Fresnel lens (both available from science hobby suppliers) to focus the lightning luminosity changes on the photocell. The signal from the photocell is then sent to the audio input of the camera. From this simple idea evolved the idea of looking at cloud tops from space to assess lightning rates and convective cloud growth quantitatively. This concept has been tested with the use of a space-worthy version of the sound movie camera aboard the space shuttle (Vonnegut et al., 1985) and in instrumentation aboard a NASA U-2 research airplane that is flown above the tops of storms (Brook et al., 1980; Goodman et al., 1984). The circuit for one instrument to measure lightning waveforms with microsecond resolution from the U-2 is shown in Fig. 8.15. Similar instrumentation has been used at the ground

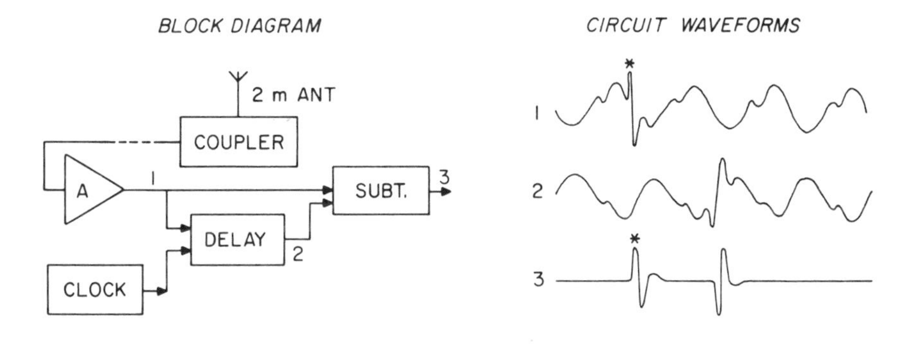

Figure 8.14. A functional diagram and waveforms of an ELF system. The delay and subtractor (SUBT.) circuits remove most of the 60 Hz, which would otherwise dominate the output signal at normal operating sensitivities. Waveform 1 is observed out of the main amplifier; this signal is mostly noise from 60 Hz and its harmonics, with a pulse from a return stroke (marked with *). The same pulse is seen in waveform 2 after delay by 1/60 cycle. When waveforms 1 and 2 are subtracted, waveform 3 is obtained. Its noise is significantly reduced and the return stroke pulse (*) and its mirror-image artifact 16.67 ms later are easily detected.

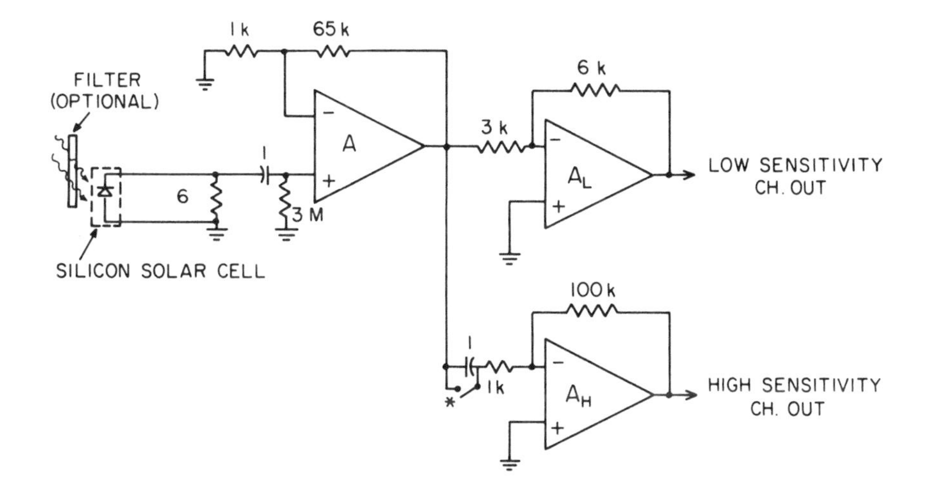

Figure 8.15. A circuit diagram of the optical detector flown on a NASA U-2 research airplane. Capacitance is in microfarads; resistance is in ohms. The optional filter can be used to reduce ambient light while allowing a spectral line prominent in lightning to be passed. The switch marked with the * allows the high sensitivity channel to be AC- or DC-coupled. Amplifier A is a fast FET input type, and A_L (low sensitivity) and A_H (high sensitivity) are line-driver amplifiers to maintain waveform quality in long 50 to 75 Ω coax lines. They instrument risetime of about 1 μs.

and on airplanes by various investigators in studies of lightning characteristics. Also technology now exists for the development of an array of optical detectors that not only provide lightning waveforms but could allow lightning to be mapped from a geosynchronous satellite. This technology is described in Sec. 8 below.

The optical sensor and camera technique has been extended to television systems. Here also optical data can be recorded on the audio track. There are several advantages in using a video camera instead of a movie camera; most notable is that data are not lost (a video field takes about 17 ms to produce, and there are no gaps between successive frames as there are in movie cameras, whose shutters are open only about half the time), data acquisition is inexpensive, and data can be viewed immediately. The use of video to document lightning was demonstrated by Winn et al. (1973) and has been expanded in various configurations by several investigators to integrate the study of cloud development and lightning occurrence. Video documentation has been applied to the study of lightning strikes to airplanes and to an assessment of site errors and detection efficiency of automatic ground-flash locating systems. In the latter study an all-azimuth video system was used (Mach et al., 1986).

Other optical methods for the study of lightning include photographic and spectrographic techniques. They are most often used for the study of lightning physics. Reviews of such instrumentation and results were given by Uman (1969) and Orville (1977).

8. Lightning-locating Techniques

a. Introduction

Knowing where lightning occurs in storms is essential to gaining an understanding of storm electrification, particularly of relationships between lightning and its storm environment. Until the early 1970s, however, the technology for locating lightning was limited to radar, to magnetic direction finding with the use of very low frequency (VLF) sferics, and to measurements of electric field changes for estimating the charge neutralized in a lightning flash. Beginning about 1970, advances in electronics and computers were coupled with significant new insights to improve existing methods and to provide new methods. Today, in addition to improvements in the previous techniques, lightning can be located from thunder by using microphone arrays and from very high frequency (VHF) sferics with antenna arrays. These advances may contribute to major strides in understanding electrification of storms, particularly in combination with significant advances in the technology for observing other meteorological parameters of storms.

b. Charge Center Analysis

The first technique developed for estimating the location of lightning inside clouds used measurements of electric field changes from lightning at several stations simultaneously. Simultaneous equations for the electrostatic field changes were solved to estimate the coordinates and net charge of an equivalent charge center lowered to ground by the flash (note that, even though the term "neutralization" is often used, only the movement of charge is required to produce the effect). Jacobson and Krider (1976) and Krehbiel et al. (1979) used this technique to study the location of charge centers, Krehbiel et al. using a network of field change sensors and Jacobson and Krider using field mills.

To infer the location and magnitude of charge from measurements of electric field changes at the ground, it is necessary to model charge geometry. Two geometries have commonly been used for this analysis. The first, a point charge model, applies only to flashes lowering charge to ground. If it is assumed that the charge to be lowered is spherically symmetric, or that its dimensions are small compared with its height aboveground, and if the charge is lowered to a perfectly conducting plane, the resulting field change for a station at coordinates x_i, y_i, z_i is

$$\Delta E_i = \frac{1}{2\pi\varepsilon_0}\frac{Qz}{[(x - x_i)^2 + (y - y_i)^2 + (z - z_i)^2]^{3/2}}, \quad (8.9)$$

where x, y, z are the coordinates of the charge center, Q is the charge being lowered, and ε_0 is the permittivity of free space. Since there are four unknown variables (x, y, z, and Q), measurements of ΔE_i at four locations are sufficient to determine the coordinates and charge.

The second model, a dipole model, is assumed to describe field changes from intracloud processes. If an intracloud process neutralizes equal but opposite point charges above a perfectly conducting plane, the electric field change is the superposition of the field changes from each point charge and is given by

$$\Delta E_i = \frac{Q}{2\pi\varepsilon_0}\left[\frac{z_p}{[(x_p - x_i)^2 + (y_p - y_i)^2 + (z_p - z_i)^2]^{3/2}} - \frac{z_n}{[(x_n - x_i)^2 + (y_n - y_i)^2 + (z_n - z_i)^2]^{3/2}}\right], \quad (8.10)$$

where x_p, y_p, z_p are the coordinates of the positive charge, and x_n, y_n, z_n are the coordinates of the negative charge. In this situation seven measurements of ΔE_i are needed to determine the six coordinates and the charge. If only six measurements are available, the expression for ΔE_i can be reformulated in terms of the dipole moment, **p**:

$$\mathbf{p} = Q\,[(x_p - x_n)\mathbf{i} + (y_p - y_n)\mathbf{j} + (z_p - z_n)\mathbf{k}], \quad (8.11)$$

where **i**, **j**, and **k** are unit vectors, although this provides less information about the charge.

When there are more measurements than the minimum required for a solution, it is necessary to reconcile the resulting redundant solutions. Two optimization techniques have been used: a scatter plot of redundant solutions and nonlinear, least-squares fitting of the measurements (first applied by Jacobson and Krider, 1976). Krehbiel et al. (1979) found that results from the scatter plot and least-squares analysis are in good agreement with each other. Since the least-squares technique gives a centroid more directly and objectively, it is now preferred over the scatter plot.

In least-squares fitting, the analysis seeks to minimize the chi-squared statistic, which measures the deviations between the measured and derived values of the field and is given by

$$\chi^2 = \sum_i \frac{(\Delta E_i - \Delta\xi_i)^2}{\sigma_i^2}, \quad (8.12)$$

where $\Delta\xi_i$ is the measured field change at the ith location, ΔE_i is the field change calculated from the model charge distribution, and σ_i^2 is the variance in the measurement of $\Delta\xi_i$. Since ΔE_i is a nonlinear function of the coordinates and charge of the centroid, an iterative, nonlinear algorithm, such as the Marquardt algorithm described by Bevington (1969), is needed for the least-squares analysis.

The success of the charge centroid analysis in finding solutions has been cited in arguing that a point charge can model the actual geometry of charge in lightning flashes reasonably well. However, the charge center analysis is relatively insensitive to the assumption that charges are spherically symmetric, as long as there is some degree of symmetry in the charge geometry. In a computer simulation of a thin disk of charge, for example, it was found that the charge analysis determined Q and the horizontal coordinates of the charge center within errors that would be expected in the measurement of ΔE (E. Williams, 1984, personal communication). The vertical coordinate was the most strongly affected, and even it was within 0.5 km of the actual centroid.

The constraints imposed on electric field change measurements by the technique are fairly strenuous. The instruments should be spaced only a few kilometers apart and in a relatively level region. Krehbiel et al. (1979) indicated that to measure the location of a charge centroid to within 100 m it is necessary to measure electric field changes to within 1% at the different stations. Because of the dependence of the form factor on the position of the sensor relative to its surroundings, this corresponds to variations in the position of the sensor of about 2.5 cm. Other sources of error include the electronic circuitry, the telemetry, and the recording system. Environmental effects such as space charge and charging from splashing rain also introduce errors, though the errors can be subtracted if their period is much slower than the field change being analyzed. Krehbiel et al. (1979) estimated that a typical error in their measurement of field changes was approximately 2%.

As mentioned previously, similar information on the distribution of charge lowered to ground by lightning can be provided by measurements of the electric field change by a network of several field mills. The frequency response of a typical field mill is inadequate for resolution of the field change of return strokes, but a typical mill does measure the total field change from the flash. From the total field change one can find the total charge neutralized during a flash. Constraints on the network are the same as those on the network of field change sensors. A network of field mills has been operating for a number of years at Kennedy Space Center. Analysis of network data by Jacobson and Krider (1976) and comparison with independent data have shown the technique to be useful in ascertaining the heights of charge centers relative to ambient temperatures.

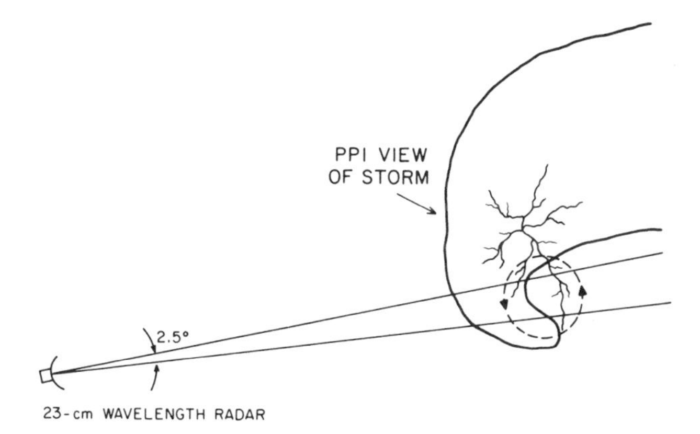

Figure 8.16. Observing lightning with radar. Lightning is being observed in the mesocyclone region (dashed circle) of a severe thunderstorm. The antenna is stationary except for occasional azimuthal changes to follow the moving storm. The radar antenna pattern is cosecant squared in the vertical, so flashes from within a large elevation sector are detected and located in range.

c. Lightning Location by Means of Radar

Although the observation of lightning with weather radars was first reported in the 1950s, it was not until recently that radar techniques were enhanced with the specific goal of locating lightning, determining physical characteristics of lightning channels, and relating lightning to storm evolution. Radar offers unique opportunities to study lightning out to very long ranges (several hundred kilometers), as well as very close. The concept of observing lightning with radar is shown schematically in Fig. 8.16. Radar wavelengths of 10 cm and longer are required for easy identification of lightning in the presence of precipitation.

With a 70-cm-wavelength radar, Mazur et al. (1984a) observed lightning strikes to an instrumented airplane. Not only can the lightning echoes be related to lightning channel characteristics, but also these observations show that the airplane usually initiates the lightning strikes rather than intercepting existing flashes as previously thought. Of about 500 strikes to the airplane, more than 100 have been analyzed at this writing; all but three show lightning echoes that generally begin in the same radar resolution volume as the airplane and propagate outward, often bidirectionally. In addition, flashes seen visually near the airplane by onboard observers and observed with the radar to propagate past the airplane have almost never struck the airplane. Using the same 70-cm-wavelength radar and a 10-cm wavelength weather radar, Mazur et al. (1984b) determined gross features of the vertical distribution of lightning relative to precipitation structure.

Recently vertically pointing Doppler radars have been used to study lightning and also to examine possible interactions between precipitation and electricity. An example of such a radar configuration is shown in Fig. 8.17; note

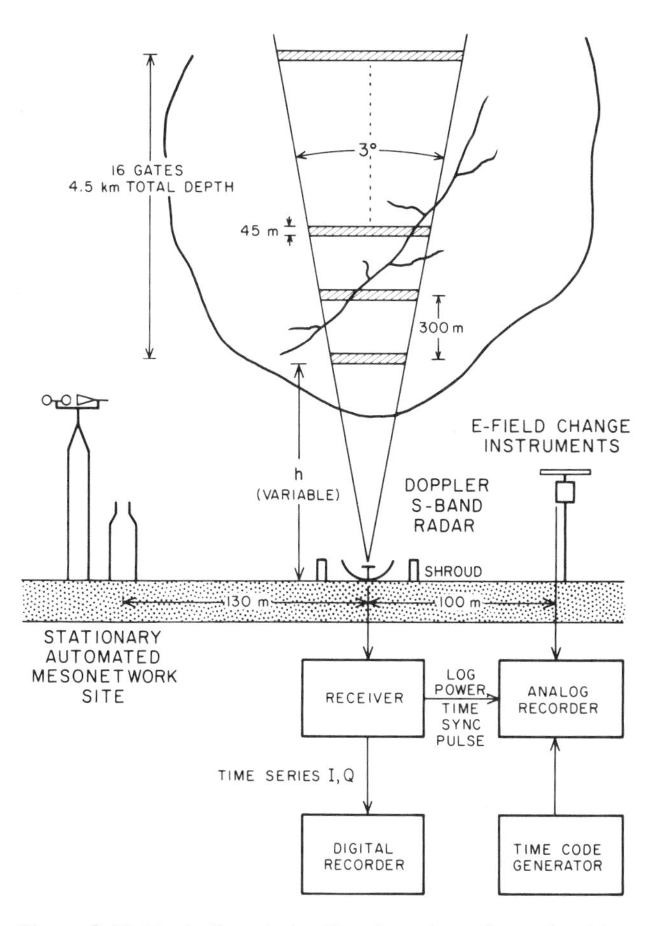

Figure 8.17. Vertically pointing Doppler radar and associated instrumentation to study lightning and the interactions of lightning and precipitation.

Table 8.1. Characteristics of Vertically Pointing NSSL Doppler Radar In Mode for the Study of Lighting

Wavelength	10.52 cm
One-way 3-dB beam width	3°
Peak transmitter power	20 kW
Pulse repetition time	768 μs
Pulse width	0.25 μs
Antenna one-way gain	35 dB
Receiver bandwidth	10 MHz

that the radar antenna is shrouded to reduce significantly (40 dB) the antenna beam side lobes. In the NSSL arrangement the Doppler radar observational volume consists of 16 layers, i.e., gates, within the conical antenna beam. Each layer is only 45 m deep and is separated by 300 m from the adjacent layers; the total range extent of the observational volume is 4.5 km. This range "window" can be moved in height by controlling the time delay to the first range layer. Several different recording schemes have been used. The output of the radar's logarithmic video amplifier is recorded in analog form, and time series data from each of the 16 layers are recorded digitally for subsequent processing. Also recorded on analog tape with the reflectivity are electric field changes, optical transients from lightning, and timing and radar synchronization information that allows comparison among all variables. The radar characteristics used in several experiments and indicative of suitable values are given in Table 8.1.

Because the reflecting lightning channel moves with the air velocity, it is possible to measure vertical air velocities as well as the terminal fallspeed of the precipitation from analysis of Doppler spectra of lightning echoes (Zrnić et al., 1982). Summaries of results from several investigators are given in Mazur et al. (1984a, b), Zrnić et al. (1982), and Rust and Doviak (1982).

d. Acoustic Mapping

Techniques for tracing the propagation of sound, originally applied to underwater uses (sonar) and oil exploration, can be adapted to locate lightning channels from analysis of recordings of the thunder that they generate. Since the development of a lightning flash is very fast on the time scale applicable to acoustic propagation, the collection of mapped acoustic sources derived from the thunder recording of a flash gives basically a snapshot of the hot, explosive channels producing thunder, with no information about their sequence of development. Although acoustic techniques are limited to a range of roughly 20 km for audible frequencies, and the analysis of data is complicated by the sensitivity of sound propagation to normally encountered atmospheric variations, the required instrumentation is relatively simple and can be deployed easily.

To locate the source of a thunder pulse, means are required for determining the pulse's range and direction. Range can be determined because electromagnetic signals travel at the speed of light whereas acoustic signals travel at one-millionth that speed; thus the time delay between a flash's electric field change and a thunder pulse defines the range to the source of the pulse. Sensors for recording electric field changes are described above. Any rugged outdoor microphone with a reasonably flat frequency response between a few hertz and a few hundred hertz can be used to record the arrival of a thunder pulse.

There are two techniques, thunder ranging and ray tracing, for determining location of an acoustic source. Thunder ranging, the simpler of the two, was evaluated first by MacGorman (1977) and later, in greater generality, by Bohannon (1978). Ray tracing was developed by Few (1970) and evaluated by Few and Teer (1974). Each technique has particular advantages and disadvantages.

Thunder ranging requires at least three microphones separated by 1–3 km in a noncollinear array. The ranges of the source of a thunder pulse to each of the microphones are computed as described above. The intersection of the three ranges then gives the location of the acoustic source in three-dimensional space. The equations for the coordinates of the intersection are simplest if the microphones are arranged in a right triangle with one of the legs aligned with north as shown in Fig. 8.18. In this arrangement the coordinates are given by

$$x = \frac{R_3^2 - R_2^2 - D_{23}^2}{2D_{23}}, \tag{8.13}$$

$$y = \frac{R_2^2 - R_1^2 + D_{12}^2}{2D_{12}}, \tag{8.14}$$

and

$$z = (R_2^2 - x^2 - y^2)^{\frac{1}{2}}, \tag{8.15}$$

where D_{ij} is the distance between microphones i and j, R_i is the range from the ith microphone to the acoustic source, x and y are the coordinates east and north from microphone 2, and z is the height above the plane that contains the three microphones.

In the thunder-ranging analysis, the thunder waveforms from all microphones are plotted along with the electric field change record on a common time base at a scale on the order of 1 cm s^{-1} and are examined by eye to find features common to all. Because the thunder waveform is influenced by propagation effects and by the different position and orientation of a lightning flash relative to each microphone, the analysis should be based only on features that are fairly prominent in waveforms from all the microphones. To determine the ranges from a thunder source to each microphone, the propagation time is measured as the time interval from the lightning field change until the corresponding feature in each thunder signature; the speed of sound is derived from atmospheric data. The resulting ranges are then used in Eqs. 8.13–8.15 to compute coordinates of the source of the thunder feature.

Ray tracing, on the other hand, uses an array of three or

more microphones separated by only a few tens of meters. A cross-correlation analysis such as that described by Bendat and Piersol (1971) is used to identify corresponding thunder pulses at all microphones and to compute the propagation time of the pulse from one microphone to another. Each cross correlation uses a 0.1–0.5-s segment from the data record. To avoid computational problems, the segment should be about twice the maximum expected propagation time but needs to be as short as possible to minimize confusion caused by thunder from multiple sources arriving at about the same time. The maximum expected propagation time is set either by the time required to traverse the baseline being analyzed or by the lags between a clearly defined feature in the thunder waveform.

The values found by the cross-correlation analysis for the propagation time between each pair of microphones are used to compute the direction from which the thunder pulse arrived. For the same geometry of baselines as in Fig. 8.18, but with the legs of the triangle equal in length, the azimuth ϕ and elevation θ from which the pulse arrived are given by

$$\phi = \arctan \frac{\tau_{23}}{\tau_{12}} \tag{8.16}$$

and

$$\theta = \arccos \frac{c}{L\tau_{23}^{-1} \sin \phi - u \cos (\phi' - \phi)} \tag{8.17}$$
$$= \arccos \frac{c}{L\tau_{12}^{-1} \cos \phi - u \cos (\phi' - \phi)},$$

where c is the speed of sound, L is the length of the north-south and east-west baselines, τ_{ij} is the propagation time between microphones i and j (positive when the pulse arrives at microphone i first), u is the surface windspeed, and ϕ' is the azimuth from which the wind is blowing.

To locate the source of the thunder pulse, an acoustic ray with this initial azimuth and elevation must be propagated backward through a model atmosphere. The total path length is determined from the speed of sound and the measured time between occurrence of the lightning flash and arrival of the pulse at the microphones. If the atmosphere had uniform temperature and humidity and no wind, the path would be a straight line, and the computation of the coordinates of the thunder source would be simple. However, in the real atmosphere paths are curved, and this must be approximated in the computation. For a continuously varying, stratified atmosphere with wind, a form of Snell's law describes how the direction angles change along the ray path:

$$\cos \alpha = \cos \alpha_0 \left[\frac{1 + \Gamma_c dz}{1 - (\Gamma_{u_x} + \Gamma_{u_y})\, dz} \right],$$

$$\cos \beta = \cos \beta_0 \left[\frac{1 + \Gamma_c dz}{1 - (\Gamma_{u_x} + \Gamma_{u_y})\, dz} \right],$$

and

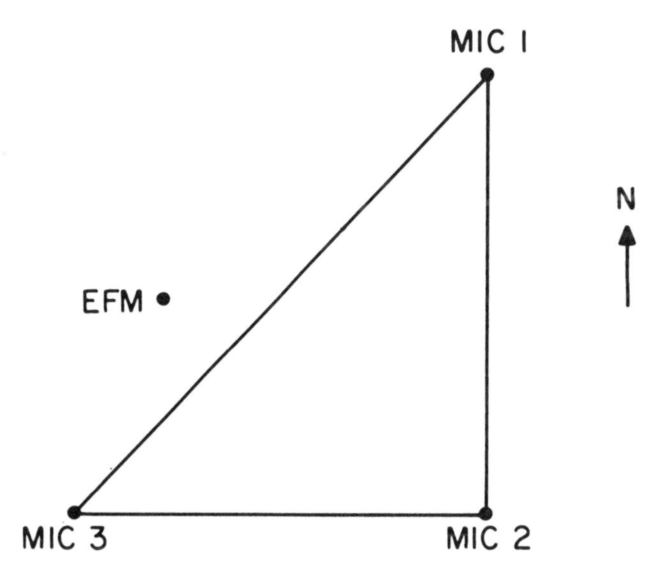

Figure 8.18. Arrangement of a microphone array. EFM denotes the location of an electric field change meter to record the flash times.

$$\cos^2 \gamma = 1 - \cos^2 \alpha + \cos^2 \beta, \tag{8.18}$$

where α, β, and γ are the new direction angles to the x, y, and z axes, respectively; α_0 and β_0 are the previous direction angles; and

$$\Gamma_c = \frac{1}{c_0} \frac{\partial c}{\partial z},$$

$$\Gamma_{u_x} = \frac{\cos \alpha_0}{c_0} \frac{\partial u_x}{\partial z},$$

and

$$\Gamma_{u_y} = \frac{\cos \beta_0}{c_0} \frac{\partial u_y}{\partial z}, \tag{8.19}$$

where u_x and u_y are the components of the local wind, c is the local speed of sound, and c_0 is the previous value of c. The ray path can then be approximated as a series of steps of length ds, the components of each step being given by

$$dz = \cos \gamma \, ds,$$

$$dx = \frac{c \cos \alpha + u_x}{c} ds,$$

and

$$dy = \frac{c \cos \beta + u_y}{c} ds. \tag{8.20}$$

The time increment for each step is given by

$$dt = \frac{ds}{c}. \tag{8.21}$$

Steps are added until the sum of the time increments equals the measured propagation time.

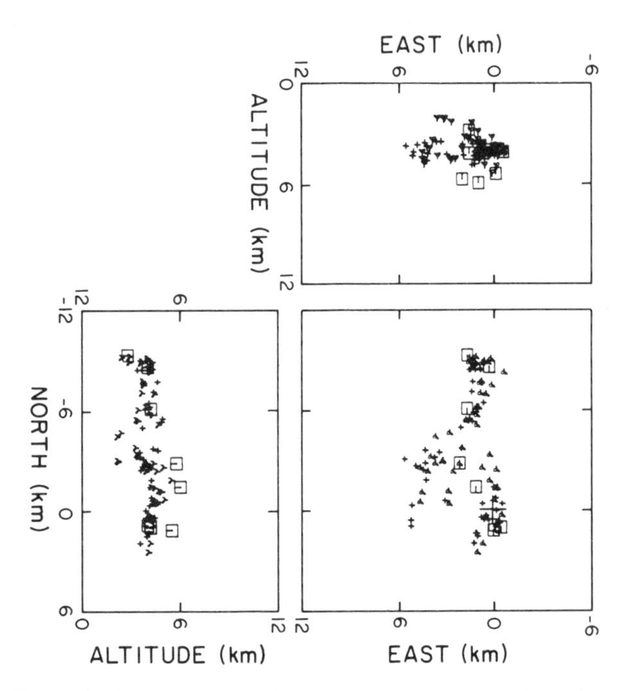

Figure 8.19. Comparison of three acoustic reconstructions of the same lightning flash. Acoustic sources mapped by ray tracing from two arrays are here shown as computer-generated plots of +'s and *Y*'s to indicate the array used. (Some *Y*'s are upside down, and some have a loop or circle on one side.) Sources mapped by thunder ranging are plotted as squares.

Inherent errors in the two acoustic techniques are of comparable magnitude. The source of largest errors in ray tracing is the difficulty in modeling wind shear. Few and Teer (1974) estimated that differences in wind shear between the model and real atmospheres would cause errors typically within 10% of the range to the source.

Thunder ranging is much less sensitive to wind shear than ray tracing, but it is more sensitive to other sources of error. MacGorman (1977) found that when lightning is within about 5 km of the microphone array errors in reconstructed locations from thunder ranging are dominated by the different orientation and range of lightning structure relative to each site, typical errors in this region being 15% of the range. When lightning is more than 5 km outside the microphone array, errors in thunder ranging are dominated by the uncertainty in identifying a particular short pulse within bursts of pulses in the thunder signature; errors in this region are typically 10% of the range. Much larger errors can result from errors in subjective choices of corresponding pulses in the thunder signatures from different microphones.

MacGorman (1977) examined whether results from the two techniques were consistent. Three flashes were reconstructed by ray tracing with two independent arrays and by thunder ranging with three microphones separated 2–3 km. Agreement between the two ray-tracing analyses and between the ray-tracing and thunder-ranging analyses was generally within the predicted errors, as shown by the example in Fig. 8.19.

Thunder ranging is most useful for focusing attention on interesting data before ray tracing or for studying large numbers of lightning flashes to determine the general location of lightning channels. Thunder ranging is much faster than ray tracing, requiring a total analysis effort of a few hours per flash versus 20–40 h for ray tracing. However, thunder ranging provides much less detail in the reconstructed lightning structure, and the analysis is somewhat more subjective, since the choice of corresponding features in the thunder signatures is made by eye. Therefore, ray tracing is more appropriate when greater detail or more reliable locations are needed. If the microphone array is small enough (probably a few meters) that thunder signals are highly coherent between microphones, it may be possible to automate the calculation of direction angles for ray tracing (Eqs. 8.16 and 8.17); the time required for analysis might then be less than that required by thunder ranging and possibly less than an hour.

e. VHF Radio Mapping

Several techniques have been developed to map the location and development of lightning by using the radio frequency signals it generates. These techniques use frequencies in the VHF band (30–300 MHz) or higher for more sensitivity to intracloud processes. The effective range of these techniques is generally 60 km or less because of signal attenuation and decreased spatial resolution in the mapping analysis at longer ranges. Beyond these common factors, the techniques vary considerably. Each technique can be described in terms of the size of its antenna array and its method of determining locations. Small arrays have baselines on the order of 10–100 m and determine only direction angles to the radio sources. Range can be determined if there are two such arrays for triangulation. Large arrays have baselines on the order of 10 km and determine both direction angles and range. Short-baseline arrays determine direction angles either by measuring the difference in times of arrival of the radio signal between pairs of antennas or by using an interferometric technique. Long-baseline arrays determine locations only by measuring differences between times of arrival.

(1) Long-Baseline, Time-of-Arrival Technique: The first VHF system to locate noise sources on lightning channels in three dimensions was a long-baseline, time-of-arrival system developed by Proctor (1983). The chief initial disadvantages of the system were that data could be recorded continuously for only 250 ms and analysis of the data was extremely laborious, requiring six months to determine source locations for a single 250-ms record. With the addition of new devices as described by Proctor (1983), these problems have been largely overcome, although analysis is still fairly tedious.

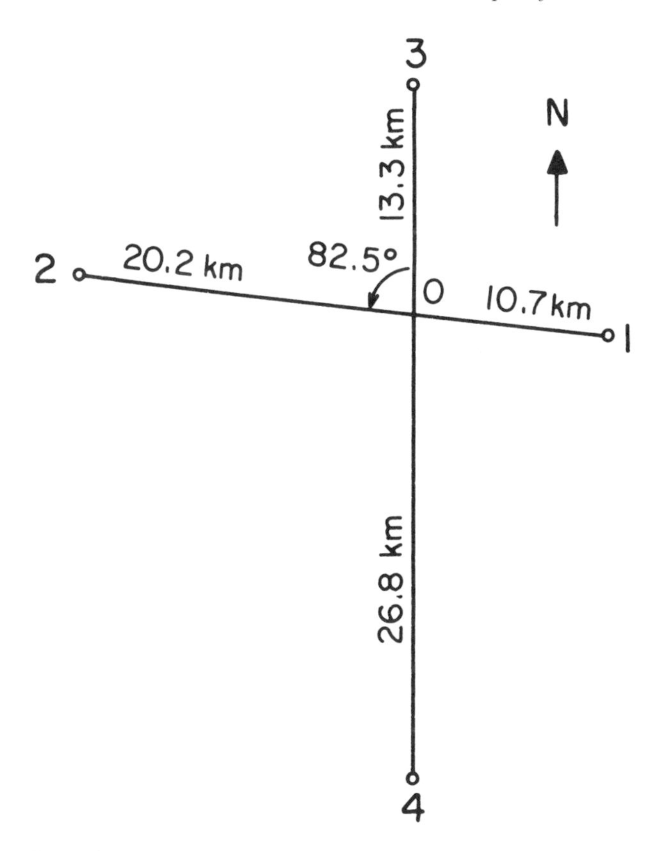

Figure 8.20. Long-baseline VHF array (after Proctor, 1981). The numbers 0–4 denote locations of receiving antennas.

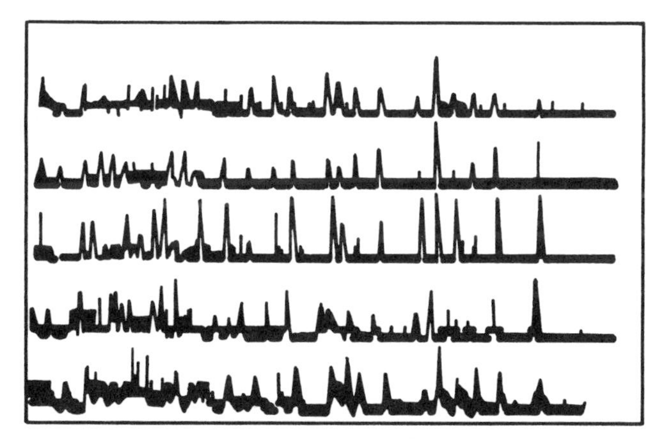

Figure 8.21. Tracing of outputs from five VHF receivers for a portion of a lightning flash (after Proctor, 1983). The time across the plot is approximately 60 μs.

In Proctor's present system radio signals are sensed by five broad-band, vertically polarized antennas arranged to form two crossed baselines (Fig. 8.20). The signal from each antenna is fed into a crystal-controlled receiver operating at a center frequency of 355 MHz, and the outputs of the outlying receivers are transmitted to the central station over frequency-modulated 3-cm-wavelength microwave links. Precautions are taken in the telemetry to avoid signal contamination by sferics or by cross coupling of signals from different stations. Coordinates of the outlying stations relative to the central station are determined to within 10 cm, which provides the propagation time between stations much more accurately than required for the time-of-arrival measurements. Time of arrival is given by the time the telemetered receiver output arrives at the central station minus the propagation time of the telemetry between the two stations. Data from the system are recorded on long strips of photographic film by an analog, laser-optical recorder. Up to 20 minutes of continuous data from five channels can be recorded with an interchannel rms timing error of 82 ns and a bandwidth of 6 MHz per channel.

For analysis the data are slowed down by a factor of 6,000 during playback from the film. Each channel of data is sampled and digitized every 62.5 ns of recorded time and transferred into memory capable of storing 1 ms of data from all five channels. A custom-built analyzer is then used to determine time differences between signals from different stations. All five channels are displayed on the screen of a cathode-ray tube (Fig. 8.21). The operator of the analyzer can shift each channel to the left or right to line up a series of corresponding pulses, and the analyzer automatically keeps track of the relative delays between channels. By displaying segments of the data with increasingly better time resolution, signals can be lined up in 62.5-ns increments, and delays can be read automatically.

The resulting delays are sent to a minicomputer that calculates the coordinates of the source of the signal and checks that the calculated location is consistent with all the redundant time delays (any two of the three measured delays along each baseline are sufficient to determine the coordinate of the source along that baseline). To calculate the source location, equations are determined for three hyperbolic surfaces corresponding to the measured time delays between stations, and these equations are solved to find the point where the surfaces intersect. For a coplanar array with orthogonal baselines, and with stations arranged as shown in Fig. 8.20, one of the redundant solutions can be written as

$$x = \frac{A_1L_1 + A_2L_2}{2(A_2 - A_1)}\left\{-1 \pm \left[1 - \frac{c^2(A_2 - A_1)(\tau_2^2A_2^2 - \tau_1^2A_1^2)}{(A_1L_1 + A_2L_2)^2}\right]^{1/2}\right\},$$

$$y = \frac{A_3L_3 + A_4L_4}{2(A_4 - A_3)}\left\{-1 \pm \left[1 - \frac{c^2(A_4 - A_3)(\tau_4^2A_4^2 - \tau_3^2A_3^2)}{(A_3L_3 + A_4L_4)^2}\right]^{1/2}\right\},$$

and

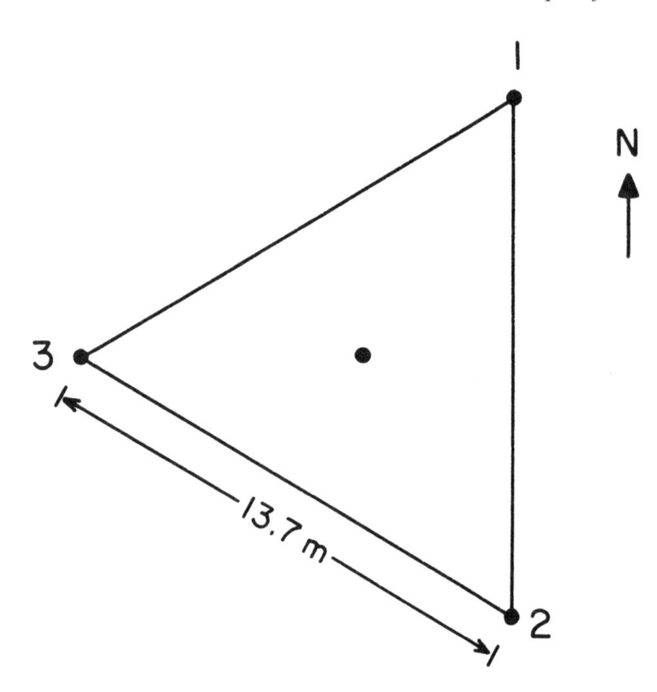

Figure 8.22. A full-hemisphere VHF antenna array layout. The interior dot indicates the location of the vertical baseline antennas, i.e., one antenna on a pole 13.7 m above another antenna on the ground.

$$z = \left[\frac{1}{4}\left(L_1^2 - c^2\tau_1^2\right)\left(\frac{4x^2}{c^2\tau_1^2} - 1\right) - y^2\right]^{1/2}, \quad (8.22)$$

where $\tau_i = T_i - T_0$, T_i is time of arrival at the ith station in seconds, L_i is the length of the baseline between station i and station 0, c is the speed of light, and

$$A_i = \frac{L_i^2}{c^2\tau_i^2} - 1. \quad (8.23)$$

The choices of sign in the expressions for x and y are positive when $\tau_1\tau_2 > 0$ and $\tau_3\tau_4 > 0$, respectively. See Proctor (1971) for additional information about the computations.

Because there is no vertical baseline, the standard error in z is much larger than standard errors in x and y. Within a plan quadrilateral having the four outlying stations as vertices, the error in x and y is approximately 25 m. The error in z, however, is a function primarily of elevation angle, varying from approximately 100 m directly over station 0 to 300 m at $(-3, -3, 4)$ km and to 1 km at $(-6, -6, 2)$ km.

(2) Short-Baseline, Time-of-Arrival Technique: Taylor (1978) developed a time-of-arrival system for locating radio sources radiating in the 20–80 MHz band, as described in Chap. 7 of this volume. The system was upgraded in 1984. The upgraded system determines direction to VHF impulse sources over the entire hemisphere above the horizon, instead of being limited to selectable sectors that cover approximately 60° in azimuth by 45° in elevation. Five antennas are used, three at the vertices of an equilateral triangle, 13.7 m on a side, and two forming a vertical 13.7-m baseline at the center of the triangle. Coverage is divided into six sectors, each spanning 60° in azimuth by 60° in elevation; a seventh sector overhead covers the upper 30° of elevation at all azimuths. Each VHF impulse that is accepted, i.e., that has a sufficiently fast risetime and an amplitude above the system threshold, is assigned to the appropriate sector from the sequence of arrival at the antennas. This sector is encoded along with the time, azimuth, and elevation of the source on nine-track digital magnetic tape.

The cycling time of the system has been decreased so that impulse sources can be processed every 16 μs instead of every 40 μs (peak rates of 64,000 impulses per second instead of 25,000). Continuous-data rates are still limited to 12,000 impulses per second by the time required to buffer and record the data. However, the lower continuous rate did not appear to cause data loss when the peak rate was 25,000 impulses per second.

Equations given in Chap. 7 for calculating azimuth and elevation from the measured propagation times are valid in the improved system except in the overhead sector. In the overhead sector the VHF system measures time lags of a signal along two horizontal baselines of the array, and these time lags can be used to determine azimuth and elevation. For the array geometry shown in Fig. 8.22, azimuth is given by

$$\phi = \arctan\left(\frac{2\tau_{13} - \tau_{12}}{\sqrt{3}\tau_{12}}\right), \quad (8.24)$$

where τ_{ij} is the lag between antennas i and j, positive when the signal reaches i first, and elevation is given by

$$\theta = \arccos\left(\frac{c\tau_{12}}{D\cos\phi}\right) = \arccos\left[\frac{2c}{\sqrt{3D}}(\tau_{12}^2 - \tau_{12}\tau_{13} + \tau_{13}^2)^{1/2}\right], \quad (8.25)$$

where c is the speed of light, and D is the distance between antennas.

With two of the VHF systems separated by 15–50 km, it is possible in later analysis to triangulate the real-time directions to locate sources of VHF signals. To do this, however, it is necessary first to synchronize time by finding the timing offset in the clocks for the two arrays. During data acquisition clocks at the two arrays are set to within 1 ms of the WWV time standard. The offset is determined more accurately by finding the difference in time between clocks that maximizes the number of time-coincident impulses in the data from the two arrays. This offset is then added to the time from the slower clock, the data are searched for coincident signals, and locations are computed. If η is the azimuth of the baseline from array A to array B, L is its length, and array A is the origin, the coordinates of the source are given by

$$x = L \frac{\tan \phi_B \cos \eta - \sin \eta}{\tan \phi_B - \tan \phi_A} \tan \phi_A$$
$$= y \tan \phi_A,$$
$$y = L \frac{\tan \phi_B \cos \eta - \sin \eta}{\tan \phi_B - \tan \phi_A} \tag{8.26}$$
$$= \frac{x}{\tan \phi_A},$$

and

$$z = (x^2 + y^2)^{1/2} \tan \theta_A.$$

With a baseline of 40 km it is usually possible to locate only 20–30% of the signals detected by a single array. The percentage is low for three reasons. First, the logic in the circuitry for the mapping system has thresholds that must be exceeded before a signal can be accepted. Since the amplitude of electromagnetic radiation decreases linearly with range, pulses that satisfy thresholds at one array may fall below the thresholds at the other array. Second, the lightning channel segments that generate the signals are anisotropic radiators. Thus again there can be different amplitudes at the two arrays, even if ranges are similar, because the relative orientation of the radiating segment is different. Third, one of the arrays is occasionally processing another signal at the time a signal detected by the other array arrives. Failure to locate the source of a radio signal is usually the result of one of the first two causes.

It is impossible to place absolute confidence in any single, three-dimensional source location, because it is always possible that signals coincident within a reasonable time window are not actually from the same source. This problem increases as longer baselines are used between arrays. However, clusters of located sources have greater reliability because it is unlikely that many locations in a cluster are identified with wrongly associated impulses. In addition, greater confidence can be gained in a particular source by comparing the measured time delay between arrays (assuming that the synchronization analysis is correct) with the time delay that would result from the calculated location.

(3) Short-Baseline, Interferometric Technique: Interferometric techniques have been used for a number of years in radio astronomy and have been adapted by Warwick et al. (1979) to map centroids of VHF sources generated by lightning. In its simplest form an interferometer consists of two identical antennas some distance apart, each connected to a receiver. The outputs of the two receivers are sent to a phase detector that produces a voltage level that is a function of the phase difference between the signals at the two antennas. The phase difference, in turn, is a function of the direction angle of the VHF source relative to the baseline between the antennas. To determine the azimuth and elevation

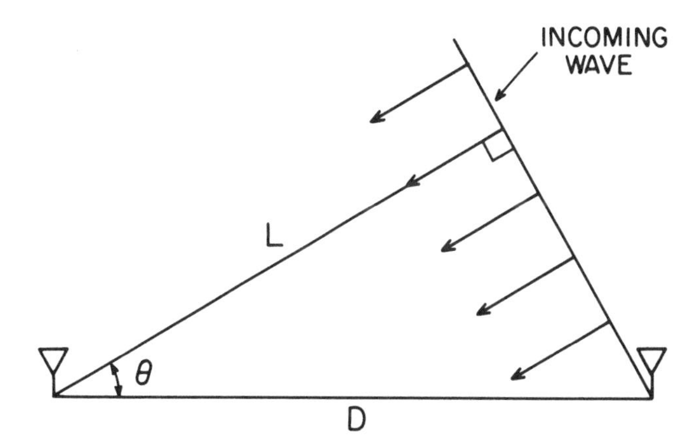

Figure 8.23. Geometry for determining the dependence of phase difference on the direction angle in an interferometer system. D is the baseline; L is the distance that the wavefront propagates while traversing the baseline; θ is the direction angle along which the wavefront propagates.

to a VHF source, an interferometer with two orthogonal baselines is needed, and to locate the source in three dimensions, two separated interferometers are needed.

The dependence of the measured phase difference on the direction angle is determined by the geometry shown in Figure 8.23. For a signal of wavelength λ arriving from a direction angle θ, the phase difference α between two antennas separated by a distance D is given by

$$\alpha = \frac{2\pi L}{\lambda}, \tag{8.27}$$

where

$$L = D \cos \theta. \tag{8.28}$$

Then the direction angle θ is given by

$$\theta = \arccos \left[\frac{\alpha \lambda}{2\pi D} \right]. \tag{8.29}$$

Measurements of α for one to tens of microseconds are averaged to reduce random noise fluctuations.

Since the output of the interferometer is a trigonometric function of the phase difference with a cycle of 2π radians, α cannot be uniquely determined if it varies by more than 2π as L varies from $+D$ to $-D$. Thus Eqs. 8.27 and 8.28 imply that the measurement of α and, hence, of θ is ambiguous for any $D > \lambda/2$. This can be shown explicitly if we rewrite Eq. 8.29 as

$$\theta = \arccos \left[1 - (\alpha_0 - \alpha) \frac{\lambda}{2\pi D} \right], \tag{8.30}$$

where $\alpha_0 = 2\pi D/\lambda$ is the value of α at $\theta = 0°$. As θ in-

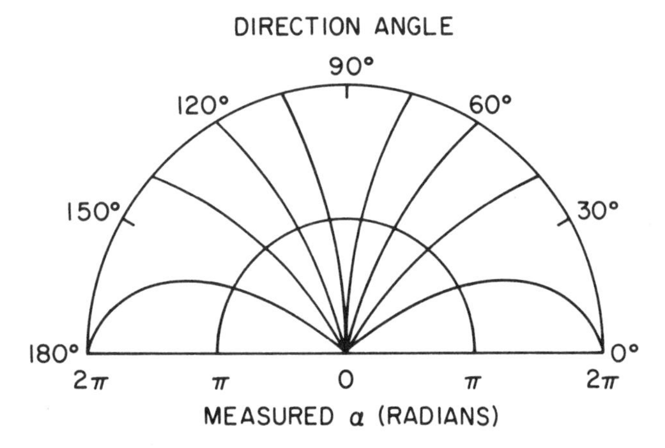

Figure 8.24. Direction angle as a function of measured phase difference α for an interferometer with a 4λ baseline. Note that there are eight direction angles corresponding to a given value of α.

creases from 0°, the measured phase difference α changes until $\alpha_0 - \alpha$ reaches an integral multiple of 2π, at which point the system measures the same phase difference as at $\theta = 0°$. Thus all values of θ such that

$$\theta = \arccos\left(1 - \frac{n\lambda}{D}\right) \qquad (8.31)$$

correspond to the same interferometer output as $\theta = 0°$. Figure 8.24 shows the multiple values of θ corresponding to any given value of measured phase difference for an interferometer with a baseline 4λ long. These cycles in the phase difference are often referred to as fringes. For a given baseline, there are $2D/\lambda$ fringes.

Although the discussion above suggests that a short baseline is desirable, the lessening in number of ambiguities for shorter baselines is countered by a corresponding increase in angular errors. Warwick et al. (1979) assumed that a lightning source generates band-limited white Gaussian noise, and they estimated the resulting error in the direction angle to be

$$\delta\theta = \frac{\sqrt{2}\,\lambda}{\pi D\,(B\tau)^{1/2}}, \qquad (8.32)$$

where B is the bandwidth of the system, and τ is the time over which the phase measurement is averaged (examinations of interferometric data and other unpublished estimates of the error indicate that this equation overestimates the error; however, the functional dependence on D, B, λ, and τ is similar in all estimates). Thus the error is inversely proportional to the baseline length. To circumvent balancing angular ambiguity versus error, two colinear baselines can be used, one with $D = \lambda/2$, which has no ambiguities but poor resolution, and one with a value of D such that the interval of direction angles over which the phase difference makes a complete cycle equals $\delta\theta$ for the $\lambda/2$ baseline.

There are a number of techniques for determining the phase difference of a signal between antennas, and interferometers can be classified by the techniques they use. The interferometer designed by Warwick et al. (1979) is a simple multiplying interferometer. In this system the radio frequency signal received by each of the antennas is amplified and mixed with the output of a local oscillator to produce a signal at an intermediate frequency; the amplitude of the intermediate frequency signal is proportional to the amplitude of the radio frequency signal. The signals from the two antennas are then amplified and multiplied together to produce a voltage level that is directly proportional to the power of the combined signals. This voltage is a function of the relative phase difference and the direction angle, such that for a point source

$$V = K\cos\alpha \qquad (8.33)$$
$$= K\cos\left(\frac{2\pi D\cos\theta}{\lambda}\right),$$

where K is a constant of proportionality.

If analog recording is used with the interferometer, it is expedient to output a time-varying sinusoidal signal instead of a DC voltage level. Warwick et al. (1979) accomplished this by offsetting the frequencies of the local oscillators mixed with each of the antenna outputs by f_0. The relative phase and the output to the recorder then beat at a frequency equal to f_0; i.e.,

$$V = K\cos\left(2\pi f_0 t + \frac{2\pi D\cos\theta}{\lambda}\right), \qquad (8.34)$$

where t is time. As can be seen from this relationship, a result of offsetting the two mixed frequencies is that a stationary point source gives a time-varying output equivalent to that of a moving source for an interferometer in which the mixed frequencies are identical. This apparent motion of a stationary source can also be viewed as equivalent to moving the fringe pattern of an interferometer across the sky along the direction of its baseline. Since the value of t is unknown, the measured phase differences and hence the direction angles are only relative; it is no longer possible to determine direction angles from the output of the system and the length of the baseline alone, although changes in the direction angles can still be determined. To determine the direction to a source, it is then necessary to provide a source of periodic signals at a known location to provide a reference direction for the relative direction angles.

The interferometer described by Hayenga and Warwick (1981) operated at 34.3 MHz and had two orthogonal base lines, each 2λ long and having four fringes. There was not a second site, so all VHF sources were located in only two spatial dimensions. Data were recorded on analog magnetic tape. The local oscillators for each pair of antennas were offset 200 kHz, so the output of the system had a center frequency that was also 200 kHz. The bandwidth was 3.4 MHz, and the signal averaging time was 2.5 μs. From Eq. 8.32 the resulting angular uncertainty for the direction angle to each baseline is 4°. An estimate that appears to be more consistent with the measured data is 2.5°, based on an error of $(B\tau)^{-1}$ radians in the measurement of phase.

Recently the system was redesigned, and two interferometers were built so that three-dimensional locations could be determined. The revised operating frequency of the interferometers is 274 MHz, and the bandwidth is 6 MHz. Each interferometer has baselines along two orthogonal axes. Three colinear antennas are connected on each axis to give two baselines, one $\lambda/2$ long and one 4λ long, so that there is no angular ambiguity in the measured direction angles. The output is averaged and sampled every 1 μs. With these values the angular error calculated from Eq. 8.32 is 2.5°. A more optimistic and, as mentioned in the preceding paragraph, apparently more reasonable estimate of the angular error is 1.5°. Data from each system are recorded on a 28-track, digital, high-density recorder capable of recording 6.25 Mbytes s^{-1} for 15 min.

f. Comparison of Three-Dimensional Lightning-locating Techniques

Each technique for locating lightning can contribute something unique to our understanding of lightning. By the same token, each has shortcomings. The equivalent charge center analysis, for example, is the only technique that estimates directly the net charge involved in lightning processes. If the analysis is performed on sequential segments of the field change, as by Krehbiel (1981), it is even possible to study the development of lightning by studying how the net charge and the coordinates of the charge centroid vary. However, the required electric field measurements are difficult to make, and the analysis gives no indication of lightning channel structure or of the geometry of the charge involved in a particular short segment of the field charge.

On the other hand, the acoustic technique can define in considerable detail the structure of the volume permeated by explosive lightning channels, and measurements for the analysis are relatively easy to make. Furthermore, in the limited number of instances in which an infrasonic thunder signal is received, the location of the infrasonic source corresponds to the location of charge neutralized by the flash (the theoretical basis for this has been discussed by Few, 1985, and Bohannon et al., 1977). However, the acoustic technique generally gives no direct indication of charge magnitude or of the development of a lightning flash, the analysis is fairly time-consuming, and errors are sensitive to atmospheric conditions. There is also some indication that the acoustic technique may be biased against thunder sources at higher altitudes. Although high-altitude sources have been reported by MacGorman et al. (1981) and Rust et al. (1981), Holmes et al. (1980) reported that they observed few thunder sources in the upper region of storms, where there were many strong radar echoes from lightning.

The radar technique is the only technique that actively probes lightning channels instead of passively receiving signals. It provides the location and development of lightning channels directly out to long ranges with accuracy defined by the resolution of the radar being used. Furthermore, since radar responds to ionization, it is possible to study how the ionization of a channel increases and decays. The primary disadvantage of radar is that it observes lightning channels only within its beam. Radars with mechanical antenna drives cannot scan a storm within the time interval that ionization is typically maintained by lightning channels (0.1 – 1 s) unless the angular field of view of the radar encompasses the entire storm. Phased-array antennas used with UHF radars for lightning studies would be capable of extremely fast scanning, but practical design constraints make them marginal for mapping lightning flashes at ranges less than about 100 km. Even if entire lightning flashes cannot be reliably mapped by radar, however, important details on portions of individual flashes, as well as statistical data indicative of characteristics in the whole storm, can be obtained.

Techniques for locating sources of radio noise have a unique capability as a class to observe how lightning develops in time and space throughout a storm. However, some uncertainty persists about precisely what is being observed. Anything that creates a suitable change in the current moment of a lightning channel generates a signal. Proctor (1981) and Hayenga and Warwick (1981) observed that radio sources usually occur in new regions, instead of continuing to be generated in approximately the same place, and so suggested that the sources are created by breakdown processes.

Although the three radio systems for mapping lightning respond to the same parameter of lightning and so have many similarities in their data, there are significant differences too. Proctor's long-baseline system is capable of beautifully detailed maps with excellent accuracy directly over the network. However, the manual correlation of signals that makes the detailed maps possible also makes the analysis very laborious, even with modern instrumentation. Also, because there can be no vertical baseline in Proctor's system, errors become large as elevation angles become small.

Taylor's short-baseline system has good accuracy over the entire hemisphere above the horizon, and useful data can be collected out to a range two to three times farther than with the other two systems. Furthermore, Taylor's system gives direction angles to sources in real time and requires less analysis to determine the location of sources than Proctor's system requires. The chief disadvantage of Taylor's system is that the automatic processing allows only well-defined impulses to be accepted. The system requires that impulse risetimes be fast enough for accurate determination of differences in times of arrival, and it will not process subsequent impulses until the signal level has dropped below the system threshold. Thus the system can process only a single impulse from a train of impulses that has a sustained level above the threshold, whereas the other two systems can process several impulses in the train.

Hayenga's interferometer also does some processing of received signals automatically, but it imposes no restrictions on waveforms and so can process more impulses. Furthermore, analysis of data from the system requires no manual correlation analysis like that for Proctor's system, although the data require more processing than data from Taylor's system. The interferometer system appears to be less accurate than the time-of-arrival systems because of trade-offs involving bandwidth and signal integration time. The poorer accuracy of interferometry is aggravated at low elevation angles by lack of a vertical baseline, although a vertical baseline could be added. However, the system was designed for ranges out to only about 25 km, where sources at low elevation angles are relatively infrequent and poorer angular resolution is adequate. An unpublished criticism of the interferometer has been that if two or more radio sources radiate during a single sample by the interferometer interference between them causes the measured phase and the derived direction angles to have essentially random values. Addressing this criticism, Hayenga and Warwick (1981) argued that simultaneous sources are not a problem in their data; they did not observe obvious random jumps in their reconstructions of lightning development, and it would be possible to detect cases of simultaneous sources by comparing measurements from two colinear baselines and looking for inconsistent amplitudes and direction angles. Two colinear baselines were not available in the earlier system used by Hayenga and Warwick (1981) but are available in Hayenga's new system. Simultaneous sources (occurring within 1–2 μs) may occur more often in storms larger than those that have been studied with interferometers thus far, because lightning can occur much more frequently, and flashes often span greater distances.

g. Selected Results from Three-Dimensional Lightning-locating Systems

With advances in the technology for locating lightning have come corresponding advances in our understanding of lightning and its relationships with other storm parameters. Comprehensive discussion of these advances is beyond the scope of this chapter, but some examples suggest the breadth of the contributions.

Lightning structure had been thought to be predominantly vertical, and this has been found to be the case in some storms (Proctor, 1971; Christian et al., 1980). However, the radar echoes of lightning by Ligda (1956) and more recent results from lightning mapping systems have shown that lightning structure is often predominantly horizontal, sometimes with a channel length of tens of kilometers (Teer and Few, 1974; MacGorman et al., 1981; Taylor, 1978; Proctor, 1981; Rust et al., 1982). Sequential lightning flashes often overlap in space, subsequent flashes extending activity into new regions (MacGorman, 1978; Proctor, 1983). Most measurements of flash progression speeds for regions containing VHF sources within clouds have found the speed to be $1 \times 10^4 - 5 \times 10^5$ m s^{-1} (Taylor, 1978, 1983; Proctor, 1981, 1983; Warwick et al., 1979; Hayenga and Warwick, 1981; Hayenga, 1984; Rust et al., 1982; Mazur and Rust, 1983). However, Proctor (1981), Hayenga and Warwick (1981), and Hayenga (1984) identified a second category of progression speeds of approximately 10^7 m s^{-1} that applies to progression of sources of short bursts of VHF pulses. All estimates of progression speed are average velocities between two points or regions and so underestimate the velocity of channel development, which is characteristically tortuous and may occur sporadically within the time of the average.

Lightning does not occur randomly with height in a storm but tends to occur in a layer spanning a temperature range of approximately 10–20°C. In continental storms this layer is usually somewhere between 5° and −20°C, with a second layer between −20° and −45°C in some storms (MacGorman et al., 1981; MacGorman et al., 1983; Proctor, 1983; Taylor et al., 1984). This is also true of storms along the eastern U.S. coast, though the temperature ranges for lightning activity can be 5–10°C colder (Taylor, 1978; MacGorman et al., 1981; Mazur et al., 1984b). Charge analyses have found that the negative charge lowered to ground by lightning is often distributed horizontally and occurs at temperatures of 5° to −20°C (e.g., Jacobson and Krider, 1976; Krehbiel et al., 1979; Krehbiel, 1981; Krehbiel et al., 1984). Infrasonic source locations, which, as mentioned previously, correspond to the location of charges neutralized by flashes, also tend to be in this layer (MacGorman et al., 1981). MacGorman et al. (1981) interpret these observations of lightning and charge structure as indicating that most channel structure occurs inside regions of charge, not between regions of opposite charge. Support for this interpretation is provided by Krehbiel (1981) in his study of VHF sources and charge locations in intracloud flashes and by Williams et al. (1985) in his experiments on spark propagation through Plexiglas doped in different patterns with electrons. Initial channels for an intracloud lightning flash tend to occur above or in the upper portion of the lower region of lightning activity (Lhermitte and Krehbiel, 1979; Rust et al., 1982; Taylor, 1983; Liu and Krehbiel, 1985; Lhermitte and Williams, 1985), consistent with a picture of these flashes as beginning in a region of high electric fields between the lower negative charge and upper positive charge that roughly characterize the thunderstorm charge distribution. Initial channels for cloud-to-ground flashes appear to occur lower than those for intracloud flashes (Lhermitte and Krehbiel, 1979; Rust et al., 1982; Taylor, 1983).

Lightning appears to be closely correlated with the development of sustained strong updrafts; Jacobson and Krider (1976), for example, found that peak lightning flash rates are much higher in storms whose radar tops reach 14–15 km, and Lhermitte and Krehbiel (1979) found that lightning flash rates increased significantly when updraft speeds deduced from Doppler radar increased rapidly to 15 m s^{-1}. The mapped locations from initial stages of a lightning flash are usually near reflectivity cores (Lhermitte and Krehbiel, 1979; Lhermitte and Williams, 1985; Proctor, 1983), but subsequent development may either stay in the vicinity of reflectivity cores (Taylor et al., 1984; Mazur et al., 1984b; Krehbiel et al., 1979) or move downwind into regions of moderate to weak reflectivity (MacGorman, 1978; Lhermitte and Krehbiel, 1979; Proctor, 1983; Ray et al., 1986), apparently depending on the degree of tilt in the storm structure and the amount of shear with height in the horizontal environmental wind. These observations and the observations by Krehbiel (1981) of vertical dipolar structure in the charge neutralized by intracloud lightning support the hypothesis that positive charge is carried on small cloud particles by the updraft, which transports them away from a lower region of net negative charge.

h. Cloud-to-Ground Strike Location

Systems for locating cloud-to-ground lightning flashes have been available for decades (e.g., Horner, 1954). However, early systems were cumbersome and were not accurate enough for desired studies of relationships between lightning strikes and storms. Two developments greatly enhanced the usefulness of this technology. First, advances in electronics provided fast and reliable communication from remote sites and speeded computations to the point where data could be processed in real time automatically in a compact unit at reasonable cost. Second, Krider et al. (1976) realized that, if the plan location of lightning was measured at the peak in the return-stroke waveform, the location would correspond to a point on the lightning channel within a few hundred meters of ground and nearly directly above the strike point. The accuracy and range of systems that determine locations from this well-defined point in the lightning waveform now compare favorably with weather radar capabilities.

Besides being easier to locate accurately, the ground strike point is the part of a lightning channel that is most important to those concerned with the lightning hazard to life and property. Thus in addition to being used for studies of storms, lightning-strike–locating systems have been used to manage or avoid lightning hazards in forest-fire detection, management of electric-power distribution lines, employee safety, and space-shuttle operations. Data can also be incorporated in climatological studies for engineering design to yield maps of lightning flash density similar to Fig. 6.1 in Vol. 1 (that map, however, is derived from thunderstorm-duration statistics; lightning-strike data were used to determine a relationship between strike density and thunderstorm duration, as described by MacGorman et al., 1984).

Lightning-strike locating systems utilizing either magnetic direction-finding or time-of-arrival technologies have been developed. Both technologies use a network of stations connected to a central analyzer by communication lines. In the magnetic direction-finding network each station determines direction to the strike point, and the central analyzer locates the strike point by triangulation. In the time-of-arrival system each station measures the time at which a lightning signal arrives, and the central analyzer solves hyperbolic equations derived from the measured propagation times between stations to determine the location of the strike point.

(1) Magnetic Direction-Finder System: The primary sensor for the magnetic direction-finder system, described by Krider et al. (1980), is a crossed-loop antenna; it consists of two vertical loop antennas housed in metal tubing to shield them from electric field changes and mounted perpendicular to each other; one loop is oriented north-south, and the other east-west. The ratio of the signal induced by a flash in the east-west loop to the signal induced in the north-south loop is then proportional to the tangent of the azimuth to the flash, if the lightning channel generating the signal is vertical.

A problem with this type of system is that there are large errors when the lightning signal has significant components with a polarity that would result from horizontal current flow in the source. These errors are called "polarization errors" and can be caused by horizontal components in the orientation of lightning channels, by ionospheric reflections of the signal, and by the effects of terrain and structures on signal propagation (referred to as "site errors"). For several years techniques have been available to discriminate against signals reflected by the ionosphere, because they arrive later than the directly propagating signal. Site errors are also usually manageable. They can be as high as 20° for a particular installation, but they are generally less than 5–10° when reasonable care is taken in the selection of a site (e.g., Mach et al., 1986). Furthermore, since site errors are systematic and appear to be constant in time at most locations, it is possible to compensate for them once they are determined.

The remaining source of error, the orientation of lightning channels, is random and so far has been difficult to detect or correct. However, Krider et al. (1980) circumvented much of the problem with channel orientation by gating the system to measure the ratio of signals at the peak of the return-stroke waveform, as mentioned previously. Since the threshold-to-peak risetime is usually a few microseconds for ranges out to 300–400 km and the return-stroke pulse propagates up the channel at roughly 10^8 m s^{-1}, the initial peak is radiated by a part of the return-stroke channel that is about 100 m aboveground. This part of the channel to ground tends to be predominantly vertical; thus the resulting polarization errors tend to be small when measurements of azimuth are made at the peak of the waveform. Uman et al. (1980) showed that errors from nonvertical channels near the ground are generally less than 1° for lightning at ranges greater than 10 km.

The system described by Krider et al. (1980) receives

frequencies up to several hundred kilohertz but filters frequencies above 100 kHz to remove noise from radio stations. A number of tests are applied to magnetic field waveforms to ensure that accepted waveforms are from return strokes and that signals from intracloud processes are rejected. Criteria for the waveforms include a maximum time to peak, a minimum pulse width, a maximum amplitude for precursor signals, and a maximum amplitude for following peaks. For example, for flashes lowering negative charge to ground, a typical minimum width criterion would be 10 μs, and a typical maximum amplitude for subsequent peaks of opposite polarity would be 50% of the initial peak. Criteria are modified automatically, depending on the polarity and amplitude of the initial pulse and on its position as a first or subsequent return stroke.

In its present form the magnetic direction-finder system consists of two subsystems, the direction finder and the position analyzer. The direction-finder subsystem consists of a crossed-loop antenna, a flat-plate electric field antenna, and analog and microprocessor electronics. The electric field antenna is used to determine the polarity of charge lowered by a flash and so remove the 180° ambiguity that is inherent in azimuth measurements from the crossed-loop antenna when current direction is unknown. The microprocessor tests the magnetic field waveforms to reject all but return-stroke waveforms, digitizes the amplitudes at the peak of the waveform from each loop, determines azimuth to the ground strike, and transmits strike data to the position analyzer.

If a signal arrives within a preset window (about 180 ms) of the last accepted signal, the direction finder evaluates it as a subsequent stroke, and the two azimuths are compared to determine whether they are within a preselected azimuthal window. If the azimuths are consistent, the signal is considered to be a subsequent stroke of the same flash. If not, the signal is considered to be from the first stroke of a different flash and is held in memory for comparison with subsequent signals. The direction finder can simultaneously process up to five flashes with interleaved return strokes, up to a maximum of 14 strokes per flash.

The position analyzer is the central processor of the locating system and receives strike data from two or more direction finders to locate ground strikes. Two versions are available. In the so-called point-to-point version the position analyzer monitors the status of each direction finder over a dedicated communication line and accepts flash data sent in real time, assigning a corrected time to each flash when the data are received. In the second version, referred to as the "multidrop version," the position analyzer periodically polls the direction finders, and several direction finders share a single communication line. During each polling of a multidrop direction finder, the position analyzer samples the clock to check for drift and reads the flash data stored in memory since the last polling.

In both versions the direction-finder data are checked to see whether a flash has been detected by two or more direction finders within a preset time window (typically 20 ms or less), and if it has, a strike location is computed. For each ground strike that is located, the system records the time, the calculated coordinates of the strike, the amplitude and polarity of the peak in the waveform, the number of strokes per flash, and the direction-finder sites that detected the flash. In addition, the system can record parameters measured by each site, including amplitude, number of strokes, polarity, and the azimuth from the site to the strike.

A common installation of the magnetic direction-finder system consists of a single-position analyzer receiving data from three or more direction-finder subsystems in an array with baselines of 10–250 km, depending on the sensitivity chosen for the units. In a normal calculation of strike location the position analyzer uses spherical trigonometry to triangulate azimuths from the two direction finders with the largest signal amplitudes, which should correspond to the two sites closest to the flash. If a ground strike lies near the baseline of these two sites and a third station also detects the flash, the azimuth from the third station is used with that having the largest amplitude to calculate the strike location by triangulation. If, however, there is no third azimuth, range to the strike location near the baseline is calculated from the measured signal amplitudes by utilizing the r^{-1} dependence of amplitude on range.

The accuracy of the locating system has been examined both by comparing strike azimuths from the system with azimuths determined from television video recordings of channels to ground and by intercomparing azimuths from all strikes in multistroke flashes (recorded by a special software option for the direction finder). In both types of analysis standard deviations for the difference in determinations of azimuth have been 1–2°. The difference between the measured azimuths and the azimuth of the calculated locations appears to have a standard deviation of 1° or less if a least-squares optimization scheme is employed to make use of all the redundant measured azimuths and amplitudes instead of only the azimuths from the two sites closest to the flash.

The detection efficiency of the system has also been measured by comparing its data with television video recordings at installations in Florida, Oklahoma, and Arizona. These locating systems operated either at medium sensitivity (a 50% detection efficiency is expected at a nominal range of 200 km) or at high sensitivity (nominal range, 400 km). Within the appropriate nominal range, reported percentages of ground truth locations that are located by the system have ranged from 65% to 90%, 70–80% being typical. The decrease in detection efficiency with range, which is expected beyond a few tens of kilometers, is not sufficient to explain completely the variation in results from these studies. The variation may be caused in part by differences from installation to installation, as in background noise, or by particular storms possibly having a higher percentage of ground-flash waveforms that fail the criteria of the system.

(2) Time-of-Arrival System: A time-of-arrival system for locating lightning ground strikes was developed in the early 1980s. This system, described by Bent and Lyons (1984), locates strike points by measuring the differences in time of arrival of a return-stroke signal at three to six stations and then solving to locate the intersection of the corresponding hyperbolas on a spherical surface. If only three stations are available, the solutions are ambiguous in some areas surrounding the network; four stations are required to determine solutions uniquely everywhere. Baselines between stations vary from 10 to 35 km for low-gain systems and from 150 to 400 km for high-gain systems. Stations are connected to a central analyzer facility by communication links.

Each station consists of a vertical antenna to receive reference timing signals and sense electric field changes from lightning, a time-signal receiver, a time-signal generator, and a lightning-stroke detector. To maintain good locational accuracy, the time-signal generators from the different stations should be synchronized to within a fraction of a microsecond. For this purpose they are synchronized to a common reference timing signal, usually the ground wave from a single Loran-C navigational transmitter. Bent and Lyons (1984) reported that synchronization can be maintained between stations to within a few tenths of a microsecond.

The lightning-stroke detector appears similar in function to the corresponding part of the magnetic direction-finder system in that the peak in the received waveform is used for the measurement of location. However, the system samples a clock to determine time of arrival of the peak, instead of measuring direction to the source of the peak, and there is no waveform testing to evaluate whether the signal is from a cloud-to-ground return stroke. Instead, the system requires signals to exceed two amplitude thresholds and relies on the requirement that lightning signals must be detected at multiple stations to eliminate intracloud flashes (signals from intracloud processes tend to lie in higher-frequency bands, so their signals do not propagate as far as signals from ground flashes).

In the time-of-arrival system errors in the computed locations are introduced by anything that affects the determinations of time of peak in the return-stroke waveform (the error in location is a complicated function of range, azimuth, and timing error, and adequate theoretical treatments are not yet available in the scientific literature). Random errors are caused by noise affecting either the synchronization of the time-signal generator or determination of the peak in the return-stroke waveform, or by sampling errors in assigning a time from the time-signal generator to the peak. Such random timing errors appear to be 1 μs or less.

Systematic errors are caused primarily by propagation effects either on the reference timing signal or on lightning waveforms. Dominant propagation effects for the time-of-arrival system are deviations in path length from the length over a spherical Earth and a lengthening in risetime caused by attenuation of higher frequencies. Bent and Lyons (1984) estimated the effect of intervening mountains 1850 m and 3700 m high on path length with transmitter-to-receiver distances from 74 to 370 km. The propagation delays that resulted from the increases in path length were 0.25–1.5 μs.

Lengthening in risetime is a function of the path length and the conductivity of the terrain. A difference in the risetime in lightning waveforms received at two stations translates directly to an error in the measured time interval for propagation between them. Uman et al. (1976) reported that propagation over a path length of 200 km in Florida, where terrain is mostly swamp or farmland, resulted in an average increase in risetime of approximately 1 μs. Longer propagation paths would result in correspondingly slower risetimes. Cooray and Lundquist (1983) reported also that common conductivity differences for various land types could cause differences in risetimes of more than 1 μs for different 200-km propagation paths. Thus, if the path length to one station is longer than the path length to another station, the risetime at the more distant station will tend to be longer, and if the conductivity of terrain is poorer over one path than another (e.g., rock versus farmland, or land versus water), the risetime over the path with poorer conductivity will be longer.

If the effects of propagation were measured over a particular path, the time of the peak in the waveform could be corrected accordingly, since many of the effects are relatively constant in time. Although such measurements would be difficult, the correction for the Loran-C signal would involve only a single propagation path to each station. For lightning signals, however, the locating system cannot know the length or direction of the propagation path a priori; therefore, any such correction would have to be iterative and would have to account for significant differences in topography and conductivity in the vicinity of the network. It thus appears that it would be impractical to correct for the effects of propagation on lightning signals.

It appears that systematic errors from propagation effects are as large as or larger than random errors and would be difficult to correct. As a result, the random scatter in locations for multiple strokes in a lightning flash from a time-of-arrival system does not give an adequate indication of system accuracy, and even internal consistency of redundant solutions is not necessarily an indicator of accuracy. Evaluation will depend on experimental comparison with independently located lightning strikes and on a theoretical analysis of the locational errors that result from an appropriate mean error in time-difference measurements.

i. Possible Future Mapping Techniques

Several areas of future sensor development are indicated by present technologies. Such development is a matter of commitment and funding rather than more sophisticated technology. For example, a long-wavelength radar, such as a UHF system with a phased array antenna, would allow rapid scanning of storms and provide data for three-dimensional mapping of lightning within storms at ranges of ≥ 100 km.

Another example involves satellite-borne lightning mapping. Available optical technology can detect lightning from geosynchronous altitudes during daylight and darkness over nearly the full visible disk, with spatial resolution of 10 × 10 km and with timing resolution of 1 ms (Davis et al., 1983). Such a system would provide mapping information on intracloud lightning as well as cloud-to-ground lightning and would allow lightning and storms to be studied on large scales. As has been reported by Robertson et al. (1984), lightning measured from space could be examined relative to storm-top development, growth rates, precipitation, and storm and larger-scale wind fields. Thus were such an instrument to be flown on a satellite and the data made available routinely to the scientific and operations communities, it would increase our understanding of such topics as the global electrical circuit, relationships between storm development and electrical activity, and thunderstorms over ocean areas, while providing information for improved forecasts and nowcasts of storm and lightning activity. Undoubtedly such data would introduce possibilities for research and applications now unforeseen.

9. Observation and Measurement of Hailfall

Griffith M. Morgan, Jr.

1. Introduction

The development of adequate measurement techniques is an important early milestone in any scientific endeavor. Before that stage a science consists of little more than a collection of stories and impressions. The study of hailfall sits, at the moment, astride the measurement milestone. It is only since the last part of the preceding century that networks of specifically instructed observers have attempted organized observations of hailstorms in Europe, and less than 30 years since the first networks of hail-observing instruments were brought into use in North America.

Most modern data on hailfall have come from a few research groups that have established the necessary extensive field observing networks. Such groups have been in Alberta, Canada; Pretoria, South Africa; Illinois; and Colorado. As hopes rose that modern science and technology might develop means for reducing damage caused by hail, additional countries supported projects aimed at hail suppression, and the First International Workshop on Hailfall Measurements was held at Banff, Alberta, Canada, in October 1977 (Goyer, 1978). The major field program conducted by the National Center for Atmospheric Research during the 1970s is the subject of a major work by Knight and Squires (1982), which includes a brief history of hail studies and of the planning process that attends modern efforts involving cooperation among many agencies and (often) governments.

During the late 1970s a major program in Switzerland, including participants from France and Italy, tested during five storm seasons a method for hail suppression that had been suggested by scientists of the Soviet Union. The comprehensive report of the project (Federer et al., 1986) is highly informative with respect to all of the basic concepts, deployment of observational tools, and data analysis.

Complete description of a hailfall would include the parameters for each stone falling on a representative area: time and place of arrival, size, shape, ice and liquid-water content, content and distribution of air bubbles and foreign substances, hardness, crystal structure, velocity, rotation, and perhaps more. In most hail research, much less complete information is required, and only a few parameters suffice to describe an entire hailfall at a point. For example, the total kinetic energy of the hail integrated during a hailfall event has been used as the primary measure for hail-prevention experimentation (Schleusener and Jennings, 1960) because of a presumed relationship between kinetic energy and crop damage. The National Hail Research Experiment (NHRE) based its hailfall documentation on measurements of the time-integrated volume (or mass) of hail owing to theoretical arguments about the effect of cloud seeding on that parameter. Results of hailfall measurement studies carried out within the NHRE were presented by Morgan (1982).

2. Hail Observation by Conventional Weather-Station Networks

Hail is observed and reported by the weather-observing networks that furnish information on which the familiar weather forecasts are largely based, and much of the readily available climatology of hail is based on these observations (Stout and Changnon, 1968). Although the stations are approximately 25 to 100 miles apart, depending on the region, so many hailstorms occur each year that some of them are observed at each station. The value of the stations' hail reports lies primarily in the length of the observations (50 to 80 years at some stations). To show their importance it is necessary to explain the relationship between point hail frequency and areal hail frequency.

The average number of times hail occurs at a given weather station is the point hail frequency. Though it does not directly define the number of hailstorms taking place in the vicinity of the station, it is an indicator of generalized hailstorm frequency. By comparing point frequencies with hail data from other sources, such as hail insurance reports and observer networks, it is possible to determine the average relationship between the local point frequencies and the frequency of hail in areas of specified sizes. Figure 9.1 shows the point-area relationship based on data from Illinois and one point in Colorado, and similar relations derived from data for Alberta and South Africa. The curve for Illinois and Colorado approximates a straight line; South Africa and Alberta show higher area-point ratios for equal

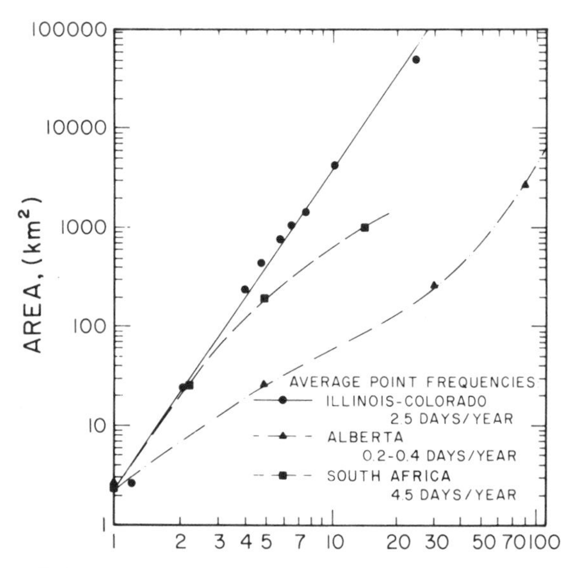

Figure 9.1. Relationship between areal hailfall frequency and point frequency for Illinois-Colorado (Changnon, 1971b), Alberta (Douglas, in Hitschfeld, 1971), and South Africa (Carte and Held, 1978).

areas. The differences between these curves should be due to differences in the characteristic sizes of hail areas (and their variations) from place to place, but probably reflect effects due to nonuniformities of the point hail frequency within the areas studied and ordinary sampling problems. The large-scale point frequency distribution in Illinois is rather uniform, with point values generally between 2 and 3 days per year. Both the Alberta and the South Africa areas contain both high and low average frequencies within the areas of study. Maps of point frequency of hailfall such as those found in Stout and Changnon (1968) are valid and useful for comparing one area with another. Point frequencies reliably indicate regions of high or low hail incidence.

3. Hailfall and Crop Damage

Crop damage is the most important economic impact of hailfall. Much has been learned about hailstorms by studying damage to crops, but there are some important complications to be understood to interpret such information.

Hail damages crops because hailstones are fast-moving missiles. Each stone has some degree of potential for puncturing or removing leaves, breaking stems and branches, crushing stalk fibers, and smashing fruit. This damage potential depends on the hailstone itself (mass, shape, hardness, and speed), the plant that is exposed to it (type of plant, stage of growth, mode of planting, and condition), and other factors such as wetting by previous or concurrent rain. The wind that accompanies hail also has a strong influence on damage primarily because it can impart a velocity component to the hailstones several times that acquired through free fall alone. Added horizontal velocity increases the path length of the stone through the crop and, hence, the probability of its striking a plant and increases the number of stones that have enough energy to do damage. The wind can also cause the plant to present a different cross section to the hail than it would in the absence of wind. A very impressive fall of hail might do little or no damage to a crop if the wind is very light; a lesser fall of hail might completely destroy it if accompanied by a strong enough wind. An observed increase in crop damage from one side of a field to another could be due to an increase in the same sense of either hail or wind.

Lack of wind information has hindered efforts to determine relationships between crop damage and hailfall characteristics such as stone sizes and energies (Changnon, 1971a; Garcia et al., 1976; Morgan and Towery, 1976a).

a. Hailfall Inferred from Kinetic Energy

Most attempts to express hailfall parameters in terms of equivalent crop loss have been based on the assumption that damage is a simple function of the sum of the kinetic energy of the hailstones falling on a unit area. That energy, for an arbitrary hailfall in the absence of wind, is

$$KE = 1/2\Sigma m_i v_i^2, \qquad (9.1)$$

where m_i and v_i are the masses and fallspeeds of individual stones striking a unit horizontal area. Since stones are assumed to fall at their terminal fallspeeds, given by $v_i = k\, d_i^{\frac{1}{2}}$, where k is a constant and d_i is the diameter of the ith stone,

$$KE = \frac{\pi}{12} k^2 \rho_i \Sigma d_i^4. \qquad (9.2)$$

This analytical approach to hail damage has been further developed by Morgan and Towery (1976b), who have shown how a windblown hailfall can be divided into its component fluxes of kinetic energy for analysis. With no wind, there is only a vertical flux of vertically directed kinetic energy. When wind blows with hail, five new flux components can be defined, each of which can be greater than the no-wind energy flux.

Even if we neglect the complications owing to wind, the damage is actually a function of more than the kinetic energy. First, the geometry of the plant must be taken into account. Second, although the largest stones carry the most energy, they are few (and for that reason alone may not play a strong role in crop damage), and their numbers are not well estimated. The damage will be more dependent on some intermediate-sized stones that are both great in numbers and large enough to do damage (Morgan and Towery, 1976a).

An example of an attempt to relate hail kinetic energy to

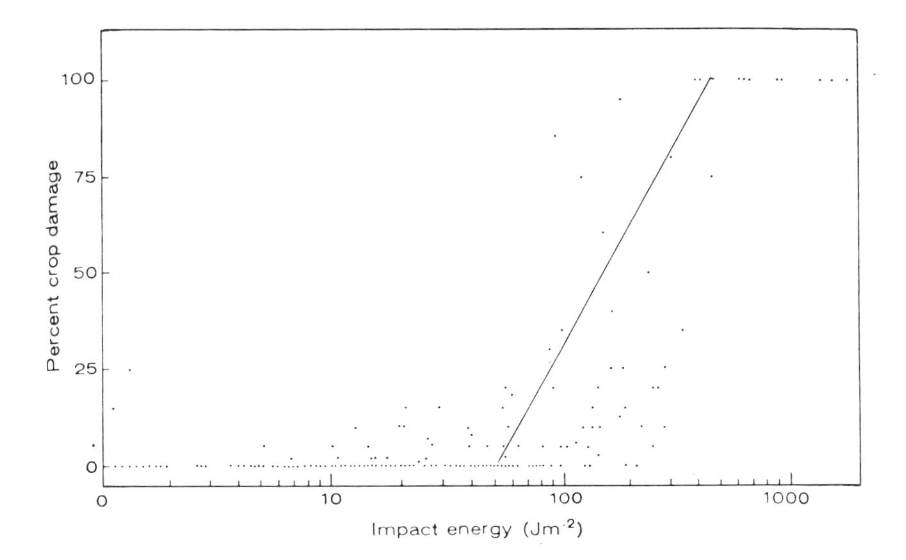

Figure 9.2. Crop damage in Alberta as a function of hailfall kinetic energy (Strong and Lozowski, 1977).

crop damage is seen in Fig. 9.2. This shows that there is a threshold of about 50 J m^{-2} below which little or no damage is done to crops and that above about 450 J m^{-2} the crops are totally destroyed. Between these limits, however, the actual data show no regular relationship between hail kinetic energy and crop damage.

Vento and Morgan (1976) used hailcube (see Sec. 5 below) measurements to estimate the enhancement of hailfall kinetic energy by wind. The kinetic energy contributed by wind exceeded that owing to fallspeeds alone in 33% of 524 observations made in the Po Valley of Italy and in almost all 88 measurements made in hailstorms in Nebraska.

Wojtiw and Renick (1973) used both hail and wind observations to determine the probability of damage. From the reports of wind (strong, moderate, etc.), hailstone size, ground coverage, and crop damage, they estimated the total (fallspeed plus wind) impact energy (KE) received by a unit horizontal surface. Table 9.1 shows their results for the probability of damage of four classes of intensity as a function of the total impact energy.

Table 9.1. Relationship Between Total Impact Energy and Hail Damage in Alberta

Total Impact Energy (J m^{-2})	Probability (%) of Damage Category Occurring			
	Nil	Slight	Moderate	Heavy-Severe
$<10^{-2}$	98.8	0.3	0.0	1.2
10^{-2}–$<10^{-1}$	94.2	4.3	1.4	0.0
10^{-1}–$<10^{0}$	81.2	16.1	1.1	1.7
10^{0} –$<10^{1}$	67.8	19.1	10.1	3.0
10^{1} –$<10^{2}$	46.2	29.6	14.8	9.4
10^{2} –$<10^{3}$	23.0	23.3	27.8	25.9
10^{3} –$<10^{4}$	12.7	13.0	23.9	50.4
10^{4} –$<10^{5}$	10.6	6.4	6.4	76.6

b. Hailfall Inferred from Insurance Records

Insurance records of crop loss are an invaluable source of hailfall information. CHIAA (1978) gives a detailed account of the terminology and technical characteristics of crop hail reports. Without a full understanding of these, one could acquire some very misleading ideas about hailfall from using insurance information. Not all hail produces damage, and the hail area shown by insurance data will be smaller (at times considerably) than the area over which hail actually fell. Insurance against hail normally includes a deductible clause whereby no payment is made for damage less than a certain percent, and this can further reduce the inferred hail area.

Crop hail insurance reports represent an almost continuous distribution of observers at times when all farms have hail-sensitive crops, but there must be crops to experience damage for insurance claims to reveal the occurrence of a hailfall.

Figure 9.3 shows the average number of haildays per month in a 26,500-mi^2 (67,800-km^2) area of central Illinois from 1967 to 1973 both with and without the incorporation of hail insurance data. In March, when no crops are exposed and susceptible, the insurance data add nothing, but in June, when crops are susceptible, they double the number. This illustrates simultaneously the effect of high observation density on the number of observed hail days and a shortcoming of insurance data as a source of information for the researcher. When insurance data are most effective, it seems (in the example of Fig. 9.3) that the monthly number of hail days is doubled. On this basis one may deduce true peaks of about 15 haildays per month in April and May.

An insurance adjuster generally makes estimates of one

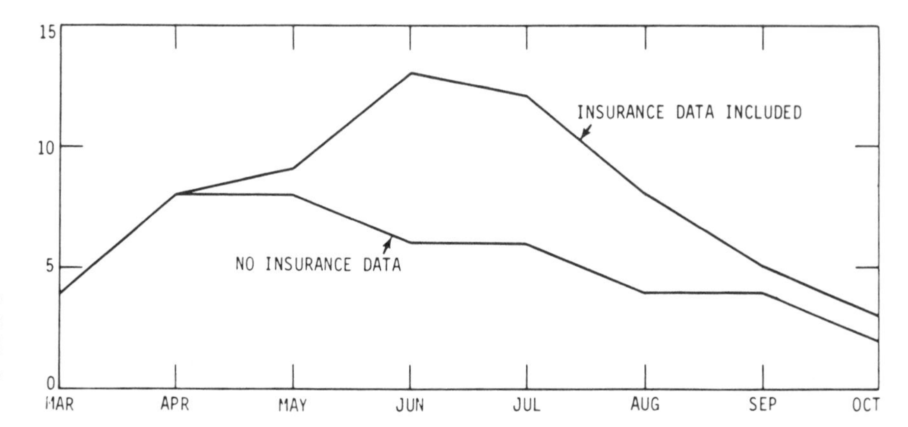

Figure 9.3. The average number of haildays per month in a 26,500-mi² (67,800-km²) area of central Illinois, 1967–73, with and without reports of insurance losses (Changnon and Morgan, 1976).

or two types of physical damage to the crop, such as percentage of leaf loss and percentage of plants completely destroyed and then, with the aid of empirically derived tables and graphs, estimates the percent loss of crop yield. These forecast losses are at times seriously in error. The relationship between these forecast values and the actual physical damage depends on factors (potential for plant recovery, for example) not related solely to the physical damage. Plants go through stages in their development during which they are sensitive in varying degrees to damage by hail. For instance, large leaf loss to corn may have a strong influence on net yield at one stage of its life and none at all at a later stage.

c. Hailfall Inferred from Aerial Photography of Crop Damage

An interesting observation technique, which within very few trials led to some surprising new insights about hail patterns, is aerial photography of damaged crops. Changnon and Barron (1971) employed aerial photography with both infrared and conventional color film in an exploratory study and found that with careful interpretation damage areas could be well delineated. A study by Towery et al. (1976) examined the utility of aerial photography using false-color infrared film as an aid to hail insurance companies in storm assessments. The results were encouraging enough that the sponsoring group uses the technique on a quasi-routine, trial basis. Some of the insights about surface hailfall that have come from studies of aerial photographs are described in Vol. 2, Chap. 11.

4. Hail Data from Observers' Reports

Collation of witnesses' accounts probably gave the first indications of the strange and variable patterns of hailfall. Studying such accounts is still an exciting and valuable way to learn about hailfall patterns, as well as people's attitudes and perceptions of hailstorms. The use of large networks of trained observers, however, has produced greater scientific knowledge of hailfall.

Networks of observers have been organized over large areas since late in the last century. In 1874 a network of postal-reporting observers, augmented by hail insurance adjusters, military volunteers, and railroad employees, was established to cover northern Italy (Pini, 1885). Until World War I, it furnished valuable documentation of thunderstorm motions (speeds and directions), times of occurrence, sequences of events, hailfall characteristics (stone sizes, types, numbers), and hailfall patterns. Figure 9.4 shows reconstructions of some thunderstorm tracks and isochrones in 1879 based on the postal reports. A similar effort was made in England in 1887 (Marriot, 1892), and still later a notable observer network involving weather stations and school principals was used to study hailstorms in Austria (Prohaska, 1905, 1907). Large networks of postal-reporting observers were established in Canada (Douglas and Hitschfeld, 1958), Illinois (Wilk, 1961), Colorado (Beckwith, 1960), South Africa (Carte and Held, 1978), Italy (Morgan, 1973), and Oklahoma (Nelson and Young, 1978). Figure 9.5 shows a representative reporting card.

Careful monitoring and studying of the reports can determine the reliability of each observer (see Carte et al., 1963). The observer reports of hailfalls are subjective, and certain average biases must be considered for them to be useful. Douglas (Carte et al., 1961) noted that observers tend to round off times to the nearest 5 minutes. Changnon (1971a) also showed that observers tend to overestimate sizes of small hailstones but, on the average, estimate the size of the larger stones well. Durations were greatly overestimated (typically by 10 minutes) by observers, compared with durations derived from rain gages that indicate the times of hailfall (Changnon, 1966).

5. Hailfall Measurement Instruments

Difficulties encountered in surface hail measurements derive from these: (1) rarity of hail occurrence, so that highly

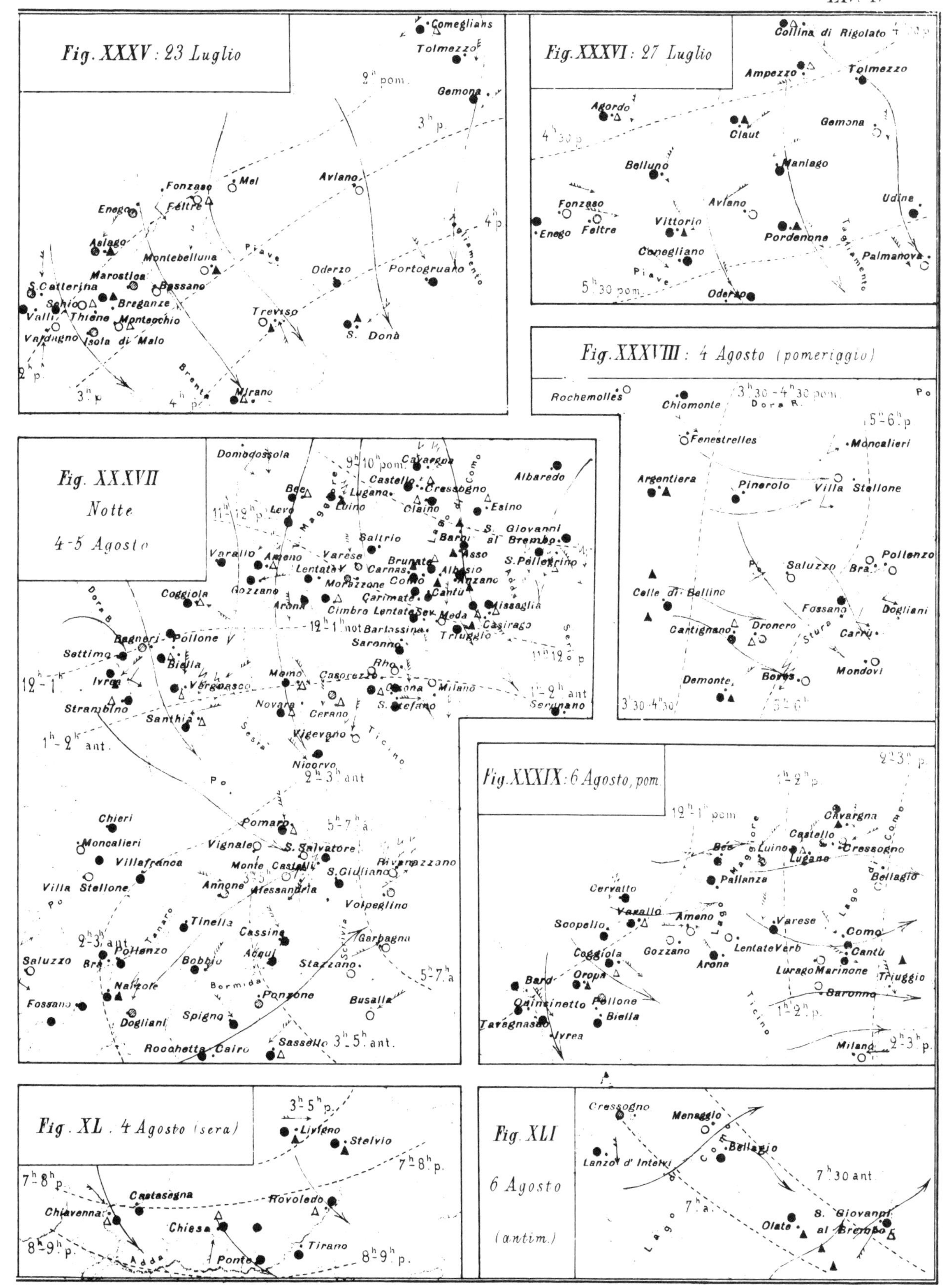

Figure 9.4. Analyses of postal observers' reports of thunderstorms and hail in northern Italy in 1879. Dashed lines are isochrones; arrows indicate direction of propagation; lightning, rain, and hail symbols are like those in current usage. Barbed arrows indicate winds with storms.

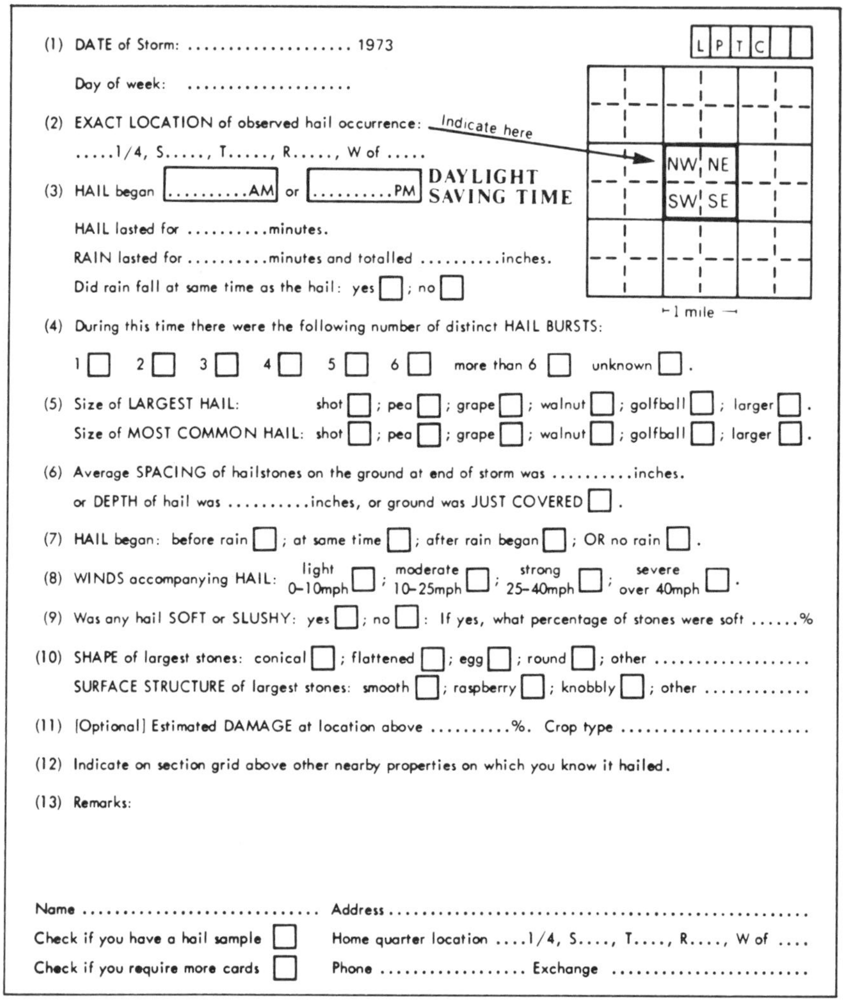

(1) DATE of Storm: .... 1973 L P T C

Day of week: ....

(2) EXACT LOCATION of observed hail occurrence: Indicate here

.....1/4, S....., T....., R....., W of .....

NW NE
SW SE

← 1 mile →

(3) HAIL began ....AM or ....PM DAYLIGHT SAVING TIME

HAIL lasted for ....minutes.

RAIN lasted for ....minutes and totalled ....inches.

Did rain fall at same time as the hail: yes ☐; no ☐

(4) During this time there were the following number of distinct HAIL BURSTS:

1 ☐ 2 ☐ 3 ☐ 4 ☐ 5 ☐ 6 ☐ more than 6 ☐ unknown ☐.

(5) Size of LARGEST HAIL: shot ☐; pea ☐; grape ☐; walnut ☐; golfball ☐; larger ☐.
Size of MOST COMMON HAIL: shot ☐; pea ☐; grape ☐; walnut ☐; golfball ☐; larger ☐.

(6) Average SPACING of hailstones on the ground at end of storm was ....inches.
or DEPTH of hail was ....inches, or ground was JUST COVERED ☐.

(7) HAIL began: before rain ☐; at same time ☐; after rain began ☐; OR no rain ☐.

(8) WINDS accompanying HAIL: light 0–10mph ☐; moderate 10–25mph ☐; strong 25–40mph ☐; severe over 40mph ☐.

(9) Was any hail SOFT or SLUSHY: yes ☐; no ☐: If yes, what percentage of stones were soft ......%

(10) SHAPE of largest stones: conical ☐; flattened ☐; egg ☐; round ☐; other ....
SURFACE STRUCTURE of largest stones: smooth ☐; raspberry ☐; knobbly ☐; other ....

(11) [Optional] Estimated DAMAGE at location above ....%. Crop type ....

(12) Indicate on section grid above other nearby properties on which you know it hailed.

(13) Remarks:

Name .... Address ....

Check if you have a hail sample ☐ Home quarter location ....1/4, S...., T...., R...., W of ....

Check if you require more cards ☐ Phone .... Exchange ....

Figure 9.5. A typical observer's hail-reporting form, from Alberta.

dependable instruments and large study areas are required, (2) great horizontal variability of hailfall parameters, which places demands on the sensing areas and numbers of instruments, (3) departures of hailstone properties (sphericity, hardness) from the ideal, and (4) complications introduced by wind.

The instruments for recording and measuring hailfall now in use are all of quite recent origin. Reviews of hail instruments have been presented by Towery et al. (1976) and Nicholas (1978).

a. Integrating Sensors

There are several types of integrating hail sensors. The most widely used sensor, the hailpad, has been deployed in field programs in Montana, Colorado, Illinois, Alberta, Italy, France, Switzerland, and South Africa.

Hailpads are very simple instruments that can be fabricated in large numbers from inexpensive materials. In most field studies the hailpads have been slabs of Styrofoam wrapped in aluminum foil, which are mounted horizontally on special supporting frames that rest on the ground or are fixed to fence posts (Fig. 9.6). The mechanical properties of the materials used have varied somewhat from one project to another, and each user has had to perform his own calibration. The problems and minor controversial points regarding calibration (Vento, 1976; Lozowski and Strong, 1978) are not essential to this discussion. Descriptions of various combinations of materials used in hailpads can be found in Schleusener and Jennings (1960), Towery et al. (1976), Roos (1978), and Lozowski et al. (1978).

G. G. Goyer (personal communication) experimented with pressure-sensitive paper backed with a hard Masonite surface as a potential replacement for the hailpad. This sensing surface worked well for hard, plastic spheres but is useless for recording real hailstones because of their variable shapes and hardness. He also constructed a device intended to sort hailstones by size, which functioned well with spherical test objects but poorly with actual hailstones.

There are distinct advantages and disadvantages to the use of hailpads. The advantages are that they are inexpensive, easy to service and work with, and provide the following information:

1. Hail occurrence (yes-no).
2. The number of hailstones ($\geq$0.1-cm diameter) per unit area, determined by counting dents. The upper limit on number is about 10^4 stones per square meter; beyond this number it becomes difficult to distinguish individual stones.
3. An estimate of stone size, made by measuring dent size.
4. An estimate of the vertical kinetic energy and momentum imparted by hailstones (based on observed dent size and theoretical values of fallspeed).
5. An estimate of the accumulated mass of hail.
6. The areal extent of hail, usually on a hailday basis.

If data are collected over a period of years, the above information can provide valuable climatological information about hail occurrences, which is important in evaluating hail-suppression projects.

Disadvantages of hailpads include the following:

1. All hail falling between pad changes will be recorded. Should more than one hailfall occur, the hail events cannot be separated.
2. Stone-frequency errors occur if stones cause overlapping dents, a major problem with very large numbers of stones.
3. No information is obtained about windblown stones.
4. No information is obtained on the time of hail, a distinct problem in the study of individual storms. An attempt has been made to overcome this problem by installing hailpads near weighing-bucket recording rain gages that have been altered (evaporation funnel removed) so that hailstones fall directly into the bucket, deflect the pen arm, and leave a spike on the rainfall trace. This spike can give a reasonably accurate estimate of time and duration of hail (Changnon, 1966), but some judgment is required to identify spurious signatures that occasionally occur because of strong, gusty winds.

Another variation of the hailpad had been adapted for quantitative analysis of windblown hail. Figure 9.7 shows a hailcube, a five-sided hail sensor, which has been distorted for ease of viewing. It is an adaptation of a device extensively employed in Italy. Vertical and azimuthal angles of hailfalls are obtained by means of a relationship between the numbers of stones that occur on the vertical and horizontal sensing surfaces (Morgan and Towery, 1976b; Vento, 1972). The horizontal components of hailstone velocity, momentum, and kinetic energy can also be calculated. A minor disadvantage is transporting five times as many hailpads as required by the single hailpad technique. Hailcube analysis, described in detail by Morgan and Towery (1976b), typically requires about 20% more time than that required to analyze single horizontal hailpads. Hailcubes have been employed in field projects in the United States, Canada, and Western Europe; in the late 1970s a network of 300 hailcubes was installed in support of a hail-prevention project near Pecs, in southern Hungary (Endre Wirth, director, Hungarian Hail Prevention Project, Pecs, Hungary, personal communication).

Figure 9.6. A typical hailpad installed in the field.

b. Recording Instruments

The need for more sophisticated hail data such as hailstone arrival times has led to the development of several recording hail gages.

The South Dakota School of Mines and Technology developed a geophone hail gage (Fremstad, 1968; Johnson and Smith, 1978). It consists of a 15-cm diameter plate that transfers the hailstone impact momentum to a sensor of large mass. Electronic circuits derive an electrical pulse from a geophone attached to the mass. The pulse amplitude is a function of the hailstone momentum.

The geophone hail gage gives information on size, number, and time of occurrence of hailstones, covers a reasonable range of vertical momentum, gives a unique and reproducible output of momentum values, operates on batteries, and records the data on magnetic tape (resulting in reduced analysis time and cost). Time distribution and size provide information useful in assessing hail-suppression efforts and hailstorm characteristics. The primary disadvantage to this hail gage is the small sensing area, which may yield a some-

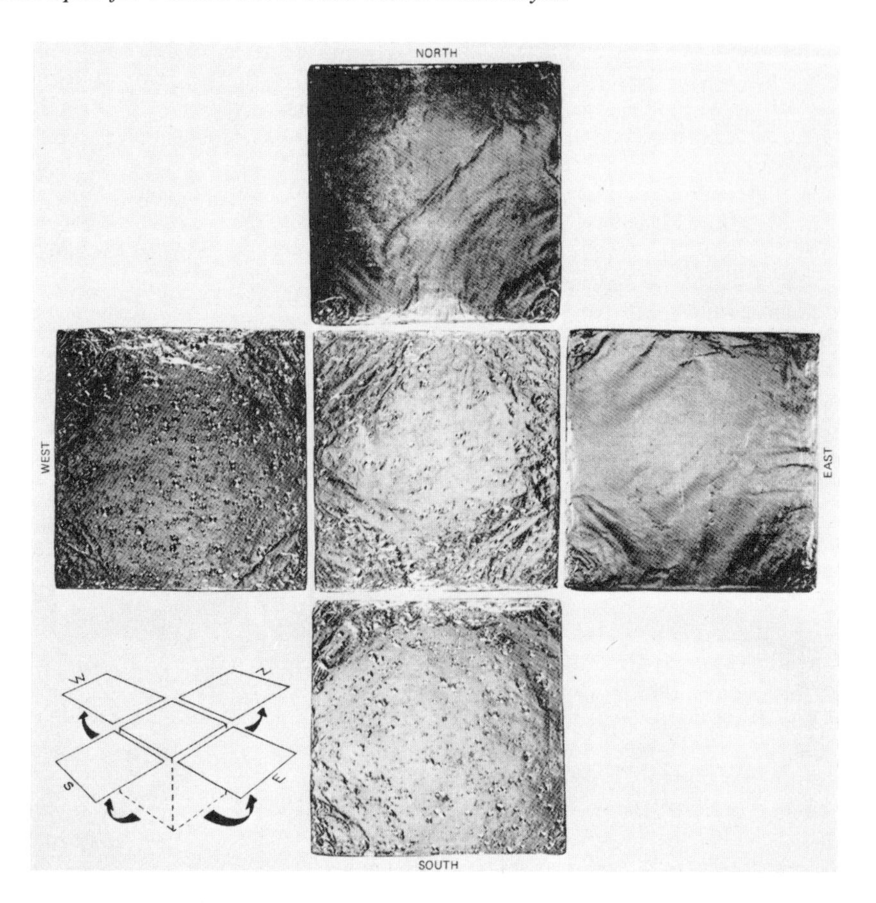

Figure 9.7. The five-sided hailcube. Example from a hailstorm in southwestern Nebraska on 22 July 1973.

what unrepresentative stone size distribution (Changnon and Staggs, 1969).

The Illinois State Water Survey hail gage (Mueller and Changnon, 1968) is a mechanical gage based on ballistic pendulum theory. Vertical displacement of a 36-cm-diameter sensing platform is mechanically recorded as an amplitude deflection on a motor-driven recording chart. The downward movement of the platform activates (with a start magnet) a constant-speed motor and drum drive to provide accurate timing for a hailfall record. The gage was designed to record three separate 10-min (or shorter) hail occurrences between servicings, but in field trials the start mechanism tended to become activated by strong winds alone, with loss of ability to record subsequent episodes of hail (E. A. Mueller, Illinois State Water Survey, personal communication).

NHRE has used, since 1972, a hail-rain separator (Nicholas, 1978) very similar to an instrument described by Kozminski and Bac (1964). The separator was designed to obtain a measurement of the mass of rain and hail. Recording and nonrecording types have been designed, with a 36-cm^2 opening that receives the rain and hail. The rain passes through a screen into a container, and the hailstones roll or slide down the screen into a second container. The nonrecording hail separators can be automatically closed at a specified time by a spring-loaded lid. The precipitation from the rain and the hail containers is measured daily. In the recording hail separator, the hailstones fall into a water-filled container, and the amount of water displaced is equal to the mass of hail. The displaced water flows into a calibrated, automatic siphon device; the amount is recorded digitally on 35-mm film, allowing estimation of the hail rate and duration. The rain-receiving portion of the separator allows similar recording of rainfall.

Admirat developed a time-resolving hail recorder (Pierre Admirat, Institute of Glaciology, University of Grenoble, France, personal communication). The instrument consists of several conventional hailpads mounted on a drum that rotates in steps on command from a timing device. One hailpad at a time is exposed to the hail through an opening in the top of the instrument's housing. A moving-foil hail recorder was developed in Canada (Koren, 1969). Also, the Illinois State Water Survey examined a device that successively exposes two hailpads for 24 hours each. A device of this general type has been used successfully in the field for controlling the exposure of rain sampling containers. These devices are very promising candidates for employment in large networks.

A hailstone spectrometer (Federer and Waldvogel, 1975) is a research tool at the Swiss Federal Institute of Technology. It consists of a 1,650-cm^2 platform covered with foam rubber. The area is exposed for 27 seconds, and then the stones are photographed by an automatic 35-mm camera; during the next 3 seconds the platform is automatically swept clean to expose a fresh surface for another 27 seconds. The spectrometers are portable and require the presence of an operator during observations.

Ezio Rosini (University of Rome, personal communication) constructed and tested in Italy a device consisting of a collecting funnel and a clock-driven (stepwise) rotating elbow spout that would deliver all hailstones falling in fixed time periods to receptacles filled with chilled kerosene; this procedure would thus preserve all stones entering the collector in each period. This device suffered from problems with the collection funnel, such as bouncing of stones, but otherwise functioned well.

An electrooptical hail recorder considered during the National Hail Research Experiment consists of an array of photodiodes illuminated by a collimated light beam. The operating principle of the device is similar to that of the Knollenberg probe discussed in Chap. 3. Hailstone size and fallspeed are measured by the number of elements of the array that are shaded by the passage of each stone and by the stone's time of transit past the array.

A hailstone sensor consisting of an electrically shielded Butyl rubber sheet was developed at the National Center for Atmospheric Research by Buck and Goyer (Changnon and Staggs, 1969). Impact of a hailstone on the sheet produces a pulse whose amplitude is a function of the size of the hailstone. Though the sensor had many desirable properties such as ruggedness, it was insensitive to stones less than 0.5 in (1.25 cm) in diameter, and the output response to impact was not uniquely related to input.

The Illinois State Water Survey tested two devices that recorded the time and duration of hailfall, using a piezoelectric sensing element attached to a 30-by-30-cm Plexiglas impaction surface horizontally supported above the ground. Proper thresholding of the output of this sensor permitted a very reliable distinction between hail and rain. In one of the devices a standard weighing-bucket recording rain gage with an additional pen arm was used for recording the hail-occurrence times. Although tests with the device in Illinois and Colorado were partly successful, the developers were never able to produce a surface both sufficiently large and uniformly sensitive to provide representative measures of hailfall (E. A. Mueller, Illinois State Water Survey, personal communication).

c. Remote-sensing Techniques

McNeil and Houston (1966) discussed their preliminary investigation of interesting ideas for remotely sensing hail. They reported that exploratory studies of seismic and acoustic techniques indicated that, though sound in concept, they would be usable only to ranges of about 0.25 miles (0.4 km) and therefore were not worth pursuing. They also discussed and rejected techniques based on radar sensing of the changes in reflectivity of a passive target owing to hail impacts. Another technique considered was allowing the impact of hail to switch on an inexpensive VHF transmitter whose output could be sensed by a remote receiver.

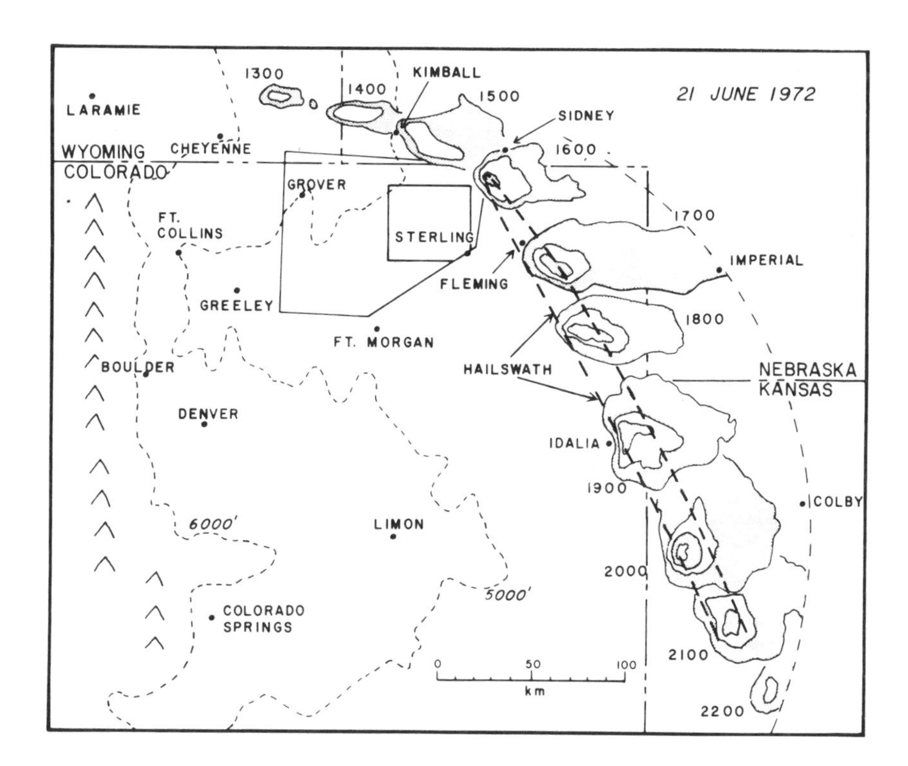

Figure 9.8. Hourly positions of a large hailstorm as determined by radar. The approximate limits of the hailswath are indicated by the bold dashed line. Outer contours are 20 dBZ. Successive contours are at intervals of 12 dB (i.e., 32, 44, etc.) (Browning and Foote, 1975).

They believed that the main problem with such an approach would be that of frequency multiplexing an array of several hundred such sensors to know which of them was transmitting at a given time.

An interesting remote sensing approach to hail detection and measurement pursued by Weickmann (1969), Goyer (1970; 1971), and Roads (1973) is the mapping of melting hail on the ground by airborne infrared radiometric devices. Under ideal circumstances this technique can produce a good estimate of the extent of hail on the ground; in general, however, it is of limited applicability.

Weather radar has been an indispensable remote sensing tool in the study of hailstorms and other forms of severe weather. Radar allows rather approximate localization of hail areas; Fig. 9.8 shows radar echo tracings of a large, severe hailstorm that took place on the high plains of Colorado and Kansas on 21 June 1972. Weather radar research has included developing methods of measuring both rainfall and hailfall at the ground (in terms of rainfall rate, rate of hailfall mass deposition, or measures of hailfall damage potential such as kinetic energy flux). The detection and measurement of hail by radar is fraught with complications, and radar cannot yet be recommended for routine monitoring and measurement of hailfall.

The correspondence between location of hail at the ground and low-angle radar echo has been investigated by many, and no clear picture has emerged. Waldvogel et al. (1978) have reported some success at estimating hail kinetic energy with radar, the basic assumption being that hail, when it occurs, is associated with the strongest radar echo at the lowest angle of scan.

In Illinois (Changnon and Morgan, 1976) strong hailfall at the ground was associated with strong low-altitude gradients of 3-cm wavelength reflectivity. Carte and Held (1978) reported 86% of surface hail occurrences were in the strong gradient (10-cm wavelength) regions, 38% occurred on the storm's left flank (N.B.: these are Southern Hemisphere observations), and 39% occurred on the leading edge, with hail occasionally occurring at the rear.

The tendency of hail to appear in the strong-gradient portions of the echoes might be explained by wind displacement below the radar beam; it has been shown in Vol. 2, Chap. 11, that hail tends to be blown to the right of the direction of hailshaft motion (in Illinois, Northern Hemisphere). However, aircraft penetrating hailstorms (Musil et al., 1976) have found hail aloft on the edges of updrafts in regions of strong gradient of radar reflectivity.

A major effort in remote hail detection was the dual-wavelength radar hail-detection technique proposed by Eccles and Atlas (1969). This method, based in the different scattering properties of hail at different radar wavelengths (Atlas and Ludlam, 1961), matured during the 1980s. Its use has entailed some practical difficulties, mainly involving mismatch of beam patterns at the different wavelengths (Rinehart and Tuttle, 1982, 1984; Jamison and Heymsfeld, 1984). In the mid-1980s this technique saw application in research projects, but it does not seem promising for routine operations, because the radars of different wavelengths must be maintained in rather precise calibration relative to one another. Other radar hail-detection techniques utilize measurements of polarization, since this property of incident radiation is altered according to the shape and orientation of targets (Atlas et al., 1953). In particular, large raindrops appear as anisotropic scatterers because of their systematic oblateness; large hail is usually irregular and tends to tumble as it falls, indicating no systematic orientation to the radar. Radar methods that utilize polarization characteristics were proposed and tested in the 1960s and 1970s (Barge, 1970; Seliga and Bringi, 1976), and have been developed quite effectively during the 1980s (Torlaschi et al., 1984; Bringi et al., 1984; Aydin et al., 1986). According to R. Carbone (National Center for Atmospheric Research, personal communication), greatest precision in delineation of hail regions is attained through use of research radars that bring dual wavelength and multiple polarization capabilities to bear simultaneously.

Reviews of radar sensing of hailfalls and hail areas have been provided by Changnon (1972), Srivastava and Jameson (1977), and Doviak and Zrnić (1984). See also Chap. 10 in this volume.

d. Limits to Accuracy of Point Hailfall Measurements with Instruments

Instruments such as hailpads are calibrated under controlled laboratory conditions, where uncertainties of measurement can be evaluated and are usually quite small (about 10%; Lozowski et al., 1978). Other errors or sources of uncertainty, which occur in real field conditions, probably dwarf the basic measurement errors.

Since ideal properties of hailstones are assumed in calibrating instruments, departures from ideal result in errors. Large departures from sphericity (see Vol. 2, Chap. 11) are infrequent, and the normal range of shapes does not suggest that this is a large source of error. Deviations of fallspeeds at impact from the theoretical terminal fallspeeds, however, seem likely to occur in the turbulent airflow near the ground, and for this reason there have been programs (in Canada and Colorado) to observe fallspeeds of hailstones in the field.

Most hailstones are hard enough to be considered close to the ideal (see Vol. 2, Chap. 11). Soft hail is not rare, but it does not appear to be a frequent source of measurement uncertainty.

These problems and others unforseen produce a net uncertainty (or instrumental error variance) in the measurements. Attempts in Europe and the United States to evaluate this uncertainty (for hailpads) in the field involve catching or photographing all the stones that strike a hailpad and comparing their dimensions with the dents on the pads (Matson et al., 1978).

Another source of error, characteristic of any instrument of small surface area, is purely statistical. It is reasonable to expect the statistics of hailstone concentration in the air (per cubic meter) at any instant to vary from one small volume to another in a random fashion that can be described by the Poisson probability distribution, which has found validity in problems similar to hailfall, such as sedimentation of particles, fluctuations in the concentration of molecules of a gas, and, more closely, fluctuations in raindrop concentrations (Joss and Waldvogel, 1969). The distribution law is given by

$$p(x) = \frac{m^x e^{-m}}{x!}, \qquad (9.3)$$

where $p(x)$ is the probability of finding x hailstones in any of many equal volumes, and m is the mean of x over many volumes. The same distribution should be equally valid for the total number of hailstones per unit area on the ground, or, for that matter, for the number of hailstones in any selected size interval.

Confidence limits for the Poisson distribution have been tabulated (e.g., Crow [1974], for values of x up to 100), and approximation formulas exist for large values of x. The 95% limits for selected values of x are given in Table 9.2. These limits are those within which many close-by measurements would be found to lie 95% of the time. It is clear that the number of the generally more numerous small hailstones is estimated more reliably ($\pm 10\%-20\%$) than that of the large or largest hailstones, only a few of which are detected. To obtain reliable information on the frequency of occurrence of larger stones it is necessary to put out enough instruments to achieve a net sampling area large enough to receive 100 or so stones. This area could amount to tens or hundreds of square meters, and thus an observer's report of the size of the largest hailstone is, in spite of its lack of precision and objectivity, probably better than the estimate from a 900-cm^2 hailpad.

Table 9.2. The 95% Confidence Limits for Hailstone Concentration Samples on Surfaces of Fixed Size

Measured Number of Hailstones	Upper Confidence Limit	Lower Confidence Limit
2	7.2	0.242
4	10.2	1.09
8	15.8	3.5
16	26.0	9.1
30	43	20.2
50	66	37.1
75	94	59
100	122	81
484	~528	440
900	~960	840

6. Collection of Hail for Scientific Analysis

The collection of hail for careful analysis of structure and content is of relatively recent origin. However replete the relevant literature may be with references to the results of such activities, it contains little mention of the means by which the analyzed hailstones have been obtained. Fortuitous presence of bystanders, usually farmers, or of the authors themselves has been stated or implied in most of what has been published.

There have been, however, in various hail-infested areas of the world, some efforts by scientists during the past two decades directed toward catching hail as it falls. Probably the most persistent of this small company are Knight and Knight (1968), who have been collecting samples of falling hail in northern Colorado, Oklahoma, and the lowveld of South Africa for 20 years. One of the reasons for the Knights' work has been to remove elements of fortuity from studies of hailstone structures. Bystanders tend always to collect the largest stones in any given samples. More important than this size bias, however, is the lack of other information to which the results of the analyses can be related. All organized attempts to collect hail, therefore, have been connected with projects in severe-storm research wherein data from radar, aircraft, and ground networks are also available.

Hail chasing and attendant problems were described by Browning et al. (1968), who intercepted hailstorms in the vicinity of, and in cooperation with, the National Severe Storms Laboratory in Norman, Okla. (see also Nicholas, 1978).

Browning and his colleagues examined hailstones in the field, but in most cases the stones were simply collected, stored, and later examined in the laboratory. Because the crystal structure of a hailstone changes rapidly at temperatures close to 0°C in a process known as "recrystallization," the stones must be collected in a manner that will preserve their original structure (Knight and Knight, 1968). This is usually done by letting the stones fall into liquid hexane at dry-ice temperature (−78°C), which freezes, immediately and in situ, any liquid water contained in the hailstone.

A number of vehicles have been designed to intercept severe storms and to sample various types of precipitation. The more elaborate of these, as noted in Chap. 2 of this volume, have also contained instruments to measure and record the standard meteorological parameters such as temperature, pressure, windspeed and direction, humidity, and rainfall rate. Ambient air filter samplers, raindrop disdrometers, hail cameras, and even pibals are also sometimes carried on the vehicles.

When they first began chasing hailstorms in northeastern Colorado, Knight and Knight used an ordinary delivery van and set out catchers on the ground. In time radios were added to permit communication with a radar center at Grover, Colo., and project aircraft. A simple device was later mounted on top of the van to funnel the hailstones inside, at once allowing them to observe the storm from inside the vehicle and providing time-resolved samples.

The complexity of the instrumentation contained in any one vehicle depends primarily upon its mission. One of the

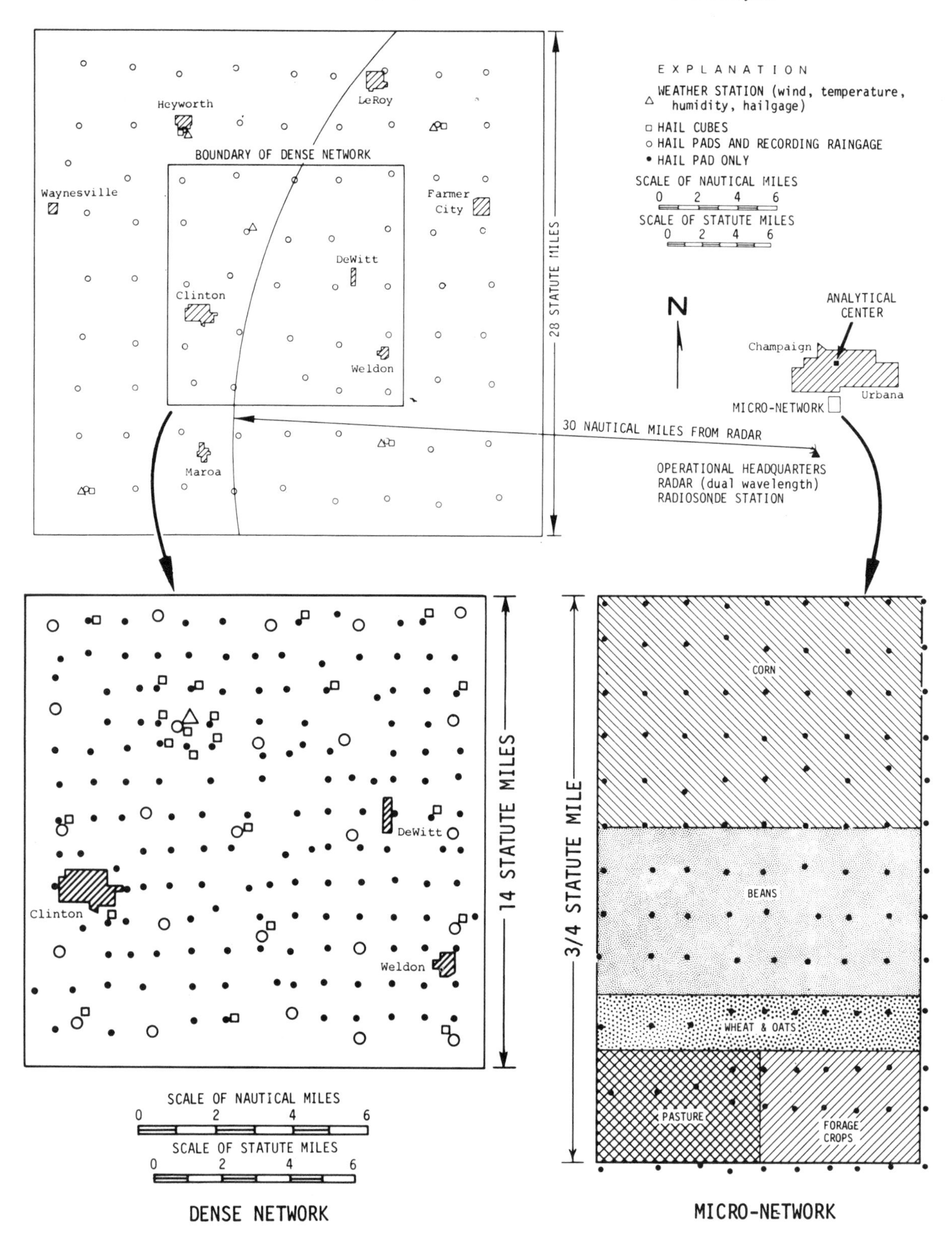

Figure 9.9. Networks of three different sizes and densities operated by the Illinois State Water Survey in 1974 (Changnon and Morgan, 1976).

first vehicles designed to sample both rain and hail was used in the Alberta Hail Studies Program for several years (Summers, 1968). In 1969, Morgan designed a similar but more complicated vehicle for a research program conducted in Verona, Italy. A still more elaborate vehicle was designed by the University of Wyoming (Veal, 1970) and used in cooperation with the National Hail Research Experiment (NHRE) in northeastern Colorado from 1971 to 1976, as were a number of more spartanly equipped vehicles designed by the Desert Research Institute to collect rain and hail samples for silver analysis in connection with weather-modification projects. A more recent addition to the vehicle fleet was built by Federer and Waldvogel for use in Grossversuch IV, the hail research program conducted in Switzerland (Federer and Waldvogel, 1975).

Successful interception of a severe storm requires a great deal of cooperation and communication between the mobile units and those who are monitoring the progress of the storm by radar or aircraft. In an ideal situation storm occurrence and motion must be forecast in sufficient time to allow mobiles to be deployed into the area where the storm is expected to occur. As the storm materializes and grows, its development is relayed to the mobile units, which maneuver into a position in front of the storm, where the core of highest radar reflectivity may be expected to pass overhead.

In reality, nature rarely cooperates in such a straightforward way, and hail chasing is often a game of "catch-up." There are hazards involved in chasing hail, not least of which is the necessity of driving over miles of wet, unpaved farm roads that, during storms, turn into rivers of slithery gumbo. There are meteorological hazards as well: gust fronts, blowing dust, and lightning. In areas such as Oklahoma, where hail is often associated with tornadoes, one must always be especially aware of the vehicle's position with respect to the storm. Large hail often falls just to the north of a tornado. In such circumstances it is obvious that the technique employed for catching hail differs somewhat from that employed in less violent storms. One approaches from the north, drives into the precipitation shaft but never through it, collects hail, and returns to the north again, turning away from the storm. Clear escape routes, preferably in more than one direction, are extremely important (see also Chap. 2 of this volume).

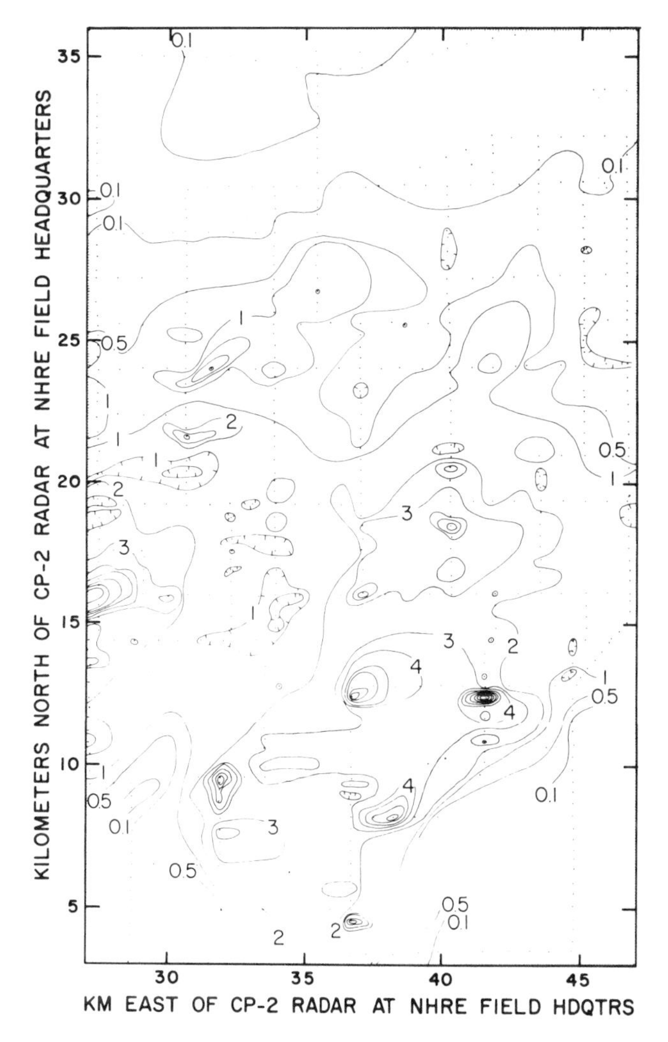

Figure 9.10. Pattern of hailfall mass deposition, 22 June 1976, over the NHRE network, based on all 551 damaged hailpads.

7. Design of Networks for Hailfall Observation

The principal characteristic of a network is its station density, or the number of stations per unit area (sometimes expressed as its inverse, the area represented by each station, or as the mean separation between stations). This must be great enough that a number (to be specified) of observation points will fall within any significant hail entity (e.g., a hailstreak) that occurs within the network. Figure 9.9 shows networks of three different station densities operated by the Illinois State Water Survey in 1974.

Clearly, the perfect network would be dense enough to allow a description of every hailfall event to any desired degree of detail. The network density actually required to achieve an acceptable network sampling variance depends on (1) the statistical characteristics of the hailfall in the network area, described by the distribution of sizes (lengths, widths, areas) of hailfall entities and the typical scales of variability within entities of the desired measures of hailfall; and (2) the purpose the measurements must serve. For example, a network intended to estimate the total mass of hail falling from an ensemble of 50 hailstorm days need not be as dense as one intended to describe in detail a single hailfall.

a. An Example of the Effect of Sampling Density on Estimates of Areal Extent of Damaging Hail

Two areas in central Illinois, one comprising 4,000 square miles and the other 1,000 square miles, located inside the

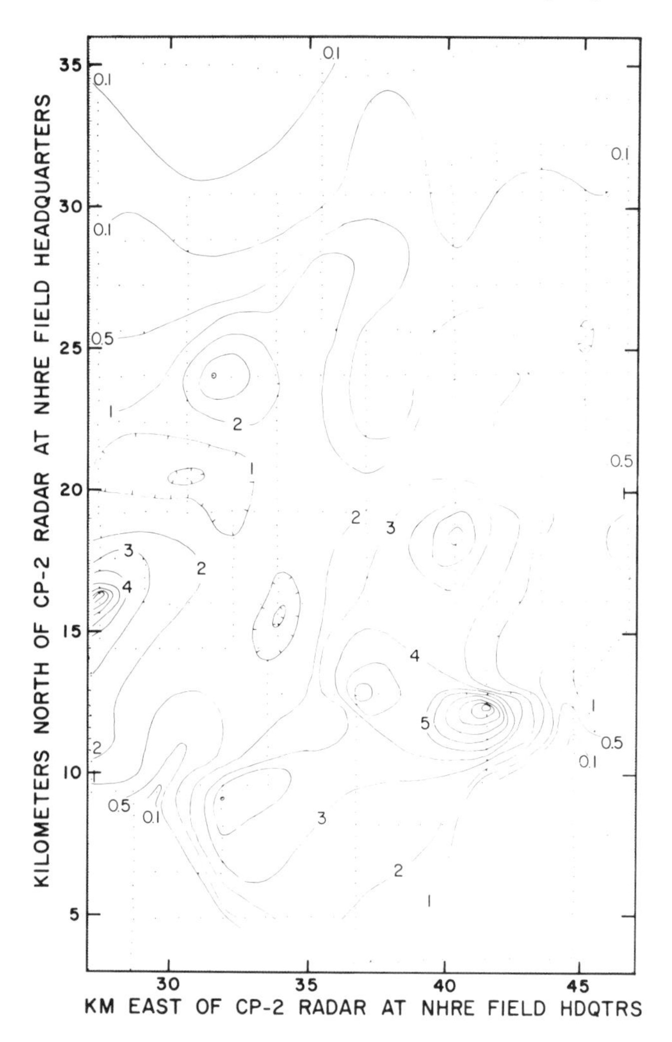

Figure 9.11. Pattern of hailfall mass deposition based on one-quarter of the observations in Fig. 9.10.

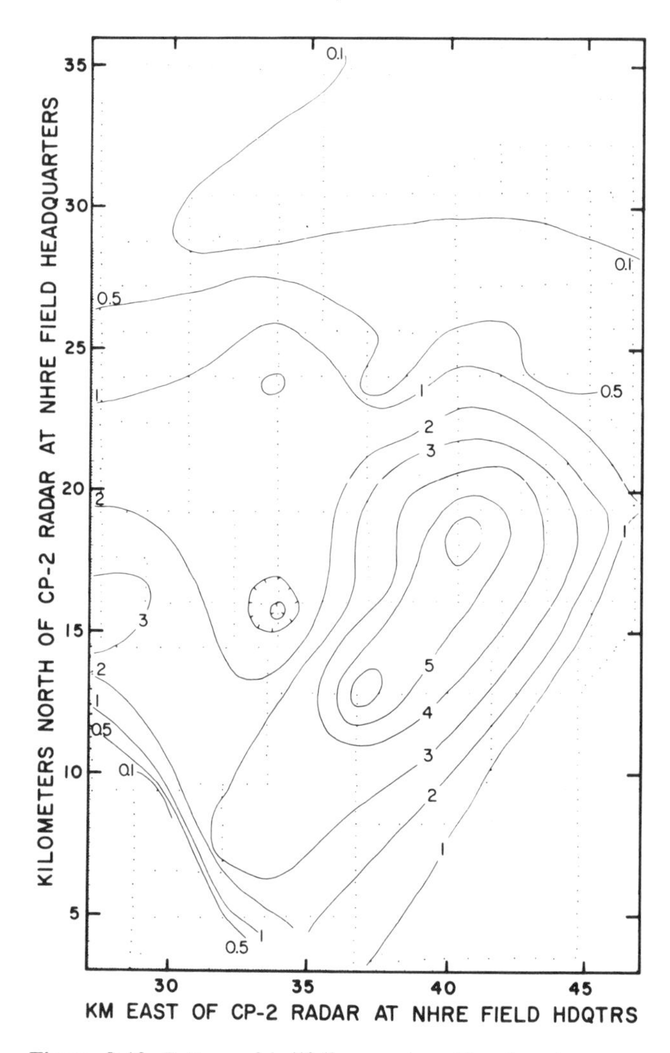

Figure 9.12. Pattern of hailfall mass deposition based on one-sixteenth of the observations in Fig. 9.10.

larger area, were chosen for study by Changnon (1968). More than 85% of the total area was covered by insurance over an 11-yr period. Maps for each storm day were prepared, showing the actual areas of paid losses using all the available insurance data that represented a sampling density of four observation points per square mile. Transparent overlays at the same map scale with evenly distributed points representing observation sites with densities of 0.5, 1, 3, and 9 square miles per point were placed on each damage map to determine the area of damage per storm day for each density.

The average values for the areal extent of damaging hail on storm days showed a considerable variation with sampling density. In the 4,000-mi^2 area, 16,000 sampling points (36 points per 9 square miles) indicated that the average daily areal extent of damaging hail was 36.1 square miles, whereas 445 points (1 point per 9 square miles) measured only 6.3 square miles of damaging hail. Underestimation of hail area with the sparser measuring network in these cases was partly a consequence of patchiness in the hail distribution and partly a consequence of the analysis method, which tended to assign a fixed area to each (farm) report, regardless of station density. In other words, with fewer stations, there were fewer hail reports and a smaller area defined.

The seasonal average areal extents of damaging hail defined by different densities were expressed as percentages of the total damaged hail area, as defined by four points per mile (see Table 9.3). On the average, a hail network with a density of one point per 9 square miles in a 1,000-mi^2 area measures 11% of the actual area of damaging hail in a season. A network with this density measured a 1-yr high of 13% and a 1-yr low of 8%. These results indicate that a network comprising one or two observation sites per square

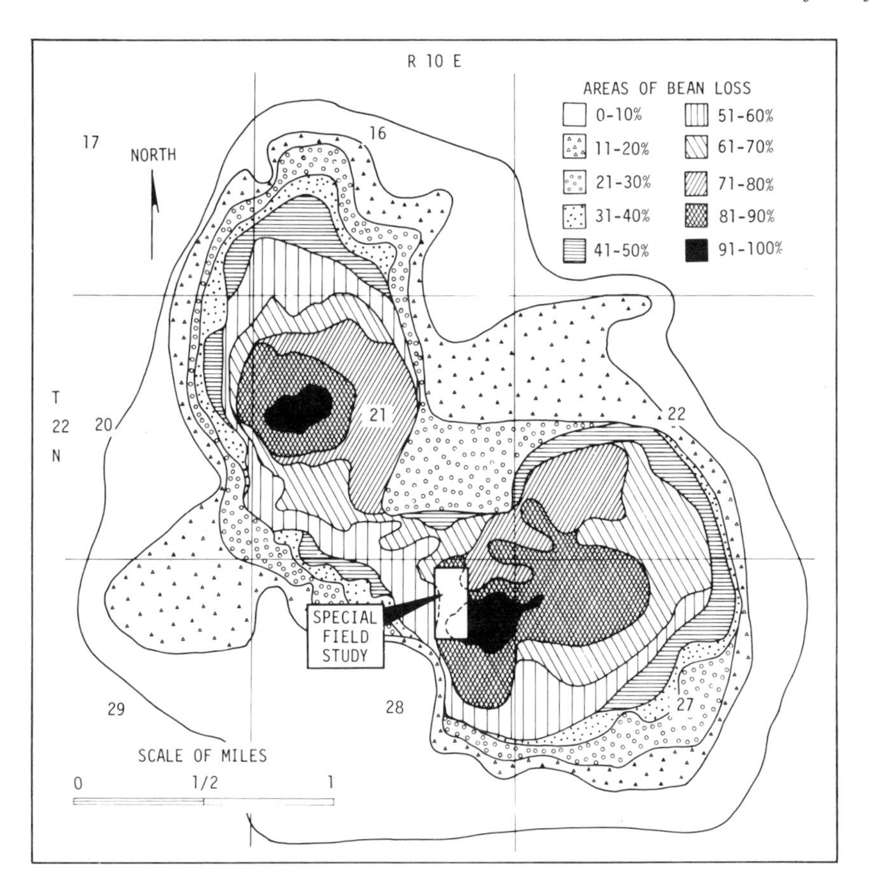

Figure 9.13. The distribution of damage (in percent) in a 5.4-mi hailstreak in central Illinois, 11 July 1969. Loss to soybeans is defined by 198 assessments.

Table 9.3. Percent of Total Area of Damaging Hail in 1000-m^2 (2560-km^2) Area, Measured by Different Densities of Observation Points During a Crop Season

	Density (mi^2/Sampling Point)			
	9	3	1	0.5
Average	11	25	67	85
Highest	13	30	76	91
Lowest	8	20	62	80

mile is necessary to measure adequately the areal extent of damaging hail.

b. An Example of Degradation of Hailfall Pattern with Decreasing Network Density

Long (1978) studied the variation of estimates of total hailfall mass for a storm that occurred on 22 June 1976 over the NHRE hail network in northeastern Colorado. On that day 551 hailpads registered hail, and the topography of mass of hail (in millimeters of rainfall equivalent) based on all of them is shown in Fig. 9.10. The dots represent the hailpad locations. Figure 9.10 is an example of the realistic compromises that must be made between the ideal network configuration and the road system that is available. In this case the roads are very widely separated. An attempt has been made to offset this by spacing the stations very closely, as close as 400 m, along the roads. The mean spacing turns out to be 1 km.

In Figs. 9.11 and 9.12 stations are removed from the map, and new analyses are performed. Figures 9.10–9.12 show how the pattern of hailmass becomes more and more degraded as the number of stations (N) decreases.

c. An Illustration of the Evaluation of Network Sampling Performance

Area and pattern are only two aspects of hailfall. For example, predicting the results of trials of hail-suppression techniques requires quantitative estimates of hailfall intensity and amount.

Changnon and Barron (1971) studied the patterns of crop damage within single hailstreaks, making as many as 45 point estimates in a single 20-acre field and in one case 198 estimates in a 5.4-mi^2 streak (Fig. 9.13). These measurements show a damage pattern with large variations over

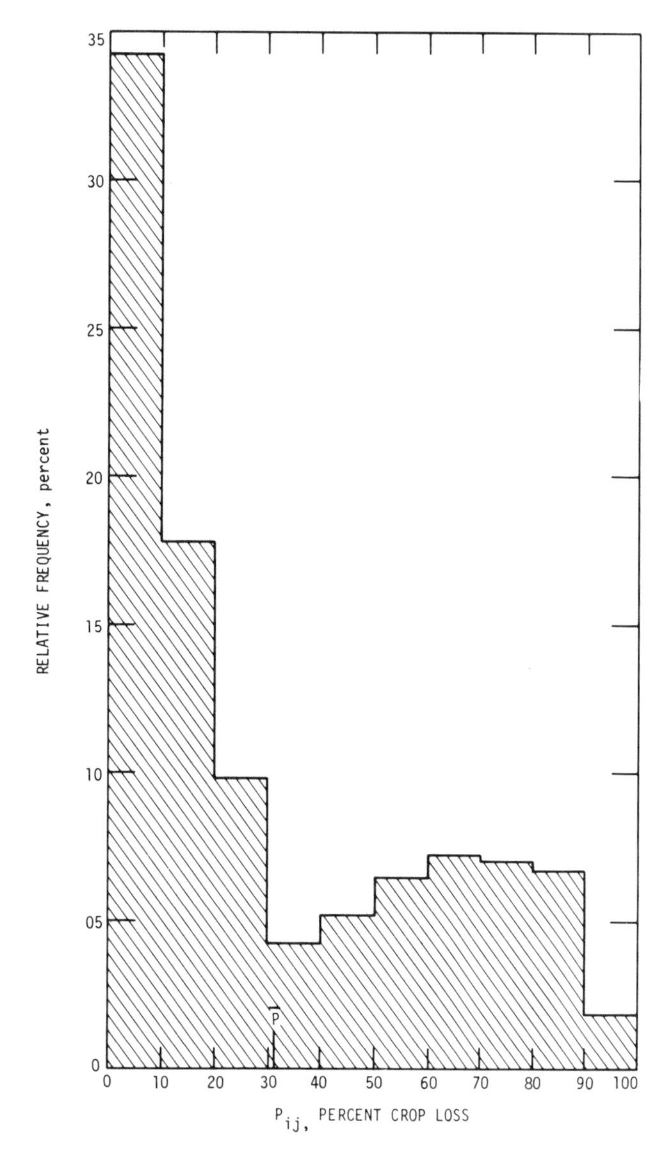

Figure 9.14. Histogram of damage values (in percent classes) for the hailstreak of Fig. 9.13.

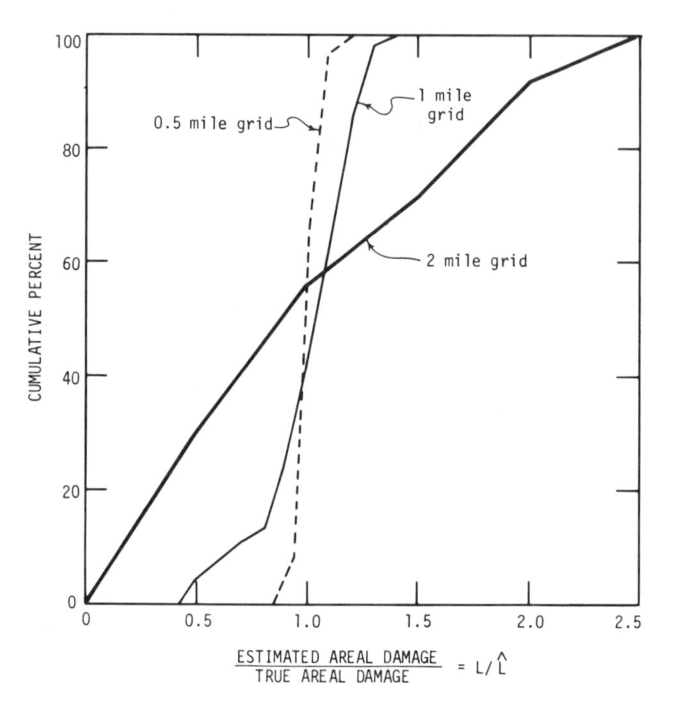

Figure 9.15. Cumulative curves of the estimates of areal total damage divided by the true value for three network grid sizes for the hailstreak of Fig. 9.13.

small distances. Morgan and Towery (1975) determined the distribution of these loss estimates and the average gradients of loss over this streak. There was a large number of points (Fig. 9.14) with small loss and a broad peak, representing a fraction of all the points, with values between 40% and 90%. The average N-S loss gradient was 6.5% per 150 meters. A Monte Carlo simulation was performed in which the pattern was sampled 100 different times by a 1-mi-grid network and also by 0.5- and 2.0-mi-grid networks. The process can be visualized by imagining the grid formed by the intersections of the four straight lines in Fig. 9.13 to be moved around on the damage pattern to 100 different positions. The variance of the areal estimates in this kind of simulation is called "network sampling variance." This simulation showed (Fig. 9.15) that the total damage (the integral over area of the percentage loss) would be estimated to within 20% of the true value 80% of the time by a 1-mi-grid network. This is the only available evaluation of the performance of a given network in estimating the areal total damage from a single hailstreak. The result is highly specific to the damage distribution that was used and is not necessarily representative.

An important parameter in determining the network performance is undoubtedly the ratio between the area of the hailstreak and the area represented by one grid point. This ratio should be obtained for a sample of hailstreaks that represents the range and distribution of hailstreaks to be expected in the network. Such ratios have been used by Silverman (1981) for designing a network to measure thunderstorm rainfall in the high plains. This approach can be applied to any areal quantity (e.g., total mass, kinetic energy, or the area itself) and need not be restricted to square or uniform grids.

d. Network Size and Shape

The size of the network required for a given purpose rests on information derived from Fig. 9.1, which shows the number of hail days per year as a function of area. The main importance of area size is in determining the number of study events available each year and the number of instru-

10. **Weather Radar**

Richard J. Doviak, Dale Sirmans, and Dusan Zrnić

1. Introduction

Radars (Radio Detection and Ranging) were originally developed to detect and determine range of aircraft by radio techniques. As they became more powerful, their beams more directive, receivers more sensitive, and transmitters more coherent, radars were successfully applied in mapping the Earth's surface and weather; their signals have even reached out into space to explore surface features of our planetary neighbors. The radar beam penetrates thunderstorms and clouds to reveal, like an X-ray photograph, their inside structure. Pulsed Doppler radar techniques have been applied to map severe-storm structures with some astounding success, particularly in showing, in real time, the development of incipient tornado cyclones (Donaldson, 1970; Doviak et al., 1974; Burgess, 1976). Such observations should enable weathermen to provide better forecasts and warnings, and researchers to understand the life cycle and dynamics of storms.

The most comprehensive treatment of radar techniques is found in the collected works compiled by Skolnik in his *Radar Handbook* (1970). Battan's text (1973) on weather radar applications is probably the most widely used by meteorologists, and Atlas (1964) also gives a concise and informative review of many weather radar topics. Both emphasize electromagnetic scatter and absorption by hydrometeors. Nathanson (1969) emphasizes the total radar environment as well as radar design principles. The radar environment causes unwanted reflection (clutter) from sea and land areas (precipitation produces clutter when aircraft are the targets of interest); Nathanson treats precipitation echoes comprehensively. Anomalous propagation of radar signals enhances ground clutter. A good general reference on propagation of electromagnetic waves through the stratified atmosphere is Bean and Dutton (1966). A comprehensive treatment of the techniques used in Doppler weather radar and applications to observe both clear and stormy weather is given in Doviak and Zrnić (1984).

We provide here a foundation of weather radar principles and relate the radar parameters and signal characteristics to the target's meteorological properties. We lightly touch upon subjects comprehensively treated elsewhere (e.g., scattering properties of hydrometeors), present techniques used in extracting a target's properties from its echoes, and relate radar parameters and echo power to the weights given to a hydrometeoric target's backscatter cross section. To accomplish this, we tour the radar signal path starting from the transmitter, through the antenna, and along the beam to the target and return to the receiver, highlighting along the way the important signal properties.

We discuss the pulsed Doppler radar and consider a discrete target to develop radar principles and the radar equation. The theory for non-Doppler radar is naturally included. These principles are then extended to the more complex weather target, which is a conglomerate of discrete targets producing a continuous stream of echoes with rapidly fluctuating amplitude and phase. The origin of these fluctuations is shown, and the weather radar equation is developed.

We relate the weather Doppler spectrum and its principal moments to the spatial distribution of target reflectivity and its velocity. Finally, we show example analyses of radar data from severe storms.

2. Principles of Radar

Weather radar illuminates the atmosphere around it with a series of pulses of focused electromagnetic radiation, directed with an antenna mechanically rotated in azimuth and elevation. Alternately transmitting and receiving, a radar is equipped to display echoes from precipitation particles and other targets in relation to the time delay between pulse transmission and echo arrival, hence indicating range. Echo amplitude is a measure of precipitation intensity or, in the case of a discrete target, its size. Coherent radars can go one important step further: they gage the phase of successive echo signals to determine the Doppler shift and hence radial velocity of the hydrometeors. Pulsed-Doppler radars with advanced signal-processing equipment using solid-state digital hardware can estimate precipitation intensity, radial velocity, and atmospheric turbulence at about 1,000 contiguous but independent locations every second. Although massive amounts of data are gathered and need to

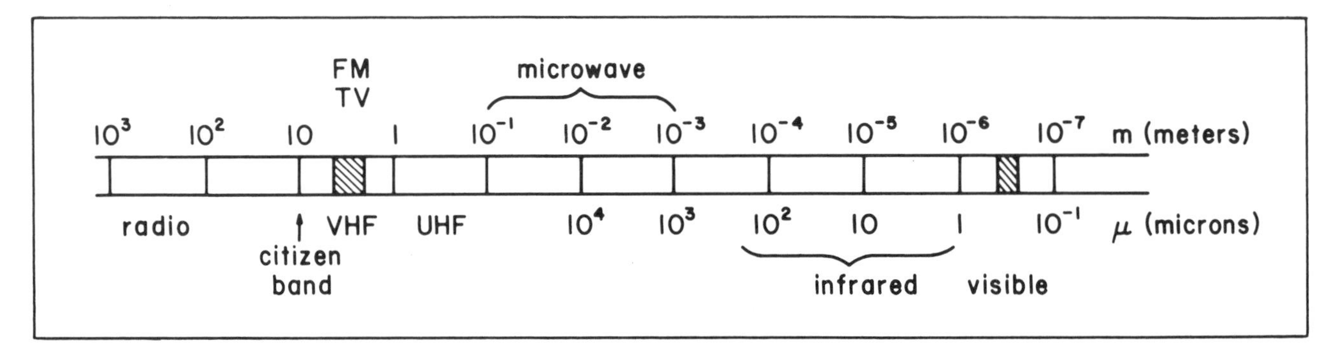

Figure 10.1. Electromagnetic spectrum showing location of microwave band of radar wavelengths relative to radio frequency and optical bands.

be processed in relatively short periods of time, much of the data is redundant and contains few bits of meteorologically significant information. This redundancy facilitates processing and assimilation of large data sets both by dedicated computers and by human observers.

a. Electromagnetic Waves

Electromagnetic or radio waves from radar are electric, E, and magnetic, H, force fields that propagate through space at the speed of light and interact with matter along its path. These interactions cause the scattering, diffraction, and refraction common to visible electromagnetic radiation (light). The waves, focused into beams by the antenna system, have sinusoidal spatial and temporal variations—the distance and time between successive wave peaks of electric (magnetic) force defines the wavelength, λ, and wave period (i.e., the reciprocal of frequency, f, in hertz). These two important electromagnetic field parameters are related to the speed of light c:

$$c = \lambda f = 3 \times 10^8 \text{ m s}^{-1}. \qquad (10.1)$$

Microwaves are electromagnetic forces having wavelengths between 10^{-3} and 10^{-1} m—visible radiation has a wavelength of about 6×10^{-7} m. The upper end (0.01 to 0.1 m) of the microwave band is used in weather and aircraft surveillance radars (Fig. 10.1).

The electric field wave far from the transmitting antenna has a time t and range r dependence, generally given by

$$E(r,\theta,\phi,t) = \frac{A(\theta,\phi)}{r} \cos [2\pi f(t - r/c) + \psi] \text{ V m}^{-1}, \qquad (10.2a)$$

where A depends on θ, ϕ, the direction of r from the radiation source, and ψ is usually an unknown but constant phase. The dependence of E on r, t, θ, and ϕ is characteristic of all electromagnetic waves propagating in space devoid of matter, be they radio or light. Equation 10.2a approximates well, at weather radar frequencies, the properties of waves propagating through our atmosphere. Because force has a direction, E is a vector quantity, and Eq. 10.2a represents one component of the electric field. The waves propagate in the direction of r—that is, an observer's range r must increase at rate c in order to stay on a wave crest ($t - r/c$ = const). The vectors E, H are always perpendicular to r and to each other when r is large (see Sec. 2b(2)).

Because the principal factors characterizing periodic electric fields are amplitude, $A(\theta,\phi)/r$, and phase, $2\pi f(t - r/c) + \psi$, it is convenient to use complex number or phasor notation to describe these parameters (Jordan, 1950, p. 116). The electric field (Eq. 10.2a) is then expressed as

$$E = \frac{A(\theta,\phi)}{r} e^{j2\pi f(t-r/c) + j\psi}, \qquad (10.2b)$$

which, according to Euler's formula, can be represented by a two-dimensional phasor diagram (Fig. 10.2). This diagram clearly presents the time and space dependence of amplitude and phase to within an integral multiple of 2π. E and H are real, measurable quantities, and it is understood that we need to take the real part of Eq. 10.2b, or of any other phasor we may introduce, in order to have the real time and space dependence. The time dependence of phase is of paramount importance in understanding the principles of Doppler radar.

Another important electromagnetic field quantity is the time-averaged power density, $S(r,\theta,\phi)$:

$$S(r,\theta,\phi) = \frac{1}{2}\frac{EE^*}{\eta_0} = \frac{A^2(\theta,\phi)}{2\eta_0 r^2} \text{ W m}^{-2}, \qquad (10.3)$$

where * is the complex conjugate operation. It can be shown that $EE^*/2$ is the time average of Eq. 10.2a squared. Factor η_0 is the wave impedance (in space or Earth's atmosphere at radar wavelengths, η_0 is a constant equal to 377 ohms) and is the ratio of the electric to the magnetic field intensity. It simply relates the electric field intensity required to advance the power density S. Time averages represented by Eq. 10.3 are averages of power over a cycle or period of f^{-1} of the wave, and Eq. 10.3 is usually called peak power density of an electromagnetic packet of waves if power is pulsed (i.e., transmitted in a burst of energy). The significance of $S(r,\theta,\phi)$ is that it represents the power density flowing outward from a source, either continuously or in bursts, and products of S with areas, later to be specified, represent power that is received, absorbed, scattered, and so on.

Spatial resolution of observations, i.e., discrimination between two adjacent similar objects, is dependent on wavelength and antenna size. The angular resolution or diameter of the first null in a circular diffraction pattern is well approximated by

$$\Delta\theta \sim \frac{140\lambda}{D} \text{ (degrees)}, \tag{10.4}$$

where D is the diameter of the antenna system (Born and Wolf, 1964, p. 415). Thus radio waves (i.e., $\lambda = 10^2$–10^3 m) require huge antenna installations to achieve an angular resolution of a few degrees—rather poor for optical antenna systems. For example, the human eye has an angular resolution of about 0.02°. It is evident that remote radar sensing at microwave frequencies with practical-size antennas is characterized by poor resolution compared to optical standards.

The essential distinguishing feature favoring microwaves for weather radars is their ability to penetrate rain and cloud and thus provide a view inside showers and thunderstorms, day or night. Rain and cloud do attenuate microwave signals, but only slightly (for $\lambda \geqslant 0.05$ m) compared with the almost complete extinction of optical signals. Raindrops scatter electromagnetic energy, and the portion scattered constitutes the signal whose characteristics are diagnosed to determine storm properties. Scattered signal strength measures rain intensity, and time rate of signal phase change (Doppler shift) measures raindrop speed in the r direction.

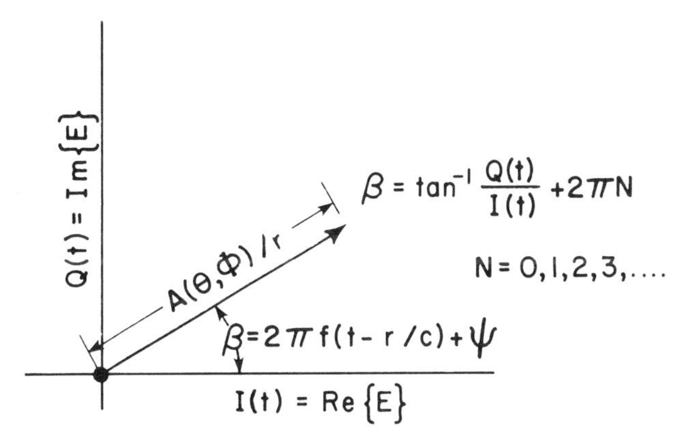

Figure 10.2. The phasor diagram. Re $\{E\}$ is the real part of the complex electric field E, and Im $\{E\}$ the imaginary part.

b. Doppler Radar Transmitting Aspects

Figure 10.3 shows in block diagram the principal components of a simplified but representative pulsed-Doppler radar. The Klystron amplifier, turned on and off by the pulse modulator, generates a train of high peak power microwave pulses having duration τ of about 1 μs (10^{-6} s), with spacings at the pulse repetition time (PRT) T_s, the sampling time interval. The pulse of peak power density can be represented as $S(r,\theta,\phi)U(t - r/c)$, where

$$U(t - r/c) = \begin{cases} 1, & r/c \leqslant t \leqslant (r/c + \tau) \\ 0 & \text{otherwise.} \end{cases} \tag{10.5}$$

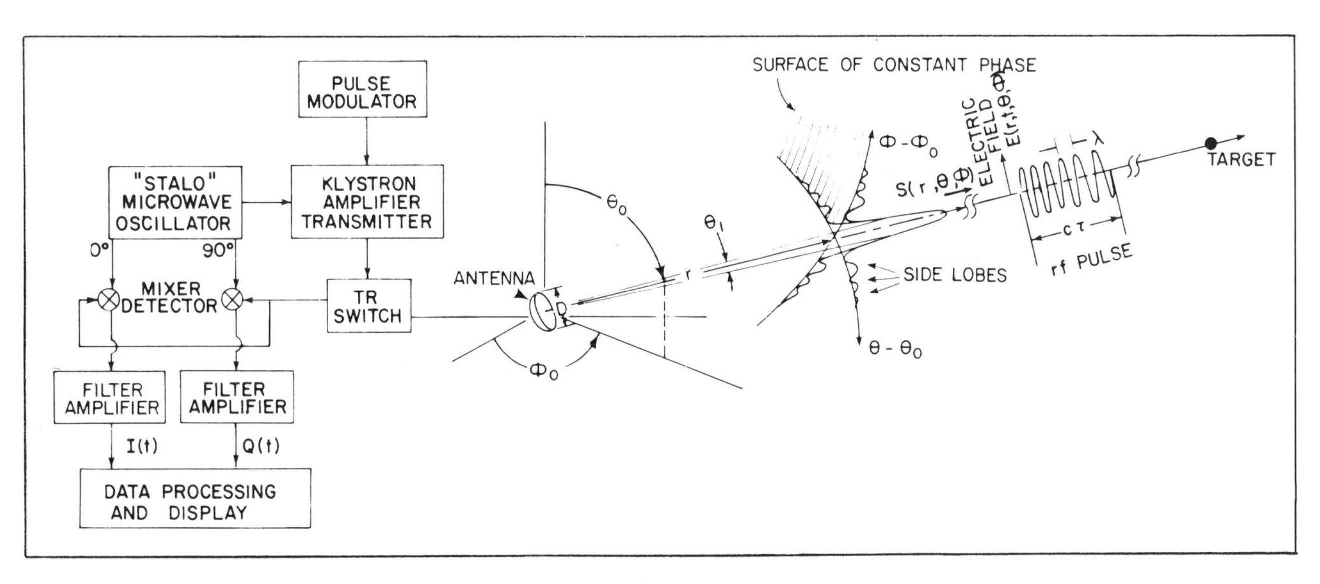

Figure 10.3. Simplified Doppler radar block diagram.

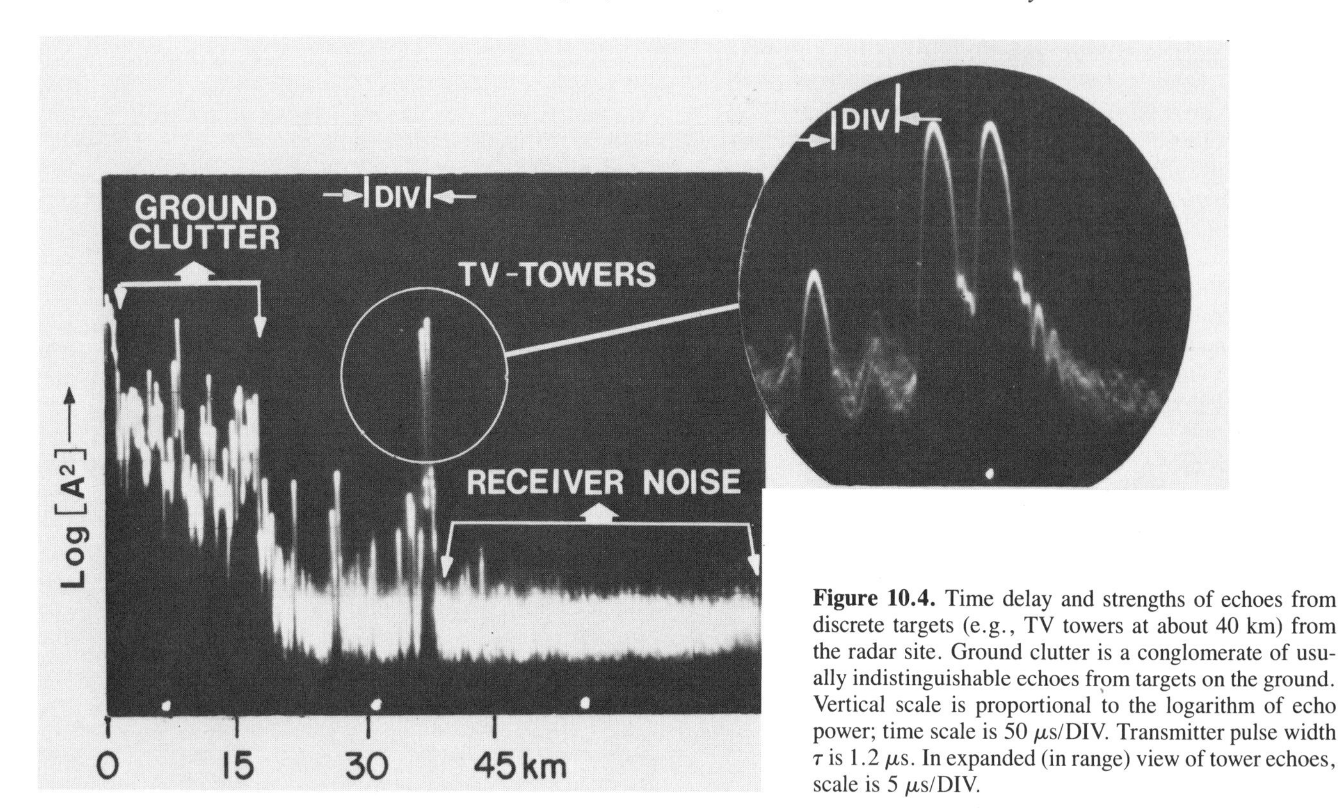

Figure 10.4. Time delay and strengths of echoes from discrete targets (e.g., TV towers at about 40 km) from the radar site. Ground clutter is a conglomerate of usually indistinguishable echoes from targets on the ground. Vertical scale is proportional to the logarithm of echo power; time scale is 50 μs/DIV. Transmitter pulse width τ is 1.2 μs. In expanded (in range) view of tower echoes, scale is 5 μs/DIV.

This power illuminates objects (targets) as it propagates within a narrow beam, and a tiny fraction of this radiation is scattered by these targets toward a receiver located in most cases at the transmitter site. Furthermore, for economic reasons, the same antenna is shared by the transmitter and receiver.

The transmit-receive (TR) switch connects the transmitter to the antenna during the period τ, whereas the receiver (i.e., mixer-detector plus filter-amplifier) is connected during the time interval $T_s - \tau$, the listening period. The switching is not performed instantaneously, and there is a period of time (usually a few tens of microseconds) when the receiver does not have full sensitivity for detection.

When targets are small (compared with $c\tau/2$) and well separated (no two targets within $r \pm c\tau/4$), the returned signals or echoes are replicas of the transmitted pulse. These replicated pulses are shaped by the receiver, which also adds electronic noise. The echoes are then displayed on video equipment to show their relative strengths and delay-referenced to the time of the transmitted pulse (Fig. 10.4).

The stretch of echoes in the range between 21 and 25 km mainly shows atmospheric echoes which are added to the receiver noise, whereas at ranges inside 21 km there is a mixture of echoes from ground targets and from refractive index irregularities whose physics is discussed fully by Doviak and Zrnić (1984, Chap. 11). Note that, because the vertical scale of Fig. 10.4 is logarithmic, atmospheric echoes (i.e., fluctuations superimposed on ground clutter) may actually have amplitudes larger than receiver noise, though the fluctuation on top of the stronger ground-clutter echoes appears to be less than in regions void of clutter. To further illustrate this point, note that the receiver noise and atmospheric echoes are not evident on the expanded view of echoes from the TV towers because on this logarithmic scale the vertical displacements owing to noise and atmospheric echo fluctuations are much smaller than fluctuations seen at ranges without strong ground target echoes.

Echoing principles were first applied in the late 1920s to measure remotely the properties of the ionosphere, and in late 1934 a 10-m wavelength radar was first used by the Naval Research Laboratory to locate an airplane. The use of microwaves in radars did not become practical until early in 1940, when a powerful and efficient transmitting tube, the magnetron, was developed. The first detection of storms by microwave radar was made in England in early 1941. There is an excellent historical review by Atlas (1964) of early developments in radar meteorology.

The development of high-power and high-gain Klystron amplifiers in the 1950s made practical the generation of microwaves that are phase-coherent pulse to pulse, a requirement for pulsed-Doppler radars if velocity of targets is to be measured. Radar signals are phase-coherent from pulse to pulse if the distance (or time) between wave crests of successive transmitted pulses is fixed or known. Magnetrons are not phase-coherent from pulse to pulse, but if the phase of each transmitted pulse is measured and stored, these inherently incoherent oscillators can be used in radars to measure velocity of targets.

(1) The Electromagnetic Beam: The microwave pulse leaves the antenna in an essentially collimated beam of diameter D equal to that of the antenna reflector (Fig. 10.5). However, because of diffraction, the electromagnetic beam begins to spread at a range $r = D^2/\lambda$ into a conical one having an angular spread or beamwidth, θ_1, given approximately by Eq. 10.4. The beamwidth θ_1 is commonly specified as the angle within which the microwave radiation is at least one-half its peak intensity (3-dB width).

The radiation pattern $S(\theta,\phi)$ specifies the angular distribution of power density (W m^{-2}) that emanates from the antenna. It is impossible to confine all the energy in a narrow conical beam, and some of it inevitably falls outside the main beam lobe into side lobes (Fig. 10.3). The main lobe is that region where S monotonically decreases, though other definitions have been proposed (Kraus, 1966, p. 154). Usually the power density in any side lobe is less than 1/100 of the peak density in the main lobe. Furthermore, the sum total of power in side lobes can often be held to just a few percentage points of that transmitted within the main lobe (Sherman, 1970).

The antenna reflector is usually a paraboloid of revolution and is illuminated by a source located at its focal point. The illumination is made to be nonuniform across the reflector in order to reduce the side-lobe levels, and the 3-dB beamwidth θ_1 is then

$$\theta_1 = 1.27\ \lambda/D \text{ (rad)}, \tag{10.6}$$

and the width between the first nulls becomes $3.27\lambda/D$ (rad).

The microwave energy is within a $c\tau$-thick spherical shell that expands (i.e., propagates) at a speed c. Thus the power density $S_i(\theta,\phi)$ incident on targets must decrease inversely with r^2, although power P_t' transmitted through any enclosing sphere is constant. This is the reason for the $1/r^2$ dependence of S in Eq. 10.3. Because of losses in the antenna, its transmission lines, and its protective radome, the power P_t delivered to the antenna's waveguide port is larger than P_t'.

Figure 10.5. The National Severe Storms Laboratory's weather radar antenna-reflector inside its protective radome. The reflector, a paraboloid of revolution, has a 9.14-m diameter. The radiation source is the "horn" at the end of the curved (black) waveguide. The tubes extending to the right support the source and waveguide.

(2) Antenna Gain: If P_t' were radiated equally in all directions, S would be equal to $P_t'/4\pi r^2$. However, the antenna focuses radiation into a narrow, angular region where the peak power density S_p is much stronger than $P_t'/4\pi r^2$. The ratio

$$\frac{S_p}{P_t'/4\pi r^2} = g_t' \tag{10.7}$$

defines the maximum directional gain of the antenna. Measurement of P_t' is difficult, and the antenna engineer therefore measures power P_t delivered to the antenna's waveguide port and S_p at distances far (i.e., $r \geqslant 2D^2/\lambda$) from the antenna. In this case the computed gain g_t accounts for losses of energy associated with the antenna system (e.g., radome, waveguide). Then the incident radiation power density at range r is given by

$$S_i(\theta,\phi) = \frac{P_t g_t f^2(\theta,\phi)}{4\pi r^2}, \tag{10.8}$$

where $f^2(\theta,\phi)$ is the normalized power gain function (i.e., $f(\theta,\phi) = 1$ at θ_0,ϕ_0; Fig. 10.3), and g_t is simply the antenna gain along the beam axis.

c. The Target Scattering Cross Section

The cross section σ of a target is an apparent area that intercepts a power σS_i, which is assumed to radiate isotropically to produce at the receiver a power density,

$$S_r = \frac{S_i\sigma(\theta',\phi')}{4\pi r^2}, \tag{10.9}$$

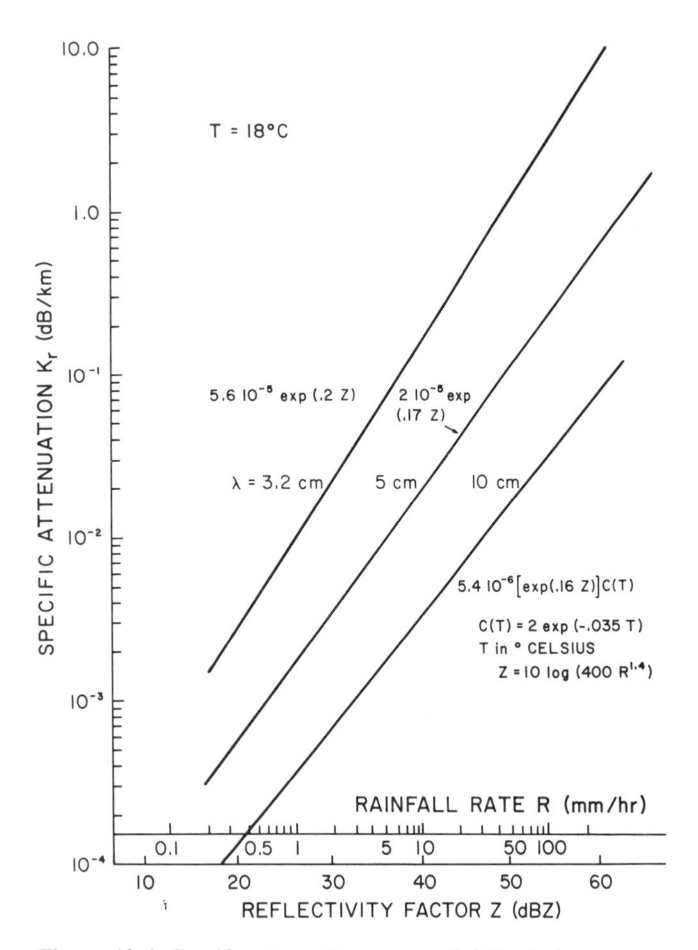

Figure 10.6. Specific attenuation versus rainfall rate for propagation through rain showers ($T = 18°C$). The Laws and Parsons (1943) drop-size distribution is assumed. $C(T)$ is a temperature adjustment factor; $Z = 10 \log(400R^{1.4})$, and the units of Z are dBZ (from Doviak and Zrnić, 1984).

equal to that scattered by the actual target. The area σ is sometimes called "differential cross section" to distinguish it from total cross section, which is proportional to the total power scattered by the target. The polar angles θ', ϕ' to a receiver are referenced to an axis drawn from the transmitter to the target with the target at the origin. Equation 10.9 suggests that scatterers do not echo power isotropically, and hence the target cross section $\sigma(\theta', \phi')$ can depend on the relative location of the transmitter and receiver.

It is easy to deduce that a target's scattering cross section may have no resemblance to the target's physical cross section. In fact, thin metallized fibers (chaff) of real cross section that is small compared with λ^2 have backscatter cross sections many times larger than their physical area. On the other hand, a large metallic sheet can have nearly zero backscatter cross section when it is oriented to minimize the energy scattered to the receiver.

The backscatter cross section σ_b for a water drop of diameter D that is small compared with λ (i.e., $D \leq \lambda/16$) is

$$\sigma_b = \frac{\pi^5}{\lambda^4} |K_w|^2 D^6, \qquad (10.10)$$

where $|K_w|^2$ is a parameter, related to the refractive index of the water, that varies between 0.91 and 0.93 for wavelengths between 0.01 and 0.10 m and is practically independent of temperature (Battan, 1973, p. 39). Ice spheres have $|K_w|^2$ values of about 0.18 (for ice density of 0.917 g cm^{-3}) independent of temperature as well as wavelength in the microwave region.

The backscatter cross section (Eq. 10.10) is called the "Rayleigh approximation" because its wavelength dependence is the same as the scatter cross sections of atmospheric molecules whose diameters are small compared with the optical wavelengths, a scattering condition first accurately elucidated by Lord Rayleigh. If the target is small compared with wavelength, the scatter energy is radiated nearly isotropically. Equation 10.10 shows that waves at shorter λ are more strongly scattered—a fact that Rayleigh used to explain why the sky is blue. Large targets have a much more directive scatter radiation pattern, and often more radiation flows in directions other than back to the transmitter. For example, scattered power density in the direction of r (forward scatter can be 100 to 1,000 times larger than that returned in the direction of the source (known as the "Mie effect"; Born and Wolf, 1964).

Battan (1973) and Atlas (1964) have reviewed the abundance of experimental and theoretical work that relates a particle's scatter cross section to its shape, size relative to wavelength when $D \geq \lambda/16$, temperature, and mixture of phases (e.g., water-coated ice spheres).

d. Attenuation

Were it not for electromagnetic energy absorption by water or ice drops, radars with shorter wavelength radiation would be in much greater use because of their superior spatial resolution (Eq. 10.6). Short-wavelength (e.g., $\lambda = 3$ cm) radars suffer echo power loss that can be 100 times larger than that of radars operated with $\lambda \geq 10$ cm.

Each drop absorbs an amount of power that can be expressed as

$$P_L = \sigma_a S_i, \qquad (10.11)$$

where σ_a is the absorption cross section that intercepts a power equal to the power dissipated as heat in the drop. In contrast with Eq. 10.10, there is no simple formula for σ_a that can be used for different wavelength radars and for different drop temperatures; we must resort to numerical evaluations of complicated solutions (Herman et al., 1961) or measurements. In addition, the incident wave suffers power loss from scatter. Analogous to the backscatter cross section is a total scatter cross section σ_s that accounts for the total power scattered by a drop. Total scatter cross section is the area that, when multiplied by the incident power density, gives a power equal to that scattered by the target in all directions. σ_s is proportional to the integral, over a sphere enclosing the target, of the scattered power density $S_r(\theta',\phi')$; for small spherical drops, $\sigma_s = 2\sigma_b/3$. Thus the total power extracted from a wave is proportional to $\sigma_a + \sigma_s$.

If we assume that the presence of drops within an elemental volume ΔV does not significantly alter the incident power density S_i within this volume (i.e., we can neglect scattering of the scattered field—the Born approximation), then the power density change in a wave propagating a short distance Δr through the volume is

$$\Delta S_i = -\frac{\Delta r}{\Delta V}\sum_{n=1}^{N}(\sigma_{sn} + \sigma_{an})S_i, \qquad (10.12)$$

where the summation extends over all N drops within ΔV, and σ_{an} and σ_{sn} are the absorption and scatter cross sections of the nth particle. The rate of power density loss is then

$$\frac{dS_i}{dr} = \lim_{\Delta r\to 0}\left(\frac{\Delta S_i}{\Delta r}\right) = -kS_i, \qquad (10.13)$$

where

$$k = \lim_{\Delta r\to 0}\sum\frac{(\sigma_{an} + \sigma_{sn})}{\Delta V}$$

is the specific attenuation, or, as denoted by some authors, the attenuation coefficient. In the limit $\Delta r \to 0$, S_i can be considered a constant within ΔV and thus can be placed outside summation. Then the power density at any range r is the integral solution of Eq. 10.12:

$$S_i(r_2) = S_i(r_1)\exp\left(-\int_{r_1}^{r_2}kdr\right). \qquad (10.14)$$

The specific attenuation expressed in decibels per kilometer is

$$K = \frac{d}{dr_2}\left[10\log\frac{S(r_1)}{S(r_2)}\right]$$
$$= 4.34 \times 10^3 k \text{ dB km}^{-1} \qquad (10.15)$$

Values of K have been tabulated by Bean and Dutton (1966, pp. 292–96), and Fig. 10.6 plots K for various rainfall rates and radar wavelengths as well as for various reflectivity factors (see Sec. 3c[3] for explanation of Z).

Data on cloud attenuation rate versus wavelength and temperature are given by Gunn and East (1954). Easy-to-use formulas are given by Doviak and Zrnić (1984). Although attenuation through ice clouds is usually negligibly small, it should not be ignored for water clouds, where the attenuation rate is generally an order of magnitude larger.

When drop diameters are small compared with wavelength, attenuation becomes independent of drop size distribution and depends only on cloud liquid mass density. Clouds between the radar and a distant storm can produce measurable attenuation even at $\lambda = 10$ cm (e.g., 1.8 dB for two-way propagation through 100 km of 1 g m^{-3} liquid water) and thus could cause underestimates of rain in the distant storm.

Besides attenuation owing to rain and/or cloud droplets, there is attenuation owing to energy absorbed by the atmosphere's molecular constituents, mainly water vapor and oxygen. This gaseous attenuation rate k_g is not negligible when storms are far away ($r \geqslant 60$ km) and beam elevation is low, even at $\lambda = 10$ cm if accurate cross-section measurements are required. In such a case a correction of 1–2 dB is typically required. Brandes and Wilson (Chap. 11 of this volume) give expressions for attenuation owing to water and gas as well as rainfall and cloud.

Combining Eqs. 10.8, 10.9, and 10.14, we deduce that the echo power density at the radar antenna is

$$S_r(r,\theta,\phi) = \frac{P_t g_t f^2(\theta,\phi) l^2 \sigma_b}{(4\pi r^2)^2}\text{ W m}^{-2}, \qquad (10.16)$$

where $l = \exp - \int(k_g + k)dr$ is the one-way loss factor owing to gaseous and droplet (both cloud and precipitation) attenuation.

e. Doppler Radar Receiving Aspects

The echo power P_r collected by the antenna system from a wave scattered by a target at r,θ,ϕ, is

$$P_r = S_r(r,\theta,\phi)\, A_e(\theta,\phi), \qquad (10.17)$$

where A_e is the effective aperture area of the antenna for radiation from direction θ,ϕ. Theoretically (Silver, 1949; Sherman, 1970, pp. 9–12),

$$A_e = \frac{g_r\lambda^2}{4\pi}f^2(\theta,\phi), \qquad (10.18)$$

where g_r, the gain of the receiving antenna, is equal to g_t if the transmitting antenna is used for echo reception and P_r is measured at the same location as P_t. For large-aperture antennas and targets along the beam, A_e is roughly the physical area of the antenna reflector. For example, the Doppler radar antenna (Fig. 10.5) of the National Severe Storm Laboratory (NSSL) has an $A_e = 42$ m^2, but its physical area is 65.7 m^2. The significantly smaller A_e is caused principally by tapering the illumination of the reflector, which is equivalent to weighting the power density reflected from the parabolic surface. Furthermore, radome and waveguide losses (omitted from our example) cause A_e to be smaller than the physical area.

(1) The Radar Equation: Combining Eqs. 10.16–10.18 gives the radar equation for a discrete target having backscatter cross section σ_b:

$$P_r = \frac{P_t g^2 \lambda^2 l^2 \sigma_b f^4(\theta,\phi)}{(4\pi)^3 r^4}\text{ (W)}, \qquad (10.19)$$

where we have substituted $g_t = g_r = g$. This equation relates echo power P_r to the radar parameters and target location.

As an example of the tremendous sensitivity that radar has for detection of some extremely small targets, consider that NSSL's 10-cm ($\lambda = 0.1$ m) radar receivers can detect echo power as weak as 10^{-14} W, its transmitted peak power is 10^6 W, and its antenna gain is 4×10^4. Thus, using Eq. 10.19, the minimum detectable cross section at a range of 20 km is $\sigma_b(\min) = 2 \times 10^{-7}$ m^2. That is, a single water drop with diameter 6.3 mm could be detected at 20 km. Thus it would not take a very large number of drops to provide a detectable signal at longer range (e.g., 200 km)—and, at that, the transmitted power and receiver performance of the NSSL radars are only moderate in relation to advanced systems in use today (as in deep-space probes, for example).

(2) The Received Waveform (Inphase and Quadrature Components): Consider a single target. If receiver bandwidth is large enough (see Sec. 2*f*[1]), the echo signal voltage $V(t)$ replicates the transmitted electric field waveform E and is proportional to it:

$$V(t,r) = A\{\exp[j2\pi f(t - 2r/c) + j\psi]\}\, U(t - 2r/c), \quad (10.20)$$

where $2r$ is the total path traversed by the incident and scattered waves, A is now the complex voltage (containing the phase shift ψ_i produced by the scatterer) at the input to the mixer detector (Fig. 10.3), and, as before, $U = 1$ when its argument is between 0 and τ, zero otherwise. Range time $\tau_s \equiv 2r/c$ specifies, in units of time, the target location. The phase of the echo,

$$\beta = 2\pi f(t - 2r/c) + \psi, \quad (10.21)$$

is dependent on t as well as r.

An important receiver component is the STALO (for stabilized local oscillator), which oscillates at the transmitted frequency f plus (or minus) a small but fixed offset frequency f_Δ. (For simplicity Fig. 10.3 shows homodyning wherein STALO frequency is the same as the transmitted frequency; i.e., $f_\Delta = 0$.) A portion of its signal is summed with the echo signal, whose strength is weak relative to the STALO's. For this condition the sum of the two can be well approximated by an expression in which the STALO signal is modulated by another that replicates the echo signal in both phase and amplitude shape except that its frequency is f_Δ (Rideout, 1954, p. 311). The summed signals are applied to a nonlinear (usually square law) device (mixer-detector) whose output contains sum and difference frequencies and harmonics of these. Filtering is used to remove the STALO plus harmonic signals while retaining the modulating signal. This summing, nonlinear operation and filtering is called heterodyning if $f_\Delta \neq 0$. Thus the voltage $V_0(t)$ at the filter output (Fig. 10.3) is

$$V_0(t) = Ae^{-j(4\pi r/\lambda - \psi)} U(t - 2r/c) e^{j2\pi f_\Delta t}, \quad (10.22)$$

if losses are ignored. Heterodyning serves only to convert (shift) the carrier frequency (i.e., f) without affecting the modulation envelope. Usually f_Δ is selected to be an intermediate frequency (e.g., 30 MHz) much higher than the frequencies contained in the spectrum of $U(t - 2r/c)$. However, in order to have signals that can be handled by digital data-processing and video-display equipment, it is necessary to convert the intermediate frequency to the baseband at which $f_\Delta = 0$.

We assume for now that the filter rejects all harmonics of $f, f + f_\Delta$, as well as the harmonics of the sum and difference frequencies, but has sufficient bandwidth so that we can ignore loss of frequencies that constitute Eq. 10.22 (see Taylor and Mattern, 1970). The STALO signal is cw (continuous wave) so that whenever an echo arrives a STALO signal is mixed with it. A Doppler radar usually has two channels (without which direction of target motion toward or away cannot be determined); in one the STALO signal is shifted by 90° before mixing so that its detected and filtered output is (assuming $f_\Delta = 0$; Fig. 10.3)

$$V_0(t) = Ae^{-j(4\pi r/\lambda - \psi - \pi/2)}\, U(t - 2r/c). \quad (10.23)$$

Now the actual signals from each mixer are the real parts of Eqs. 10.22 and 10.23, which are designated

$$I(t) = \frac{|A|}{\sqrt{2}} U(t - 2r/c)\cos\left(\frac{4\pi r}{\lambda} - \psi - \psi_i\right) \quad (10.24a)$$

and

$$Q(t) = \frac{-A}{\sqrt{2}} U(t - 2r/c) \times \sin\left(\frac{4\pi r}{\lambda} - \psi - \psi_i\right), \quad (10.24b)$$

the inphase $I(t)$ and quadrature $Q(t)$ components (see Fig. 10.2) of the complex signal $V(t)$. Echoes from stationary targets have complex signals in which phase, $\gamma = -(4\pi r/\lambda) + \psi + \psi_i$, is time-independent. If r increases with time, the phase decreases, and the time rate of phase change,

$$\frac{d\gamma}{dt} = -\frac{4\pi}{\lambda}\frac{dr}{dt} = -\frac{4\pi v_r}{\lambda} = 2\pi f_d, \quad (10.25a)$$

where

$$f_d = -\frac{2v_r}{\lambda} \quad (10.25b)$$

is the Doppler shift (in cycles per second).

It is relatively easy to see from Eq. 10.24 that, for usual radar conditions (i.e., $\tau \sim 10^{-6}$ s) and meteorological target velocities on the order of tens of meters per second, the change in the trigonometric functions is extremely small during the time when $U(t - 2r/c)$ is nonzero. Thus we measure echo phase change during the longer time, $T_s \sim 10^{-3}$ s, from echo pulse to echo pulse rather than during a pulse period. Because of this the pulsed-Doppler radar behaves as a phase- as well as an amplitude-sampling device; samples are at $t = \tau_s + (n - 1)T_s$, where τ_s is the time delay between the nth transmitted pulse and its echo. τ_s is called "range time" because it is proportional to range (i.e., $\tau_s = 2r/c$). It is convenient to introduce another time scale, "sample time"; that is, time is incremented in discrete steps of length T_s, the sample time, after $t = \tau_s$ (see Fig. 10.28 in Sec. 3). It is important to realize that echo phase and amplitude changes are examined in sample-time space at the discrete instants $(n - 1)T_s$ for a target echo at range time τ_s. Figure 10.7 shows how samples of $I(t)$ and $Q(t)$ change for stationary and moving targets. It is quite evident that, if $Q(t)$ is negative and increasing in a positive direction while $I(t)$ is positive, the Doppler shift is positive (i.e., the angle γ increases in a counterclockwise direction), and the target is moving toward the radar.

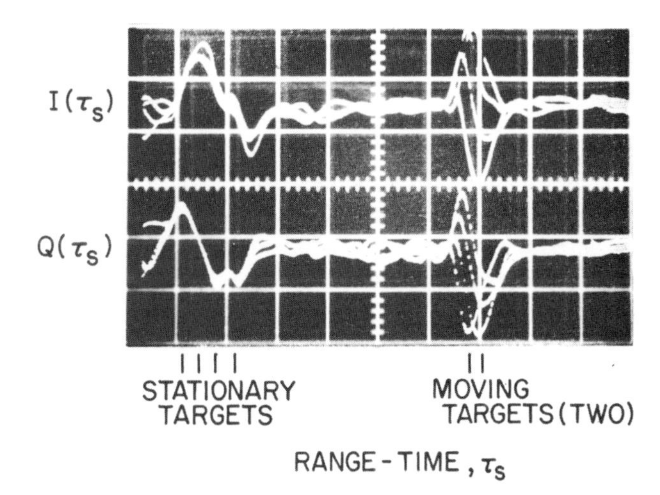

Figure 10.7. Samples of inphase $I(\tau_s)$ and quadrature phase $Q(\tau_s)$ components of the echo signal for moving and stationary targets. Four successive sampling intervals T_s have been superimposed to show the relative change of amplitude from echo sample to echo sample of targets at range-time τ_s. Vertical scale is linear in voltage, and range-time scale, τ_s, is 2 μs/DIV.

f. Radar Measurement Limitations

The discussion so far treats an ideal radar to facilitate understanding of radar principles, but it omits some important limitations in radar measurements.

(1) Receiver Noise, Bandwidth, and Signal-to-Noise Ratio: In an ideal noiseless radar receiver, there is no lower limit in detecting the weakest echo. However, all receivers have a power detection limitation usually imposed by the random electrical signals associated with thermal agitation of electron motion in each receiver's front end (in Fig. 10.3 it is the mixer). Some high-performance receivers with parametric amplifiers or other low-noise amplifiers before the mixer have so small a thermal noise that microwave emission from the Earth's surface and outer space can be the more significant contributor to the overall noise power referred to the receiver's input. Furthermore, emission from lightning and signals from other extraneous sources are noise to the radar. However, in most weather radar systems, thermal noise power N limits receiver performance, and its value,

$$N = kT_{sy}B_n, \qquad (10.26)$$

is a characteristic constant of the radar, where k is Boltzmann's constant, (1.38×10^{-23} W s K^{-1}), B_n is the receiver's noise bandwidth, and T_{sy} is the effective system temperature at the receiver's front end. The system temperature includes thermal-power contributions from environmental noise received by the antenna and thermal radiation owing to losses in the antenna's transmission line and protective radome, as well as thermal noise within the receiver itself (Blake, 1970, pp. 2–29; Doviak and Zrnić, 1984).

The echo pulse, as well as any signal, can be synthesized from its Fourier spectral components. The receiver amplifies only a band of these spectral or frequency components; i.e., the receiver has a finite bandwidth. The receiver's filter (Fig. 10.3) response $G(f)$ is usually a monotonically decreasing function of frequency, and its width, B_6, is best specified as the frequencies within which $G(f)$ is larger than one-fourth of its highest level—its 6-dB width (Taylor and Mattern, 1970). Noise bandwidth B_n is not as simply specified because it depends on $G(f)$ as well as noise power spectrum, but in practice it is nearly equivalent to the 3-dB width if noise power is white, i.e., independent of frequency (Kraus, 1966, p. 265).

The larger the B_6, the better the fidelity of the echo pulse shape, but noise power increases in proportion to B_6. We also run the risk of accepting unwanted products of the mixer output. Usually it is not important to detect the echo with nearly perfect fidelity, nor is it advisable to reduce noise excessively by decreasing B_6 because, as we show below, smaller filter bandwidth coarsens resolution. The ratio of peak echo power to noise power can be shown to be (Doviak and Zrnić, 1979, 1984)

$$\text{SNR} = \frac{(I_0^2 + Q_0^2)\,\text{erf}^2\,(aB_6\tau/2)}{kT_{sy}B_n}, \qquad (10.27)$$

where I_0 and Q_0 are the prefilter echo amplitudes $|A|/\sqrt{2}$ (Eq. 10.24), τ is the pulse width, $a = \pi/2\sqrt{ln 2}$, and erf is the error function (Abramowitz and Stegun, 1964). Equation 10.27 is the SNR for a receiver having a Gaussian-shaped $G(f)$ which often approximates the actual $G(f)$. If the receiver's $G(f)$ is matched to the spectrum of the detected echo pulse (assumed to replicate the rectangular shape of the transmitted pulse), then the output of the filters would be a triangular pulse whose base has a duration 2τ.

An important measurement made with radar is range to the target. Furthermore, radar often needs to resolve targets that are closely spaced and have largely different backscatter cross sections. For a noiseless receiver of infinite band-

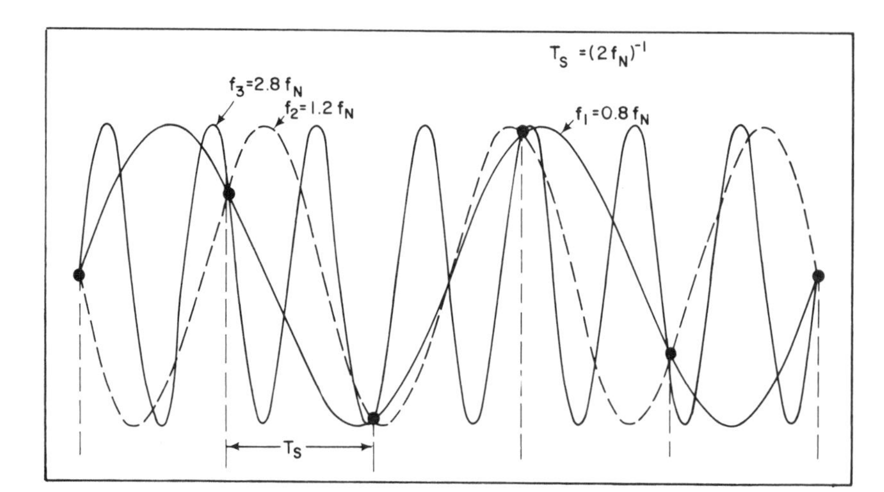

Figure 10.8. Signals at three different Doppler frequencies that yield, when sampled, the same set of data. These Doppler frequencies are aliases of each other. If both I and Q signals were sampled, it can be shown that the signal with f_2 would be an alias not of that with f_1 but of that having a frequency $-2f_N + f_1$. $T_s = (2f_N)^{-1}$.

width, targets are resolved if their spacing is wider than one-half the spatial pulse width (i.e., $c\tau/2$), and targets so spaced return echoes that arrive at separate and distinct time intervals. Finite bandwidth receivers distort the echo pulse by reverberating echo power within the receiver in amounts decreasing with time after the peak echo signal (for example, see Fig. 10.4). This causes the receiver output signal to have widths larger than τ, and weak echoes could then be masked by strong ones.

(2) Ambiguities: Two kinds of ambiguity, range and velocity, plague measurements with Doppler radars. When Doppler radars are operated with uniform PRT (i.e., T_s) and targets have range r larger than $cT_s/2$, their echoes for the nth transmitted pulse are received after the $(n + 1)^{th}$ pulse is transmitted. Therefore, these echoes are received during the same time interval that targets at $r < cT_s/2$ return echoes from the $(n + 1)^{th}$ pulse. Thus range r to the distant target has an apparent value $r' = r - (N_t - 1)r_a$ and is ambiguous with targets at $r = r'$. (N_t is the trip number, and $N_t - 1$ designates the whole number of $cT_s/2$ intervals between the target and receiver; $r_a = cT_s/2$ is that range within which all targets must lie to have their ranges unambiguously measured.) Targets for which $N_t = 2$ produce what are called "second-trip" echoes. We emphasize that r_a does not necessarily limit the range to which the pulsed-Doppler radar can achieve useful measurement. If its STALO (Fig. 10.3) is phase-coherent over many T_s intervals, the radar can accurately measure velocities of targets far beyond r_a, but interference occurs when echoes from targets in different trips are received simultaneously.

The second ambiguity relates to measurement of target velocity. As discussed in Sec. 2e(2), the target's phase γ is sampled at intervals T_s, and its change $\Delta\gamma$ over the interval T_s is a measure of the Doppler frequency $f_d = \Delta\gamma/2\pi T_s$. Unfortunately, given a set of sampled phases (computed from I, Q samples), we cannot relate them to a unique Doppler frequency. Fig. 10.8 shows the same set of samples that could have resulted from any one of three signals having different Doppler frequencies. Signals that fit the sample data set are called "aliases," and $f_N = (2\,T_s)^{-1}$ is the Nyquist (or folding) frequency. All Doppler frequencies between $\pm f_N$ are the principal aliases, and any frequency higher than f_N is ambiguous with those between $\pm f_N$. Thus target radial velocities must lie within $\pm v_a$ ($v_a = \lambda/4T_s$) to avoid velocity ambiguity. However, transmission of pulses at two interlaced pulse repetition frequencies (PRF's) and signal processing can extend the unambiguous velocity (Doviak, 1978; Doviak and Zrnić, 1984).

3. Weather Echo Signals

Section 2 delineated the radar signal properties of waves scattered from discrete targets. It has been assumed that no more than one target lies in the range interval $r_i \pm c\tau/4$, where r_i is the range to the ith target, and that the target dimensions are small compared to $c\tau/2$. Targets satisfying these conditions are called "point targets." Weather echoes are composites of signals from a very large number of hydrometeors each of which can be considered a point target. Collectively they are designated distributed targets or clutter. After a delay (the round-trip propagation time between the radar and the near boundary of the scatter volume) echoes are continuously received over a time interval equal to twice the time it takes the microwave pulse to propagate across the volume containing scatterers. Because one cannot resolve the individual targets, we resort to sampling, at discrete range time delays τ_s, the composite continuous echo that forms a complex voltage $V(t) = I(t) + jQ(t)$ (Fig. 10.9).

Each voltage sample is a weighted composite of discrete echoes from all the scatterers, with weights determined by the radiation pattern $f^2(\theta,\phi)$ and a range weighting function $W(r)$ dependent on the product of receiver bandwidth and transmitted pulse width $B_6\tau$. Figure 10.10 illustrates an

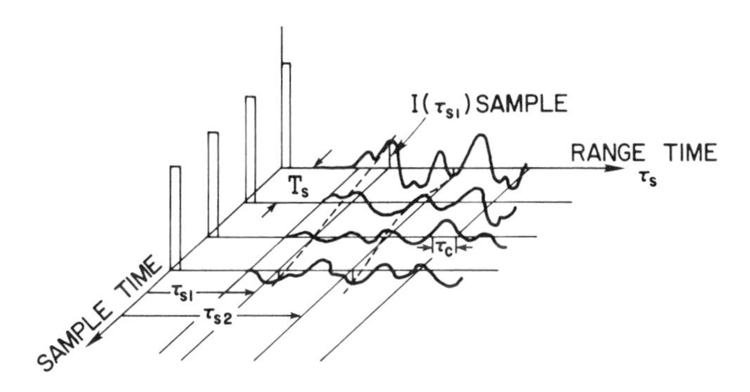

Figure 10.9. Idealized traces for the inphase component, $I(\tau_s)$, echoes from distributed targets. Each trace represents echoes from a single transmitted pulse. Instantaneous samples are taken at τ_{s1}, τ_{s2}, etc. Dashed line indicates probable time dependence of sample at τ_s if sampling rate T_s^{-1} were increased and ambiguities resolved. τ_c depicts signal correlation time along τ and is related to $B_6\tau$. Samples at fixed τ_s taken at T_s intervals are used to construct Doppler spectra for scatterers located about the range $c\tau_s/2$.

echo from a single point target at the output of the receiver filter. The weighting functions determine a resolution volume in space where targets contribute significantly to the echo sample at τ_{s1}.

The echo signal from each scatterer in the resolution volume constructively or destructively (depending on the phase of the signal) interferes with signals from other scatterers to produce a composite echo sample $V(\tau_{s1})$, and the random size and location of scatterers cause the amplitude and phase of $V(\tau_{s1})$ to be random variables. The echo sample at another range time delay τ_{s2} is a composite signal from scatterers in a different resolution volume, and hence we expect $V(\tau_{s1})$ to differ from $V(\tau_{s2})$. Thus $V(\tau_s)$ fluctuates as τ_s increases (Fig. 10.9) even when the scatterer density is spatially uniform. The correlation between $V(\tau_s)$ samples taken at different τ_s is related to the radial dimensions of the resolution volume and the spacing $\delta\tau_s$, between samples.

In this section we determine the statistical properties of the echo samples $V(\tau_s)$ and instantaneous power $P(\tau_s)$, relate the average of successive $P(\tau_s)$ samples to the backscatter cross section of targets within the resolution volume, and develop a form of the weather radar equation that accounts for range weighting.

a. The Echo Sample

As mentioned previously, the echo sample $V(\tau_s)$ is a composite

$$V(\tau_s) = \frac{1}{\sqrt{2}} \sum_i A_i W_i e^{-j4\pi r_i/\lambda} \qquad (10.28)$$

of signals, where $|A_i|/\sqrt{2} = (I_i^2 + Q_i^2)^{1/2}$ is the prefilter echo amplitude (see Eq. 10.24) of the ith scatterer located at r_i, ϕ_i, and θ_i. W_i is the corresponding range weighting function. We are not yet specific about the functional form of W_i. For now it suffices to note that the range dependence of W_i is such that it principally weights those targets residing near a range r determined by τ_s (i.e., $r = c\tau_s/2$). Furthermore, the antenna weighting factor $f^2(\theta,\phi)$ is assumed to be part of A_i. We can ignore the constant phase ψ given in Eq. 10.22. We also assume target velocity slow enough that W_i is independent of it. This is tantamount to assuming that the Doppler shift is small compared to B_6; otherwise, fast-moving targets would return signals at frequencies that fall outside the bandwidth of the receiver. Weather targets never move at speeds to shift echoes outside the receiver bandwidth.

The sample, at range time τ_s, of echo power averaged over a cycle of the transmitted frequency f is proportional (see Eq. 10.3) to

$$P(\tau_s) = VV^* = \frac{1}{2} \sum_{i,k}^{N_s} A_i A_k^* W_i W_k^* \exp[j4\pi(r_k - r_i)/\lambda]$$

$$= \frac{1}{2} \sum_i |A_i|^2 |W_i|^2 + \frac{1}{2} \sum_{i \neq k} A_i A_k^* W_i W_k^* \exp[j4\pi(r_k - r_i)/\lambda]. \qquad (10.29)$$

We have summed the power in the I and Q channels, and therefore the factor of 1/2 in front of . . .VV^* required for the power in either the I or Q channels of real signals is not present in Eq. 10.29. The instantaneous echo power $P(\tau_s)$ is for one transmission, and N_s is the number of scatterers. An instantaneous power is considered to be the power averaged over any one cycle of the radar frequency. When scatterers move relative to one another, the echo sample at range time τ_s differs for each transmitted pulse, and the amount of change depends on the sampling time interval T_s (the interval between transmitted pulses) and the relative velocity of the scatterers. The equation defining $P(\tau_s)$ for successive transmission will have the same form, but the r_i and r_k will have changed owing to scatterer motion. If scatterers within a sample volume move randomly a significant fraction of a wavelength (e.g. $\lambda/4$) between successive transmissions, each successive echo sample $V(\tau_2)$ (spaced T_s apart) will be uncorrelated. In order to make

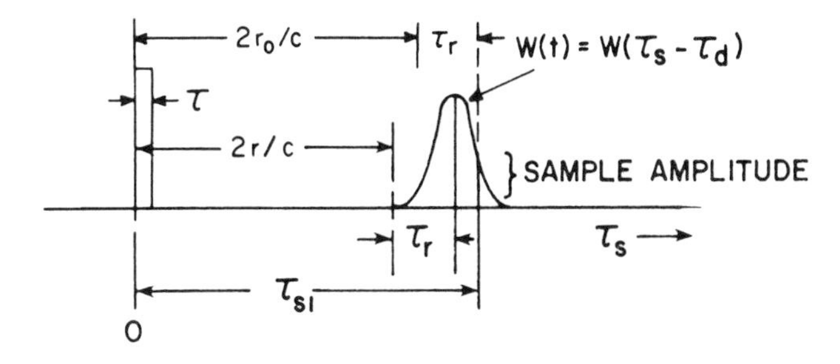

Figure 10.10. Realistic receiver response to point target echo of a rectangular transmitter pulse. Time delays relate to target range r, receiver response delay τ_r, and the sample range-time delay τ_{s1}. τ_s = 0 is the time at which the transmitter signal is emitted from the antenna.

coherent Doppler measurements of the scatterer's mean radial speed, the time T_s between successive samples must be small enough that successive echoes, at fixed delay τ_s, are correlated.

Whereas the second sum in Eq. 10.29 represents a rapidly fluctuating contribution to the instantaneous power $P(\tau_s)$, the first sum is relatively constant if scatterer's displacement is small compared to the distances over which the weighting functions change appreciably. Hydrometeor displacements of the order of a wavelength can cause large changes in the second sum, whereas they need to be displaced 100 m or more (for typical weather radar operating parameters) before significant change occurs in the first term. Fluctuations are caused by target displacements, which change the phase of each echo sample. Although the second sum can be significantly larger than the first (it has $N_s(N_s - 1)$ contributions, compared to N_s for the first term) for some echo samples, its average over many successive samples (i.e., its sample-time average) approaches zero as the number of samples increases without limit, because the average of the complex exponential term tends to zero. The first sum is then the sample time mean power $\overline{P}(\tau_s)$ if, as stated earlier, the scatterer's displacement on the average is small compared to the distances over which the weighting functions change significantly, or, from another point of view, if, during the sample-time average, scatterers displaced far from the region where $f^2(\theta_i,\phi_1)\ W(r)$ has significant weight are replaced by others having the same statistical properties. In other words, we assume that the statistical properties (e.g., the mean scatter cross section per unit volume) of the scattering medium are stationary. In practice sample-time averaging intervals are much less than 1 second, and hence scatterers are displaced much less than a few tens of meters during the period of average power estimation. However, we stress that $\overline{P}(\tau_s)$ estimates can be made only because hydrometeors are moving relative to one another. An accurate estimate of $\overline{P}(\tau_s)$ is important because it relates to the estimates of liquid water in the resolution volume.

b. Signal Statistics

The I and Q components of the echo sample are random variables if the scatterer's positions are unpredictable. Let us consider two consecutive echo samples spaced T_s seconds apart. The first is given by Eq. 10.28, and the second can be written

$$V(\tau_s,T_s) = \frac{1}{\sqrt{2}}\sum_i |A_iW_i|\cos\gamma_i - j\frac{1}{\sqrt{2}}\sum_i |A_iW_i|\sin\gamma_i, \quad (10.30a)$$

where

$$\gamma_i = \frac{4\pi r_i}{\lambda} + \frac{4\pi v_i T_s}{\lambda} - \psi_i, \quad (10.30b)$$

$v_i = \Delta r_i/T_s$ is the average radial velocity needed to move the ith scatterer by Δr_i, and ψ_i contains the phase owing to scattering and the phase of W_i. Because the range extent of the resolution volume is much larger than the wavelength ($c\tau/2 > \lambda$), and there are many scatterers, it is natural to expect the first term of Eq. 10.30b to be uniformly distributed between $-\pi$ and π. Even though the distribution of $4\pi r_i/\lambda$ need not be uniform, its width usually spans many intervals of 2π so that multiple aliasing (of phases into the unambiguous 2π interval) causes the distribution across 2π to be, for all practical purposes, uniform. Therefore, regardless of the v_i or ψ_i distributions, the phases γ_i are also uniformly distributed. This follows because the distribution of the sum of random variables is obtained after convolving (on a circle from $-\pi$ to π) the individual distributions. Because one of them is uniform, the distribution of the sum will always be uniform.

Now we are in a position to apply the central limit theorem to the real and imaginary parts of Eq. 10.30a. The theorem states that a sum of independent random variables tends to have a Gaussian distribution if their number is large and none of the variables is dominant (i.e., much larger than the rest). Both conditions are certainly true for hydrometeors, and thus the $I(\tau_s,T_s)$ and $Q(\tau_s,T_s)$ have Gaussian distribution with zero mean. It is worth noting that

$$I(\tau_s,T_s) = \frac{1}{\sqrt{2}}\sum_i |A_iW_i|\cos\gamma_i = |V(\tau_s,T_s)|\cos[\theta(\tau_s,T_s)], \quad (10.31a)$$

$$Q(\tau_s,T_s) = -\frac{1}{\sqrt{2}}\sum_i |A_iW_i|\sin\gamma_i = |V(\tau_s,T_s)|\sin[\theta(\tau_s,T_s)]. \quad (10.31b)$$

The equality in Eq. 10.31 follows because a sum of sinusoids can always be expressed as a sinusoid with a phase $\theta(\tau_s,T_s)$ and an amplitude factor (envelope) $|V(\tau_s,T_s)|$. However, this does not mean that I and Q have pure sinusoidal variation with time.

In addition to being Gaussian, the in-phase and quadrature components are independent random variables (Doviak and Zrnić, 1984). Therefore, the joint probability distribution of I and Q is the product of the individual probabilities:

$$p(I,Q) = (1/2\pi\sigma^2)\exp(-I^2/2\sigma^2 - Q^2/2\sigma^2), \quad (10.32)$$

where σ^2 is the mean square value of I (equal to that for Q).

From Eqs. 10.31 and 10.32 one can obtain, using well-established procedures, the distributions of $|V|$, θ, and the power $P(\tau_s)$ (Papoulis, 1965, p. 418). It can be shown that the phase θ is independent of the amplitude $|V|$ and is uniformly distributed, while the amplitude $|V| = (I^2 + Q^2)^{1/2}$ has Rayleigh probability density:

$$p(|V|) = (|V|/2\sigma^2)\exp(-|V|^2/2\sigma^2). \quad (10.33)$$

Because the power $P(\tau_s)$ is proportional to $I^2 + Q^2$, it follows from Eq. 10.32 that $P(\tau_s)$ is exponentially distributed with density

$$p(P) = (1/2\sigma^2)\exp(-P/2\sigma^2) \qquad (10.34)$$

and a mean value $\overline{P}(\tau_s) = 2\sigma^2$. Constants of proportionality and impedance factors that make the transition from Eq. 10.33 to Eq. 10.34 dimensionally correct have been ignored and will be ignored henceforth. Radar receivers have gains and losses, and therefore the receiver needs to be calibrated with a known input power to relate accurately the output $I^2 + Q^2$ to the echo power.

We emphasize that, although I and Q are independent random variables, the stochastic processes $I(\tau_s, nT_s)$ and $Q(\tau_s, nT_s)$ are not independent. This means that in general the expected value $E[I(\tau_s, mT_s)Q(\tau_s, kT_s)] \neq 0$ for $k \neq m$ (see Doviak and Zrnić, 1984, for the proof). The correlation between two successive samples of the complex signal will be appreciably different from zero only if the distribution of $4\pi v_i T_s/\lambda$ in Eq. 10.30b is narrow compared to 2π, which is equivalent to saying that the distribution of the v_i is narrow compared to $\lambda/2T_s$ (i.e., the radar's Nyquist velocity).

c. The Weather Radar Equation

We now relate the sample-time average of echo power $\overline{P}(\tau_s)$ to the radar parameters and target cross section. The contribution to the average echo power from each scatterer is, from Eq. 10.29,

$$P_i = \tfrac{1}{2}|A_i|^2|W_i|^2, \qquad (10.35)$$

where $\frac{1}{2}|A_i|^2$ is the prefilter echo power and hence can be directly expressed in terms of radar parameters and target cross section through use of Eq. 10.19. Thus the sample-time average power at delay τ_s is

$$\overline{P}(\tau_s) = \frac{P_t g^2 \lambda^2}{(4\pi)^3} \sum_i \frac{l_i^2 \sigma_{bi} f^4(\theta_i, \phi_i)|W_i|^2}{r_i^4}, \qquad (10.36)$$

where σ_{bi} is the expected or average value of the backscatter cross section of the ith hydrometeor. It is necessary to use the expected value because hydrometeor scatter cross sections are not constant (e.g., drops vibrate; snowflakes wobble). Now consider an elemental volume ΔV and the sum of σ_{bi} over this volume divided by ΔV which defines the reflectivity

$$\eta \equiv (\Delta V)^{-1} \sum_{\Delta V} \sigma_{bi}, \qquad (10.37)$$

which is the scatter cross section per unit volume. The terms $l_i, f(\theta_i, \phi_i), r_i$, and W_i are assumed not to vary significantly over this elemental volume. Because the parameters η, f, and W are well-behaved functions, we replace the sum by an integration to obtain the following form of the weather radar equation:

$$\overline{P}(\mathbf{r}_0) = \frac{P_t g^2 \lambda^2}{(4\pi)^3} \int_0^{r_2} \int_0^{\pi} \int_0^{2\pi} \frac{\eta(\theta, \phi, r) l^2}{r^4} f^4(\theta, \phi)|W(r)|^2 dV, \qquad (10.38)$$

where

$$dV = r^2\, dr \sin\theta\, d\theta\, d\phi,$$

and θ, ϕ are angular positions relative to the mainbeam axis, whose vector radius $\mathbf{r}_0$ is directed to the location of maximum weight (i.e., the center of the resolution volume). The upper limit for the r integration does not extend to infinity because targets beyond some range r_2 cannot return echoes soon enough to contribute to the echo sampled at delay τ_s. In the next section we are more specific about the range of the limits of integration for r and the functional form of $W(r)$. Here we need only recognize that Eq. 10.38 does not hold for $r < 2D^2/\lambda$.

In general η, l are functions of r, but let us assume that $f^4(\theta, \phi)|W(r)|^2$ has a scale (resolution volume dimensions) such that the reflectivity and attenuation can be considered constant over the region that contributes most to $P(\tau_s)$. We also assume insignificant contribution from regions outside. Soon we shall be more precise as regards resolution volume size. Furthermore, we assume that the range to the resolution volume is large compared to its range extent. Thus we can approximate Eq. 10.38 by

$$\overline{P}(\mathbf{r}_0) \simeq \frac{P_t g^2 \lambda^2 l^2 \eta}{(4\pi)^3 r_0^2} \int_r |W(r)|^2\, dr \int_0^{2\pi} \int_0^{\pi} f^4(\theta, \phi) \sin\theta\, d\theta\, d\phi. \qquad (10.39)$$

When antenna patterns are circularly symmetric and with Gaussian shape, it can be shown (Probert-Jones, 1962) that

$$\int_0^{\pi} \int_0^{2\pi} f^4(\theta, \phi) \sin\theta\, d\theta\, d\phi = \pi\theta_1^2/8\ln 2, \qquad (10.40)$$

where θ_1 is the 3-dB width (in radians) of the one-way pattern.

(1) The Range-weighting Function: The range-weighting integral in Eq. 10.39 causes a loss in weather echo strength. This loss, owing to finite receiver bandwidth, was first enunciated by Nathanson and Smith (1972) to explain residual discrepancies between measurements and theory. Probert-Jones's solution accounted for most of the discrepancy and, in the case $B_6 >> \tau^{-1}$ often used in conventional non-Doppler radar systems, for practically all differences. If the receiver's filter has Gaussian frequency response, it can be shown (Doviak and Zrnić, 1979) that a form of the weather radar equation,

$$\overline{P}(r_0, \theta_0, \phi_0) = \frac{P_t g^2 \lambda^2 l^2 \eta}{(4\pi)^3 r_0^2} \left[\coth(2b) - \frac{1}{2b}\right] \frac{\pi\theta_1^2}{3ln2} \frac{c\tau}{2}, \qquad (10.41)$$

accounts for finite bandwidth filter loss where $2b = \pi B_6 \tau / 2\sqrt{ln2}$. Equation 10.41 shows the dependency of averaged

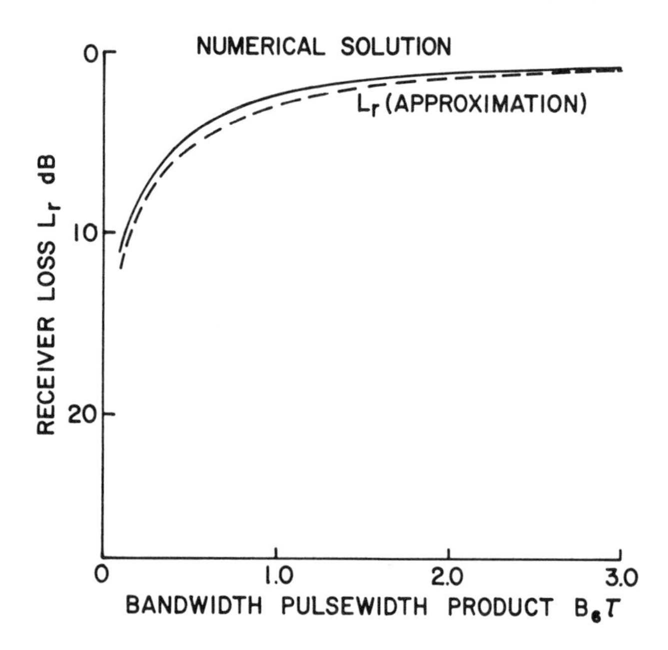

Figure 10.11. Receiver signal power loss L_r (dB) because of finite bandwidth. Receiver frequency transfer is Gaussian, and echo pulse rectangular. The solid curve is a numerical solution of the exact response function, and the dashed curve is obtained from an analytical approximation.

echo power $\overline{P}$ upon measurable quantities for the radar, and the term in the brackets is the loss l_r owing to finite receiver bandwidth, and $L_r = -10 \log l_r$ is plotted in Fig. 10.11. This loss is a function of both transmitted pulse width τ and receiver bandwidth B_6 analogous to the two measurable parameters g and θ_1, of the antenna. For radar receivers having B_6 nearly equal to τ^{-1} or less, l_r is significant and needs to be accounted for in the weather radar equation.

(2) Resolution Volume: It is useful to define a resolution volume V_6 as that circumscribed by the 6-dB contour. Thus the resolution volume is a complicated axially symmetric region determined by the product of range and angular weighting functions. The maximum 6-dB diameter of this region is $r\theta_1$, and it has a maximum range extent given by Doviak and Zrnić (1984):

$$r_6 = \frac{c\tau/2}{aB_6\tau}\cosh^{-1}(2 + \cosh aB_6\tau), \qquad (10.42)$$

where $a = \pi/2\sqrt{ln2}$. When the receiver's bandwidth B_6 is matched to the transmitted pulse width τ (i.e., $B_6\tau = 1$), r_6 is about 20% larger than $c\tau/2$, the resolution for $B_6 \rightarrow \infty$. From Eq. 10.42 we note that both B_6 and τ determine the range distribution of weights, as g and θ_1 determine the angular distribution. The resolution volume is less than 0.1 km^3 for $\theta_1 = 1°$, $r = 50$ km, $B_6 = 1$ MHz, and $\tau = 1$ μs. Outside this volume, reflectivity is weighted by factors less than one-fourth, because of the antenna pattern function and the finite bandwidth receiver response.

(3) Reflectivity Factors: Radar meteorologists need to relate reflectivity (η) to factors that have meteorological significance. If scattering particles are known to be spherical and have diameters that are small compared with wavelength (i.e., the Rayleigh approximation), then we can substitute Eq. 10.10 into Eq. 10.37 to obtain

$$\eta = \frac{\pi^5}{\lambda^4}|K_w|^2 Z, \qquad (10.43a)$$

where

$$Z \equiv \frac{1}{\Delta V}\sum_i D_i^6 \qquad (10.43b)$$

is the reflectivity factor. Whenever the Rayleigh approximation does not apply, and this is usual for short-wavelength ($\lambda < 10$cm) radars measuring thunderstorm precipitation, it is accepted practice to write

$$Z_e \equiv \frac{\lambda^4}{\pi^5}|K_w|^{-2}\eta \qquad (10.43c)$$

where Z_e is the equivalent reflectivity factor. Because values of Z commonly encountered in weather observations span many orders of magnitude, radar meteorologists use a logarithmic scale $10\log_{10}Z$ (where Z is in units of mm^6/m^3). Precipitation produces Z values ranging from near 0 dBZ to values somewhat larger than 60 dBZ in regions of heavy rainfall and hail.

Reflectivity factors do not relate radar echo power to meteorologically significant parameters such as rainfall rate or liquid-water content because one more essential parameter needs to be known in addition to substance phase (i.e., $|K_w|^2$): the drop size distribution (see Chap. 11). Although η or Z_e could be used in the radar equation, radar meteorologists have chosen Z_e.

(4) A Useful Form of the Weather Radar Equation: Incorporating Eq. 10.43c into Eq. 10.41, we have the weather radar equation that gives average echo power in terms of Z_e (m^3):

$$\overline{P}(\mathbf{r}_0) = \frac{\pi^3 P_t g^2 l^2 l_r \theta_1^2 c\tau |K_w|^2 Z_e(\mathbf{r}_0)}{2^{10}(ln2)\lambda^2 r_0^2}, \qquad (10.44a)$$

where all units are in the MKS system (i.e., meters, seconds, and watts). θ_1 is in radians, g is dimensionless, and r_0 is range to the center of the resolution volume.

However, it is common to express Z_e in units of mm^6 m^{-3}, θ_1 in degrees, r_0 in kilometers, λ in centimeters, τ in microseconds, P_t in watts, and $\overline{P}$ in milliwatts. Thus, using units conventional to the radar meteorologist, the mean weather echo power for resolution volume at r_0, θ_0, ϕ_0 is

$$\overline{P}(\text{mW}) = \frac{\pi^5 10^{-17} P_t(\text{W}) g^2 l^2 l_r \tau(\mu\text{s}) \theta_1^2(\text{deg}) |K_w|^2 Z_e(\text{mm}^6/\text{m}^3)}{6.75 \times 2^{14}(ln2) r_0^2(\text{km})\lambda^2(\text{cm})}. \qquad (10.44b)$$

d. Signal-to-Noise Ratio for Weather Targets

The signal-to-noise ratio (SNR) for weather echoes is Eq. 10.41 divided by kTB_n. A filter having a Gaussian-shaped frequency response $G(f)$ has $B_n = 1.06\, B_6/\sqrt{2}$, and then SNR depends on B_6 as

$$\mathrm{SNR} = \frac{C_0}{r^2}\left[\coth(aB_6\tau) - \frac{1}{aB_6\tau}\right]\frac{\tau^2}{B_6\tau}, \qquad (10.45a)$$

where C_0 contains constants pertaining to the radar. If $B_6\tau$ is a constant (a reasonable selection could be 1 because it "matches" the receiver bandwidth to the transmitted pulse width, producing a maximum SNR for point targets), SNR is proportional to the square of the transmitted pulse width. But it can be shown that maximum SNR is obtained as $B_6 \rightarrow 0$, in contrast to point target measurements, where we obtain a maximum SNR at $B_6\tau = 1$ (see Eq. 10.27). Even though SNR increases monotonically as B_6 decreases, resolution r_6^{-1} worsens. Zrnić and Doviak (1978) define an optimum B_6 as the value that maximizes SNR for a given resolution.

4. Doppler Spectra of Weather Echoes

The Doppler spectrum is a power-weighted distribution of the radial velocities of the scatterers that lie mostly within the resolution volume. The power weight depends not only on the reflectivity of the scatterers but also on the weighting given to them by the antenna pattern, the transmitted pulse shape, and the receiver's response to it. A derivation leading to a relationship among the velocity and reflectivity fields, the resolution volume weighting functions, and the Doppler spectrum was first put forward by Sychra (1972). Doviak and Zrnić (1984) derive identical results using a different approach. They show that the power spectrum density $\overline{S}(v)$ (i.e., power per unit velocity) of the Doppler velocities is

$$\overline{S}(\mathbf{r}_0,v) = \iint_A \eta(\mathbf{r}_1)I(\mathbf{r}_0,\mathbf{r}_1)\,|\mathrm{grad}\ v(\mathbf{r}_1)|^{-1}\,dA, \qquad (10.46a)$$

where $\mathbf{r}_0$ and $\mathbf{r}_1$ are the radius vectors to the resolution volume and the elemental scattering volume, A is an area on a surface of constant Doppler velocity $v(\mathbf{r}_1)$, and $I(\mathbf{r}_0, \mathbf{r}_1)$ is the illumination function,

$$I(\mathbf{r}_0, \mathbf{r}_1) = C_1 f^4(\theta - \theta_0, \phi - \phi_0)|W(r_0,r_1)|^2/r_1^4, \qquad (10.46b)$$

where C_1 is a constant.

Equation 10.46 is fundamental and worthy of more discussion. The area A consists of all isodop surfaces (surfaces of constant Doppler velocity); i.e., it is a union of such surfaces. At each point $\mathbf{r}_1$ on the surface, the reflectivity is multiplied with the corresponding illumination function. The gradient term adjusts the isodops' contribution according to their density; i.e., the closer the isodop surfaces, the smaller the weight applied to the spectral components in the velocity interval between two isodops.

For calculation of the mean velocity and the spectrum width, the normalized $\overline{S}_n(\mathbf{r},v)$ version of Eq. 10.46 is used:

$$\overline{S}_n(\mathbf{r}_0,v) = \frac{\overline{S}(\mathbf{r}_0,v)}{\int_{-\infty}^{\infty} \overline{S}(\mathbf{r}_0,v)\,dv}. \qquad (10.47)$$

If the reflectivity and velocity distributions within and about V_6 are uncorrelated random variables, then it can be shown that $\overline{S}_n(\mathbf{r}_0,v)$ is proportional to the probability distribution of radial velocities. Note that the integral in the denominator is the total power and equals the volume integral, Eq. 10.38. The weighted space averaged reflectivity $\overline{\eta}(\mathbf{r}_0)$ is

$$\overline{\eta}(\mathbf{r}_0) \equiv \frac{\iiint \eta(\mathbf{r}_1)I(\mathbf{r}_0,\mathbf{r}_1)\,dV_1}{\iiint I(\mathbf{r}_0,\mathbf{r}_1)\,dV_1}. \qquad (10.48)$$

The integral value in the denominator of Eq. 10.48 can be obtained from Eq. 10.41 for weather radar parameters usually met in practice. Now the mean Doppler velocity is defined as

$$\overline{v}(\mathbf{r}_0) \equiv \int_{-\infty}^{\infty} v\,\overline{S}_n(\mathbf{r}_0,v)\,dv, \qquad (10.49)$$

a combination of reflectivity and illumination function weighted velocity that could be quite different from the $I(\mathbf{r},\mathbf{r}_1)$ weighted velocity. The velocity spectrum width $\sigma_v(\mathbf{r})$ is obtained from

$$\sigma_v^2(\mathbf{r}_0) = \int_{-\infty}^{\infty} [v - \overline{v}(\mathbf{r}_0)]^2\,\overline{S}_n(\mathbf{r}_0,v)\,dv. \qquad (10.50)$$

The relationship between the point velocities $v(\mathbf{r}_1)$ and the power-weighted moment $\overline{v}(\mathbf{r}_0)$ is obtained by substituting Eq. 10.47 into Eq. 10.49:

$$\overline{v}(\mathbf{r}_0) = \frac{\iiint v(\mathbf{r}_1)\,\eta(\mathbf{r}_1)\,I(\mathbf{r}_0,\mathbf{r}_1)\,dV_1}{\iiint \eta(\mathbf{r}_1)\,I(\mathbf{r}_0,\mathbf{r}_1)\,dV}. \qquad (10.51)$$

Unlike the pulse volume-averaged reflectivity, this is the average of point velocities weighted by both reflectivity and the illumination function. Similarly, the width (Eq. 10.50) reduces to

$$\sigma_v^2(\mathbf{r}_0) = \frac{\iiint v^2(\mathbf{r}_1)\,\eta(\mathbf{r}_1)\,I(\mathbf{r}_0,\mathbf{r}_1)\,dV_1}{\iiint \eta(\mathbf{r}_1)\,I(\mathbf{r}_0,\mathbf{r}_1)\,dV} - \overline{v}^2(\mathbf{r}_0) \qquad (10.52)$$

and corresponds to a weighted deviation of velocities from the averaged velocity.

The mean Doppler velocity (Eq. 10.49) depends on the distribution of scatterers' cross section within the resolution volume and its weighting functions. Thus $\overline{v}(\mathbf{r})$ cannot in general be equated to a spatial mean velocity. However, if reflectivity and illumination are symmetrical about the resolution volume center and if radial wind changes linearly across V_6, then $\bar{v}(\mathbf{r})$ is the true radial component of wind at the volume center. But hydrometeor fall velocity must be either insignificant or known (see Sec. 5a[1]).

a. Estimating Doppler Power Spectra

To measure the power-weighted distribution of velocities, frequency analysis of $V(\tau_s)$ is needed and can be accompanied by estimating its power spectrum. Bear in mind that the frequency analysis is performed along the sample-time axis for samples $V(\tau_s)$, at fixed τ_s. Thus we have discrete samples $V(\tau_s, mT_s)$, spaced T_s apart, of a continuous random process. Next we make some general statements concerning spectral analysis of continuous random signals.

The power spectrum density is the Fourier transform of the signal's autocovariance function $R(T_l)$:

$$\bar{S}(f) = \int_{-\infty}^{\infty} R(T_l)e^{-j2\pi fT_l}\,dT_l, \qquad (10.53)$$

where T_l is a time lag.

The autocovariance function of a stationary signal (statistics do not change during the time of observation) is found from the time average:

$$R(T_l) = \lim_{T\to\infty} \frac{1}{T} \int_{T/2}^{T/2} V^*(t)V\,(t+T_l)dt. \qquad (10.54)$$

Because $V(t)$ has zero mean, autocorrelation and autocovariance are identical. Note that conservation of power relates the Doppler spectrum $\bar{S}(v)$ to the power spectrum $\bar{S}(f)$ through

$$\bar{S}(v) = \frac{2}{\lambda}\,\bar{S}(f). \qquad (10.55)$$

When the spectrum or the autocovariance is known, all pertinent information (e.g., mean Doppler velocity) can be readily obtained. Neither $\bar{S}(v)$ nor $R(T_l)$ is available; they must be estimated from the ensemble of samples $V(\tau_s, mT_s)$ spaced T_s apart. Because radars observe any resolution volume for a finite time, one is faced with estimating at a given range, $c\tau_s/2$, the Doppler power spectrum and its parameters from a finite number M of time samples $V(mT_s)$. We delete the range time argument τ_s, and $V(mT_s)$ designates an ensemble sample at the implicit delay τ_s and T_l increments in steps of T_s.

Samples $V(mT_s)$, often multiplied by a weighting factor $W(mT_s)$, are Fourier-transformed, and the magnitude squared of this transformation is a power spectrum estimate $\hat{S}(k)$, commonly known as the "periodogram":

$$\hat{S}(k) = \frac{1}{M}\left|\sum_{m=0}^{M-1} W(mT_s)V(mT_s)e^{-j2\pi km/M}\right|^2, \qquad (10.56)$$

where k and m are integers, and unit increments of k correspond to velocity increments of $2v_a/M = \lambda/(2MT_s)$.

The finite number of time samples from which the periodogram is computed limits velocity resolution and creates an undesirable "window effect"; i.e., one may imagine that the time series extends to infinity but is observed through a finite length window. The magnitude squared of the data window transform is referred to as the "spectral window" and is significant because its convolution with the true spectrum equals the measured spectrum (Doviak and Zrnić, 1984).

Figure 10.12 illustrates a weather signal weighted with a uniform window and one with a von Hann (raised cosine) window, showing considerable difference in the spectral domain, especially in spectral skirts. Since the von Hann window has a gradual transition between no data and data points, its spectral window has a less concentrated main lobe and significantly lower side lobes. The resulting spectrum retains these properties and enables us to observe weak signals to more than 40 dB below the main peak. This is very significant when one is trying to estimate the peak winds of tornadoes or other severe weather (Zrnić, 1975a) within the resolution volume; power associated with high velocities in the spectral skirts is rather weak and would be masked by the strong spectral peaks seen through the window side lobes unless a suitable window were applied. The apparent lack of randomness of coefficients in the spectral skirts for the rectangularly weighted data is due to the larger correlation between coefficients. This correlation is attributed to the strong spectral powers seen through the nearly constant level window side lobes (Zrnić, 1975a).

The example in Fig. 10.12 is from a tornadic circulation with translation; the broad spectrum results from high-speed circulatory motion within the resolution volume. The envelope shape $|\sin x/x|^2$ is readily apparent for the rectangular window (at negative velocities), and the dynamic range for spectrum coefficients is about 30 dB. This is in contrast to more than 45 dB of dynamic range with the von Hann window, which also defines better the true spectrum and the maximum velocity (60 m s^{-1}).

Besides the window effect, which is intimately tied to signal processing, there are several spectral artifacts owing to the radar hardware, which are discussed in Zeoli (1971), Zrnić (1975b), and Zrnić and Bumgarner (1975).

b. Velocity Spectrum Width, Shear, and Turbulence

The velocity spectrum width (i.e., the square root of the second spectral moment about the mean velocity) is a function of both radar system parameters such as beamwidth, bandwidth, and pulse width and the meteorological parameters that describe the distribution of hydrometeor density and velocity within the resolution volume. Relative radial motion between targets broadens the spectrum. For example, turbulence produces random relative radial motion of drops. Wind shear can cause relative radial target mo-

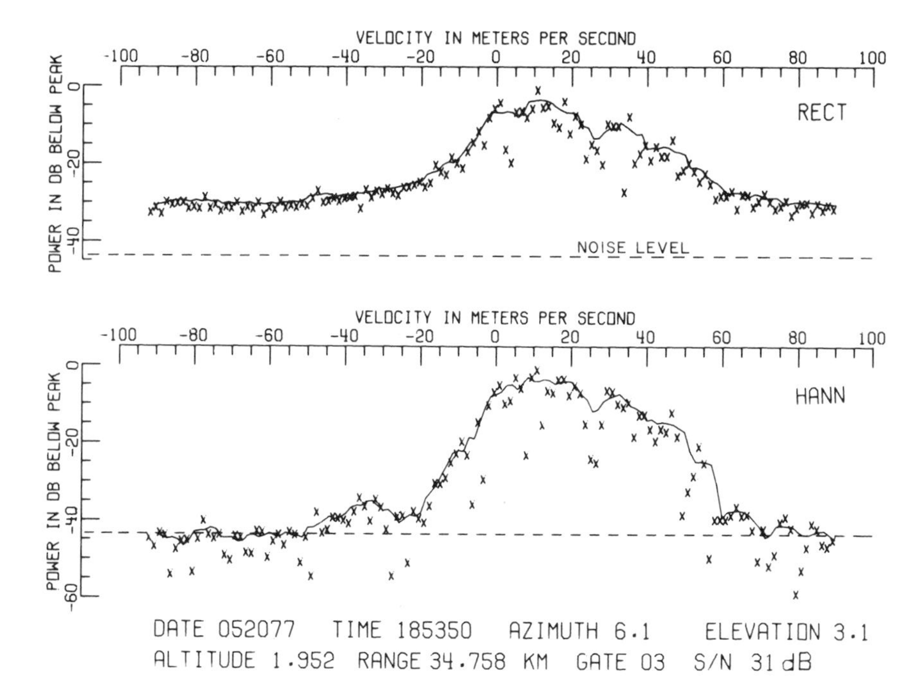

Figure 10.12. Power spectra of weather echoes showing statistical fluctuations in spectral estimates (x). RECT (above) signifies spectra of echo samples unweighted, whereas HANN (below) signifies samples weighted by a von Hann window. Solid curves are five-point running averages of spectral powers.

tion, as can differences in the speed of fall of various-size drops. Drop oscillations and/or rotation of hydrometeors produce modulation of the echo signal which increases the spectrum width. There is also a contribution to spectrum width caused by the beam sweeping through space (i.e., the radar receives echoes from targets that are not weighted identically on successive samples). This change in the location of the resolution volume V_6 from pulse to pulse results in a decorrelation of echo samples and a consequent increase in the spectrum width σ_v (Doviak and Zrnić, 1984). The echo samples will be uncorrelated more quickly (independent of particle motion inside V_6) the faster the antenna is rotated. Thus spectrum width increases in proportion to the antenna angular velocity.

Because the cited spectral broadening mechanisms are independent of one another, the square of the velocity spectrum width σ_v^2 can be considered as a sum of the variances contributed by each. That is,

$$\sigma_v^2 = \sigma_s^2 + \sigma_\alpha^2 + \sigma_d^2 + \sigma_t^2 + \sigma_h^2 \qquad (10.57)$$

where σ_s^2 is due to shear, σ_α^2 to antenna motion, σ_d^2 to different speeds of fall for different-sized drops, σ_t^2 to turbulence, and σ_h^2 to hydrometeor oscillation and/or rotation. Some add another term to account for beamwidth broadening. Even with uniform wind, v(**r**) changes across V_6. We consider this part of σ_s^2 because uniform wind produces angular shear of $v(\mathbf{r})$, and this broadening is accounted for in Eq. 10.60. Because broadening caused by the mean relative motion between the radar and the targets is small except in the case of an airborne radar, σ_α refers only to antenna rotation. The significance of the width σ_v for the weather radar design is discussed by Doviak and Zrnić (1984, Chap. 7).

The spectrum second moment σ_d^2, owing to the different radial components of fall speed of the assorted-size drops, is related to the radar and meteorological parameters:

$$\sigma_d^2 = (\sigma_{d0} \sin \theta_e)^2. \qquad (10.58)$$

The width σ_{d0} is caused by the spread in terminal velocity of various-size drops falling relative to the air contained in V_6. Lhermitte (1963) has shown that for rain $\sigma_{d0} \simeq 1\ \mathrm{m\ s^{-1}}$ and is nearly independent of the drop size distribution. The elevation angle $\theta_e = (\pi/2) - \theta_0$ is measured with respect to the beam center.

If the antenna pattern is Gaussian with a one-way half-power pattern width θ_1 and rotates at an angular velocity of α, the spectrum width owing solely to resolution volume displacement is (Doviak and Zrnić, 1984, App. C)

$$\sigma_\alpha^2 = \left(\frac{\alpha\lambda \cos\theta_e}{2\pi\theta_1}\right)^2 ln 2. \qquad (10.59)$$

Doviak and Zrnić (1984) prove that the wind-shear term can be separated into three contributions. If k_θ, k_ϕ, k_r are the shears in the directions θ, ϕ, r, then

$$\sigma_s^2 = \sigma_{s\theta}^2 + \sigma_{s\phi}^2 + \sigma_{sr}^2 = (r_0\sigma_\theta k_\theta)^2 + (r_0\sigma_\phi k_\phi)^2 + (\sigma_r k_r)^2, \qquad (10.60)$$

where σ_θ^2 and σ_ϕ^2 are defined as the second central moments of the two-way antenna power pattern in the indicated directions, and σ_r^2 is the second central moment of the weighting function $|W(r)|^2$. A circularly symmetric Gaussian pattern has

$$\sigma_\theta^2 = \sigma_\phi^2 = \frac{\theta_1^2}{16\, ln 2}. \qquad (10.61)$$

For a rectangular transmitted pulse and a Gaussian re-

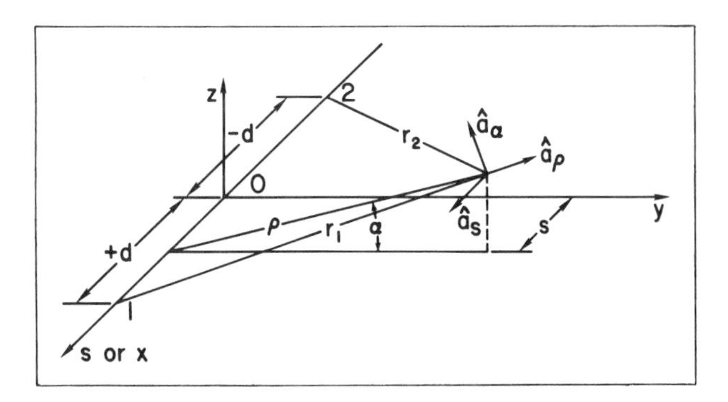

Figure 10.13. Cylindrical coordinate system used for dual radar data analysis. The radars are located at the points 1 and 2; $\hat{a}_p$, $\hat{a}_s$, and $\hat{a}_\alpha$ are the unit normals defining directions of the three orthogonal velocity components.

ceiver frequency response under matched conditions (i.e., $B_6\tau = 1$),

$$\sigma_r^2 = \left(\frac{0.35c\tau}{2}\right)^2. \tag{10.62}$$

The width σ_t owing to turbulence is somewhat more difficult to model. For turbulence that is isotropic within V_6, Frisch and Clifford (1974) have shown that σ_t^2 is related to the eddy dissipation rate ε. A detailed discussion of the contribution of turbulence and shear to the spectrum width is presented in Doviak and Zrnić (1984, Chap. 10), where examples of data fields are also given. Broadening of spectra by drop oscillation or rotation of hydrometeors has not been quantified but is expected to be relatively unimportant in most instances.

5. Observation of Weather

The great utility of storm observation with centimeter wavelength pulsed-Doppler radar derives from its capacity to map reflectivity η and mean radial velocity $\bar{v}$ inside the storm's shield of clouds. A three-dimensional picture of a single storm takes about 2 to 5 minutes of data-collection time not only because of antenna rotation limitations but also because many echoes from each resolution volume must be processed to reduce the statistical uncertainty in the η and $\bar{v}$ estimates. Although storm structure can change significantly during this period with distortion in the radar's image of the true reflectivity and velocity fields, highly significant achievements have been made in depicting the structure and evolution of the thunderstorm.

But the meteorologically interesting variables are not η or $\bar{v}$ but parameters such as rainfall rate (on the ground) and wind. Pulsed-Doppler radar measures the radial speed of hydrometeors, not air, and in certain situations (e.g., with vertically directed beams) these speeds can differ significantly from the radial component of wind. Surface rainfall rate estimates are also not easily related to η, and radar reflectivity measurements are often supplemented by surface rain gages (Brandes, 1975; Crawford, 1977; Ahnert et al., 1986).

Incoherent radars map η and, if their resolution volume is small enough and reflectivity estimates are accurate, can track prominent reflectivity structures to map the vector wind (Crane, 1979; Reinhart, 1979). However, Doppler radar measures, practically instantaneously, velocities in each resolution volume and hence can provide far better resolution of the velocity field.

a. Dual Doppler Radar

A single Doppler radar maps a field of a velocity component directed toward (or away from) the radar. A second Doppler radar, spaced far from the first, provides a field of different radial velocities, because the wind vector is projected on different radials. The two radial-velocity fields can be vectorially synthesized to retrieve the two-dimensional velocities in the plane containing the radials (Armijo, 1969). The synthesis is usually accomplished on common grid points to which radar data are interpolated. Radial velocities in each of the radar's resolution volumes surrounding a grid point are not measured simultaneously but are separated up to the few minutes required for each radar to scan the common volume. The respective resolution volumes are also usually quite different in size and orientation.

Nevertheless, useful estimates of wind can be made on scales of air motion large compared with the biggest resolution volume dimensions, if the velocity field is quasistationary during the period required for data collection. Targets such as water drops having small mass quickly respond to horizontal wind forces and faithfully trace the wind. Stackpole (1961) showed that, for energy spectrum of wind scales following a $-5/3$ law to at least 500 m, more than 90% of the rms wind fluctuations are acquired by the drops if their diameters are less than 3 mm. When radar beams are at low elevation angles, hydrometeor terminal velocity (i.e., the steady-state vertical velocity of the hydrometeor relative to the air) gives negligible error in the radial wind component. At high elevation angles terminal velocity must be considered.

(1) Reconstruction of Wind Fields: Wind field determination from two Doppler radar data is greatly simplified if the synthesis is performed in cylindrical coordinates with axis chosen to be the line connecting the two radars. That is, radial velocities at data points (centers of resolution volumes) are interpolated to nearby grid points on planes having a common axis (the COPLAN technique; Lhermitte, 1970). Cartesian wind components can be derived from these synthesized cylindrical components. Although one could solve directly for Cartesian wind components, this necessitates solving an inhomogeneous hyperbolic partial differential equation to derive the vertical wind (Armijo, 1969).

The cylindrical coordinate system is illustrated in Fig. 10.13. The mean Doppler velocity must be corrected for the scatterers' reflectivity-weighted mean (i.e., average over V_6) terminal velocity $\bar{v}_t$. Thus the estimate of the radial component of air motion is

$$v_{1,2} = v'_{1,2} + \bar{v}_t \sin \theta_{e1,2}, \qquad (10.63)$$

where $v'_{1,2}$ are the mean Doppler target velocities measured by radars 1,2 at data points, $\bar{v}_t$ is positive, and θ_e is the elevation angle. To estimate $\bar{v}_t$ for each resolution volume, one could use the empirical expression (Atlas et al., 1973)

$$\bar{v}_t = 2.65\, Z^{0.114} \left[\frac{\gamma_0}{\gamma}\right]^{0.4} \text{ m s}^{-1}, \qquad (10.64)$$

where the parenthetical term is that suggested by Foote and duToit (1969) to account for height-dependent air density γ, and Z is the reflectivity factor. This relation well represents experimental data over a large range of Z (i.e., $1 \leqslant Z \leqslant 10^5$ mm^6 m^{-3}) for regions of liquid water, but large errors up to several meters per second in $\bar{v}_t$ estimates can be caused by erroneously relating regions of hail with a $\bar{v}_t,Z$ relation appropriate for liquid water. Usually there is little or no information to identify these regions uniquely, and errors in vertical wind w_z can result. However, it has been shown for typical arrangements of storms relative to the two-radar placement that the error in w_z is significantly smaller than errors in $\bar{v}_t$ (Doviak et al., 1976).

The estimated radial velocities $v_{1,2}$ of the air can be interpolated to uniformly spaced grid points in planes at angle α to the horizontal surface containing the baseline. Interpolation filters the data and reduces variance. The cylindrical wind components in the ρ,s plane are related to $\bar{v}_1,\bar{v}_2$ as

$$w_\rho = \frac{(s+d)r_1\bar{v}_1 - (s-d)r_2\bar{v}_2}{2d\rho} \qquad (10.65)$$

and

$$w_s = \frac{r_2\bar{v}_2 - r_1\bar{v}_1}{2d}, \qquad (10.66)$$

where $\bar{v}_{1,2}$ are the interpolated Doppler velocities of air, and $2d$ is the distance between the radars, as shown in Fig. 10.13.

The wind component, w_α, normal to the plane, is obtained by solving the continuity equation in cylindrical coordinates,

$$\frac{1}{\rho}\frac{\partial}{\partial\rho}(\rho\gamma w_\rho) + \frac{1}{\rho}\frac{\partial}{\partial\alpha}(\gamma w_\alpha) + \frac{\partial}{\partial s}(\gamma w_s) = 0, \qquad (10.67)$$

with the boundary condition being that $w_\alpha = 0$ at the ground. The mass density γ is given by

$$\gamma = \gamma_0 \exp\left[-g_0 M(\rho \sin\alpha)/(RT)\right], \qquad (10.68)$$

where g_0 is the gravitational constant (9.8 m s^{-2}), M the mean molecular weight of air (29 g mol^{-1}), $\rho \sin\alpha$ the altitude, T the absolute temperature (K), and R the universal gas constant (8.314 J mol^{-1} K^{-1}). Appropriate values of γ_0 and T can be obtained from surface site and upper-air soundings. The Cartesian components of wind can then easily be obtained from w_ρ, w_s, and w_α.

(2) Observation of a Tornadic Storm: The first dual Doppler radar observations of a tornadic storm were made

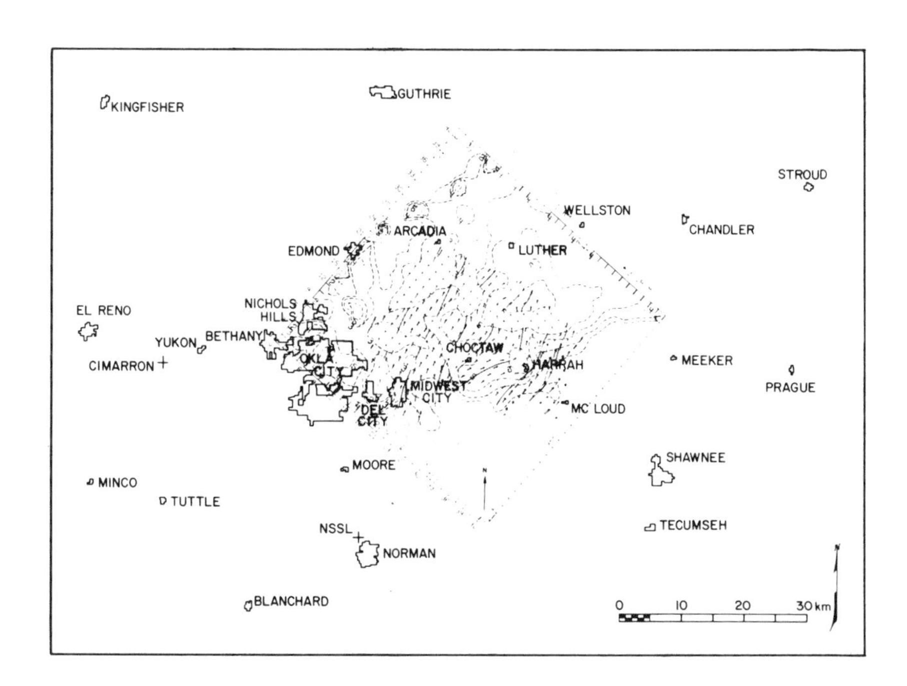

Figure 10.14. Map of central Oklahoma with reflectivity field (shaded areas) and winds relative to the ground at a height of 1 km. Radar locations are marked with +. At this height maximum velocity was 37.8 m s^{-1}. Reflectivity contours (dashed lines) are labeled log Z (Ray et al., 1975).

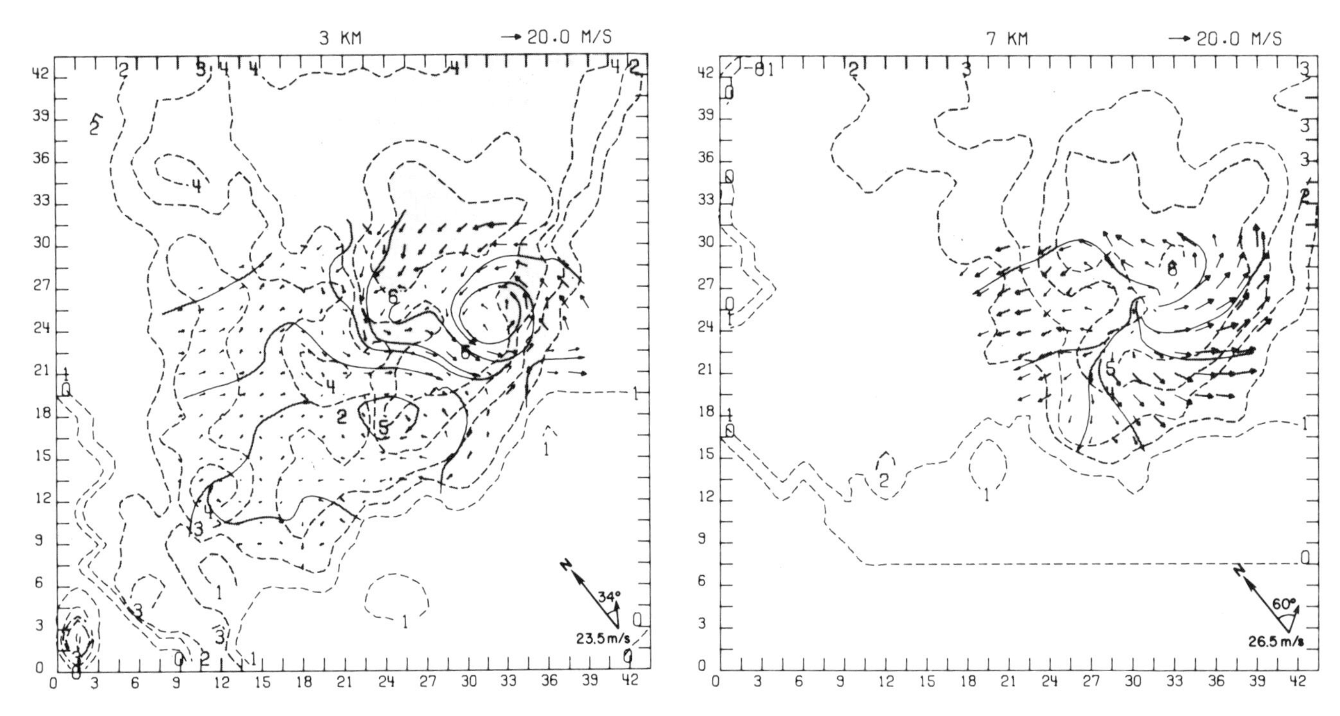

Figure 10.15a. Perturbation wind and reflectivity factor (log Z) at 3-km and 7-km altitudes. Mean wind at each altitude is given in lower right corner. Velocity scale is above upper right corner. Distances are in kilometers from an origin $s = -10$, $\rho = 15$ km ($s = 0$ is the midpoint of the base line) and abscissa is parallel to $\hat{a}_s$.

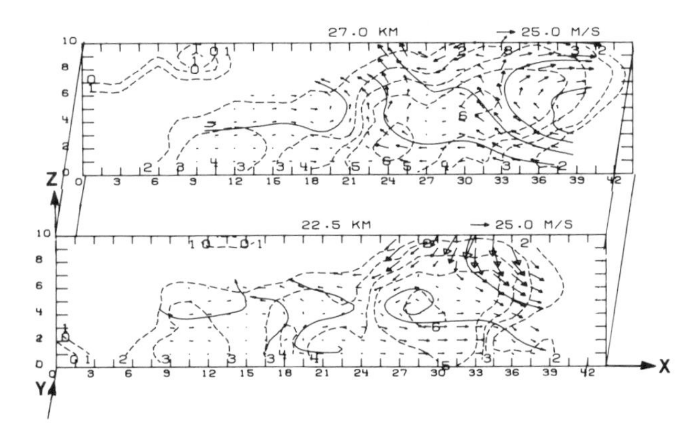

Figure 10.15b. Reflectivity and velocity fields in the X,Z plane for 27 km and 22.5 km. The mean horizontal velocity in each vertical plane was removed and is indicated in the upper right of each plane. Reflectivity contours are labeled log Z (Ray et al., 1975).

on 20 April 1974 with NSSL's 10-cm radars. These radars provide a large, unambiguous range and velocity capability well suited for observation of the large, severe thunderstorms that frequent the high plains of the United States.

Figure 10.14 locates the two radars, one at Cimarron airfield (now Page airfield) in Oklahoma City and the second at NSSL's headquarters in Norman, Okla. It also shows reflectivity contours and the horizontal wind field synthesized by use of a slight modification of the scheme outlined above (Ray et al., 1975). (Terminal velocity corrections were obtained from interpolated reflectivity factor values, and a slightly different $\bar{v}_t,Z$ relationship was used.) Streamlines have been drawn in addition to velocity vectors, whose lengths are proportional to windspeed. The curvature in the streamlines shows appreciable local vorticity in the region near the grid point (30, 24).

To discern the storm's kinematic structure at several altitudes, the mean wind at each height is subtracted from the wind vector at each grid point. This perturbation velocity field is displayed at two heights in Fig. 10.15a. Cyclonic circulation is apparent at the grid point (30, 24), where vorticity was noticed in Fig. 10.14. Inflow into the tornadic cyclone is shown at an altitude of 3 km, and divergence and outflow are apparent at 7 km. High-reflectivity factor (60-dBZ) regions are located on the downwind side. These velocity fields are in general agreement with present storm models, particularly in the weak echo region, where both imply a strong updraft northeast of the circulation (Fig. 10.15b; $X = 33$, $Y = 27$ km). A downdraft (Fig. 10.15b; $X = 33$, $Y = 22.5$ km) is found southwest of the circulation.

(3) Observation of Downbursts and Microbursts: A downburst is defined as a "strong downdraft which induces an outburst of damaging winds on or near the ground (Fujita 1981, 1985). A microburst is a downburst when the distance between diverging centers of fastest outflow is less than 4 km. Downbursts are mostly associated with intense severe storms, but they have also been observed, especially in the high plains of the U.S.A. below innocuous-looking virga (Eilts and Doviak, 1986; McCarthy and Wilson, 1984). Likewise, there is no apparent correlation between reflectivity factor and the maximum velocity difference across the flow (McCarthy et al., 1984; Rinehart and Isaminger, 1986), although observations made in regions where subcloud moisture contents are higher show many more microbursts having reflectivity factors larger than 40 dBz. Clearly evaporative cooling in dry air below cloud base reduces the reflectivity factor, and, therefore, it might be better to correlate maximum reflectivity anywhere in the storm with surface outflow intensity. Experiments indicate that intense small-scale divergent flow is more common than previously thought (Wilson et al., 1984), and, as with vortices, the smaller-size downdrafts appear to produce the strongest shear. Although it is wind speed that creates damage to surface structures, it is wind shear that is a hazard to safe flight. However, the level of threat is conditioned by the response of the aircraft to a wind-shear event. For example, a Boeing 727 aircraft has enhanced response to wind perturbations at frequencies near its phugoid frequency of 3×10^{-2} Hz (period ~30 s). Using typical values of takeoff and landing speeds, it can be deduced that wind swirls of about 3-km wavelength can have more deleterious effect on the performance of the aircraft than other wavelengths of similar amplitude. Furthermore, the most hazardous shears to safe flight are those within the first few hundred meters above the ground. Thus resolutions of a few hundred meters or less are required to resolve these small-scale diverging flows, and only radars with high angular resolution ($\leq 1°$) and observations close to the radar would resolve the shear produced by the microbursts. Compounding the difficulty in making accurate measurement of hazardous wind shear is the short lifetime of the phenomena. For example, although lifetimes of microbursts range from 5 to 15 min, the period of severe wind shear lasts from 2 to 4 min with an average velocity difference of 25 m s^{-1} across the divergent flow (McCarthy and Serafin, 1984). Even though these small-scale hazards might be difficult to detect, they are usually embedded in larger-scale phenomena (e.g. storms, gust fronts), which have longer lifetimes and thus are more easily detected. Therefore, detections of phenomena that might cause microbursts and other small-scale hazards may be as important or more important than detection of the hazard itself (Doviak and Lee, 1985).

Microburst winds derived from simultaneous observations with a 10-cm radar (the Federal Aviation Administration's transportable Doppler weather radar operated by MIT's Lincoln Laboratory; Evans, 1984) and a 5-cm radar (from the University of North Dakota) are shown in Fig. 10.16. These radars were located approximately 15 km apart near Memphis, Tenn., and the wind field displayed on Fig. 10.16 is about 5 min after the time of maximum divergence. The undulating highly turbulent front of the outflow is clearly depicted, as well as a secondary surge behind it which could be associated with another pocket of descending air.

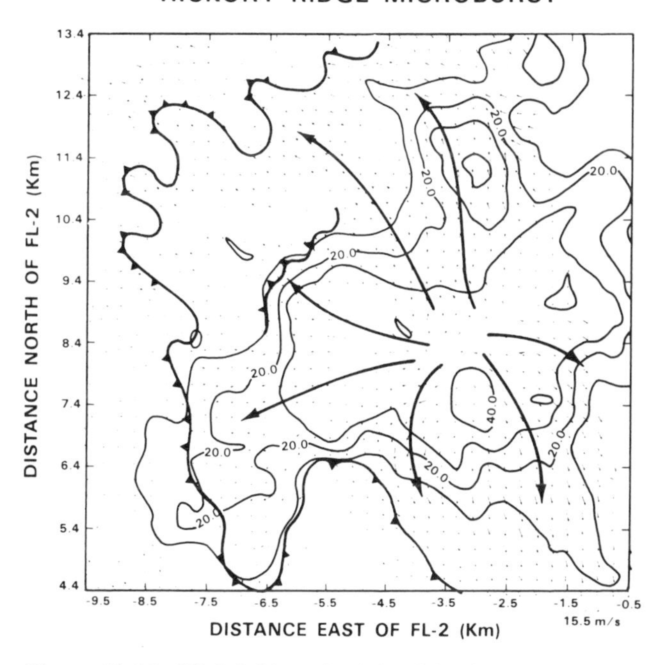

Figure 10.16. Wind fields, reflectivity fields in dBZ, selected streamlines, and the outflow front of a microburst. One grid-space length equals a wind vector of 15.5 m s^{-1}. Winds are relative to ground. Height above ground is 50 m (figure and analysis are courtesy of M. Wolfson, MIT Lincoln Laboratory).

The use of Doppler radar to predict the occurrence of a microburst along flight paths is the subject of numerous research projects. Results of recent research of the downburst phenomena reveal that there are a number of possible precursors before the formation of divergence at the surface. These include a rapidly descending reflectivity core, organized convergence near and above cloud base, and mid-altitude (2–5 km) rotation (Roberts and Wilson, 1986). Although evidence from meteorological observations and aircraft data support the idea that microbursts are a very prominent factor in most fatal wind-shear accidents, it has been proposed (Linden and Simpson, 1985) that wind shear and downdraft at the rear of a vortex ring found around the microburst and located at its leading edge (i.e., the gust front) are probably the most critically dangerous elements of flight through thunderstorm downdrafts. Because all fatal thunderstorm-related accidents to U.S. air carriers from 1970 to this writing have occurred in or near heavy precipitation, Kessler (1985) suggests that the most prudent action is to avoid high-reflectivity thunderstorms by a safe distance.

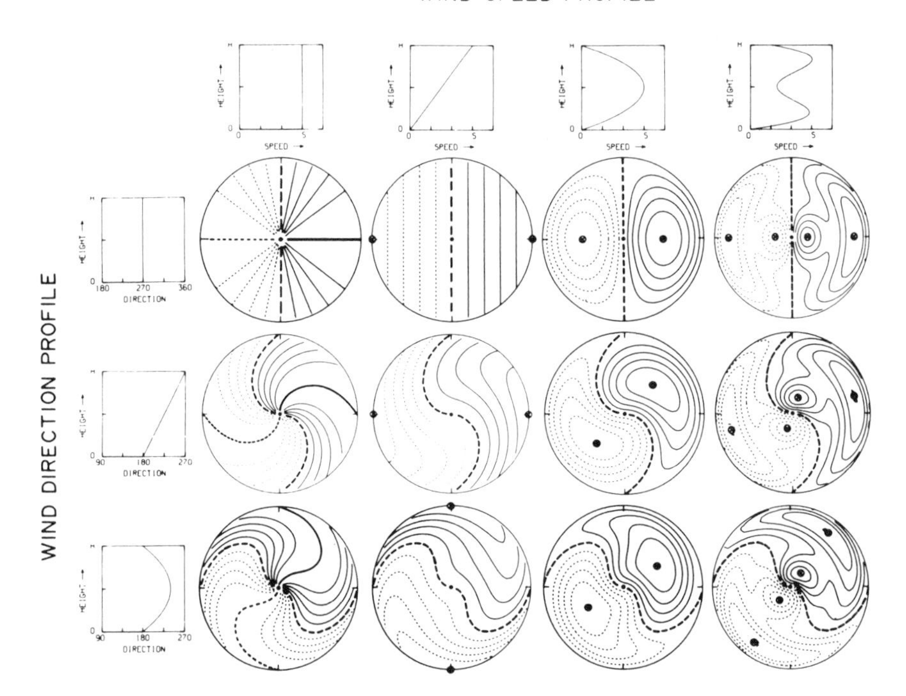

Figure 10.17. Doppler velocity patterns (plan view) for various profiles of windspeed and direction when the radar beam scans at an elevated angle θ_e. The outer circle of the display depicts Doppler velocity at height = H and range = $H/\sin\,\theta_e$. Contour interval is 0.2 S, where S is maximum windspeed (from Wood and Brown, 1986).

b. Single Doppler Radar

Although the Doppler radar measures only the radial wind component, its spatial distribution can signify meteorological events such as tornado cyclones. Moreover, uniform wind can be measured, and nonuniform wind can, with certain limitations, be mapped from data obtained with single Doppler radar (Doviak et al., 1983). There are many situations, in both clear and rainy air, where Doppler radar can detect coherent echoes over large volumes of space. In rain-free air it is the refractive index irregularities created by turbulent mixing and/or insects that are responsible for the reflectivity (see Sec. 5c). Thus a single Doppler radar offers good promise both for severe-weather warning (NSSL, 1979) and for routine wind measurements above ground.

(1) Linear Wind Measurements—Velocity Azimuth Display (VAD): When the antenna beam is scanned in azimuth ϕ while elevation angle θ_e is fixed, the radial velocity has the ϕ dependence:

$$v_r = (w - \bar{v}_t)\sin\theta_e + v_h\cos\theta_e\cos(\phi - \delta), \quad (10.69)$$

where δ is the wind direction, w the air's vertical velocity, and v_h the horizontal speed of tracers in the resolution volume. The VAD is a display of radial velocity at a single range location versus azimuth. The w and v_h velocities are readily computed from data in a VAD (Lhermitte and Atlas, 1961) under the assumption that air is in pure translation and v_t can be accurately estimated (for echoes from refractive index irregularities $v_t = 0$). Then v_r has a sinusoidal dependence on ϕ; thus amplitude and phase of the sine curve are measures of v_h and δ at the height $r\sin\theta_e$ of the sampling circle. Vertical motion produces a DC offset of the sine wave. However, when wind is not horizontally homogeneous, v_r no longer has a purely sinusoidal dependence on ϕ.

Caton (1963) showed how divergence can be determined from VAD data. Browning and Wexler (1968) carried the analyses even further by assuming that the wind field is well represented by a linear velocity field (Hess, 1959, p. 198) over the circle of measurement. With this assumption there are four basic fields of motion that convey air: translational, vortical, divergent, and deformative. Fourier analysis of v_r for linear wind reveals that only the vortical motion cannot be measured by the VAD method. The average component of v_r is proportional to mean horizontal divergence $\overline{\mathrm{DIV}\ v_h}$ plus mean $\overline{w - \bar{v}_t}$:

$$\frac{1}{2\pi}\int_0^{2\pi} v_r\,d\phi = (\overline{\mathrm{DIV}\ v_h})\,\frac{r}{2}\cos^2\theta_e + \overline{w - \bar{v}_t}\sin\theta_e, \quad (10.70)$$

where $\overline{w - \bar{v}_t}$ is the average of the target's vertical velocity on the sampling circle of radius r. The first harmonic component gives $\bar{v}_h$ and $\bar{\delta}$ averaged over the circle, and the second measures deformation. By inserting the mass continuity equation into Eq. 10.70, we can then solve for vertical wind if we have an estimate of target terminal velocity averaged over the circle. Thus a single Doppler radar can measure the three components of wind averaged over a sampling circle of radius r and produce a vertical profile of wind.

When winds are horizontally uniform while varying with height, unique patterns of Doppler velocity fields are formed on Plan Position Indicators (PPI) as the radar beam is scanned at a constant elevation angle. These patterns can be easily interpreted by trained personnel to estimate the vertical profiles of wind. PPI displays of Doppler velocity contours corresponding to a variety of vertical profiles of wind speed and direction are shown in Fig. 10.17, where the center of the PPI display represents the radar location. Winds that have a component away from the radar (termed "outbound") have positive Doppler velocity values indicated by solid contours on Fig. 10.17, whereas those toward the radar (termed "inbound") have negative values indicated by dotted contours. Winds directed normal to the beam produce zero Doppler velocities. When windspeed is constant with height (Fig. 10.17, left column), all contours pass through the center of the display. Also, maximum and minimum Doppler velocities occur along heavier (nonzero) contour lines rather than at just one point (black dots) as with other windspeed profiles. When the surface speed is zero, only the zero velocity contour (thick long dashes) passes through the center of the radar display (Fig. 10.17, three far-right columns).

If the windspeed profile has a peak within the height interval on the display, there will be a pair of closed contours 180° from each other; the azimuth of the inbound (or outbound) maximum is the direction from which the velocity jet is approaching (or departing) the radar, and the height of the peak value can be computed from the radar antenna's elevation angle and the slant range to that point.

Whereas the windspeed profile controls the overall pattern, including the spacing between contours, the vertical profile of wind direction is uniquely specified by the zero velocity contour. Note that the zero contours are identical in each row (a reflection of wind-direction profile) even though the overall patterns in each row differ significantly (a reflection of windspeed profile).

Since wind direction is perpendicular to a radial line (from display center) at the point where the line intersects the zero velocity contour, wind direction variation with height (range on the radar display) can be determined by inspection. Air motion is from the negative side toward the positive side of the zero contour. Looking at the middle row, we see that there are southerly winds at the ground, or just above the ground—the zero velocity line is oriented east-west, and air is approaching from the south and flowing away toward the north. Halfway between the center (zero height) and edge (height = H) of the display, southwesterly winds are perpendicular to a radial line. At the edge of the display wind is from the west because the radial line intersecting the zero contour is oriented north-south.

Veering winds, representative of warm-air advection, produce a striking S-shaped pattern by the zero velocity contours, whereas backing winds, representative of cold air advection, produce a backward S. These features are seen in the middle and bottom rows.

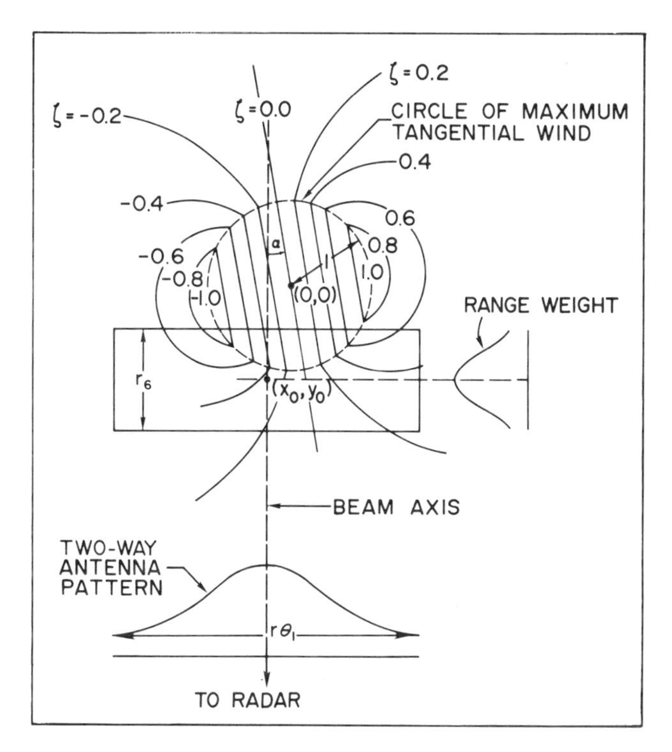

Figure 10.18. Plan view of idealized isodop pattern for a stationary modified Rankine vortex located at range large compared with vortex diameter. $\zeta = v/v_{max}$ is Doppler velocity normalized to its peak value. Resolution volume, antenna, and range-weighting functions are depicted. The angular tilt α determines radial inflow ($\alpha<0$) or outflow ($\alpha>0$).

Patterns associated with horizontal variation of windspeed and direction (fronts, mesocyclones, etc.) are presented by Wood and Brown (1983).

(2) Severe-Storm Cyclone Observations: Because radar maps spatial distribution of Doppler velocity inside storms, significant meteorological events (unseen from outside) such as tornado cyclones do produce telltale signatures. Donaldson (1970) stipulated criteria whereby a vortex can be identified from single radar observations. Briefly, there must be a localized region of persistently high, $\geqslant 5 \times 10^{-3}$ s^{-1}, azimuthal shear (i.e., the velocity gradient along an arc at constant range) that has a vertical extent equal to or larger than its diameter.

It can be shown that nontranslating cyclones have isodops forming a symmetric couplet of closed contours with equal number of isodops encircling positive- and negative-velocity maxima (Fig. 10.18). If the inner portion of the vortex is a solidly rotating core, its tangential velocity increases linearly with radius to a maximum; outside this maximum the velocity decreases (roughly) inversely with

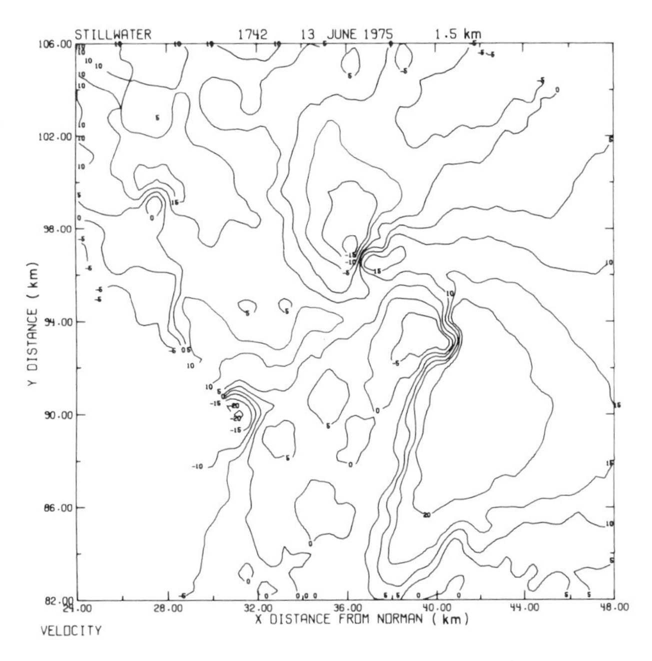

Figure 10.19a. Doppler velocity field for the Stillwater, Okla., storm at 1.5-km height. Velocities are in meters per second; contours (isodops) are in 5-m s^{-1} steps. The mesocyclone is centered at $Y = 97$ km, $X = 36$ km. The other shear region from 94 km north, 40 km east to the bottom of the field was identified from dual Doppler data to be the low-level boundary (gust front) between storm inflow to the east of the shear line and outflow to the west.

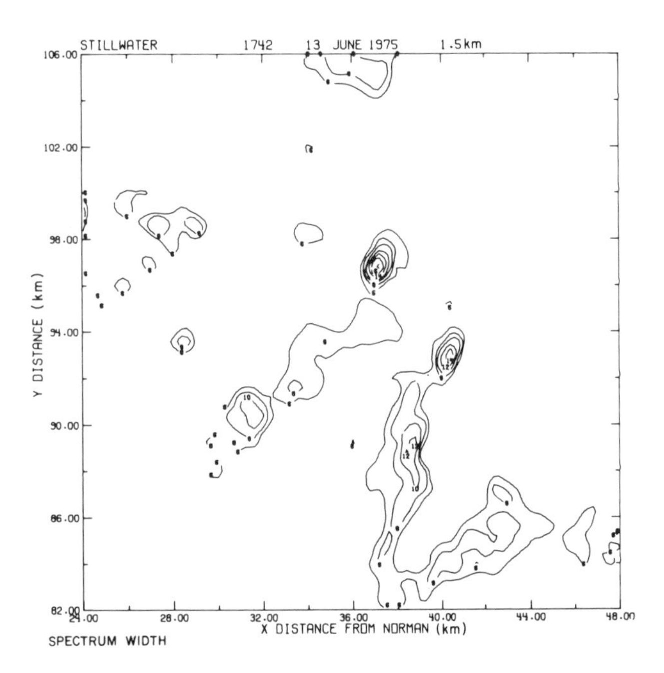

Figure 10.19b. Contours of constant spectrum width for the Stillwater, Okla., storm at 1.5 km aboveground. Values equal to or larger than 6 m s^{-1} in steps of 2 m s^{-1} are displayed. Large widths farther north coincide with the tornado mesocyclone; the other region of large width is embedded in the gust front.

the radius. The isodop contours of such a combined Rankine vortex are circular sections connected with straight lines (Fig. 10.18). This pattern has been observed many times, and an example is shown in Fig. 10.19a for a tornadic storm that did considerable damage to Stillwater, Okla., in 1975 (Burgess, 1976; Brown et al., 1978; Zrnić et al., 1977). Figure 10.19b shows contours of estimated spectrum width σ_v for the same storm, and we immediately see the striking correlation of σ_v with significant radial velocity shear. Regions of large σ_v may also indicate the presence of strong turbulence.

The National Weather Service radars routinely display reflectivity factors on Plan Position Indicator (PPI) scopes, and some television stations, using their own radar, show the radar images on a color PPI display. The PPI projects any radar data field (but usually reflectivity factor) from the conical surface on which it is acquired onto the plane surface of the indicator. Although target reflectivity cannot be reliably used for tornado detection, it has proved valuable for hydrological studies and severe-weather warnings. Those warnings are primarily based on reflectivity values, stormtop heights, and sometimes on circulatory features or "hook" echoes. However, PPI displays of Doppler velocity and spectrum width do vividly show signatures of circulations.

Color displays of reflectivity, velocity, and spectrum width allow, in real time, easier quantitative evaluation and better resolution of cyclones. Figure 10.20 is an example of Doppler radar data from a tornadic thunderstorm in central Oklahoma during the evening of 30 April 1978. These data were acquired with NSSL's 10-cm Doppler weather radar, and the pictures depict the fields of the three principal moments of the Doppler spectra.

At the time of the display (1825:32), there was a tornado on the ground at 334° and 50-km range. Although the reflectivity display showed no evidence of a tornado cyclone, the mean Doppler velocity display (Fig. 10.20b) depicted it remarkably well. The velocity scale is in meters per second with + values (red) indicating motion away from the radar. The signature of a couplet of closed contours of opposing Doppler velocities (bright green next to bright red) indicates presence of the strong circulation associated with the tornado. The display (Fig. 10.20c) of spectrum width (σ_v) is an indicator of turbulence and shear within the radar's resolution volume. We note high spectrum width values in the vicinity of the circulation and one pinpointing the exact tornado spot.

Figure 10.20d is the multimoment display (MMD) (Burgess et al., 1976); each arrow is a speedometer indicating the Doppler speed at its respective location. Zero Doppler is an arrow pointing at the 3 o'clock position, and positive velocities (away) cause the vector to rotate counterclockwise (+17 m s^{-1} velocity is at the 12 o'clock position). Negative velocities cause arrows to point toward the bottom of the diagram (i.e., toward the radar). Reflectivity is indicated by arrow length, and spectrum width by arrowhead size. The tornado circulation is depicted quite nicely by this graphic display, and the presence of wide spectra is evident from the broad arrowhead width, indicating strong shearing winds or turbulence.

(3) Doppler Spectra of Tornadoes: In 1961, Smith and Holmes reported a tornado spectrum that was obtained with a CW Doppler radar (Smith and Holmes, 1961); 12 years later a tornado was first observed by a pulsed Doppler radar (Zrnić and Doviak, 1975).

The Doppler spectrum is principally influenced by the portion of circulation that lies within the resolution volume, so that tracers moving with the same velocities contribute to a spectral component according to their reflectivity, isodop density, and antenna pattern illumination (see Fig. 10.18). Only those targets whose spectral power is above the receiver noise level may be positively identified. However, it has been our experience that tornadoes in a resolution volume offer enough reflectivity, owing to debris and hydrometeors, that a large velocity span can be observed. It is not known at this writing whether tracers moving at the peak tornado windspeed can be routinely resolved.

When centered on the beam axis, the Rankine vortex model predicts a bimodal spectrum that was observed experimentally several times (Zrnić and Doviak, 1975; Zrnić et al., 1977). Shown in Fig. 10.20 are Rankine model vortex spectra matched to data by a least-squares fit. The spectrum examples are from the Stillwater maxitornado. Maximum velocity of 92 m s^{-1} and tornado diameter of 300 m are deduced from the fitting. The deduced maximum velocity is larger than the radar's unambiguous velocity (± 34.4 m s^{-1}); therefore, aliasing has been introduced in the model spectrum, and the estimates are indirect.

Two spectra closest to the tornado were simultaneously least-squares-fitted. Simulated model vortex spectra and real spectra show very good agreement not only for the two where the fit was made but also for adjacent gate locations (Fig. 10.21). Resolution volumes corresponding to any of these simulated spectra are assumed to have uniform reflectivity within the volume. Differences in echo power from each resolution volume are accounted for by forcing each simulated spectrum to have power equal to its matching data spectrum. Asymmetry of spectral peaks (azimuth 21.1°; range 104.136 km) about zero velocity suggests that targets were centrifuged outward at a velocity of 13 m s^{-1}.

(4) Gust Fronts: Strong wind shear may also be associated with gust fronts, which are the leading edge of the mass of precipitation-cooled air, flowing first downward and then outward from the center of thunderstorms. Gust fronts harbor both vertical and horizontal shear as well as turbulence, and even after propagating several kilometers from the parent thunderstorms and away from the rain-laden air, they may retain shear at potentially hazardous levels. This large separation makes it unreliable to discount the presence of gust fronts merely because of the absence of thunderstorms in close proximity.

Doppler radar offers great promise to measure shears of horizontal wind in gust fronts void of precipitation (Zrnić et al., 1983). But owing to the nature of the instrument, measurements are usually representative of heights at beam center, and care must be exercised when comparing these measurements with surface anemometers. Results from a recent study indicate that Doppler radar estimated shears measured at heights between 50 and 600 m in Oklahoma gust fronts are stronger than shears measured at the surface by an average ratio of 1.6:1 (Eilts, 1987). Tower wind profiles measured during gust frontal passage confirm that winds (and shears) near the surface are weaker than winds (and shears) aloft.

As far as downbursts are concerned, the relationship between shear measured at the ground and radar-derived shear has not yet been established. But because surface friction has similar effects on any flow, we expect shear in microbursts to be smaller at or near the ground than aloft.

The determination that shears are stronger aloft than at the surface can be used to better detect wind shear in the area surrounding airports, where aircraft are most susceptible to it. Shears detected by a ground-based sensing device such as the Low-Level Wind Shear Alert System (LLWSAS) will underestimate shears aloft. However, if they are "corrected" with an appropriate factor (1.6 was found to be the average ratio of shear aloft to shear at the surface in gust fronts), then the surface estimate of shear along a glide path would be more comparable to the shear aloft, where aircraft are flying.

Vortices are potentially hazardous because of shear associated with them. The presence of a vortex can produce a decrease in headwind as large as that found in downbursts if the flight path is tangent to the circle of maximum wind (Doviak and Lee, 1985). Intense circulations are often found along gust fronts, and that is one more reason to avoid them.

Because of a gust front's typically long lifetime (sometimes 1–3 h) and large horizontal extent (10–90 km), the tracking of gust fronts is of practical importance. A gust front appears in a Doppler velocity field as a low-altitude convergence line which extends horizontally for tens of kilometers. A gust front algorithm has been developed (Uyeda and Zrnić, 1983) which locates areas of convergence along a radial and then pieces them together to locate lines of convergence, which are then identified as gust fronts. The algorithm can identify gust fronts and then extrapolate their location up to 20 min in advance (Fig. 10.22). This 20-min forecast, when compared with the actual location at the same time, shows good agreement for a number of cases.

(5) Turbulence: Both sheared flow in stratified environments and convection can generate turbulence. In either case the turbulent energy input is at larger-size eddies which break up into successively smaller ones until turbulent energy is converted into heat by viscosity. If the sizes of the largest eddies of isotropic turbulence (i.e., scales within the inertial subrange of eddy wavelengths where energy is

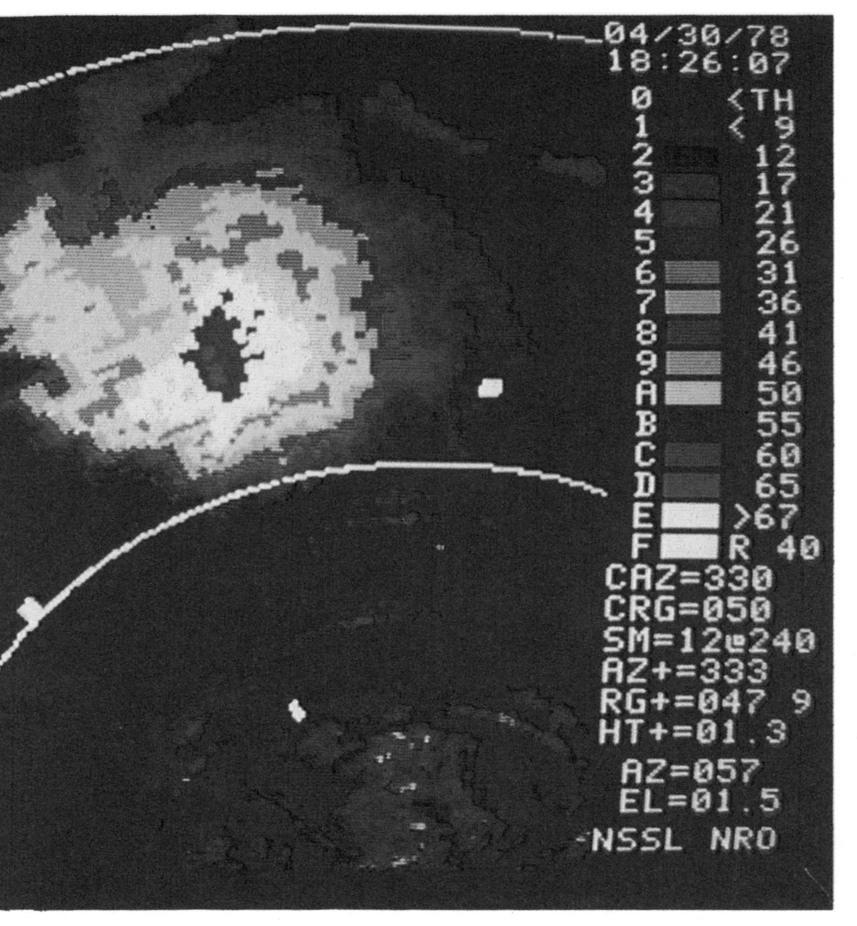

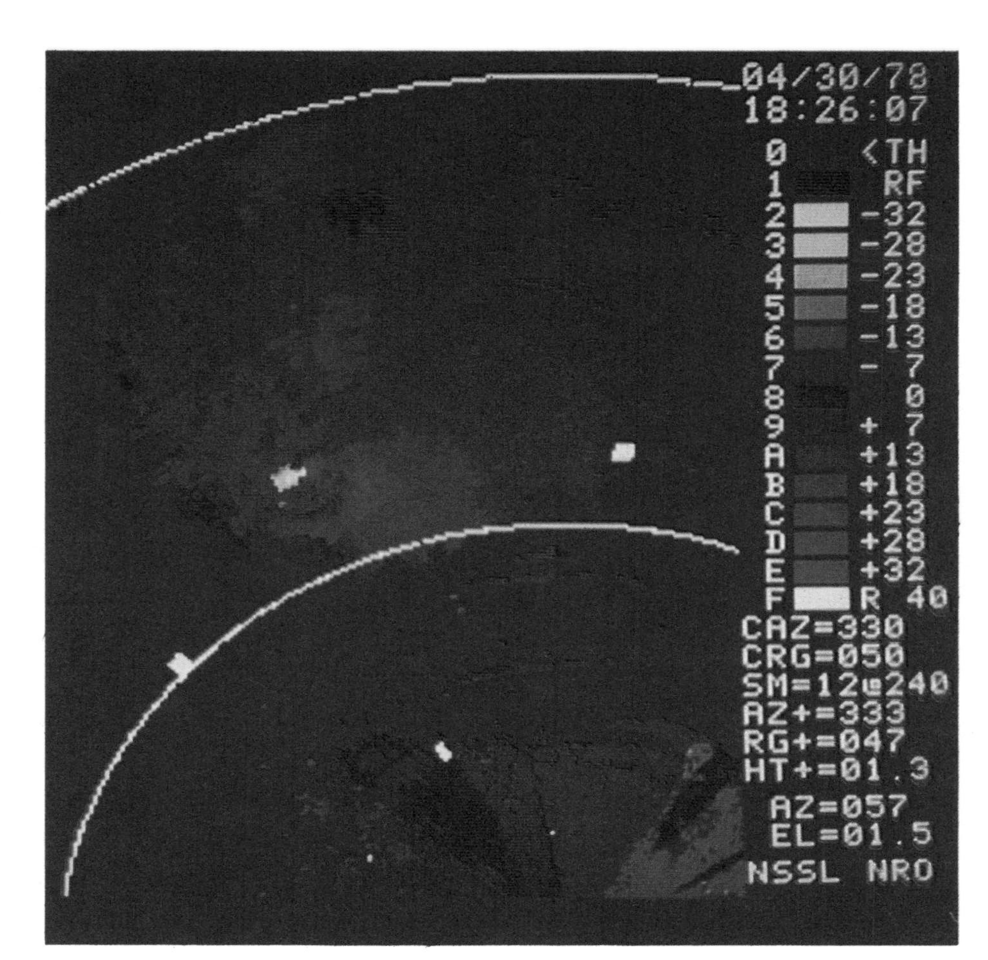

Figure 10.20. (a) Display of storm reflectivity factor Z. Color categories are in units of dBZ, i.e., of 10 log Z (mm^6 m^{-3}). (b) Doppler velocity. Color categories are in meters per second.

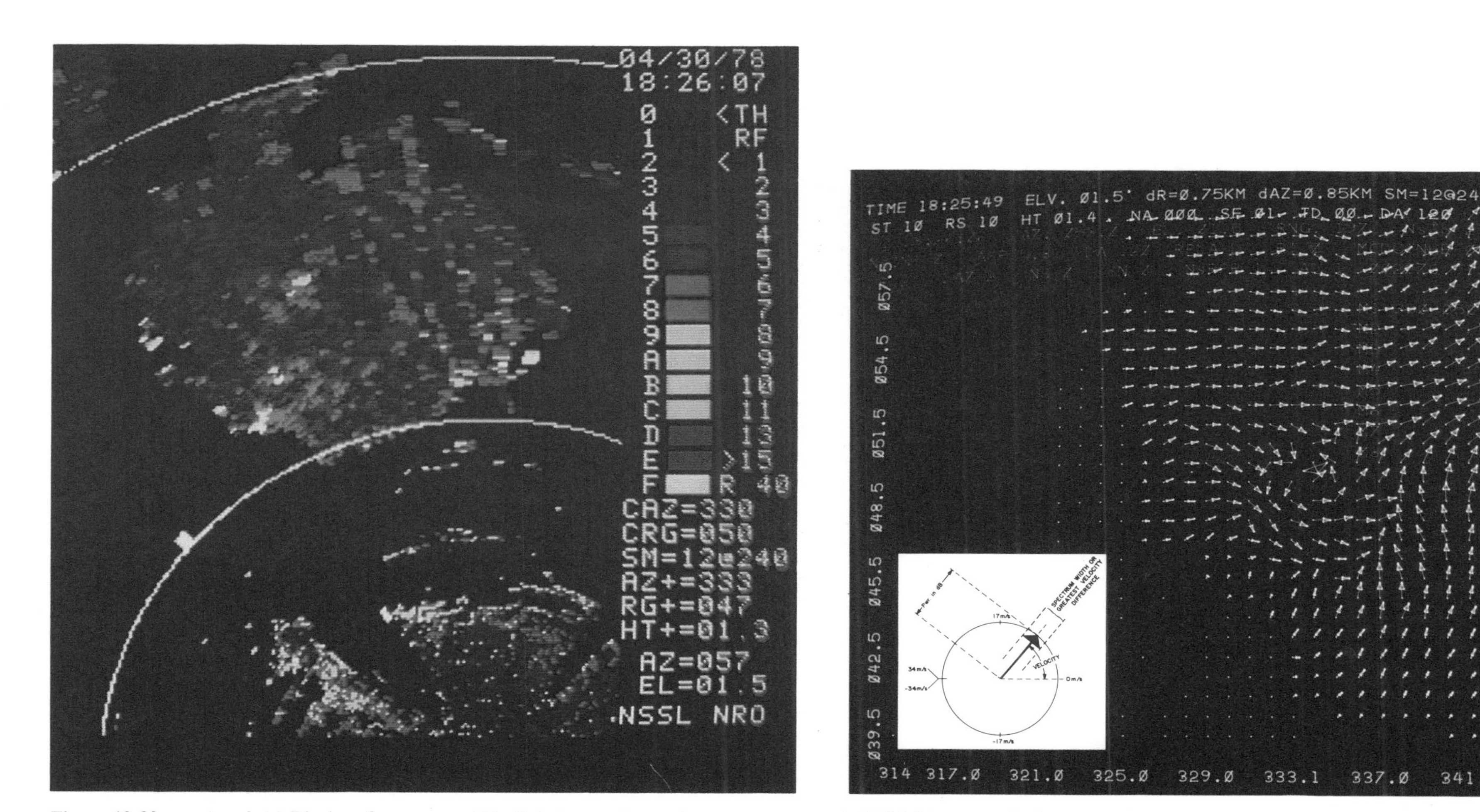

Figure 10.20. *continued.* (c) Display of spectrum width. Color categories are in meters per second. (d) Multimoment display that uses arrows to show composites of the three principal Doppler spectral moments for each resolution volume. In this B-scan format, azimuthal bearing (degrees) is indicated along the abscissa and range (kilometers) along the ordinate.

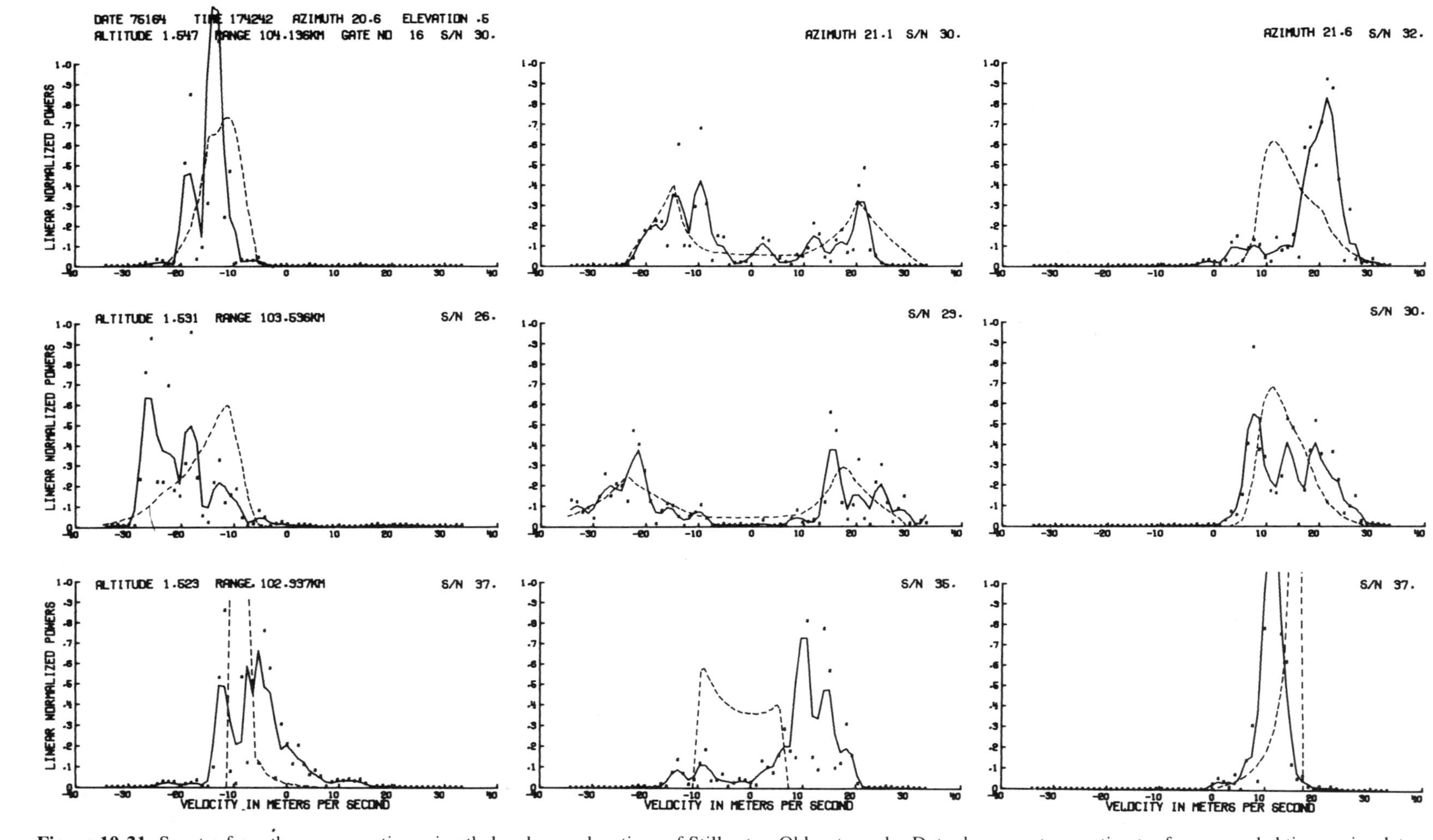

Figure 10.21. Spectra from three consecutive azimuthal and range locations of Stillwater, Okla., tornado. Dots show spectrum estimates from recorded time series data weighted with a von Hann window. Solid lines are three-point running averages. Dashed lines are simulated spectra. The mean square difference between data and simulated spectra is simultaneously minimized for two spectra (Az = 21.1° and range of 103.5 and 104.1 km). Resolution volume depth is 150 m, range gate spacing 600 m, and antenna beamwidth 0.8°. The tornado is located between the two upper middle gates. Tornado parameters obtained from these two fitted spectra are used to compute the remaining seven simulated spectra. Height aboveground for these spectra is 640 m.

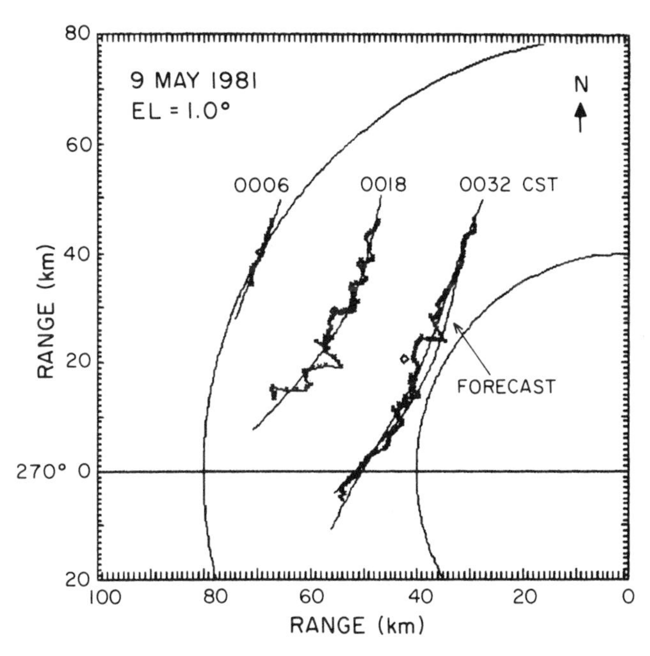

Figure 10.22. Three consecutive positions of the front on 9 May 1981. The smooth curves are least-square fits to data, and the forecast position is indicated.

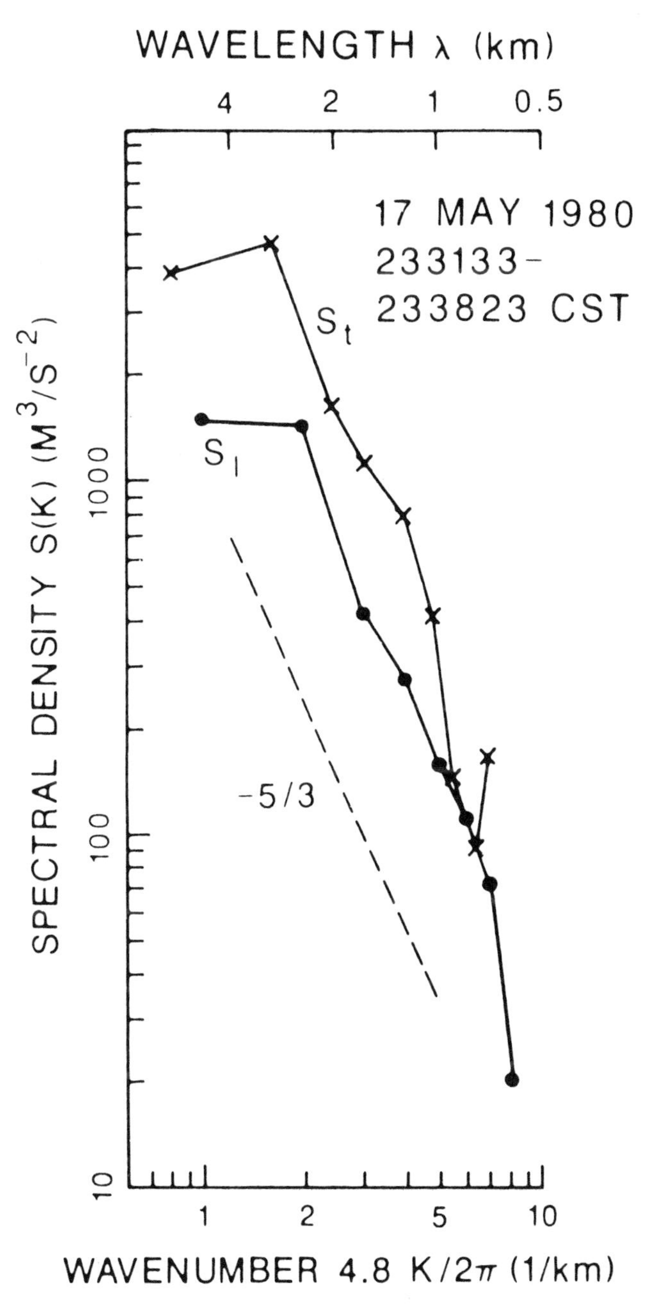

Figure 10.23. Spatial spectral densities. S_t is the transverse, and S_l the longitudinal spectrum. The dashed line represents the $-5/3$ law expected from isotropic turbulence in the inertial subrange.

transferred from larger-scale eddies to smaller ones without net energy loss) are larger than the radar resolution volume, then turbulence can be quantified from Doppler spectral width measurements (Doviak and Zrnić, 1984). So far, there have been few radar measurements to establish how well Doppler spectral width in storms characterizes isotropic turbulence. In the next paragraph we present such evidence.

Brewster and Zrnić (1986), observing a thunderstorm with a vertically pointed Doppler radar, showed excellent agreement in turbulent energy dissipation rates ε (a parameter whose values are popularly used to indicate turbulence intensity and, consequently, a measure of its hazard to aircraft) computed from spatial spectra of velocity fields with those ε computed from Doppler spectral widths. Furthermore they determined that longitudinal (S_l) and transverse (S_t) spatial spectra of velocities followed a theoretical relationship (expected for the inertial subrange) up to scales of 2.4 km (Fig. 10.23).

Thus it can be argued convincingly that the 2.4-km and smaller-scale eddies are within the inertial range of isotropic turbulence in which kinetic energy is transferred to smaller-scale eddies and to eventual dissipation when viscous forces act on these eddies. However, at long ranges the radar's resolution volume may be larger than the outer scale of isotropic turbulence; then neither the mean Doppler velocity nor spectral width fields may give accurate estimates of eddy dissipation rates. If we accept the fact that the outer scale of turbulence in storms may be 1 or 2 km, then measurement of ε from σ_v may be accurate for narrow-beam radar (i.e., $\leq 1°$) for ranges to 60 km.

c. Application of Doppler Radar in the National Weather Service

In view of the variety of displays and signatures associated with tornadoes, it is natural to ask which detection technique is most promising. A project involving the Environmental Research Laboratories, the National Weather Service, the Air Weather Service, and the Air Force Geophysics Laboratory was established to conduct experiments that would provide some answers (NSSL, 1979). Operations were conducted during the spring of 1977 and 1978. It became apparent that mesocyclone circulations showed very nicely on the color display of radial velocities. The Multimoment display (Burgess et al., 1976) used in conjunction with the color display proved most suitable for tornado and cyclone recognition. Using criteria discussed in Brown et al. (1978), the project scientists were able to detect all tornadoes that occurred within a range of less than 115 km (Burgess et al., 1978). The average lead time was about 20 min. It was established that circulation starts at middle-level altitudes (6–8 km) and works its way toward the ground. At ranges beyond 115 km, signatures of small tornadoes are lost because of poor resolution; however, large, destructive tornadoes were detected up to 240 km away through the use of the 0.8° beam of NSSL's Doppler radar. Spectrum width has not been a reliable indicator of tornadoes because turbulence in storms can produce large values of spectrum width and thus could be easily mistaken for tornadoes. On the basis of those experiments it is believed that an operational tornado detection system will involve interaction between human operators and an automated scheme in which the velocity pattern of a tornado cyclone is recognized.

d. Pulsed-Doppler Observation of Clear-Air Wind Fields

Whenever turbulence mixes air in which there are gradients of potential temperature and water vapor density, the turbulence causes spatial fluctuations in the refractive index n. The fluctuations are small (e.g., 1 ppm). Nevertheless, sensitive microwave radars detect the very faint echoes returned from these irregularities in what otherwise (without turbulence) would be a smoothly changing n with negligible backscatter.

Echoes from clear air have been seen almost from the inception of radar observations. These "angel echoes" were at first mystifying, but often were associated with birds and insects. Clear-air echoes not related to any visible object in the atmosphere were conclusively proved to emanate from refractive index fluctuations through use of multiwavelength radars at Wallops Island (Hardy et al., 1966). Simultaneous in situ measurements of refractive index fluctuations and reflectivity corroborated this finding (Saxton, 1964; Doviak and Berger, 1980). The radar studies were preceded by many measurements using tropospheric-scatter communication links that often depend on clear-air refractive index fluctuations to provide reliable wide-band circuits between distant points (Rice et al., 1967; IEEE, 1955).

In the 1960s ultrasensitive incoherent radars were used to detect remotely and resolve clear-air atmospheric structure, and these studies were well reviewed by Hardy and Katz (1969). The radars showed meteorological phenomena such as convective thermals (Katz, 1966; Konrad, 1970), sea-and-land breezes (Meyer, 1970), and Kelvin-Helmholtz waves (Hicks and Angell, 1968).

Doppler processing of coherent radar echoes can improve target detection by at least an order of magnitude (Hennington et al., 1976), and hence medium-resolution weather radars could have a detection capability often associated with large-aperture (18- to 27-m diameter) antennas used with incoherent radars. Furthermore, a coherent radar provides a way in which ground clutter can be distinguished from moving atmospheric targets and allows data acquisition at closer ranges, thereby taking advantage of the r^{-2} dependence in echo power.

Chadwick et al. (1978), using a low-power 10-cm FM-CW radar, monitored reflectivity values associated with refractive index fluctuations in the planetary boundary layer and concluded after one year of observation that winds can always be measured to a height of several hundred meters with moderately sensitive radars. Clear-air wind measurement has practical significance because pulsed-Doppler radars under development by the FAA also could measure wind-shear hazards near airports for all weather conditions (Lee and Goff, 1976).

Harrold and Browning (1971) found that radar can delineate the upper limit of convection before precipitation, showing that it is deeper in some regions than others. Some of these areas of deep convection persist for several hours, and, if showers develop, they occur within, and only within, such regions. Important economic advantage can be achieved if radars can locate aircraft weather hazards and predict the location of incipient showers.

Mapping boundary layer wind over large areas has double significance for severe-storm studies because it allows early observation of thunderstorm development—hence storm genesis can be followed from the very beginning of cumulus development—and the capability of Doppler weather radars to map clear-air wind makes it possible to monitor thunderstorm outflow, which is usually precipitation free.

Strongest fluctuations in refractive index occur where turbulence mixes large gradients of mean potential temperature and specific humidity (Yaglom, 1949). Gage et al. (1973) measured the height distribution of forward scattered signal strength to show good correlation between it and gradients of potential temperature at high altitudes, where water-vapor contributions can be ignored. The tropopause is a region where potential temperature always increases with height; it was first detected by radar in 1966 (Atlas et al., 1966). Later VanZandt et al. (1978), using backscatter results from a vertically pointed VHF radar, obtained consistent agreement between rawinsonde-inferred gradients and radar-measured ones for the clear air above the moist boundary layer.

For many years ionospheric scientists have been employing high-power VHF and UHF (Fig. 10.1) radars to observe the ionosphere. During the 1980s these radars at Jicamarca, Peru (Woodman and Guillen, 1974); Arecibo, Puerto Rico; Chatanika, Alaska (Balsley, 1978); and Lindau, Germany (Röttger and Liu, 1978), lowered their sights to examine echoes in the stratosphere and troposphere. A VHF ($\lambda = 7$ m)

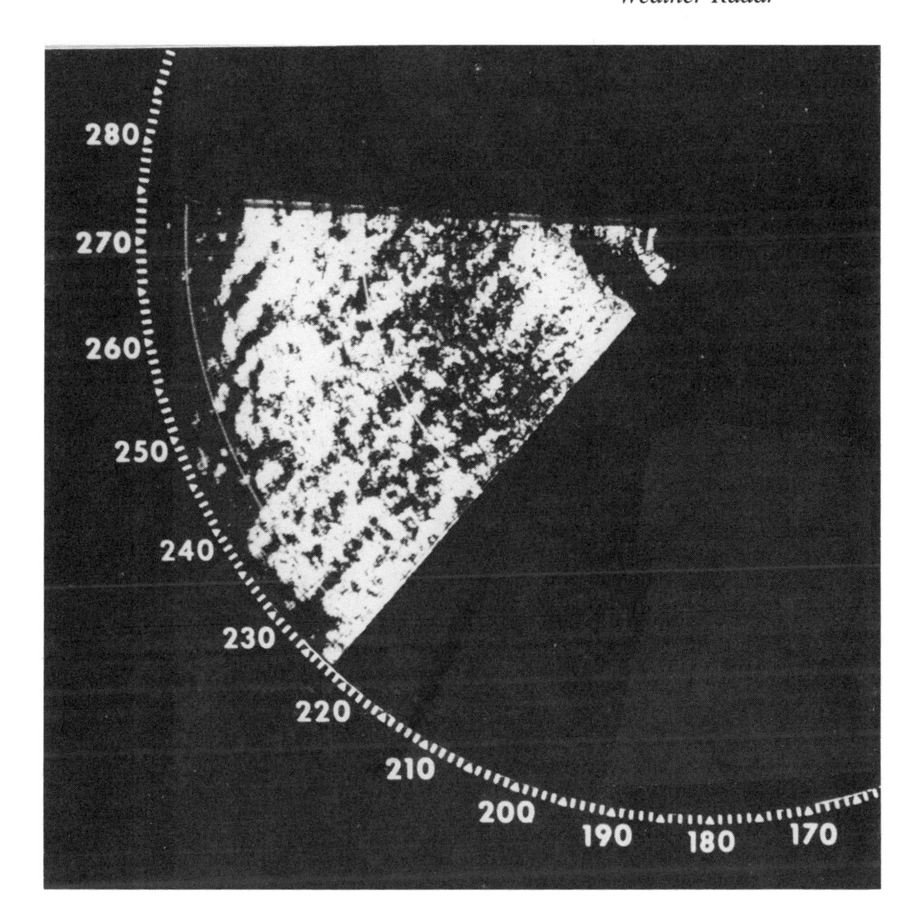

Figure 10.24. PPI sector scan of echo power from clear air as seen by the NSSL Doppler radar. The bright area of higher power is aligned roughly parallel to mean wind, and the bands are spaced about 4 km apart. Range marks are 20 km apart. $EL = 1.6°$ (27 April 1977, 1447 CST).

pulsed-Doppler radar specifically designed for tropospheric studies was assembled at Sunset, Colo., to make continuous observations of winds in the troposphere (Green et al., 1975). Some advantages of the VHF radar are that it can sometimes differentiate between scatter from hydrometeors and refractive index fluctuations (Green et al., 1978) and that it can detect coherent scatter from mean refractive index gradients at inversion layers (Gage and Green, 1978; Fukao, et al., 1985; Fukao et al., 1986).

Clear-air winds in the planetary boundary layer (PBL, i.e., surface to heights of about 1.5 km) have been synthesized from dual Doppler radar measurements (Gossard and Frisch, 1976; Kropfli and Kohn, 1978), wherein chaff was dispensed over large areas to provide suitable echo levels for detection and processing. Doviak and Jobson (1979) showed first results of two Doppler radar-synthesized wind fields in the PBL clear air, where only the diffuse and intrinsic scatterers of the medium were used as targets. They observed mean wind fields at low height to have qualitative agreement with mean wind measured (with conventional anemometers) near the surface.

On 27 April 1977, a day marked by strong nondirectional shear and curvature in the wind profile, NSSL's Doppler radar echo power measurements showed evidence of clear-air convective streets (Fig. 10.24), an observation that should signify the presence of roll vortices. The first radar detection of clear-air "thermal streets" was reported by Konrad (1968).

Figure 10.25 locates the Cimarron (CIM) and Norman (NRO) radars and a 444-m meteorologically instrumented tower. Reasonably accurate winds can be synthesized from Doppler data in the entire area enclosed by the dashed lines except for the overlapping areas. The winds were fairly uniform from the southwest on this day, but there were small perturbations from the mean wind having a magnitude about one order less than the mean wind itself. The synthesized wind field and its perturbations from the mean are shown in Figs. 10.26a and b.

Berger and Doviak (1979) examined the spatial spectra of the synthesized winds on this convectively dry day and have compared results with those spectra obtained from anemometers located on the tall tower. The spectra follow a 5/3 power law in the wavelength range of 1–8 km in agreement with spectra of tower winds. However, O'Ban-

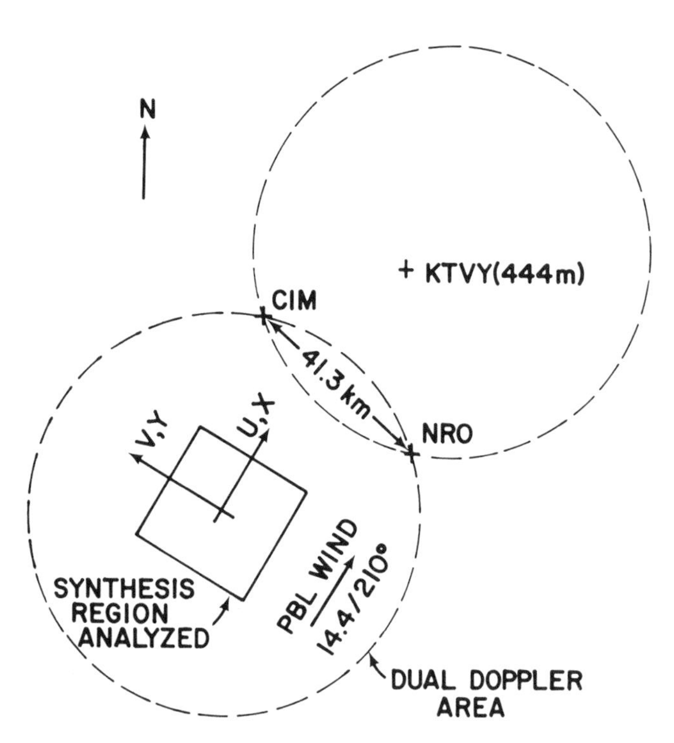

Figure 10.25. The dual Doppler-radar area (dashed lines) within which the angle subtended by the radials from the Cimarron (CIM) and Norman (NRO) radars lies between 30° and 150°. The outlined box is the region for which the PBL Doppler velocities were synthesized for detailed wind analyses. The windspeed and direction are a mean over 1.25-km depth of the PBL.

non (1978) showed spectra on another day when such a power dependence was not evident. Spectral analyses of clear-air longitudinal wind fluctuations using chaff and a single Doppler radar were reported by Chernikov et al. (1969). Their spectra, extending to 3-km wavelengths, also showed a 5/3 power law dependence in agreement with results from simultaneous measurements with airborne in situ instruments.

In addition to the 5/3 spectral form for velocity fluctuations shown in Fig. 10.26, the v component spectrum along the y direction had a peak, at the 4-km wavelength, that persisted over the one hour of data collection (Doviak and Berger, 1980). Some theories (e.g., Kuettner, 1971; Gossard and Moninger, 1975) suggest that roll vortices tend to form parallel to the mean wind in a strongly heated PBL having a large, unidirectional shear. Kuettner (1971) predicted rolls to have a horizontal spacing of 2.8 times their depth when the vertical profile of mean horizontal velocity has curvature. The strong 4-km wave in the y direction of the v component observed on 27 April is interpreted to be the convective rolls predicted by theory. Figure 10.27 depicts the observed perturbation winds after a band-pass filter had been applied in the y direction to emphasize the 4-km wave feature seen in the spectral displays (a low-pass filter was applied in the x direction along which no dominant wavelength was noted). Figure 10.27a is a vertical cross section at $X = -25.5$ km perpendicular to the mean wind. Vertical velocities were derived by integrating the mass continuity equation using wind fields from the six horizontal surfaces with vertical separation of 250 m. Readily apparent are counterrotating vortices (roll vortices), having approximately 4-km wavelength, whose maximum vertical velocities are approximately 1 m s^{-1}. Furthermore, the ratio of roll spacing to height is 2.6, in good agreement with that predicted by theory. Reinking et al. (1981) analyzed aircraft gust probe data collected on this day during the time of the radar observation. They too detected a prominent peak in power density at a wavelength of about 4 km in the y direction.

6. Conclusions

The introduction of Doppler frequency shift measurement capability into weather radars has opened new horizons for exploration by atmospheric scientists. The astounding success achieved with these radars in thunderstorm cyclone detection well in advance of tornado formation has caused incorporation of coherent systems in new radars for operation by the national services. Advances in digital signal processing and display techniques have allowed economical development of real-time presentation of the three principal Doppler spectral moments. The techniques are constantly improving, and new scientific disclosures and new operational applications in wind measurement and tornado detection are forthcoming.

Doppler radars at centimeter wavelengths do not have a large enough velocity-range ambiguity product, $r_a v_a$, to match that required to observe, without ambiguity, severe convective storms. No comprehensive data and documentation show the full extent of the problem, nor is there any foreseeable solution; it appears that storm observers will have to accept some limitations in Doppler radar weather measurements.

Dual Doppler radar observations of the kinematic structure of severe storms and the planetary boundary layer agree with theoretical models, but much investigation is still required. The Doppler weather radar shows promise of greatly increasing our knowledge of thunderstorms and the planetary boundary layer on scales not previously accessible. Furthermore, we can monitor significant mesoscale phenomena that are of importance to air traffic safety and air pollution control, and we may be able to see the triggering impulses of severe storms that each year cause substantial destruction. Increased power and sensitivity of weather radars may soon result in meteorologists being able to ob-

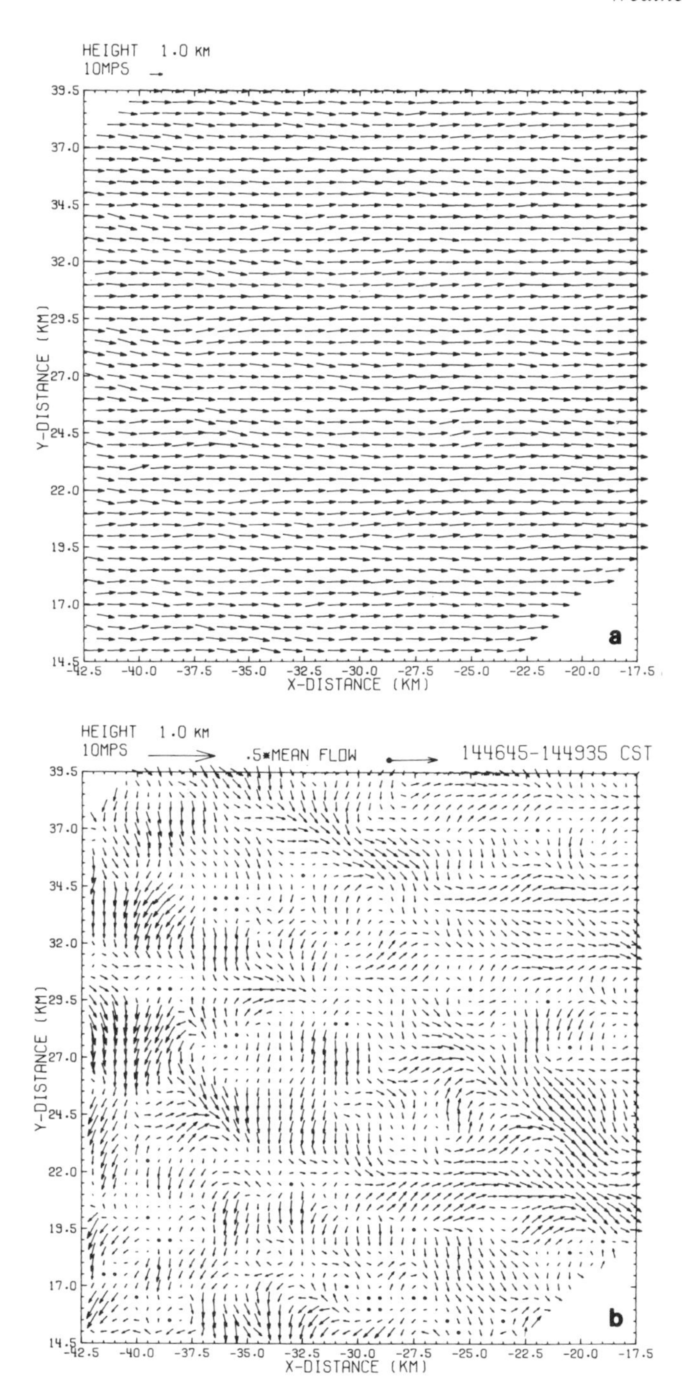

Figure 10.26. (a) Dual radar synthesized wind field at 1 km aboveground for 27 April 1977. (b) Wind field with mean removed. Synthesized wind fields were low-pass-filtered once in *X* and *Y* direction with a three-point Shuman (1957) filter.

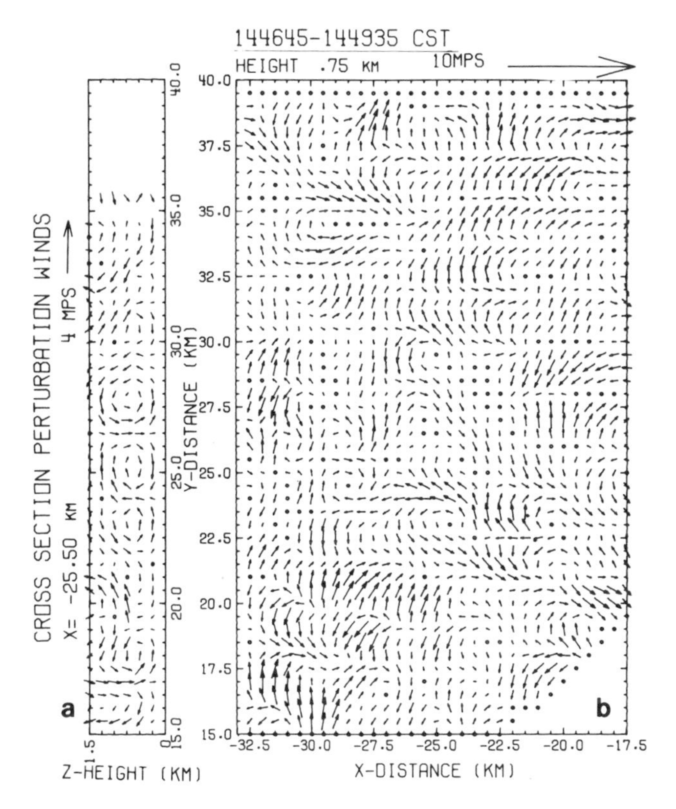

Figure 10.27. Vertical (a) and horizontal (b) cross sections of band-pass-filtered wind data that highlight the clear-air roll structure seen in the spectra displays of unfiltered data for 27 April 1977.

serve the wind structure and its evolution throughout the troposphere.

The important advances in meteorological observations brought forth by the application of Doppler techniques to weather radars will continue in the future. There is room for further improvement in the radar system to reduce the effect of ambiguities while lessening data acquisition time for observation of severe-storm convection, shear, and turbulence in clear or precipitation-laden air.[1]

[1] The authors appreciate the support they have received from their colleagues in the Environmental Research Laboratories, the Air Force Geophysics Laboratory, the National Weather Service, the Energy Research and Development Administration, the Nuclear Regulatory Commission, and the Federal Aviation Administration, not only in the preparation of this paper but also for the continuing development of Doppler weather radar technology. We are indebted to Ms. Joy Walton and Ms. Michelle Foster for efficient and accurate typing and editing of the manuscript.

11. Measuring Storm Rainfall by Radar and Rain Gage

Edward A. Brandes and James W. Wilson

1. Introduction

The measurement of thunderstorm rainfall has import for hydrology, agriculture, weather modification, climatology, and weather forecasting. Although details of requirements vary, there is always concern for rainfall distribution in space and time. For many agricultural applications and for streamflow in large river basins, rainfall data are needed over large areas for long periods (days or weeks) and a relatively sparse network of rain gages reporting once a day is adequate. In contrast, measurements from many closely spaced stations are required to forecast flash floods on small streams, where data are required for short time intervals (<10 h) and over small areas (<1000 km^2) or to evaluate weather-modification experiments. However, it is difficult and expensive to install and to maintain observation sites sufficiently numerous and capable of transmitting data in a timely manner to a central processing facility. Hence there has been considerable interest in using weather radar to measure precipitation.

Radar overcomes the primary difficulties associated with rain gages; i.e., the radar measurements are over an area, available in real time, and from a single location. The disadvantage with radar is the lack of a unique relationship between the energy returned to the radar and the precipitation reaching the ground. Although radars are expensive, their cost is similar to that for a dense rain-gage network that telemeters data to a central location. Also, when radar is used for other weather surveillance and forecasting activities, the part of the radar cost attributable to the rainfall-reporting function becomes less than the cost of a dense rain-gage network.

Figure 11.1 illustrates the detailed distributions of thunderstorm rainfall that can be obtained by incoherent radar in near real time. The spatial detail in the figure is defined not by the gages but by the radar analysis. In fact, had there been no Enid observation (near the 400-mm contour), the extreme rainfall would have been virtually unrevealed by gages. The Enid gage measured 398 mm, and the radar estimate for the same location was 374 mm. Such agreement is uncommon, however, and it is not unusual for radar measurements to be in error by more than a factor of 2. Because of this uncertainty, radar cannot replace gages; it can only supplement them.

Several measuring techniques that use various properties of radar signals have been suggested. Because rainfall attenuates microwaves, it was proposed that rainfall be determined by measuring attenuation (see Atlas, 1964; Battan, 1973) or by comparing measurements obtained at attenuating and nonattenuating wavelengths (Rogers and Wexler, 1963). Ulbrich and Atlas (1975) pointed out that measurements of both radar reflectivity factor and microwave attenuation yield two parameters of exponential drop size distributions and hence rainfall rate directly. The attenuation properties of signals from a single radar can be used to estimate rainfall from an instrumented satellite in space (Meneghini et al., 1983; Atlas and Meneghini, 1983).

Seliga and Bringi (1976, 1978) propose a technique based on measurements of radar reflectivity factor at horizontal and vertical polarizations. The dual polarization method is based on the observation that raindrops fall as oblate spheroids and that echo intensity should be concentrated in preferential planes. In the atmosphere, owing to turbulence and oscillations caused by collisions between drops, raindrops are not uniformly aligned; the differential reflectivity, the logarithm of the ratio of echo intensity in the orthogonal planes, is closer to unity than expected from theoretical considerations (Jameson and Beard, 1982). The few comparisons with rain gages (Seliga et al., 1981b; Clarke et al., 1983) indicate that the dual polarization method gives slight improvement over rainfall measurements from reflectivity alone. Differential reflectivity appears to be particularly suitable for discriminating between frozen and liquid precipitation (Hall et al., 1980; Seliga et al., 1981a; Seliga and Aydin, 1983). These schemes continue in development and require additional research before they can be considered for operational application.

Problems in estimating thunderstorm rainfall arise from extreme natural variability associated with these storms. Bias errors (primarily an undercatchment) characterize individual gage measurements, and large spatial variance in depth introduces random errors in areal mean estimates deduced from limited numbers of point observations (Sec. 2). Errors in radar measurements originate with instrumentation, sampling, and meteorological factors such as the ab-

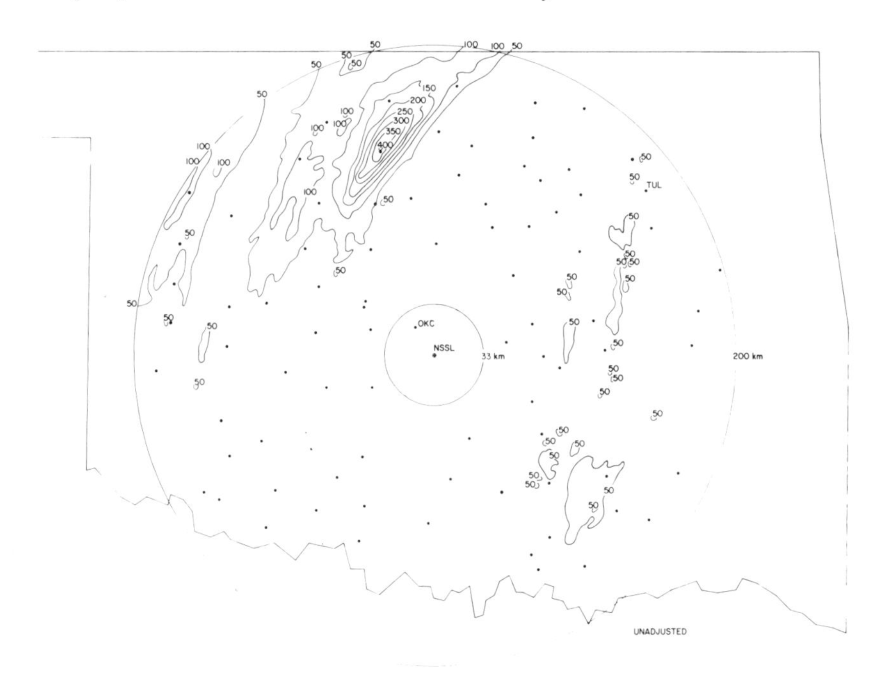

Figure 11.1. Radar-measured rainfall totals (in millimeters) for the Enid, Okla., storm of 10–11 October 1973. The rainfall analysis was derived with the National Severe Storms Laboratory (NSSL) WSR-57 radar, without adjustment by gages. The black dots indicate the location of the NWS climatological gages.

sence of a unique relationship for converting radar observations to rainfall measurement (Sec. 3). The accuracy of point and area rainfall measurements by radar is discussed in Sec. 4. Since radar and gage measurements have contrasting strengths and weaknesses, techniques have been developed (Sec. 5) to combine these two observations to produce rainfall estimates superior to those produced by either alone.

2. Gage Measurement of Rainfall

Rain gages offer a simple, inexpensive method of estimating point or areal rainfall and therefore are extensively used; however, all gage measurements are somewhat erroneous.

a. Instrument Error

Numerous papers have been published on gage-catch deficiencies in an attempt to quantify the many errors. Annotated bibliographies were prepared by Kurtyka (1953), Larson (1971), Israelsen (1967), and WMO (1973). Contributing error sources include adhesion or gage wetting, inclination of gage, splash into and out of gage, evaporation, and airflow around the gage. Fortunately, by exercising reasonable precaution, these errors (except perhaps from airflow around the gage) amount to no more than about a 2% deficiency in catch (Dahlström, 1973).

Woodley et al. (1975) compared thunderstorm rainfall measured by gages located within 2 m of each other. Although the gages were of the same type and had nearly identical exposure, differences averaged 9%. The larger percentage differences were associated with smaller storm totals. For similar comparisons, Huff (1955) found variations of only 2%, and Joss et al. (1968) reported a standard relative error of 6% for all rainfall types.

There is general agreement (Larson and Peck, 1974) that much of the total measurement error results from turbulence and increased windspeed in the vicinity of the gage orifice. As the air rises to pass over the gage, precipitation particles that would have passed through the gage orifice are deflected and carried farther downwind. A rather wide range of diminished gage catchment versus windspeed has been indicated (e.g., Koschmieder, 1934; Dahlström, 1973; Larson and Peck, 1974). Larson and Peck summarized a number of independent studies; they reported about 12% rain deficiency with a 5-m s^{-1} wind and a 19% deficiency at 10 m s^{-1}. Undercatch at higher speeds can be only estimated because of insufficient data; however, it is estimated by extrapolation that a gage may undermeasure rainfall in strong outflow regions of thunderstorms by as much as 20–40%.

b. Areal Estimate Error

Gage estimates of area-averaged rainfall reflect both errors inherent in individual measurements and random sampling errors in areas of substantial spatial variation. The sampling errors depend on the number and distribution of gages, area size, length of measurement period, and, most important, spatial rainfall variability.

The common method of estimating sampling error proceeds from a very dense network that is assumed to measure the true areal average rainfall. The difference presented by a subset of the dense network is a measure of the error resulting from sampling at a few points (Linsley and Kohler, 1951; McGuinness, 1963; Nicks, 1966; Huff, 1971; Woodley et al., 1975). In general, error decreases with increasing area size, increasing time period, increasing gage density, and increasing rainfall amount. Figure 11.2 shows sampling error versus network mean rainfall for various gage densities. Immediately apparent is the much larger error for Florida showers and thunderstorms, because rainfall gradients are considerably larger and storms are smaller in Florida than in Illinois. Huff (1971) showed there is a great variability about the average curves (Fig. 11.2), making it very difficult to predict the sampling error for a given storm.

Unfortunately, there have been no error studies with gage densities of 10–20 km² per gage, which are typically used in comparing radar and gage rainfall measurements. However, extrapolation of data used in Fig. 11.2 suggests that for Illinois-type thunderstorms the random sampling error for a watershed of 1,040 km² would be no more than 3% or 4% for rainfall amounts <10 mm and less than 1% for larger amounts. Considerably larger errors are likely for Florida-type storms.

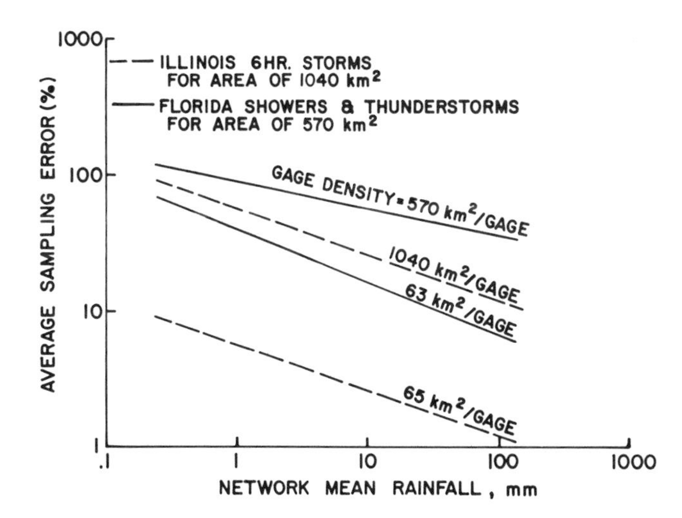

Figure 11.2. Average rainfall depth sampling error versus network mean rainfall and gage density for Illinois and Florida summer rainstorms.

3. Radar Measurement of Rainfall

Since the late 1940s there have been many radar precipitation measurement studies based upon relationships among radar reflectivity, drop size distribution, and rainfall rate. However, operational implementation of radar has been slow because of difficulties in processing large volumes of data and uncertainties in the accuracy of the measurements. Digital recording and computer technology are solving the former, and substantial progress is being made to quantify the errors.

a. Radar Equation and Calculation of Rainfall Rate

A modified version of the Probert-Jones (1962) expression relating reflectivity factor Z to the average power or backscattering energy returned to a radar, $\overline{P}_r$ (in watts), from a target volume (see Chap. 10) is

$$Z = \frac{1{,}024\ C\ ln2}{\pi^3}\ \frac{\lambda^2 r^2}{hG\theta\phi}\ \frac{l_p^2 l_g^2 l_c^2 l_a \overline{P}_r}{|K|^2\ P_t}\ \mathrm{mm^6\ m^{-3}}. \quad (11.1)$$

The other parameters are these:

λ = radar wavelength (cm).
r = radar slant range (km).
P_t = peak transmitted power (watts).
n = pulse length (m).
G = antenna gain.
θ = horizontal one-way 3-dB beamwidth (deg.).
ϕ = vertical one-way 3-dB beamwidth (deg.).
$|K|^2$ = dielectric factor.
C = 3.282×10^{23} (deg² mm⁶ rad⁻² cm⁻² m⁻² km⁻²).

Implicit in Eq. 11.1 are the assumptions that particles are significantly smaller than the radar wavelength (Rayleigh approximation) and that they are spherical. The dimensionless loss factors (l_p^2, l_g^2, l_c^2, and l_a) are larger than unity and represent precipitation attenuation (e.g., Burrows and Attwood, 1949), atmospheric gaseous absorption (Blake, 1970), cloud attenuation (Ryde and Ryde, 1945), and space-averaging loss for nonlinear receivers (Rogers, 1971; Sirmans and Doviak, 1973). Details on attenuation losses are presented in the appendix to this chapter.

The average return power per unit volume ΔV is proportional to the summation of the sixth power of particle diameters (D_i^6, mm) illuminated by the radar beam; hence the reflectivity factor is also defined by

$$Z = \frac{1}{\Delta V} \sum_{i=1}^{N} D_i^6. \quad (11.2)$$

For large scatterers ($D_i \geqslant \lambda/16$), the above relationships become unsatisfactory approximations to the more general Mie theory, and the meteorologist often defines an equivalent reflectivity factor, Z_e, i.e., the summation of an equivalent distribution of small water droplets that backscatter the same power as that observed.

Except for small diameters, raindrop size distributions have the approximate form (Marshall and Palmer, 1948)

$$N(D) = N_0 e^{-\Lambda D}, \quad (11.3)$$

where $N(D)$ is the number concentration of size D droplets per size interval ΔD. Of course, there are significant departures from this simple exponential relationship with individual drop size samples and with various types of precipitation. Thus Marshall and Palmer determined N_0 to be 8,000 m⁻³ mm⁻¹ and $\Lambda = 4.1\ R^{-0.21}$, where R is the rainfall rate (mm h⁻¹); other investigations have shown that Λ

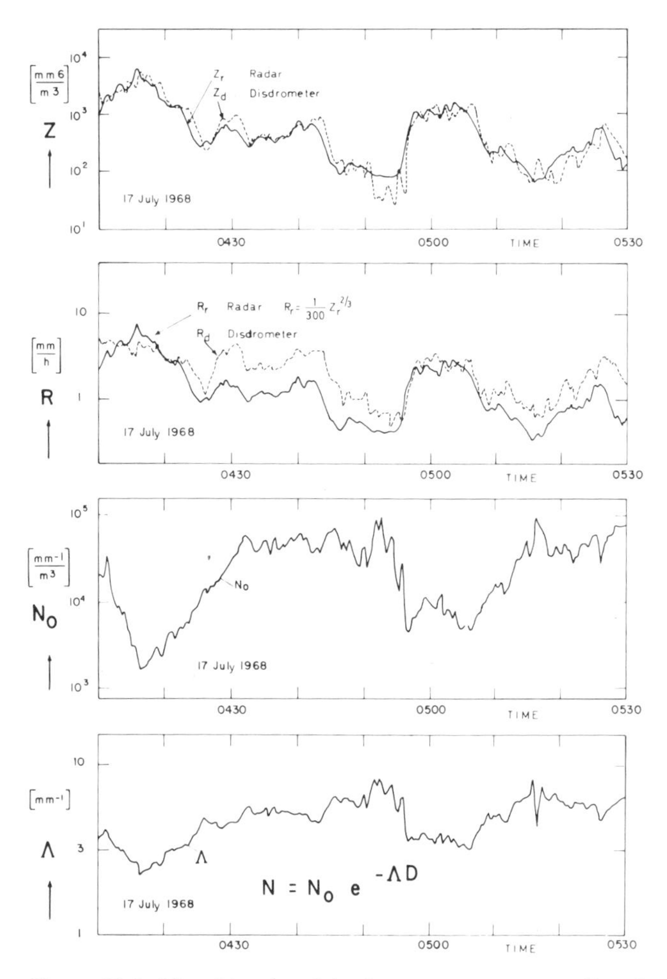

Figure 11.3. Time histories of the drop spectra parameters N_0 and Λ. Also given is a comparison of radar reflectivity and rainfall rate measured by a vertically pointing radar sampling 200 m above the ground and by drop size measurements obtained next to the radar (Joss et al., 1970).

and N_0 vary from storm to storm and that N_0 is also a function of rainfall rate; e.g., Sekhon and Srivastava (1971) found $\Lambda = 3.8\ R^{-0.14}$ and $N_0 = 7{,}000\ R^{0.37}$. An illustration of within-storm N_0 and Λ variability for stratiform precipitation is presented in Fig. 11.3.

When the drop size distribution is known, radar reflectivity factor in a unit volume can be calculated from

$$Z = \int_0^\infty D^6\, N(D) dD. \tag{11.4}$$

Similarly, when the vertical airspeed is zero, rainfall rates are given by

$$R = \frac{\pi\rho}{6} \int_0^\infty D^3 v_t\,(D)\; N(D) dD, \tag{11.5}$$

where ρ is the density of water and $v_t(D)$, the particle terminal velocity, is given approximately by $v_t = 1{,}400\ D^{\frac{1}{2}}$ (in cgs units; Spilhaus, 1948), or, more precisely, from studies by Gunn and Kinzer (1949) and Best (1950). When N_0 is constant, the equations above imply a relationship of the form (Kessler, 1969)

$$Z = AR^b. \tag{11.6}$$

Some related discussion is given in Vol. 2, Chap. 14. If the drop size distribution were known and if vertical motions were small relative to the drop terminal velocities, there would be no fundamental limitation to the accuracy of the radar rainfall estimates. But variations of drop size distribution are rarely known, and vertical air motions are frequently of the same magnitude as the particle terminal fallspeeds (particularly in thunderstorms). Hence the *Z-R* relationship is not unique (e.g., Battan, 1973), and there has been much empirical study to seek values of the coefficient and exponent in Eq. 11.6. The most commonly used relation is

$$Z = 200\ R^{1.6}, \tag{11.7}$$

following the original work of Marshall and Palmer (1948). Unless otherwise noted, all radar rainfall estimates in this chapter are derived with Eq. 11.7.

b. Sources of Error

The numerous factors causing errors in the radar measurement of rainfall can be grouped into three broad categories: (1) estimating of radar reflectivity factor; (2) time and space averaging of substantial variations in reflectivity; and (3) variations in the *Z-R* relationship and below-beam effects resulting from meteorological processes that modify the precipitation while it is falling.

(1) Estimating of Radar Reflectivity Factor: System errors (bias) in the measurement of $\overline{P}_r$ and ultimately in Z arise from incorrect hardware calibration. Even after presumably careful electronic system calibration, large, inexplicable errors in rainfall estimates occasionally remain (e.g., Harrold et al., 1974; Klazura, 1977; Saffle and Greene, 1978). A large variance (random error) in signal estimates is also associated with independent motion of the numerous target particles. This error is reduced by averaging independent range and time samples within the radar measurement volume, which reduces local measurement inaccuracy inherent in the fluctuating signals to about 0.7 dB (a standard error of ~10% in rainfall rate).

Potentially serious sources of error not associated with hardware include ground targets close to the radar site such as trees, buildings, and ridges, which can severely reduce the effective transmitted and received power and cause recurring shadows in precipitation patterns. Harrold et al. (1974) successfully corrected for partial screening of the beam by using beam patterns and topographic profiles.

Microwave propagation in the atmosphere becomes anomalous in the presence of extreme gradients of temperature and humidity. Temperature inversions with decreasing moisture with height may cause ducting (bending) of radar signals, resulting in ground target returns from extended ranges. This condition is sometimes widespread; a local condition, dubbed "thunderstorm superrefraction" by Battan (1973), occurs when local temperature inversions and specific humidity fluctuations are produced by downdraft spreading beneath thunderstorms. In regions of superrefraction and their attendant shadow zones, rainfall estimates may be significantly reduced (Brandes and Sirmans, 1976). Normally, ground returns can be identified by their high spatial reflectivity variance and reduced pulse-to-pulse fluctuation rate.

When the radome, which encloses the radar antenna, is wet, attenuation of radar signal increases. Two-way losses, a function of water film thickness and radar wavelength, may approach 2 dB (Cohen and Smolski, 1966; Wilson, 1978). Because this temporary condition is difficult to describe quantitatively and because recovery is rapid once rainfall has ended, usually no attempt is made to account for this loss.

(2) Time and Space Averaging of Radar Measurements: Radar data ordinarily are obtained by scanning in azimuth at a low elevation angle and making measurements at discrete range and angular intervals. The reflectivity factor values are converted to rainfall rate with an appropriate *Z-R* relationship and accumulated in time to yield a spatial distribution of rainfall depth (e.g., Fig. 11.1). Regardless of the *Z-R* relationship used, this procedure results in time and space sampling errors that relate to rain rate histories (i.e., storm element size, intensity, duration, and motion characteristics).

Signal averaging effectively smooths small-scale radial and azimuthal variations. For example, with the National Severe Storms Laboratory (NSSL) WSR-57 10-cm radar, on-line electronic components provide cutoff wavelengths (50% power response) of 2.26 times both the range-averaging interval of 1 km and the azimuthal sampling interval of 2°. (A technical description of the NSSL WSR-57 radar, calibration, and data acquisition procedures is given by Sirmans and Doviak, 1973.) Precipitation elements smaller than the cutoff wavelength are not resolved in the radar measurements but are detected by gages and hence decorrelate point gage-radar comparisons. Further, the increased radar sampling volume and the greater beam height at long ranges result in a higher probability that the precipitation observed by radar is not representative of that reaching the ground. Also, at greater ranges the beam is not likely to be uniformly filled by precipitation. Such problems are particularly important with stratiform rainfall at the freezing level (Wilson, 1975; Harrold and Kitchingman, 1975), because melting snow typically shows a sharp reflectivity maximum (bright band) just below the 0°C isotherm.

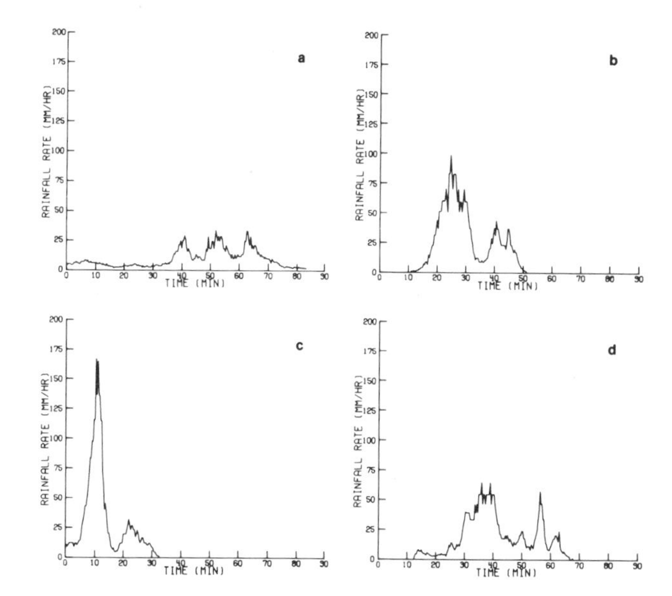

Figure 11.4. Representative point rainfall rate versus time traces for Oklahoma thunderstorms. Rainfall rates are derived from the NSSL radar data.

Typical radar-derived rain-rate histograms for a single measurement location within moderated thundershowers are given in Fig. 11.4. Some of the high-frequency fluctuation (~20 s) might be attributed to inaccuracies inherent in the basic radar measurement (Sec. 3b[1]). Such error is reduced when measurements are combined through time integration and/or by averaging over a watershed. Figure 11.4 shows that differences in the sampling interval can change the rainfall estimate.

The autocorrelation function for the four events represented in Fig. 11.5 reveals that observations spaced approximately 4 min apart are uncorrelated (coefficient of correlation ≤ 0.5), while observations with ≤ 1-min spacing have high correlation ($\gtrsim 0.9$) and therefore contain high data redundancy. When extrapolated to zero lag, random errors in the individual measurements cause the curves to intercept the ordinate at <1.

Sampling and instrumentation error (both gage and radar) combine with error in the selection of the *Z-R* relationship to further reduce the agreement between gage and radar. Figure 11.6 illustrates the effect of radar sampling interval on the ratio of gage and radar depth estimates, i.e., the parameter G/R (R refers to radar rainfall depth). The average ratio ($\overline{G/R}$) gives the systematic error, and the relative dispersion of G/R gives spatial variability of the individual comparisons. The relative dispersion or relative standard error is defined as the standard deviation ($\sigma[G/R]$) divided by the mean times 100%. For the convective storms illustrated, the dispersion among point gage-radar comparisons increases rapidly for radar sampling periods >5 min. At shorter time sampling intervals the correlation between ra-

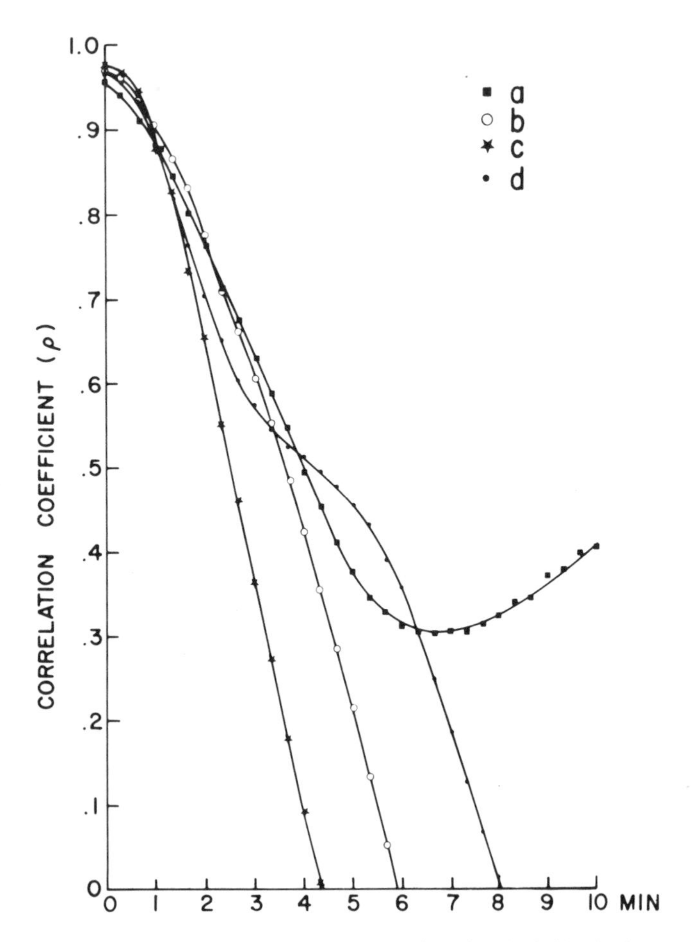

Figure 11.5. Autocorrelation function of the four individual rainfall rate traces shown in Fig. 11.4.

dar measurements increases rapidly, and there is little further error reduction. Note that considerable residual dispersion of G/R (~22%) remains even for the shortest sampling interval (20 sec). Radar bias (an underestimate in this example) increases slowly with sampling interval. Bias growth stems from an increase in radar sampling error with sampling interval. Note that the mean value of $G/(R + E_R)$ and $G/(R - E_R)$, where E_R is the radar sampling error, is greater than G/R (i.e., when the error is zero) and that the mean value grows as E_R increases. It is concluded that spatial smoothing in the basic radar measurement and integration of time-series measurements to compute the rainfall depth introduce both bias and scatter in gage-radar comparisons (Desautels and Gunn, 1970; Zawadzki, 1975). These errors have particular import when gage-radar comparisons form the basis for radar adjustment (Sec. 5).

(3) Variations in *Z-R* Relationships and Below-Beam Effects: Drop size distributions have been measured in stratiform rain, showers, and thundershowers, and characteristic *Z-R* relationships have been reported (Jones, 1955; Imai, 1960; Fujiwara, 1965; Joss et al., 1970; Jatila and Puhakka, 1973a). Ordinarily the coefficient (Eq. 11.6) increases and the exponent decreases with convection intensity. Although some improvement in radar rainfall estimates may be realized by selecting *Z-R* relationships based on storm type rather than using a single relationship for all storms, considerable error remains. For example, Jones (1966) reduced radar errors from 62% to 43% by such a technique.

Thunderstorm relationships reported by investigators typically exhibit a wide range (e.g., Fig. 11.7), comparable with storm variations observed in mean radar precipitation estimate bias (Sec. 4). Instrumental differences among drop size sampling techniques undoubtedly account for some of the total dispersion, but on the whole the reported variations in thunderstorm *Z-R* relationships as well as storm-type differences are thought to reflect vagaries in microphysical processes.

Fluctuations in drop size distributions were studied by Atlas and Plank (1953), who found that droplets in a rain shower had narrow spectra that shifted toward smaller sizes as the shower progressed. Blanchard (1953) noted the absence of small droplets at the onset of a thunderstorm and high numbers during heavy rain. Relatively high proportions of small droplets in heavy rains were also reported by Mason and Andrews (1960). For all rainfall types, Joss and Gori (1978) found that many "instant" (1-min) monodispersed samples were needed to produce an exponential distribution.

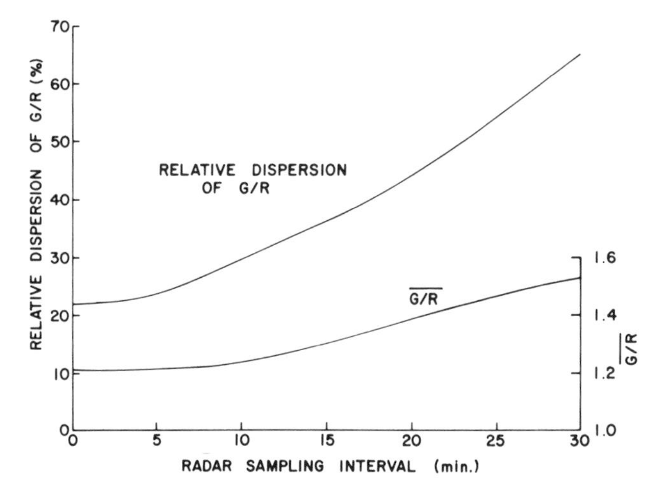

Figure 11.6. Effect of radar sampling interval on gage-radar rainfall comparisons. The lower curve illustrates the effect on the average ratio of gage and radar depths ($\overline{G/R}$), and the upper curve illustrates the effect on the relative dispersion of individual G/R ratios. Curves are based on data from 190 gage-radar comparisons of total storm rainfall from Oklahoma thunderstorms observed on 7 April 1975. Radar data obtained from the NSSL WSR-57.

Table 11.1. Microphysical and Kinematic Influences on *Z-R* Relationships and the Effect on Radar and Rainfall Estimates When No Adjustment Applied*

Process	Change in $Z = AR^b$		Probable effect on radar rainfall if *Z-R* is not adjusted	Possible region of maximum influence
	A	*b*		
Microphysical				
Evaporation (Atlas and Chmela, 1957)	Increase	Decrease	Overestimate	Inflow regions, fringe areas
Accretion of cloud particles (Atlas and Chmela, 1957; Rigby et al., 1954)	Decrease	Increase	Underestimate	Downdraft
Collision, coalescence (Srivastava, 1971)	Increase	Decrease	Overestimate	Reflectivity core
Breakup (Srivastava, 1971)	Decrease	Decrease	Underestimate	Reflectivity core
Kinematic				
Size sorting (Gunn and Marshall, 1955; Atlas and Chmela, 1957)	Increase	Decrease	Tendency to overestimate	Regions of strong inflow and outflow
Vertical motion				
Updraft	Increase	Decrease	Overestimate	
Downdraft	Decrease	Increase	Underestimate	

*Wilson and Brandes, 1979.

Drop size measurements beneath a large Oklahoma thunderstorm (Martner, 1975) yielded these relationships:

1. Leading portion: $Z = 667\ R^{1.33}$.
2. Central core: $Z = 124\ R^{1.64}$.
3. Trailing portion: $Z = 436\ R^{1.43}$.

Cloud-base spectra from Texas thunderstorms (Carbone and Nelson, 1978) show relatively high number concentrations of large droplets during growth stages, when updrafts dominate, and large total droplet concentrations with smaller median diameters during declining or downdraft stages. The mean relationships were these:

1. Growth phase: $Z = 763\ R^{1.37}$.
2. Decline phase: $Z = 400\ R^{1.56}$.

Note the similarities with the leading (updraft) and trailing (downdraft) storm regions in Martner's results. Large scatter in drop spectra over short time periods and the wide disparity among *Z-R* relationships reported by others prompted Twomey (1953) to conclude that radar rainfall estimates derived with average *Z-R* relationships were at best only approximate (a factor of 2 high or low).

Several physical mechanisms that might alter drop size distributions are listed in Table 11.1, with an indication of their probable influence on the *Z-R* relationship and of the storm region where the effect is probably a maximum. Such processes may act in combination to modify the drop size distribution to produce a complex net result. Also, droplets lose their sphericity and flatten as their size increases. Depending on the radar polarization, radar reflectivity measurements may be enhanced or reduced several decibels, causing large errors in rainfall rate (Seliga and Bringi, 1976).

Meteorological processes taking place below an elevated radar beam can further alter the precipitation before it reaches the ground. Calculations using the evaporation data of Kinzer and Gunn (1951) and terminal velocity measurements of Gunn and Kinzer (1949) indicate that rainfall rates with a Marshall-Palmer (1948) drop size distribution in an environment of 20°C and 80% relative humidity are reduced ~15% during a 300-m fall. When the horizontal windspeed and reflectivity gradient are large, precipitation drift below an elevated radar beam may negate the usual assumption that measurements are applicable at ground. Simple calcu-

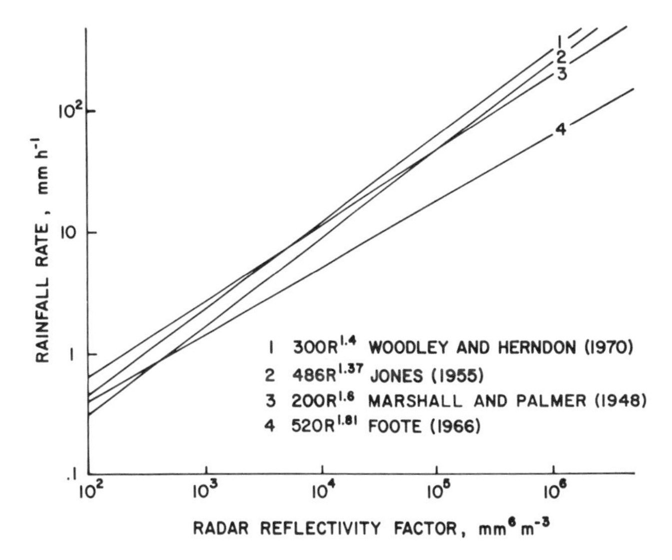

Figure 11.7. Plots of various *Z-R* relationships. Curves 1, 2, and 4 were determined for Florida, Illinois, and Arizona thunderstorms, respectively. The Marshall and Palmer relation (curve 3) represents a variety of rainfall types and is included for comparison.

Table 11.2. Summary of Gage-Radar Comparisons for 14 Oklahoma thunderstorms (Based on 5-min Radar Samples and $Z = 200R^{1.6}$)[a]

Date	Number of gages	Storm duration (h)	$\overline{G}$ (mm)	$\overline{G/R}$	Relative[b] dispersion about $\overline{G/R}$ %	Average[c] difference %	Average difference[d] (Storm bias removed) %
28 Apr. 1974	21	7	11	1.62	25	35	16
29 Apr. 1974	20	12	7	0.82	21	30	17
29 Apr. 1974	21	5	22	2.23	31	51	26
1 May 1974	21	9	24	2.41	20	57	13
20 May 1974	12	7	15	0.68	46	75	37
21 May 1974	10	3	5	0.59	36	87	27
23 May 1974	5	2	3	0.64	10	60	8
25 May 1974	22	6	25	0.91	27	30	27
25 May 1974	12	5	23	0.88	32	31	19
30 May 1974	22	7	25	1.09	34	32	33
3 June 1974	15	5	7	0.41	29	160	23
6 June 1974	13	2	16	0.68	43	79	36
8 June 1974	10	5	31	0.49	39	141	42
7 Apr. 1975	19	4	14	1.13	23	17	18
			Average 14 cases:	1.04	30[e]	63	24

[a]Wilson and Brandes, 1979.

[b]Relative dispersion (coefficient of variation) about $\overline{G/R} = 100\%\ \sigma[G/R]/\overline{G/R}$, where $\sigma[G/R]$ is the standard deviation of the gage-radar ratios.

[c]Average difference $= \dfrac{100\%}{N} \sum_{i=1}^{N} \left| \dfrac{G_i - R_i}{G_i} \right|$.

[d]Average difference (storm bias removed) $= \dfrac{100\%}{N} \sum_{i=1}^{N} \left| \dfrac{G_i - (\overline{G/R})R_i}{G_i} \right|$.

[e]The relative dispersion among all available gage-radar comparisons (14 storms) is 63%.

lations, assuming a reflectivity gradient of 2 dB km^{-1} at 300-m elevation and a mean horizontal windspeed of 20 m s^{-1}, show that rain rates in the range 2.5 to 100 mm h^{-1} can be altered 20% to 40% by horizontal winds of 20 m s^{-1}.

Depending on size and number concentration, hail may enhance reflectivity measurements, or, if the hailstones are water coated, significant attenuation may occur (see Battan, 1973, for a thorough treatment). Since hail location, composition, shape, and size distribution are rarely known, routine corrections to reflectivity measurements are not possible. Fortunately, hailfalls are usually short-lived, and their effect on storm-accumulated radar rainfall is usually small.

4. Radar-Gage Comparisons

Radar reflectivity and rainfall rates have been rather precisely related by Joss et al. (1970), who used a raindrop disdrometer and a zenith-pointing radar with a narrow-beam sampling 200 m above. Results (Fig. 11.3) show that measurements of the reflectivity factors by radar (Z_r) and disdrometer (Z_d) are very highly correlated. There is substantially less correspondence between computed rainfall rates R_r and R_d. Differences stem from time variations in the Z-R relationship owing to changes in the drop size distribution parameters N_0 and Λ. As mentioned earlier, at greater radar ranges with larger sampling volumes and with increasing beam height, the correlation between radar and ground level rainfall diminishes (Bruer and Kreuels, 1976; Martner, 1977).

Table 11.2 summarizes comparisons between radar and gage measurements for 14 Oklahoma storms observed with the NSSL WSR-57 radar. The gages were spread over 8,000 km^2 at radar ranges between 45 and 100 km. The utility of a single relationship (Eq. 11.7) for converting radar reflectivity to rainfall rate is indicated by the variability between storms in the average gage-to-radar ratio ($\overline{G/R}$). Ratios vary from 0.41 (radar overestimate) to 2.41 (radar underestimate). Storm-to-storm differences in $\overline{G/R}$, exceeding a factor of 2, are frequently observed (see also Joss et al., 1970; Woodley et al., 1974). In view of the observed range in reported Z-R relationships for thunderstorms and aberrant behavior of drop size distributions (Sec. 3b), the wanderings in $\overline{G/R}$ are not surprising and might have been expected a priori. Daily fluctuations in radar hardware calibration account for only a small portion of the total $\overline{G/R}$ variation (about 20% with the NSSL radar). Although most rainfall measurement requirements call for accuracies better than a factor of 2, radar estimates based on a single optimum *Z-R* relationship can be very useful for identifying, in near real time, regions likely to receive heavy rain and those regions that should be alerted to possible flash flooding.

The average within-storm difference between point radar and gage measurements for the 14 storms in Table 11.2 is 63%, including biases both storm to storm and within individual storms. The latter is disclosed by two statistics, defined in Table 11.2: "the relative dispersion about $\overline{G/R}$" and the "average difference (storm bias removed)". The average within-storm errors (exclusive of mean bias) vary from 8% to 42%, with an overall mean of 24%. The error reduction, i.e., 63% − 24% or 39%, represents how much of the radar error results from storm differences in the relationship between radar-received power and rainfall rate. A significant reduction in radar error could be achieved if a procedure were found to determine the average storm bias.

As might be expected, the agreement between gage and radar measurements improves when comparisons are made for areas rather than points (Wilson, 1968; Muchnik et al., 1968; Harrold et al., 1974). However, Wilson found significant error reduction with area size was possible only if the mean storm bias ($\overline{G/R}$) was removed. The Harrold et al. and Wilson studies also show error decreases as the measurement period lengthens.

Inspection of *G/R* fields for individual storms reveals that within-storm radar errors are not random but contain scales similar to rain-cell spacing. For example, Fig. 11.8a represents the rainfall pattern from two storms passing through a dense network of rain gages (one in the north half and the other in the south half). The *G/R* pattern in Fig. 11.8b shows that even contiguous storms exhibit distinctive error relationships (see also Fig. 11.9). Further, both Figs. 11.8 and 11.9 show a correlation between *G/R* and gage depth; i.e., heavy rainfalls tend to be underestimated by radar (large *G/R*) and light rainfalls overpredicted (small *G/R*). This result, mentioned by Woodley and Herndon (1970) and Desautels and Gunn (1970), is attributed to smoothing in the radar observations, variations in the drop size distribution, and vertical and horizontal air motions.

To illustrate how three-dimensional thunderstorm wind flow could create spatial errors in radar rainfall estimates, we examine horizontal wind fields (near ground) synthesized from measurements by two Doppler radars. Wind patterns for the rainfall event in Fig. 11.8 are shown for early-mature (Fig. 11.10a) and late-mature (Fig. 11.10b) stages. The two northernmost rain cells delineated by 40-dBZ contours in Fig. 11.10a moved eastward and merged to form the large single storm in Fig. 11.10b.

Light rainfall during the early-mature stage (west side of network in Fig. 11.8a) and on fringe areas of the storm was overestimated by radar ($G/R < 1$, Fig. 11.8b). On the other hand, heavy rainfall accompanying storm merger and progression to the fully matured stage was largely underestimated ($G/R > 1$). Converging inflow air in the eastern quadrant of the northernmost storm (Fig. 11.10a) is characteristic of growth and mature stages and implies upward motion. Precipitation here may be held aloft by updrafts. If the air beneath the storm is unsaturated, evaporation may also occur. Both updrafts and evaporation cause large coefficients in *Z-R* relationship (Table 11.1); unless the conversion relationship is adjusted accordingly, the actual precipitation may be overestimated, as is the case here. Large coefficients found for storm growth (Carbone and Nelson, 1978) and for leading storm edge (Martner, 1975), usually an updraft region, are in agreement with these observations (see Sec. 3b).

As storm evolution continues, updrafts and inflow give way to downdrafts and outflow (Fig. 10.10b). Note the broad divergent area that develops within the storm's reflectivity core. Downdrafts lower actual *Z-R* relationship coefficients, and rainfall estimated with a fixed relationship would be underestimated ($G/R > 1$). A smaller coefficient like the coefficients reported for a rainy storm core by Martner (1975) and for declining storm cells by Carbone and Nelson (1978), would raise the radar rainfall estimates.

Size sorting and droplet drift are possible contributors to radar underestimates beneath the south-central storm core (Figs. 11.8b and 11.10b). Strong northerly winds, opposite to the reflectivity gradient, may advect significantly higher droplet concentrations downstream below the radar beam than expected from in situ measurements. At the storm's southern boundary winds turn westerly, blowing both up and down the reflectivity gradient, and gage and radar measurements are in greater accord. It should be recalled that strong surface winds cause gage undercatchments and hence lower gage-radar ratios.

Table 11.3 summarizes nine experiments comparing areal radar and gage measurements. For each study a single relationship was used for converting reflectivity to rainfall rate. In some instances an additional constant correction was applied to eliminate any pervasive bias for the total experiment. Although each study assumes that the verification gage network provides the correct rainfall, discussion in Sec. 2 suggests that the error contribution by the gage areal rainfall estimate could be as large as 10%. The significance of the equivalent gage density results is somewhat uncertain since they seemingly relate to the size of the experimental area; i.e., the larger the area the sparser the network required to give equivalent errors.

The mean difference between radar and gage measurements (Table 11.3) varies from 23% to 67%. Excluding the two experiments with the fewest cases, the difference varies only between 49% and 67%. Note the disagreement for areal comparisons in Oklahoma thunderstorms (52% and 59%) is slightly less than the average difference (63%) given in Table 11.2 for point comparisons in similar thunderstorms. Thus, as stated earlier, measurement accuracy does not significantly improve with increasing area size if adjustments are not made for storm-to-storm variations in $\overline{G/R}$.

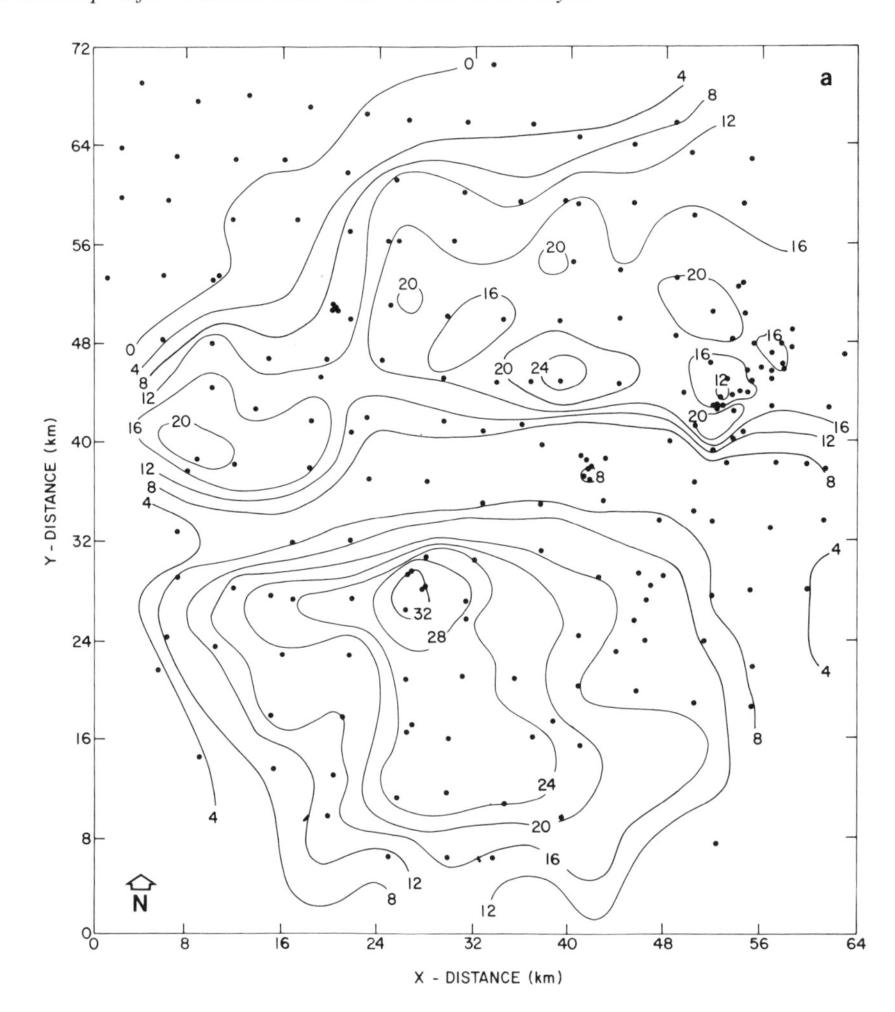

Figure 11.8. Total thunderstorm rainfall deposition in millimeters measured by (a) rain gages and (b, facing page) *G*/*R* pattern from 1500 to 1700 CST on 6 June 1974 within a dense rain-gage network. Rain-gage locations are indicated by dots.

5. Adjustment of Radar-derived Rainfall Estimates

The very large storm-to-storm and within-storm errors in radar rainfall estimates (using a single average *Z*-*R* relationship; Tables 11.2 and 11.3) dictate that estimates should be adjusted further—preferably on a within-storm basis. Adjustments usually involve either specifying the *Z*-*R* relationship by storm type or keeping the *Z*-*R* relationship fixed and using rain-gage observations to correct the radar estimates.

a. Changing the Z-R Relationship

Some improvement in radar rainfall estimates may be realized by adjusting the *Z*-*R* relationship for rain type or for synoptic situation (Jones, 1966); however, the improvement is small because of large *Z*-*R* variations within storm type classifications. Atlas (1964) and Kessler (1965) suggested that the conversion relationship be determined from radar-echo statistics. Evidence that echo coverage and reflectivity gradients from different precipitation types are correlated with *Z*-*R* relation coefficient (the exponent was not varied) was presented by Puhakka (1974). When only thunderstorms were considered, Wilson (1966) found that echo properties such as intensity, intensity variance, storm size, and orientation had little bearing on the optimum *Z*-*R* relationship. The success of the Puhakka study can probably be attributed to the usefulness of echo statistical properties for grouping storms by type (Kessler and Russo, 1963; Schaffner, 1976).

A convective rainfall study (Jatila and Puhakka, 1973a) in which drop size observations monitored at three locations within a 180-km^2 area were used to determine a single-storm *Z*-*R* relationship and ultimately to derive radar rainfalls proved unsuccessful. Errors, averaging 98%, were considerably larger than for climatological relationships based on storm type. Negative results were attributed to

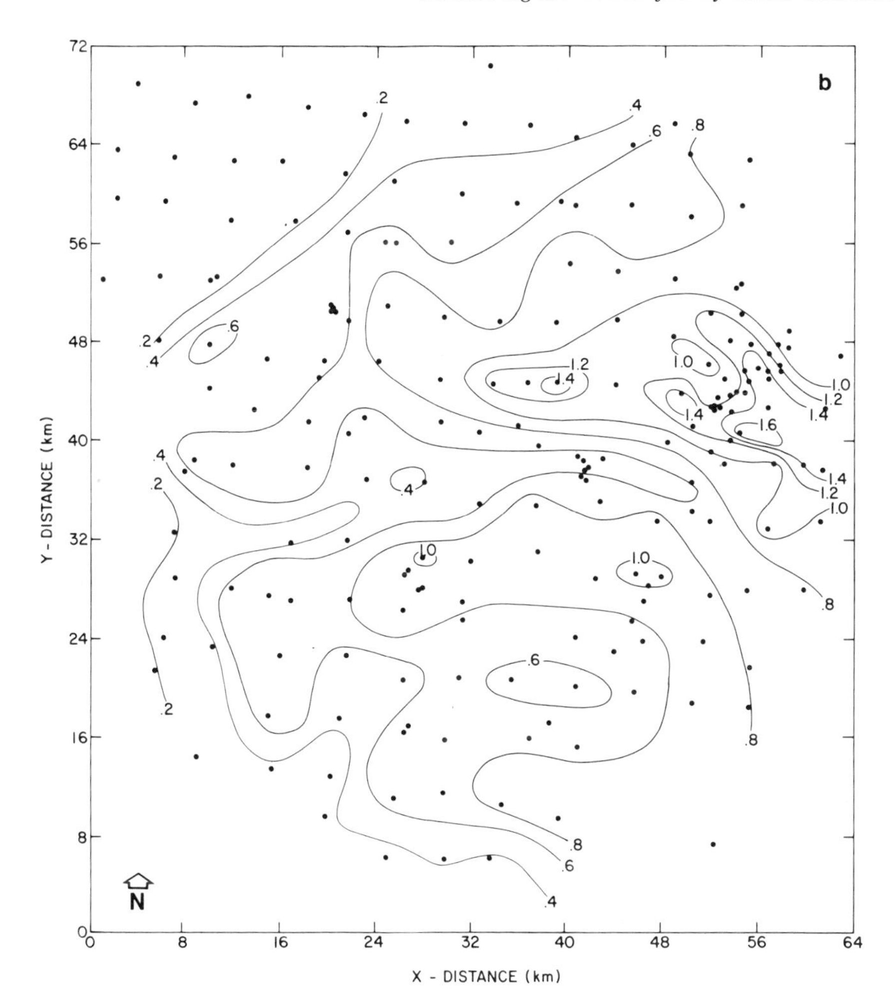

Figure 11.8b.

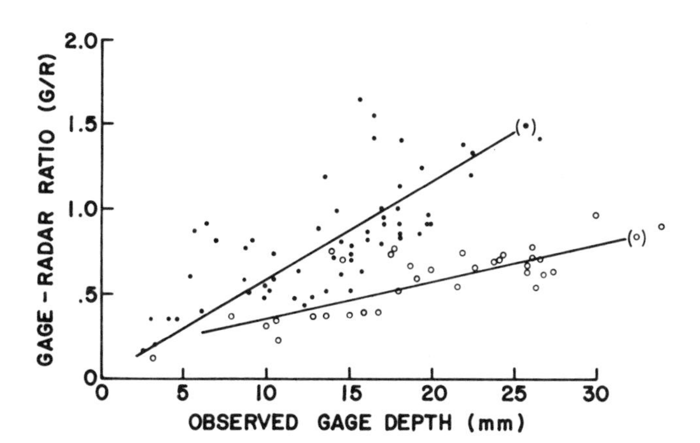

Figure 11.9. Comparison of ratios of gage- to radar-measured rainfall versus gage depth for two adjoining storms. The radar measurements are from the NSSL WSR-57 on 6 June 1974.

great temporal and spatial variability in thunderstorm drop size distributions.

b. Adjustment with Rain Gages

As early as 1954, Hitschfeld and Borden (1954) recommended that "radar used for precipitation measurements should be calibrated against rain gages." Their analysis attributes the specification of the precipitation distribution to the radar and the precipitation magnitude to the gages.

(1) Adjustment for Storm-to-Storm Radar Bias: In the simplest adjustment procedure, the mean radar bias is determined; and a correction is applied uniformly to the radar estimates. Ordinarily a number (N) of gages are used, and a multiplicative adjustment factor (F), the ratio of gage-observed (G) and radar-indicated (R) rainfalls, is computed from either

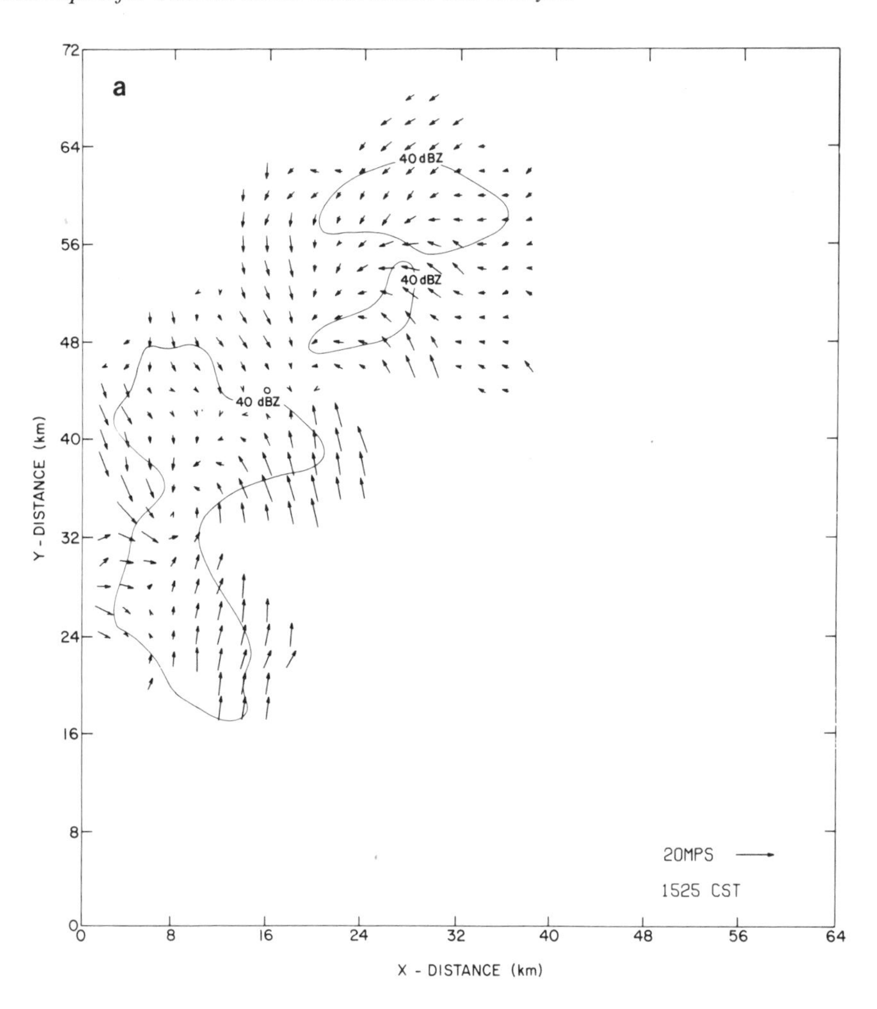

Figure 11.10. Ground relative wind fields (0.3-km elevation AGL) and radar reflectivity patterns observed at (a) 1525 CST and (b, facing page) 1625 CST on 6 June 1974. Wind fields are derived from measurements obtained with two Doppler radars.

$$F = \frac{\sum_{i=1}^{N} G_i}{\sum_{i=1}^{N} R_i} \qquad (11.8)$$

or

$$F = \frac{1}{N} \sum_{i=1}^{N} \frac{G_i}{R_i}. \qquad (11.9)$$

With Eq. 11.8 observations receive a weight proportionate to depth; with Eq. 11.9 every comparison has equal weight. (Grayman and Eagleson [1970] and Cain and Smith [1976] note log $[G/R]$ is more normally distributed than G/R; hence a more appropriate expression of Eq. 11.8 may be $F = \text{antilog}\ [\overline{\log(G/R)}]$.) Cain and Smith (1976) suggest an alternate method involving sequential analysis in which the logarithm of individual gage-radar ratios is monitored for drifts in the mean. Adjustments based on an accumulated sample mean are applied only when a statistically significant shift is detected. In each case corrections are applied for systematic bias associated with radar hardware calibration and the Z-R relationship. Indeed, the relationship selected to convert reflectivity to rainfall is of little importance and has negligible impact on the final (corrected) radar depth estimate (Brandes, 1974).

The potential of gage-calibrated radar to estimate mean watershed precipitation is elucidated by plotting natural rainfall variability as measured by the relative dispersion of gage observations against the relative dispersion of calibration ratios (G/R) computed at the same gages (Fig. 11.11). When natural rainfall has high variability, the scatter among gage observations exceeds the deviation among calibration ratios. In such storms fewer statistically independent gage-radar comparisons are needed to estimate the mean radar bias and consequently to correct the radar depth estimate

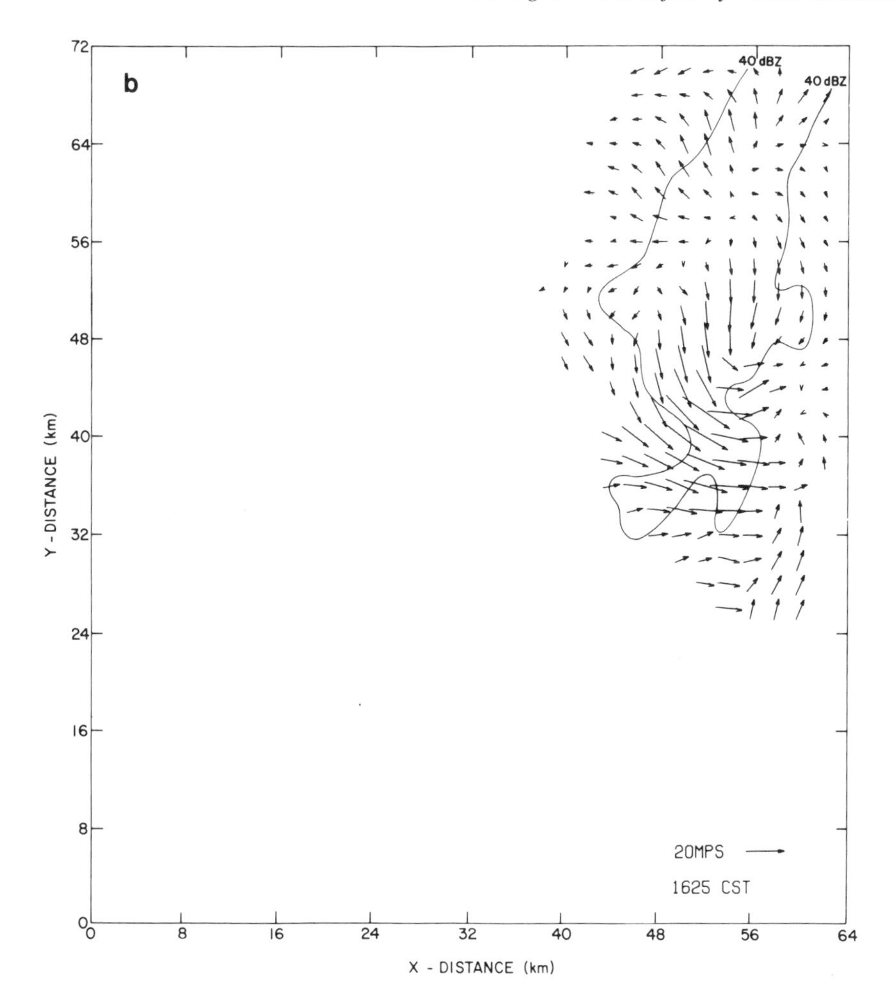

Figure 11.10b.

than to determine the average rain depth from gage observations. For more stratiform rainfalls, at least for this small sample, the advantage of gage-calibrated radar data is less apparent. On the average for individual Oklahoma storms, a single gage-radar comparison determines radar bias with an average expected error of ~30% (Table 11.2). Because this error is much less than the longer-term deviation in *G*/*R* of 63% (14 storms), even a single-gage observation can provide useful storm calibration information. If the mean radar bias is estimated from a number of gage-radar comparisons, the expected error will be smaller. Theoretically, using nine independent *G*/*R* values reduces the uncertainty in the radar bias estimate to $30/\sqrt{9}$, or 10%.

When Wilson (1970) adjusted radar-derived thunderstorm rainfalls for a 3,500-km^2 watershed measured by a single centrally located gage, the average error was reduced from 51% (unadjusted measurements) to 35%. When the single gage was used to estimate the watershed mean depth, the error was 60%. Single-gage calibration offered no improvement in the Jatila and Puhakka (1973a) study of stratiform rains, but errors in convective rainfalls dropped from 43% to 25%. These and other studies using an average storm adjustment factor are summarized in Table 11.4. Although procedural and meteorological differences preclude rigorous comparison, the common result is a significant reduction in radar rainfall estimate error when adjustments are made on a storm basis. Nevertheless, large spatial errors (e.g., the 30% error in Table 11.2) remain in the radar precipitation patterns.

(2) Spatial Adjustment: Of course, calibration by one nearby gage generally produces rainfall estimates in the neighborhood of that gage site that are superior to estimates adjusted by an ensemble average of all available calibration ratios. For example, in Fig. 11.12, when point rainfall estimates made on 7 April 1975 were adjusted by a single

Table 11.3. Comparison of Radar and Gage Measurements of Area Rainfall by Use of One Average Relationship for All Storms

Reference	Location	Rain type	Radar λ(cm)/ θ(deg)	Z-R relationship	Number of cases	Radar observation frequency (min)	Radar range (km)	Area size (km^2)	Radar observation duration	Density of verification gages (km^2/gage)	Mean difference* (%)	Gage density required to give accuracy of radar (km^2/gage)
Aoyagi (1964)	Japan	Showers	3/1	$200 R^{1.6}$	5	10	13–45	638	Storm	24	25	200
Borovikov et al. (1970)	Russia	Summer rains	3/1	$K R^{1.5}$†	59	—	60	100	Storm	1	61	400
Brandes (1975)	Oklahoma	Thunder-showers	10/2	$200 R^{1.6}$	9	5	45–100	3,400	Storm	20	52	1,600
Herndon et al. (1973)	Florida	Showers and thunder-showers	10/2	$300 R^{1.6}$†	30	5	90–125	165‡	24 h	4	67	57
Woodley et al. (1974)	Florida	Showers and thunder-showers	10/2	$300 R^{1.4}$	39	5	90–125	165‡	24 h	4	54	—
Jatila and Puhakka (1937b)	Finland	Showers	3/1.8	$360 R^{1.6}$	6	5	12–28	180	Storm	12	23	26
Jones (1966)	Illinois	Thunder-showers	10/4	$435 R^{1.48}$†	9	2	—	1,000	Storm	20	52	—
Wilson (1970)	Oklahoma	Thunder-showers	10/2	$200 R^{1.6}$†	28	10	35–100	3,400	Storm	20	59	2,000
Wilson (1975)	New York	Summer rains	5/1.7	$1000 R^{1.6}$†	121	10	95–112	170	24 h	10	49	—
Barnston and Thomas (1983)	Florida	Showers and thunder-showers	10/2.2	$300 R^{1.4}$	61	5	40–170	13,000	24 h	117	31	—

*Average percentage difference (x) computed by present authors.

$$x = \frac{100\%}{N} \sum_{i=1}^{N} \left| \frac{G_i - R_i}{G_i} \right|,$$

where G is the gage-measured rainfall, R is the radar-measured rainfall, and N is the number of cases.
†Radar rainfall amounts were adjusted to remove any average bias that existed for the total sample.
‡Five scattered areas averaging about 30 km^2 were lumped together.

calibration factor determined at a gage 20 km distant, errors in the adjusted rainfall were 17% compared with 21% when $\overline{G/R}$ was applied. The maximum useful range of any one calibration factor varies from storm to storm (7 April 1975 and 6 June 1974).

Procedures have been devised (e.g., Brandes, 1975; Collier et al., 1975; Wilson, 1975; Crawford, 1978) to combine radar observations routinely with distributions of calibrating gages. Most often local adjustments are made by interrogating nearby autoresponsive calibration sites and then assigning weights inversely by distance. (Crawford [1978] described a refined approach whereby statistical properties of both the radar and gage fields determine the combination procedure.) In essence the radar observations are molded to the gage observations by a plane-fitting technique while retaining the radar-indicated precipitation variance between gages. Depending upon the calibrating-gage density, potential source error addressed implicitly by this technique includes large-scale error patterns associated with drift in the radar hardware calibration, synoptic influences on the *Z-R* relationship, small-scale (within-storm) spatial variations in the *Z-R* relationship, and below-beam changes in the pre-

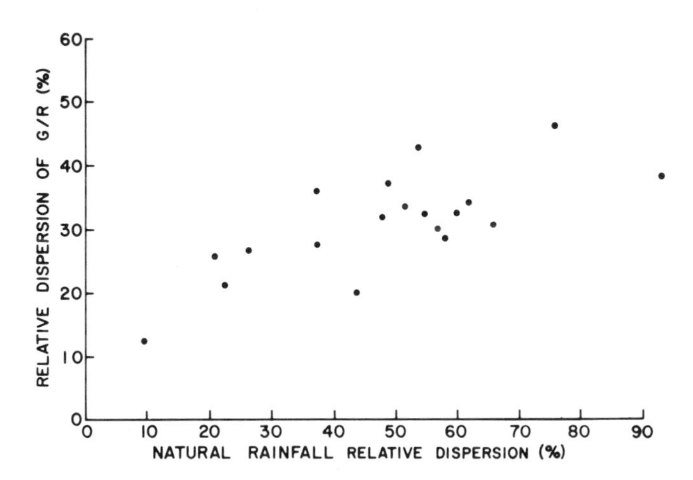

Figure 11.11. Comparison of the relative dispersion of gage-measured rainfall with the relative dispersion of the gage-to-radar ratios of measured rainfall. The gage density is 1 per 300–400 km². Data were collected with NSSL WSR-57 radar in selected spring Oklahoma thunderstorms from 1973 to 1975.

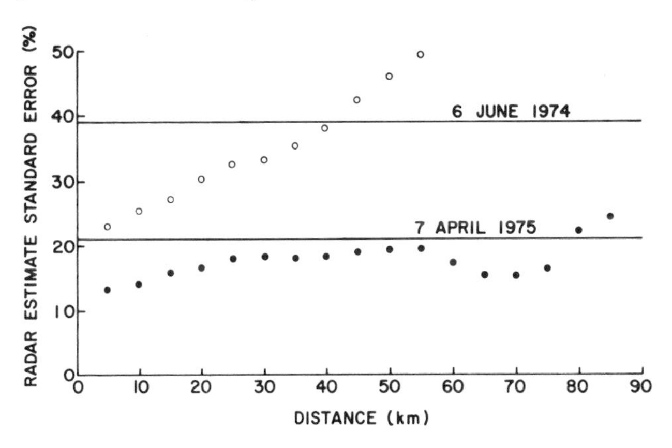

Figure 11.12. Standard error of point radar measurements (total storm accumulation) versus the distance of the measurement from the calibrating gage for two thunderstorms. The solid horizontal lines indicate the standard error for each storm after the mean radar bias ($\overline{G/R}$) is used for calibration. The curves represent the error when a single nearby gage-radar ratio is used for calibration. The radar data were collected with the NSSL WSR-57.

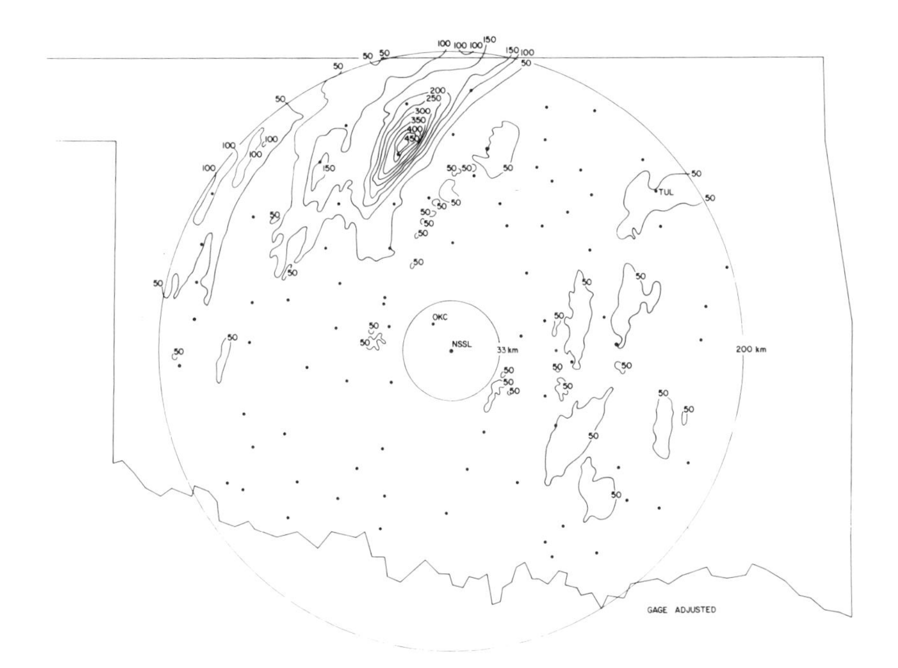

Figure 11.13. Radar-measured rainfall for Enid, Okla., storm of 10–11 October 1973, after adjustment with rain-gage observations (Brandes and Sirmans, 1976).

cipitation. Such an analysis for the Enid, Okla., storm appears in Fig. 11.13 (unadjusted radar estimates for this storm are shown in Fig. 11.1).

Although the experiments using spatial adjustment techniques (Table 11.4) differ in length of the measurement period, area size, radar range, data-collection frequency, and density of adjustment gages, the reported error range of 7% to 27% after spatial adjustment is considerably less than unadjusted estimates of 43% to 55% and somewhat less than the 18% to 35% error range for uniform adjustments.

Intuitively, we expect the radar, even if only roughly calibrated, to measure the rainfall over the entire area observed by the radar better than a single gage; if the network of calibrating gages were very dense, we would not expect the addition of radar to provide significant improvement. Data in the last three columns in Table 11.4 confirm this expectation. Results from Illinois convective storms (Hildebrand et al., 1979) indicate that the accuracy of the radar-gage estimates is matched or exceeded by the gage-only estimates when the gage density is 1 or more per 250–300 km².

Table 11.4. Radar Areal Estimates of Rainfall Utilizing Gages for Calibration[a]

Reference	Location	Rain type	Radar λ(cm)/ θ(deg)	*Z-R* relation	Number of cases	Radar observation frequency (min)	Radar range (km)	Area size (km²)	Duration	Adjustment type and calibrating gage density (km²/gage)[b]	Error before and after ( ) adjustment (%)	Percent error using calibration gages only
Wilson (1970)	Oklahoma	Thunder-showers	10/2	$KR^{1.6}$[c]	23	5–10	35–100	3500	Storm	A (3,500) A (1,200) V (1,200)	51 (35) (30) (27)	60 31 31
Brandes (1975)	Oklahoma	Thunder-showers	10/2	$200\ R^{1.6}$	9	5	45–100	3000	Storm	V (900) V (1,600) A (900)	52 (13) (14) (18)	21 24 21
Woodley et al. (1974)	Florida	Showers Thunder-showers	10/2	$300\ R^{1.4}$	39	5	85–115	570	24 hr	A (1,600)[d]	43 (30)	—
Harrold et al. (1974)	England	Showers Stratiform	10/2	$200\ R^{1.6}$	27	1	12–48	50–100	1 hr	A (500)	— (19)	48[e]
Wilson (1975)	New York	Showers Thunder-showers Stratiform	5/1.7	$200\ R^{1.6}$[f]	41	10	95–112	170	24 hr	V (275)	49 (22)	22
Collier et al. (1975)	England	Showers Stratiform	10/1	$200\ R^{1.6}$	13	1	12–48	700	3 hr	V (233)	— (7)	—
Jatila and Puhakka (1973a, b)	Finland	Showers Thunder-showers	3/1.8	$200\ R^{1.6}$	6	5	18–28	180	Storm	A (180)	43 (23)[g,h]	—
Huff and Towery (1978)	Illinois	Showers Thunder-showers	10/1	$300\ R^{1.35}$	67[i]	3	20–100	5300	0.5 hr	V (150)	55 (27)	32
Hildebrand et al. (1979)	Illinois	Showers Thunder-showers	10/1	—	56[j]	3–5	—	1,000–1,200	0.5/1 hr	V (280)	— (19)	28[k]

[a] Wilson and Brandes, 1979.
[b] A: average adjustment; V: variable spatial adjustment.
[c] Radar estimates adjusted to remove average bias for the total experiment.
[d] Density of rain-gage clusters is approximately 8 gages per cluster.
[e] Calibrating gage not within boundaries of watersheds, but is 10–20 km distant.
[f] An additional multiplicative factor of 1.7 applied to radar estimates.
[g] Varied *Z-R* coefficient to match point rainfall at central gage.
[h] Error when observed drop-size distribution used is 98%.
[i] Number of 30-min periods in four storms.
[j] Number of 30- or 60-min periods.
[k] Based on 59 situations studied.

6. Closing Remarks

It has been hypothesized that the primary sources of error in the radar measurements of rainfall reside in variations in the *Z-R* relationship caused by microphysical cloud processes and from smoothing in radar measurements. Radar rainfall error patterns suggest that discrepancies occur from storm to storm in a systematic and perhaps predictable manner. Specifically, we noted a tendency for radar to overestimate light rainfall and to underestimate heavy rainfall. The search for systematic error patterns and causes holds promise, but until this knowledge can be applied to radar measurements, we use gage data to calibrate radar data. Indeed, even simple adjustment techniques for combining sparse gage reports and radar estimates show marked improvement over either measurement system alone.

Successful implementation of radar in a precipitation measurement system requires the data user to be alert for those conditions in which the radar rainfall estimates may be erroneous or have little value, e.g., in the vicinity of ground targets, within shadow areas created by obstacles blocking the radar beam, at excessive distances where the beam becomes very wide and elevated, and at times when there is extreme variability in cloud processes and consequently in *Z-R* relationships. Such difficulties notwithstanding, radar has enormous value for flash-flood warnings. Techniques described in this chapter are not generally used by the operational weather services because they require automatic data-processing capability, which only now is becoming available. Real-time applications have been concerned with the transmission of digital radar data to users and the compositing of data from multiple radars (Greene, 1975; Collier, 1984; Cavalli, 1984).

Appendix

Adjustments to the Radar Measurement

Edward A. Brandes and Dale Sirmans

1. Attenuation by Rainfall

When radar path legends are long and precipitation echoes are highly reflective (e.g., along radially oriented squall lines), radar signals are significantly attenuated by rainfall even at 10-cm wavelengths. Thus attenuation can approach 1 dB corresponding to a diminution in rainfall estimates of about 20%. Attenuation losses, an inverse function of wavelength λ, were calculated by Burrows and Attwood (1949) for the raindrop size distribution observed by Laws and Parsons (1943). A least-squares fit applied to the logarithms of the Burrows and Attwood data yields the following two-way loss relationships:

$$L_p = 0.000687\, R^{0.97} \text{ dB km}^{-1} \quad (\lambda = 10 \text{ cm})$$

and

$$L_p = 0.0251\, R^{1.21} \text{ dB km}^{-1} \quad (\lambda = 3 \text{ cm}),$$

where R is the rainfall rate in millimeters per hour (Fig. 11.14), and $L_p = 10 \log l_p^2$, as presented in Eq. (11.1).

It is convenient to express attenuation in terms of reflectivity factor Z (mm^6 m^{-3}). Assuming $Z = 320\, R^{1.44}$, a relationship derived by Wexler (1947) for the Laws and Parsons drop size distribution, we obtain the two-way losses

$$L_p = 1.41(10^{-5})Z^{0.67} \text{ dB km}^{-1} \quad (\lambda = 10 \text{ cm})$$

and

$$L_p = 1.98(10^{-4})Z^{0.84} \text{ dB km}^{-1} \quad (\lambda = 3 \text{ cm})$$

Rainfall attenuation has the temperature dependency given in Table 11.5.

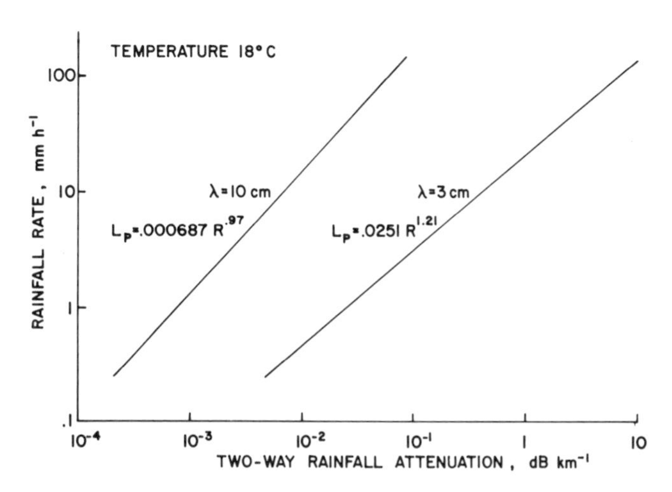

Figure 11.14. Two-way attenuation by rainfall at 18°C (Burrows and Attwood, 1949).

2. Attenuation by Clouds

Cloud and fog particles (liquid and ice) with diameters $D \lesssim 0.01$ cm generally are not detected with conventional radars but nevertheless can cause significant attenuation of radar signals. Losses, L_c ($\equiv 10 \log l_c^2$), are primarily by signal absorption and are proportional to D^3. Total loss is proportional to liquid-water content M (g m^{-3}) within the radar beam.

Ryde and Ryde (1945) determined two-way losses at $T = 18$°C to be

$$L_c = 0.009 \text{ dB km}^{-1}\, M^{-1} \quad (\lambda = 10 \text{ cm})$$

and

$$L_c = 0.100 \text{ dB km}^{-1}\, M^{-1} \quad (\lambda = 3 \text{ cm})$$

Attenuation by cloud is also temperature-dependent (Table 11.6).

Table 11.5. Multiplicative Temperature Correction Factor $K_p(T)$* for Rainfall Attenuation†

		Correction factor				
λ (cm)	T (°C)	0	10	18	30	40
10 (all rainfall rates)		2.0	1.4	1	0.7	0.6
3 (R = 2.5 mm h^{-1})		0.8	1.0	1	0.8	0.6
3 (R = 50 mm h^{-1})		0.6	0.9	1	1.0	0.8

*$L_p(T) = L_p(18°C)K_p(T)$.
†After Burrows and Attwood (1949).

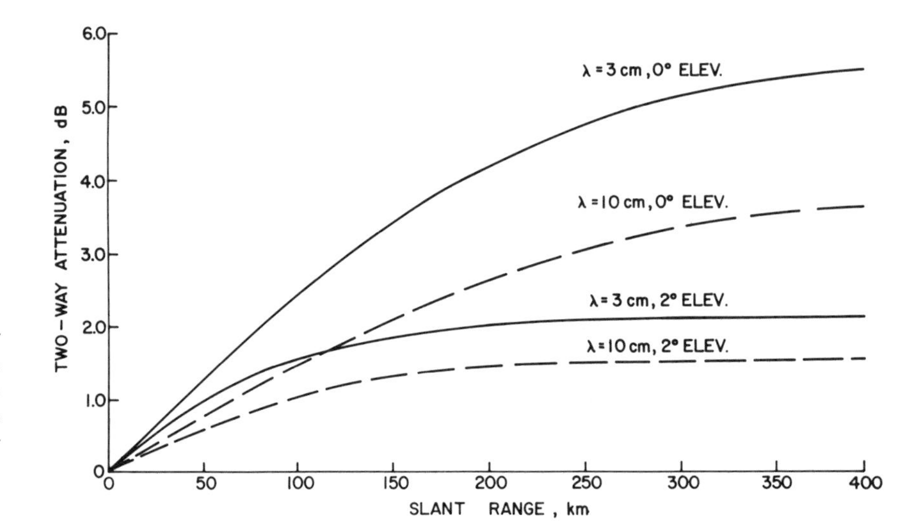

Figure 11.15. Two-way attenuation by water vapor and oxygen (From M. I. Skolnik, ed., *Radar Handbook,* Chap. 2, by L. V. Blake. Copyright © 1970 by McGraw-Hill, Inc. Used with the permission of McGraw-Hill Book Company).

3. Absorption by Atmospheric Gases

Oxygen and water vapor are the principal gaseous absorbers of microwave radiation. Losses are a function of the radar wavelength and the region of the atmosphere penetrated. Total two-way absorption by both gases L_g ($\equiv$ 10 log l_g^2) for the ICAO standard atmosphere was computed by Blake (1970) (Fig. 11.15). A formula that approximates Blake's two-way theoretical losses at λ = 10 cm for elevation angles (ϕ) less than 10° and slant ranges (r) less than 200 km is

$$L_g = L(\phi)\left[1 - \exp\frac{-r}{K(\phi)}\right]\text{dB},$$

where

$$L(\phi) = 0.4 + 3.45 \exp(-\phi/1.8)$$

and

$$K(\phi) = 27.8 + 154 \exp(-\phi/2.2).$$

At 0° elevation the relationship reduces to

$$L_g = 3.85\,[1 - \exp(-r/182)]$$

and predicts a loss of 2.6 dB (a factor of 1.45 in rainfall rate) at a range of 200 km.

4. Range-averaging Bias for Logarithmic Receivers

The radar digital integrator produces return power estimates (measurements) by averaging time and range samples. Range averaging of a logarithmic receiver output in regions of reflectivity gradient results in a bias, L_a ($\equiv$ 10 log l_a), owing to the difference between the logarithm of the spatial average of mean power (log $\overline{P_r}$) and the spatial average of the logarithm of mean power ($\overline{\log P_r}$).

Table 11.6. Multiplicative Temperature Correction Factor $K_c(T)$* for Cloud Droplet Attenuation†

		Correction factor				
λ (cm)	T (°C)	0	10	18	30	40
≥3		2	1.3	1	0.7	0.6

*$L_c(T) = L_c(18°C)K_c(T)$.
†After Ryde and Ryde (1945).

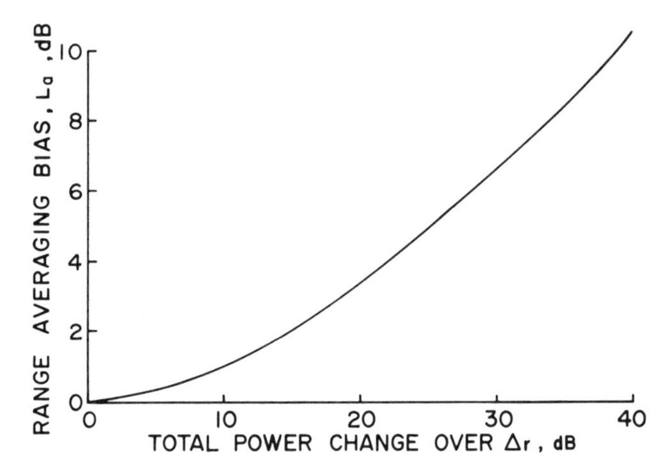

Figure 11.16. Range-averaging bias for an exponential power distribution.

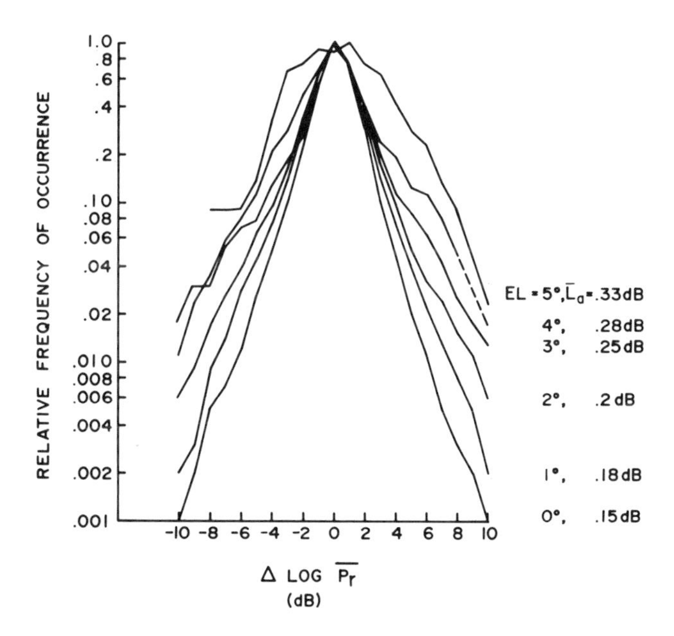

Figure 11.17. Frequency distribution of reflectivity gradients between adjacent range measurements and average bias ($\overline{L}_a$) for tilt sequence of 24 May 1973.

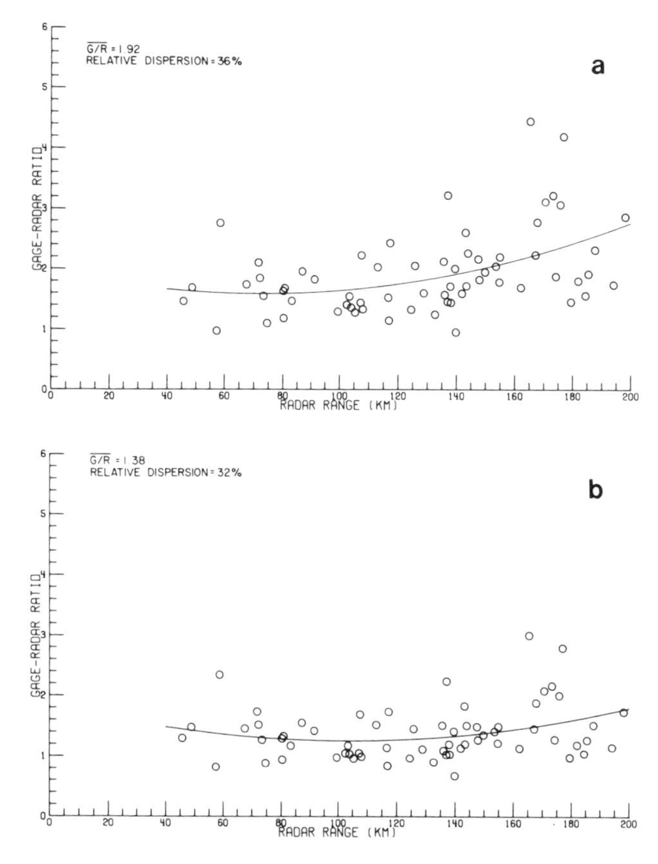

Figure 11.18. Range distribution of ratios of gage-observed and radar-estimated rainfall G/R from 10–11 October 1973 computed with radar measurements (a) unadjusted and (b) adjusted for attenuation by rainfall, attenuation by cloud, atmospheric adsorption, and signal range-averaging loss (after Brandes and Sirmans, 1976).

If the reflectivity distribution across the measurement interval (Δr) is exponential (i.e., $b = 10\ \Delta \log \overline{P}_r/\Delta r$ is constant, where b is the power gradient in dB), the measurement bias is

$$L_g = 10\left[1 - b\Delta r/20 + \log\left(\frac{10^{b\Delta r/10} - 1}{b\Delta r\ ln\ 10}\right)\right]\ \text{dB}$$

(Rogers, 1971; Sirmans and Doviak, 1973). The bias is always positive (Fig. 11.16); i.e., estimates of the mean power returned from an averaging volume are less than the true mean.

The distribution of echo reflectivity gradients between adjacent radial measurements 1 km long for a large thunderstorm is given in Fig. 11.17. Increased distribution width with elevation angle reflects the reflectivity gradient increase with height. For radar point measurements gradient bias errors can be very large compared with other errors in the reflectivity measurement.

5. Concluding Remarks

Because the spatial distribution of temperature, water vapor, cloud water, and rainfall is not measured directly and must be estimated, corrections are only approximate. The comparative parameter of importance for hydrological measurements is the ratio of gage-observed and radar-estimated rainfall depth (G/R) and the mean ratio ($\overline{G/R}$); range dependency of G/R and the dispersion among ratios are reduced if return power is adjusted for the above losses (compare Fig. 11.18a and b).

12. The Use of Satellite Observations

Robert A. Maddox and Thomas H. VonderHaar

1. Introduction

Since the advent of meteorological satellites in 1960 satellite data have been increasingly employed to improve large-scale weather analyses. Improvements in frontal analysis over data-sparse ocean areas have been dramatic. Visual-imagery and motion-picture loops provide vivid depictions of the life cycles of entire weather systems and have helped increase our understanding of large-scale meteorological processes. Many of the applications of satellite data to large-scale analysis and forecasting problems have been described by Anderson et al. (1974). As the sophistication of satellite systems has advanced (e.g., increased resolution, increased data coverage and data availability, and increased data-processing capability), the data have been used on new and small scales for studies of thunderstorms and thunderstorm-related phenomena (e.g., mesoscale convective systems and mesoscale features that lead to convective development: supercell and tornadic thunderstorms). Much theoretical background and uses of satellite data on smaller-scale research and operational forecasting problems are presented in the comprehensive work of Smith (1985).

The methods developed to interpret and use satellite data differ considerably from those used for more conventional meteorological observations. Satellite observations can provide meteorological information that bridges a continuum of scales ranging from hemispheric wave patterns to individual cumulus clouds. These data are remotely sensed from a space platform and differ in kind and especially in quantity from conventional in situ data. Therefore, we must continue to learn how to use satellite data and how to incorporate them into existing data systems, just as we are still learning to use and incorporate radar data.

The history of satellite observing systems used in the United States and an overview of the theory and practice of remote sensing from space are discussed in detail by Smith (1985). Additionally, Smith illustrates the applications of satellite data for diagnosing and studying different types of weather phenomena over a broad spectrum of scales and gives comprehensive explanations of satellite technology. This chapter, rather than attempting to review satellite systems and technology, discusses, first, applications of satellite data to research on convective storms and then the use of these data in the operational forecasting of convective weather phenomena. Explanations of the operational satellite systems of the 1980s are given by Epstein et al. (1984).

2. Satellite-based Thunderstorm Research

Early research in the satellite era demonstrated the potential value of satellite data in the detection and monitoring of intense convective storms and for applied and theoretical research on thunderstorms. Examples of the first applications of satellite data in thunderstorm studies can be found in Whitney (1961), Whitney and Fritz (1961), Whitney (1963), Erickson (1964), Boucher (1967), VonderHaar (1969), Sikdar et al. (1970), Ninomiya (1971a, b), and Hubert and Whitney (1971).

The recent development of interactive human-and-machine computer systems specifically designed for processing and analyzing satellite data has made possible quantitative thunderstorm research. Digital data transmitted from the satellite can be stored on disk and/or magnetic tape in these systems, with navigation (Earth coordinate location of features) accuracy to within a singe data element (pixel). This allows a wide variety of processing applications, ranging from sequential looping and cloud tracking to produce a wind field, to color enhancements of specific features, and to superposed displays of differing data types and analyses. For example, a multispectral display of a severe thunderstorm on NASA's Atmospheric and Oceanographic Information Processing System (AOIPS) is shown in Fig. 12.1.

Various interactive systems have been developed and are currently in use. Descriptions of some of these can be found as follows: the National Oceanic and Atmospheric Administration (NOAA)/University of Wisconsin Man-Computer Interactive Data Access System (McIDAS) in Smith (1975) and Suomi et al. (1983); the NASA Goddard Space Center AOIPS system mentioned above in Billingsley (1976) and Koch et al. (1983); the Colorado State University All Digital Video Imaging System for Atmospheric Research (ADVISAR) in Smith et al. (1979); the NOAA Program for Regional Observing and Forecasting Systems (PROFS) in

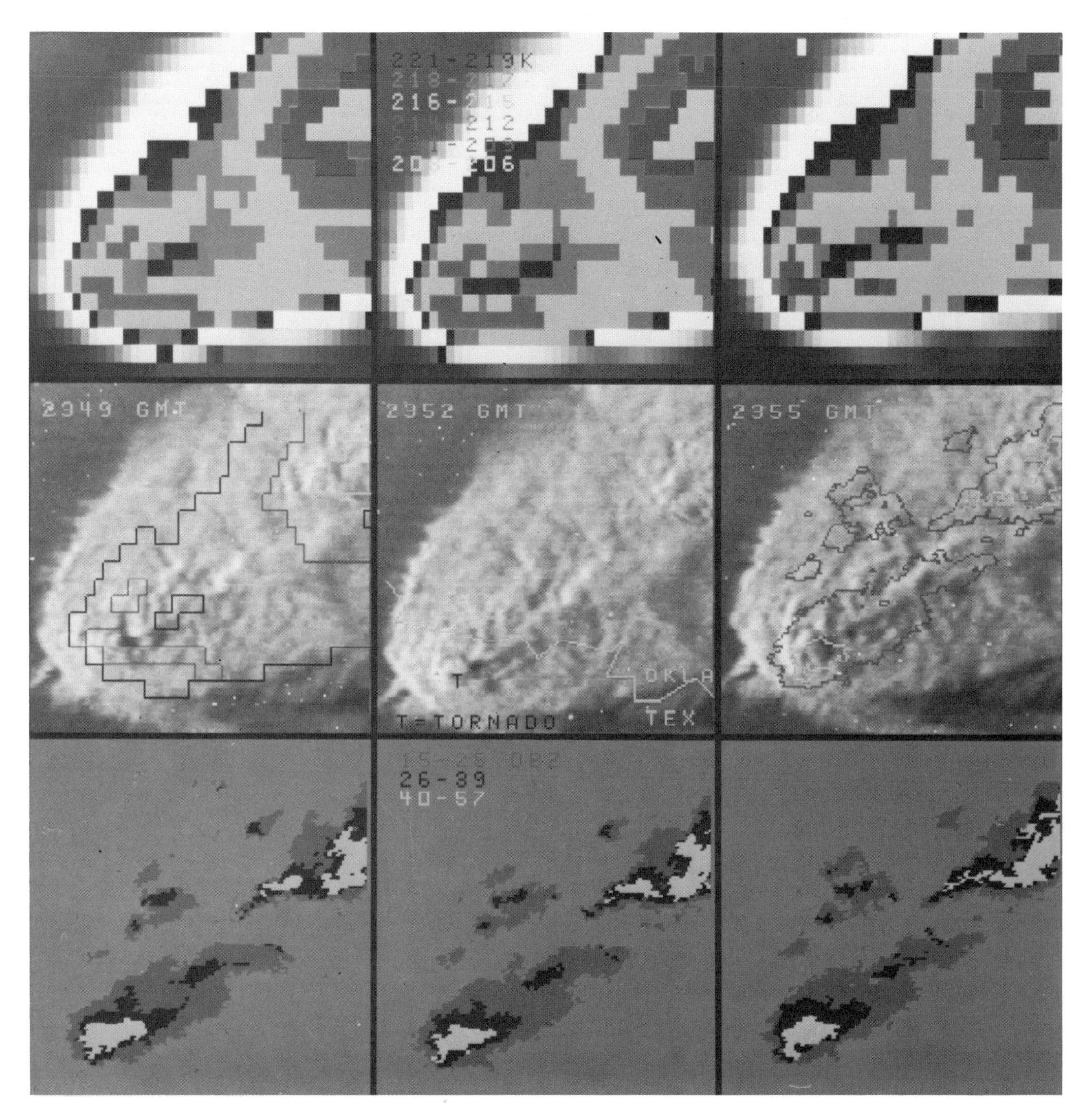

Figure 12.1. Multi-spectral display of the evolution of the Wichita Falls, Tex., tornadic storm, 2349–55 GMT, 10 April 1979. Top row: GOES infrared blackbody temperature field. Note the colder V pattern (green) and the embedded downwind warm area (red). The temperature color key is shown in the middle panel. Center row: GOES visible imagery with superimposed blackbody temperature contours (left), grid and tornado location (middle), and remapped radar reflectivity (right). Bottom row: Low-level reflectivity from the southwest quadrant of the NSSL Doppler radar, remapped to GOES coordinates. The reflectivity color key is shown in the middle panel. Note the hook echo at 2355 GMT (from Negri, 1982).

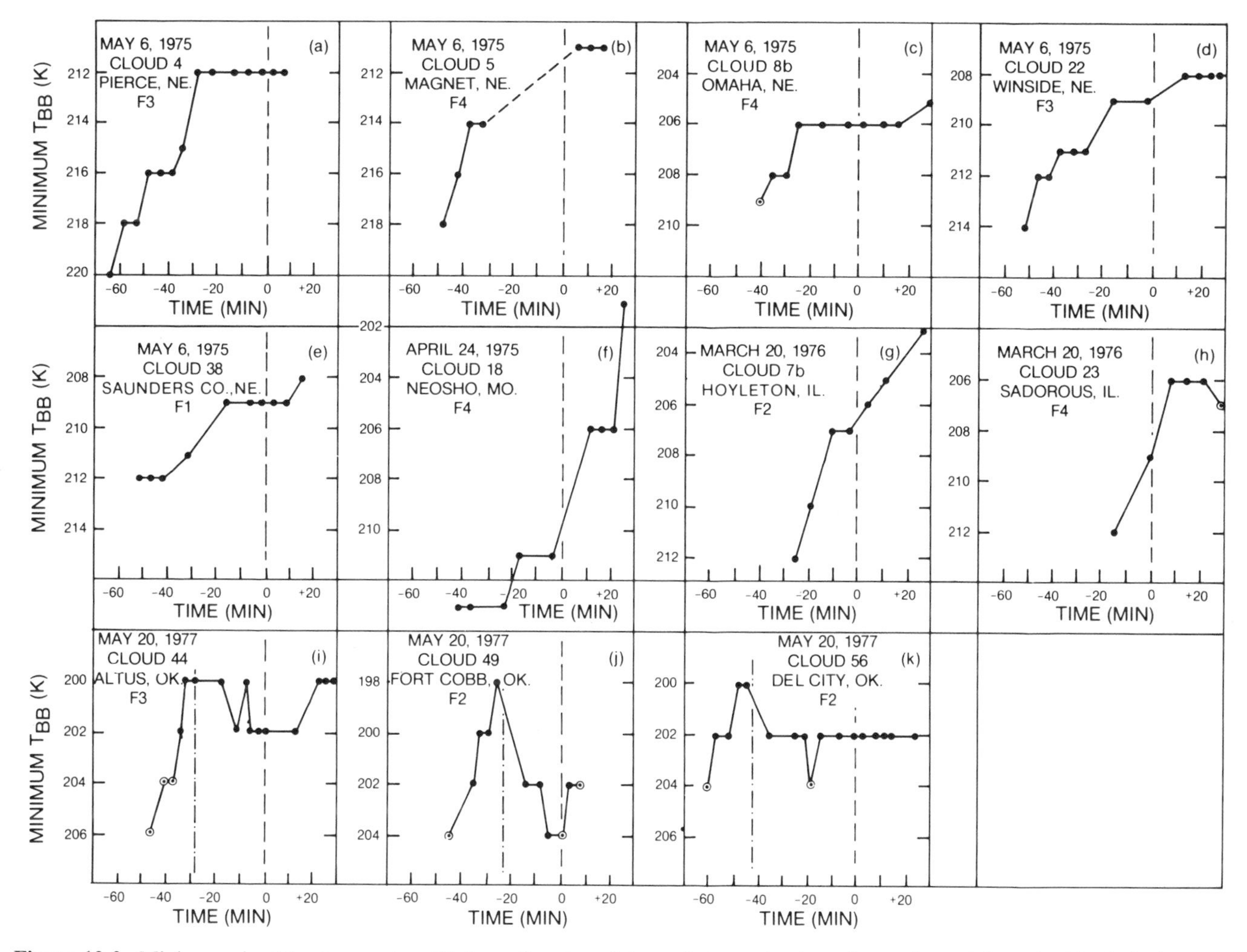

Figure 12.2. Minimum cloud-top temperature (T_{BB}) as a function of time relative to the time of tornado touchdown for 11 relatively strong tornadoes. The circled dots indicate cases in which the lowest temperature is not associated with a closed T_{BB} isotherm. The time of the tornado touchdown is marked by the vertical dashed line at time equal to zero. In (i), (j), and (k) the initial times of mesocyclones observed by Doppler radar at NSSL are shown by the vertical dash-dot line (from Adler and Fenn, 1979a).

Reynolds (1983); and the NOAA Centralized Storm Information System (CSIS) in Ostby (1984).

a. Cloud-Top Heights and Temperature Fields

Sikdar et al. (1970) showed that convective mass fluxes could be derived from satellite data; similarly, Mack and Wylie (1982) used the satellite-observed convective mass flux to estimate the condensation rate in severe thunderstorms. Adler and Fenn (1979a, 1981) and Wexler and Blackmer (1982) used infrared data and an interactive system to compute areas of cold cloud top as a function of time for many cloud systems, including several significant severe thunderstorms. Adler and Fenn also used the rate of increase of cloud-top area to develop a scheme for estimating divergence within the anvil. They found that the first report of tornadic activity often took place during or just after the rapid expansion of cold cloud area, indicating rapid ascent and growth of thunderstorm tops on the scale observed by the satellite. Figure 12.2 shows their results, which relate the minimum cloud-top temperature to the time of tornado touchdown for 11 relatively strong tornadoes.

Negri (1982) showed similar results for tornadic storms that occurred on 10 April 1979. He was able to relate the history of radar reflectivity to the cloud-top temperature evolution and the production of severe weather by the storms (see Fig. 12.3). In addition, the use of stereoscopic visual images allowed a detailed comapping of stereo heights (see Fig. 12.4) and infrared (IR) cloud-top temperature. This procedure illustrated that the highest cloud tops did not colocate with the coldest IR temperatures. Negri

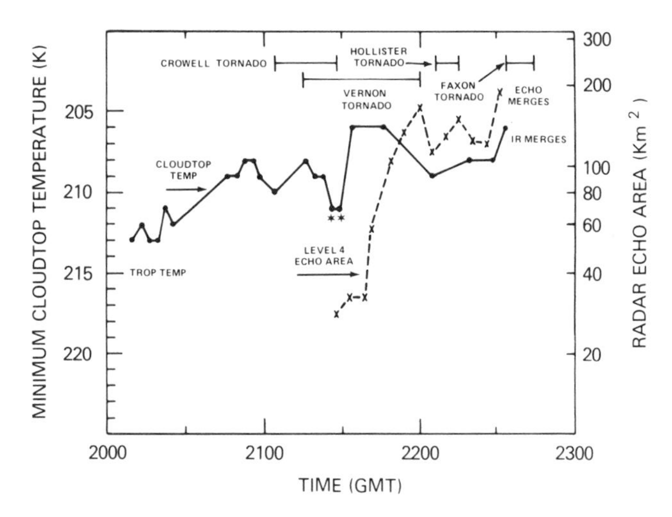

Figure 12.3. Minimum cloud-top temperature and radar-echo area v. time (from Negri, 1982). (a) The Vernon Tex., tornadic storm. The level 4 echo area includes reflectivity greater than 45 dBZ. Tropopause temperature is 215 K.

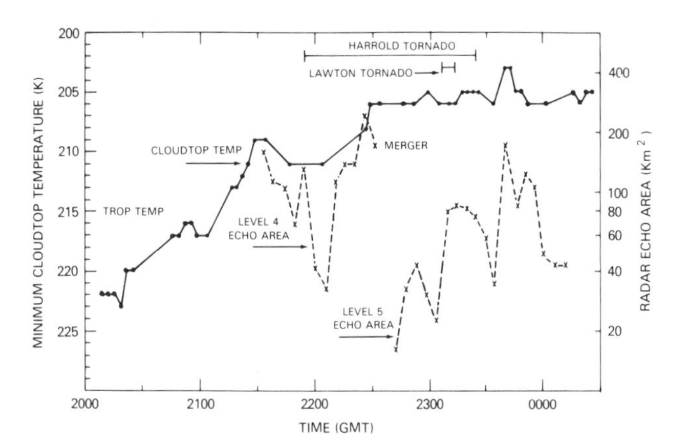

Figure 12.3 (*continued*). (b) The Harrold, Tex., tornadic storm. The level 5 echo area includes reflectivity greater than 55 dBZ.

concluded that the warm region over the high-cloud tops might be due in whole or part to several mechanisms:

1. Warming produced by subsidence of air in the lee of the ascending tower, owing solely to internal storm dynamics.
2. Obstruction of the swifter environmental flow around and above the main updraft, stripping off cirrus particles at tropopause temperature (V-shaped feature), forming a downwind warmer area by entrainment of (warmer) stratospheric air and by subsidence warming.
3. Presence of cirrus particles in the stratosphere above the anvil produced downwind in relation to the overshooting top. While the cloud top may cool adiabatically, the cirrus would assume the higher temperature of the stratospheric environment.

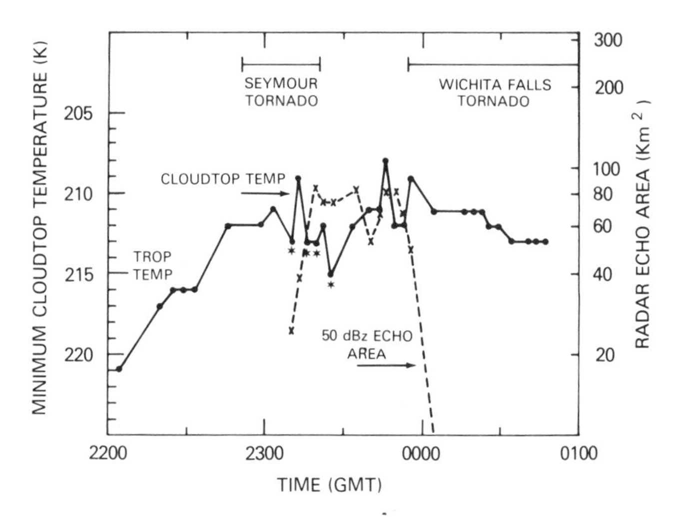

Figure 12.3 (*continued*). (c) The Seymour–Wichita Falls, Tex., tornadic storm.

4. Radiative, cloud physics, or other effects.

In this particular case he hypothesized that obstacle flow was the probable cause of the warm area.

Reynolds (1980) studied the IR cloud-top temperature fields associated with a number of damaging hailstorms. In all but one event he found that the coldest temperatures were located near the hailfall. The cloud-top temperatures were also noted to be 1° to 8°C colder than the environmental tropopause temperature. He thus identified a potential technique for identifying damaging hailstorms through variable enhancements of digital Geostationary Operational Environmental Satellite (GOES) infrared imagery that were related to the concurrent vertical temperature structure of the troposphere. However, since all his episodes occurred over the high plains, further study is required to determine whether this technique is universally applicable.

Such studies as these may lead not only to improved detection and warning of severe and/or tornadic storms but also to better understanding of storm dynamics and thermodynamics and of tornado development. However, Negri (1982) cautioned that analysis of individual severe thunderstorms in GOES imagery requires full resolution data at 3-min intervals.

b. Cloud Motions and Wind Fields

Animated film loops dramatically show cloud-motion fields, and interactive computer systems allow tracking of individual clouds and derivation of associated environmental wind fields. Since many of the important cloud features associated with thunderstorm development are convective and have relatively short lifetimes, it is important that the time interval between successive photographs in the data loop be quite short. Special projects and research efforts have been set up, and a considerable base of GOES data at intervals

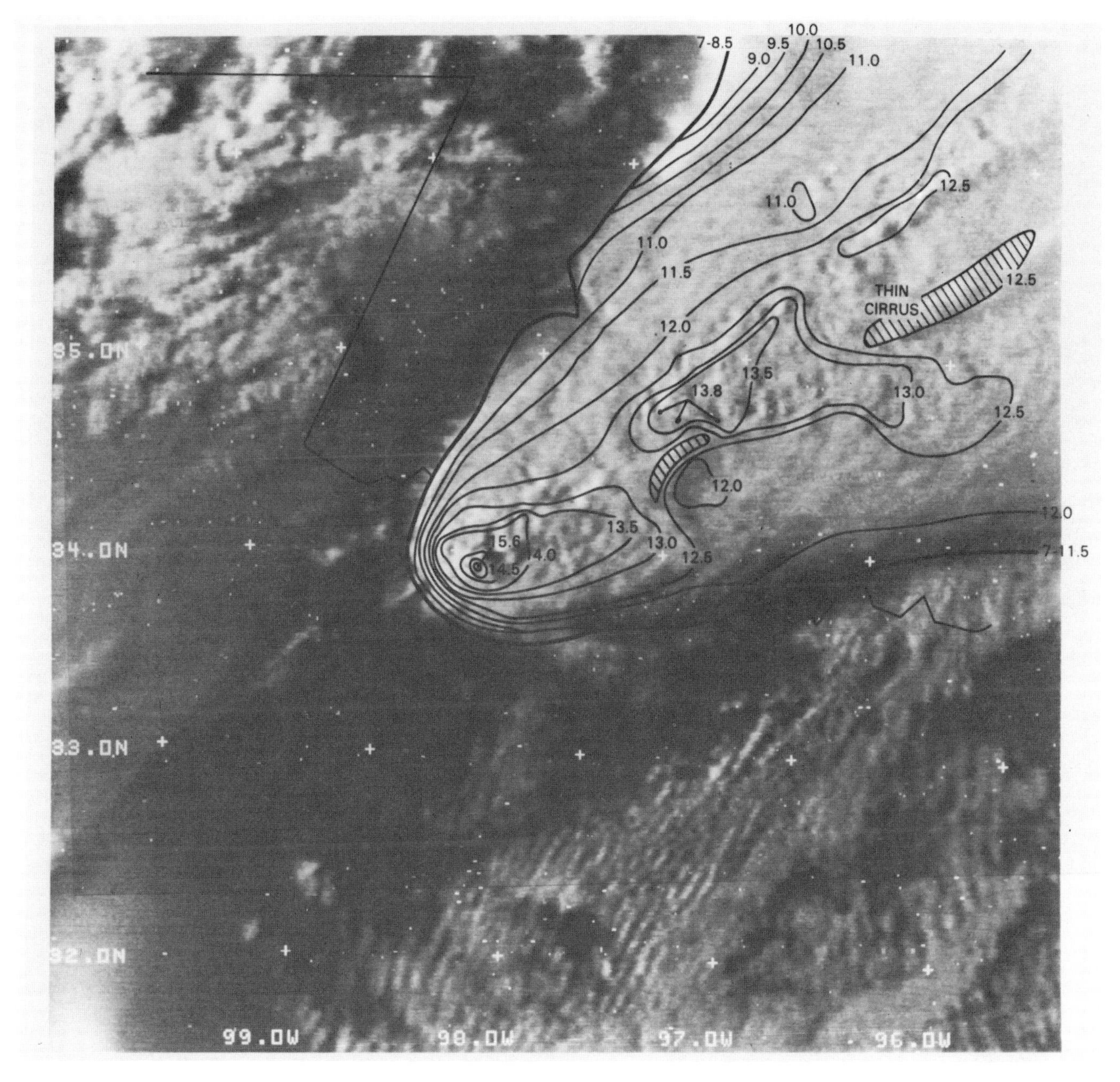

Figure 12.4. Stereoscopically derived cloud-top heights (km) at 2345 GMT, 10 April 1979, superimposed on the GOES-East visible image (from Negri, 1982).

of 3 to 15 min has been gathered. Details of several such special projects can be found in Hill and Turner (1977). A number of middle-latitude, mesoscale field programs gathered field data during the mid-1980s. Two such programs are the Genesis of Atlantic Lows Experiment (GALE) and the Oklahoma-Kansas Preliminary Regional Experiment for Stormscale Operational and Research Meteorology (OK-PRE-STORM).

Fujita et al. (1975) and Hasler et al. (1976) examined cumulus-cloud motion in relation to the environmental wind field and demonstrated that cumuli could be tracked in successive satellite images to develop a reasonable picture of low-level winds. Although low-level clouds are not perfect tracers, discrepancies (see Maddox and VonderHaar, 1979) seem similar to those for rawinsonde data. A composite flow field obtained by Tecson et al. (1977) by tracking low,

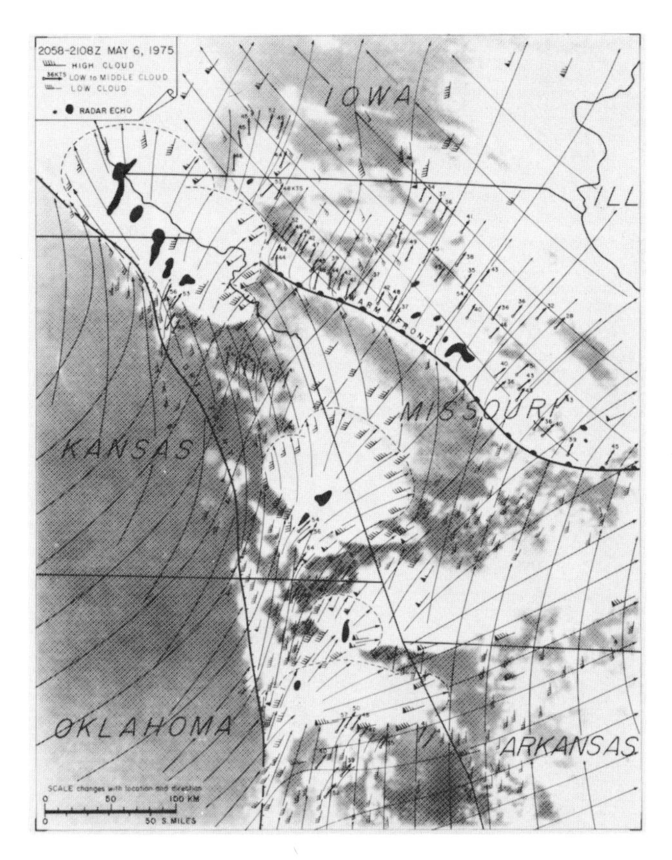

Figure 12.5. Velocities of all clouds (2058–2108 GMT) superimposed on the 2058 GMT satellite photograph for 6 May 1975 (from Tecson et al., 1977).

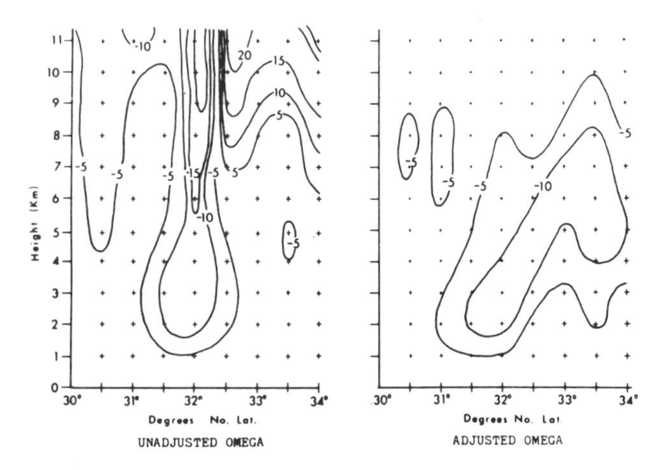

Figure 12.6. Satellite-derived vertical cross section of vertical motion, omega (X 10^{-3} mb s^{-1}), along 96° W. The original and adjusted data are shown to illustrate the effects of the adjustment procedure used (from Wilson and Houghton, 1979).

middle, and high clouds on 6 May 1975 is shown in Fig. 12.5. Several examples of research using satellite-derived wind fields from intense convective storm situations are discussed below.

Vertical Motion: Investigators at the University of Wisconsin have tracked clouds at all available levels within the troposphere and then used the three-dimensional mesoscale wind fields to compute kinematic quantities. Details of this work are explained in Wilson and Houghton (1979). The investigators believe that it is possible to obtain a sufficient number of horizontal motion vectors from the clouds to describe the three-dimensional structure of the atmospheric wind field. A vertical cross section of their omega fields (vertical air velocity in pressure coordinates), based on satellite data, is displayed in Fig. 12.6. The mesoscale, horizontal divergence and vertical motion fields appeared to be related in a consistent manner to observed clouds and to areas of subsequent severe-storm occurrence. They believe that satellite cloud tracking can enhance descriptions of atmospheric velocity fields obtained by more conventional means and perhaps provide meaningful input for initialization of numerical models. Krietzberg (1976) assessed ways that satellite data might be used in mesoscale, numerical weather prediction.

Divergence Fields: Researchers at Goddard Space Flight Center have used the NASA/AOIPS to develop many sets of low-level wind data during episodes of strong convection. Examples of divergence fields computed with the use of wind data were presented by Peslen (1980). The cloud-motion vectors and computed divergence for 1828–38 GMT on 6 May 1975 (the day of the well-known Omaha, Nebr., tornado) are superimposed on the visible GOES image in Fig. 12.7. The most intense convection built in and through the strongly convergent zone in the south-central portion of the photograph, i.e., between the thunderstorm at the southern boundary and the −44 contour. Although the availability and distribution of small cumulus cloud tracers near large storms affect the representativeness of such computations, Peslen found that satellite-derived cloud winds might be used to detect areas of intense mesoscale convergence before storm development. She also cautioned that the wind data should be derived from frequent (≤ 5 min) satellite images.

Moisture Divergence: Negri and VonderHaar (1980) combined low-level, satellite-derived wind data with an analysis of conventionally determined surface mixing ratios. The computed regions of strong moisture convergence (see Fig. 12.8) were characterized by subsequent (1–2 h later) development of severe thunderstorms. Their work indicates that operational applications making use of real-time satellite wind and lower-tropospheric moisture data may eventually play an important role in severe-storm prediction techniques, especially if extensive cumulus-cloud fields are present within the precursor storm environment.

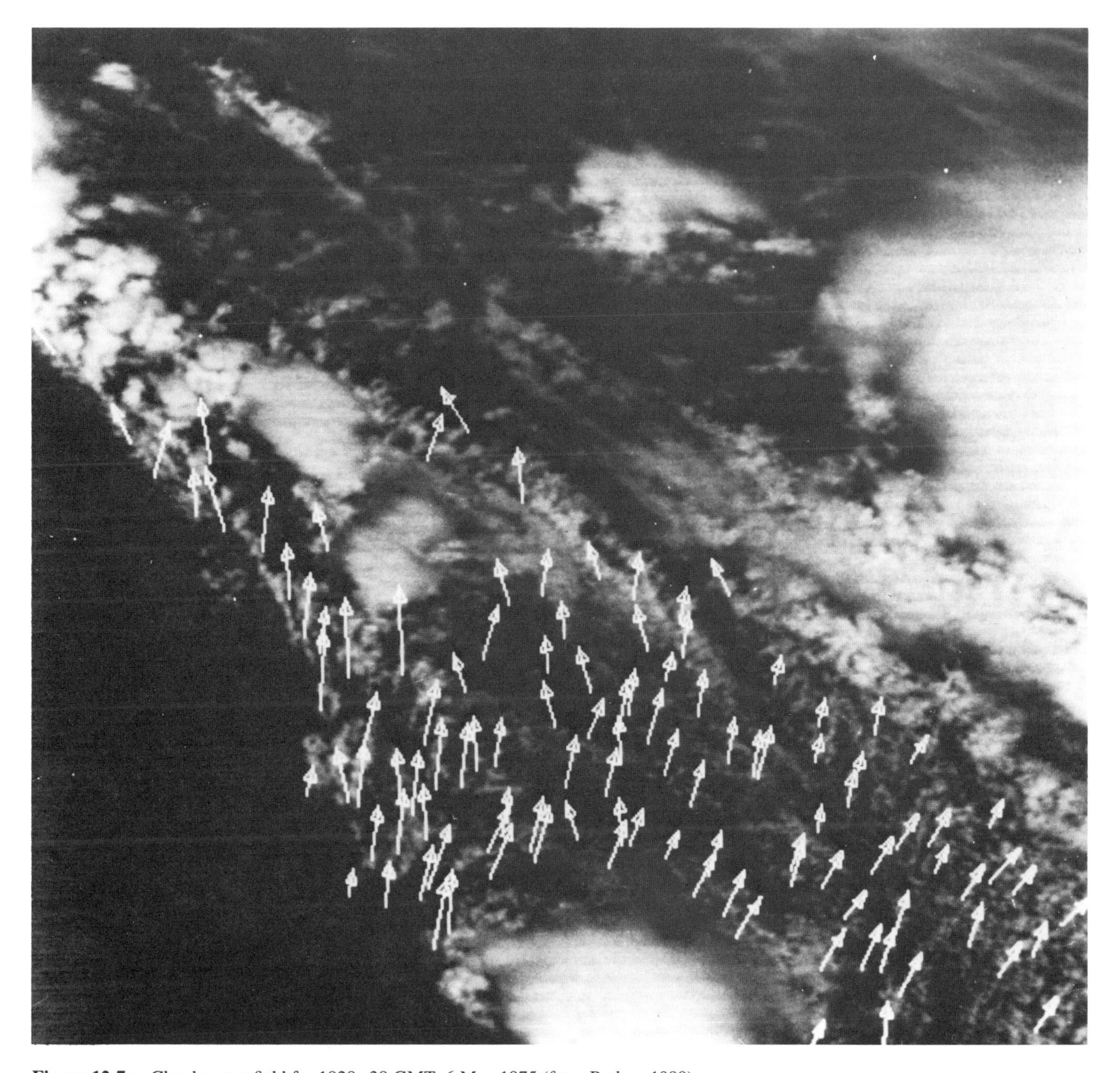

Figure 12.7a. Cloud vector field for 1828–38 GMT, 6 May 1975 (from Peslen, 1980).

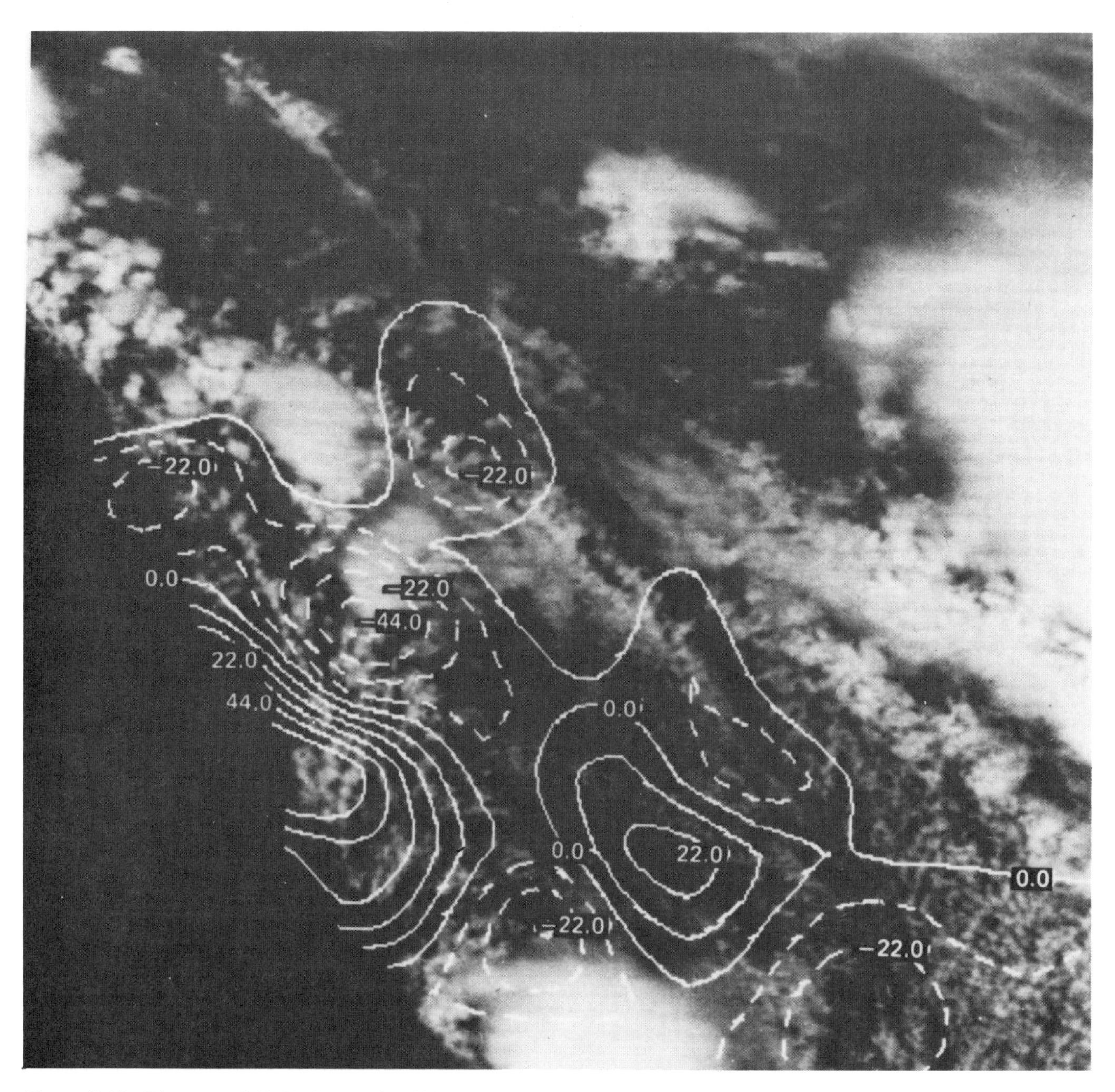

Figure 12.7b. Divergence field (X 10^{-6} s^{-1}) for 1828–38 GMT, 6 May 1975 (from Peslen, 1980).

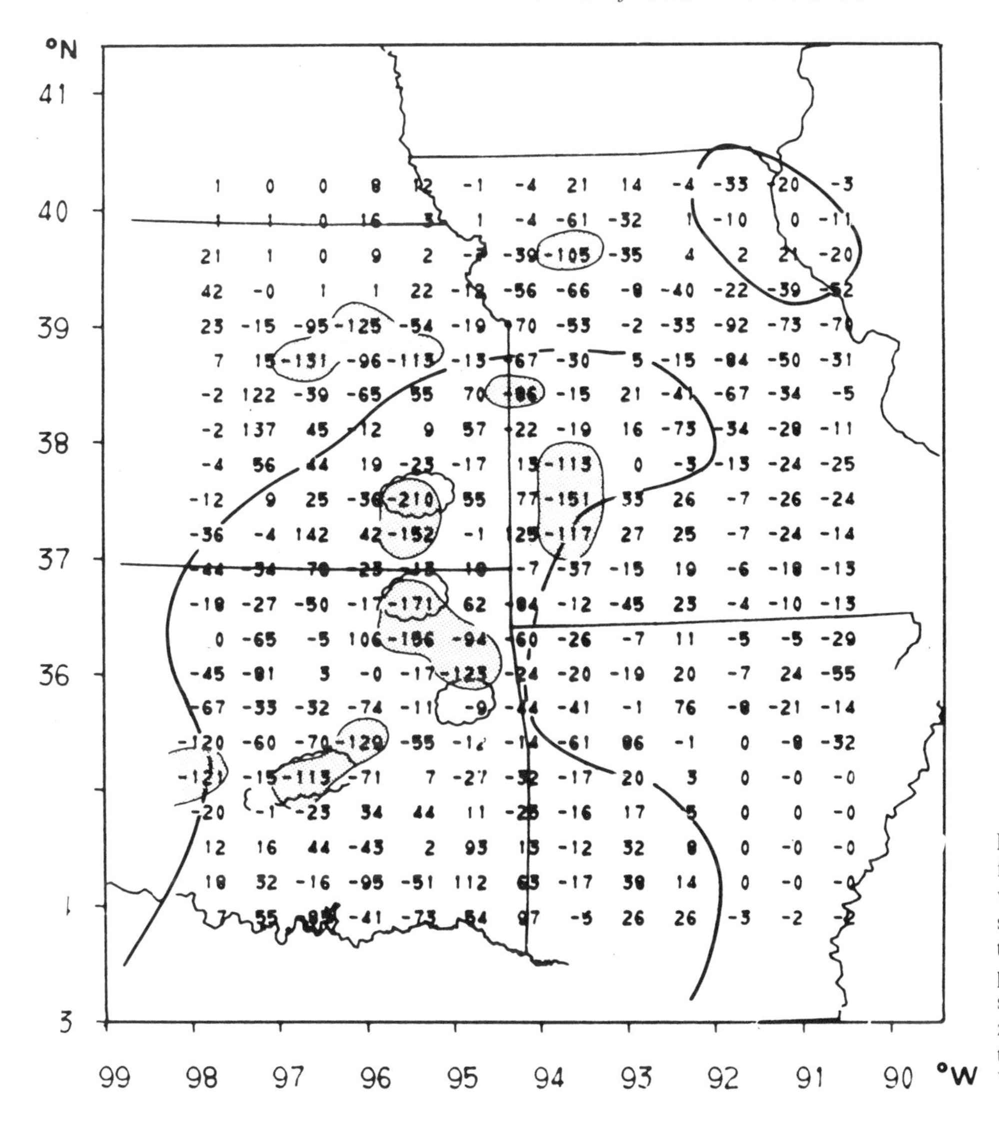

Figure 12.8. Moisture divergence field. Units are 10^{-5}g kg^{-1} s^{-1}, and values of convergence $>10^{-3}$ are stippled. In the x direction only alternate grid-point values have been plotted. Cloud outlines indicate severe-storm genesis areas, and the zone of confidence is enclosed by the solid line (from Negri and VonderHaar, 1980).

c. Thunderstorm Growth Rates

On much smaller scales Adler and Fenn (1979b) used infrared GOES satellite data (at the rapid interval of 5 min) to estimate the ascent rate of thunderstorm tops. Their computed upward vertical velocities were representative of areas approximately 10 km on a side. On this scale they found a significant difference in the mean calculated vertical velocity between storms associated with severe weather reports and storms for which severe weather was not reported (see Fig. 12.9). Their calculations indicated that GOES rapid-scan data can provide useful quantitative information on thunderstorm vertical velocities and intensities. Mack et al. (1983) followed up on Adler and Fenn's work by using stereoscopic imagery for independent computation of storm-top growth rates. Some of their results for individual thunderstorm cells in the severe squall line of 2 May 1979 are illustrated in Fig. 12.10a and b. Their comparisons of stereoscopic storm heights and ascent rates with simultaneous IR cloud-top temperatures for growing thunderstorms show the following:

1. GOES IR cloud-top temperatures grossly underestimate the actual cloud-top height observed stereoscopically, especially of immature storms.
2. Growing storms with tops below about 10 km are difficult to define in the GOES IR data.
3. The IR height trend usually lags that derived from stereo.
4. Despite the lag, IR-determined ascent rates during the rapidly growing phases are roughly the same as the stereoscopically derived rates.

These studies provide an important basis for quantitative application of satellite data to study of individual thunderstorms and their severity.

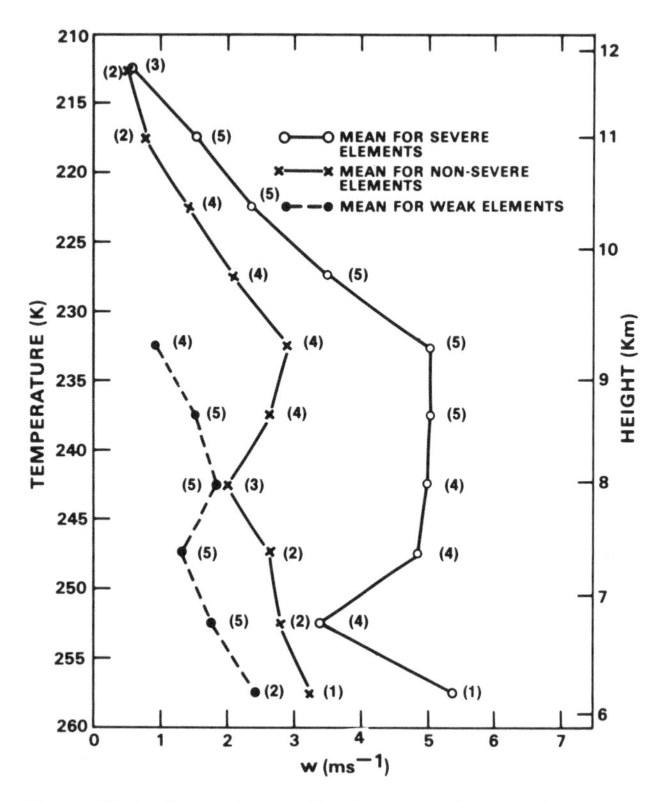

Figure 12.9. Composite satellite-derived vertical velocity profiles for 24 April 1975 (from Adler and Fenn, 1979b).

d. Areal Rain Estimates

Griffith et al. (1978) developed a technique to estimate convective rainfall over Florida and other tropical regions. The procedure (often referred to as the "Griffith-Woodley technique") was developed by relating 10-cm radar data, adjusted by rain-gage data, to satellite-observed cloud brightness. Although the rain estimation technique originally used visible satellite data digitized from negatives, changes were made (see Woodley et al., 1980) to produce total automation with IR digital data. Griffith et al. (1981) expanded the technique for application in middle latitudes. The rain estimates based on the tropical technique are adjusted with the use of output from a one-dimensional cumulus-cloud model. The model is run with the use of in situ upper-air sounding data. Sample results are illustrated in Fig. 12.11. Note that the satellite-inferred rain rates tend to lag the surface gage measurements. Areal rain amounts, however, tended to be quite comparable, which suggests that the technique may prove most useful for hydrologic and agricultural applications, since area average rainfall data for time intervals on the order of at least 3–6 h are not available over much of the globe. In related work Negri et al. (1984) suggested that a simplified version of the estimation procedure may prove to be as accurate as the Griffith-Woodley technique and more easily implemented for near-real-time, large-area rainfall monitoring.

On smaller time and space scales it is likely that the most useful rain estimation techniques will draw upon both radar and satellite data. Negri and Adler (1981) reported on a study comparing the IR cloud-top temperature history of individual thunderstorms with concurrent radar data. They found that the IR minimum temperature was well correlated with the maximum rain volume as esimated from the radar data. Reynolds and Smith (1979), Reynolds (1983), and Heymsfield et al. (1983a) have described interactive techniques for compositing simultaneous radar and satellite data. Figure 12.12 shows a display of such composited data for thunderstorms that occurred over eastern Colorado on 14 June 1981.

e. Satellite Sounding in the Thunderstorm Environment

The multichannel infrared radiation data gathered by instruments such as the Visible and Infrared Spin-Scan Radiometer Atmospheric Sounder (VAS), on the newer satellites (see Smith [1985] for technological details) can be used to infer vertical temperature structure of the atmosphere. The radiative transfer equation, through an algorithm known as "inversion" (McMillin et al., 1973; Fritz et al., 1972; Smith, 1983) provides the framework for estimating the vertical

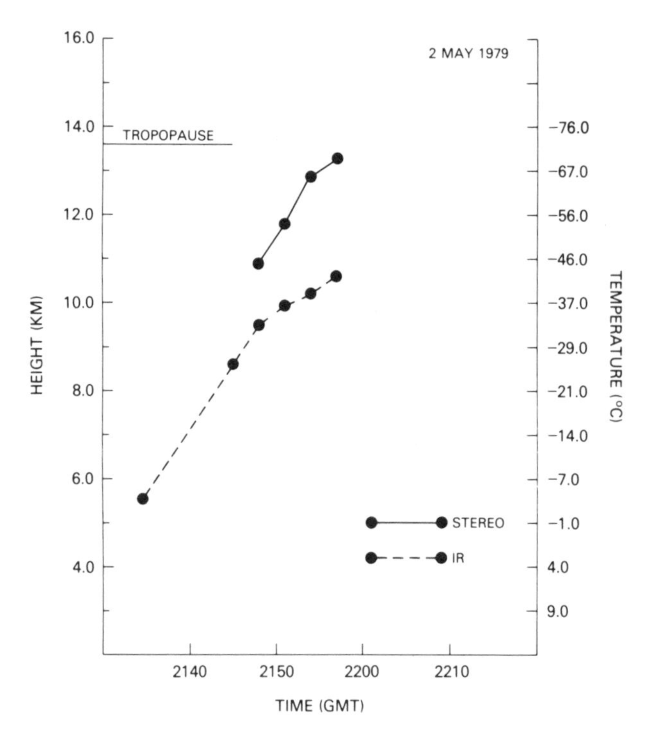

Figure 12.10a. Stereoscopically determined cloud-top heights and minimum cloud-top IR temperatures derived from GOES infrared observations for cell B3 on 2 May 1979 (from Mack et al., 1983).

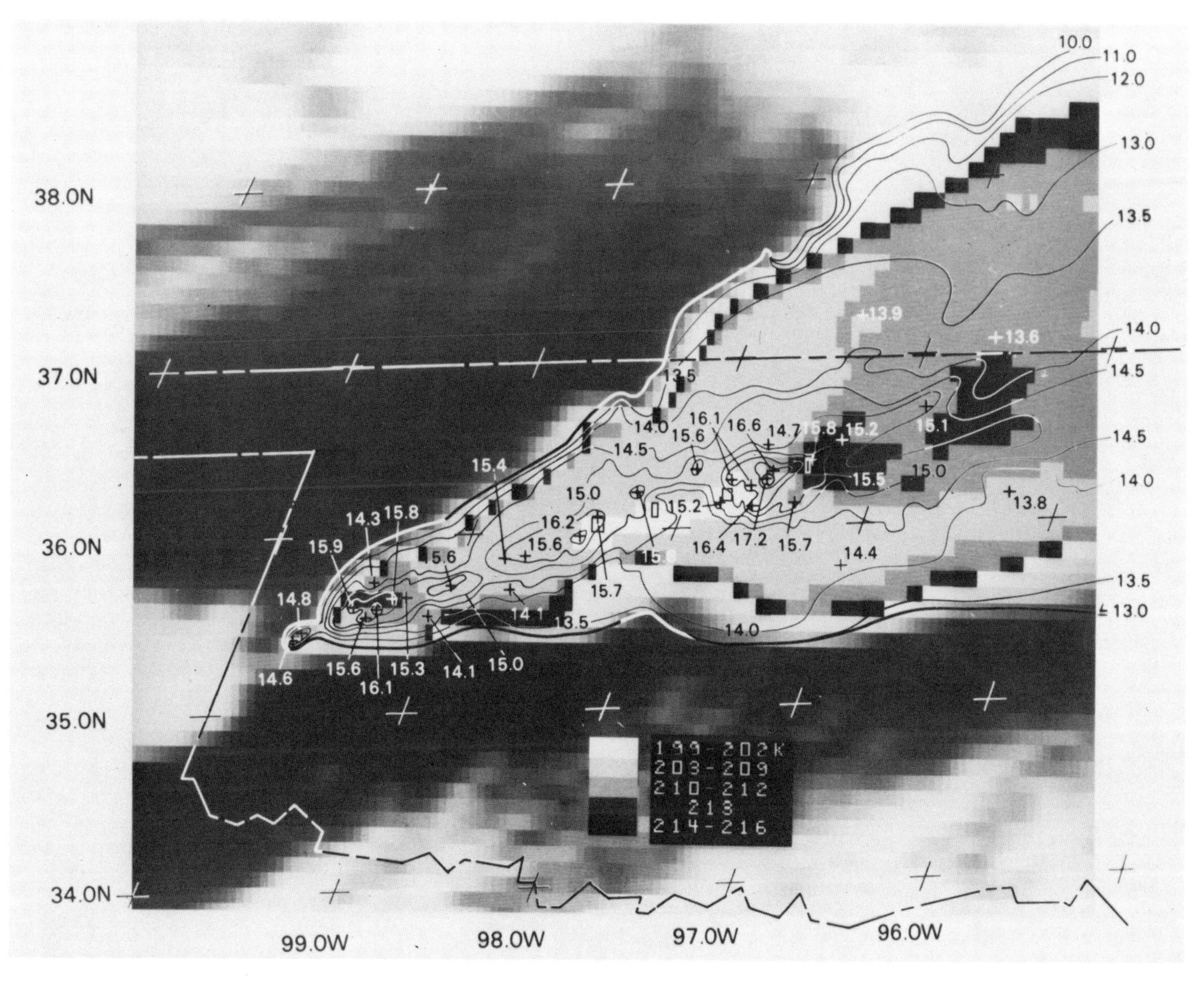

Figure 12.10b. The 0048 GMT stereoscopic cloud-top height contours overlaid on a simultaneous GOES infrared image for 2 May 1983 (from Mack et al., 1983).

distribution of temperature consistent with the observed intensity of radiation in each channel. The contribution of CO_2 presents a difficulty but can be accounted for with temperature-dependent CO_2 transmittances (Smith, 1968). The effects of varying amounts of water vapor on the CO_2 transmission functions are more significant than those of temperature variations. Cloudy situations represent a particularly difficult problem, since the transmission of infrared radiation from below clouds is highly variable and depends on cloud opacity, and since clouds may obscure only a portion of the field of view. Cloudy column radiances can be eliminated from the data sample, or an attempt can be made to account for the clouds.

Fritz (1977) presented an empirical method for retrieving temperature profiles from radiation data. In his procedure profiles known from sounding data become benchmark data to improve the absolute accuracy of retrievals at points between the benchmark locations. He applied the technique to simulated satellite radiance data for an actual severe-thunderstorm situation and found a significant increase in the accuracy of the retrieved profiles.

As improved sounders have gradually become available (see Smith et al., 1981), the use of satellite-retrieved soundings in research on the evolution of the thunderstorm environment has increased dramatically. Chesters et al. (1982) demonstrated that significant moisture gradients could be seen directly in images of the water-vapor radiance channels; that temperature and moisture profiles could be re-

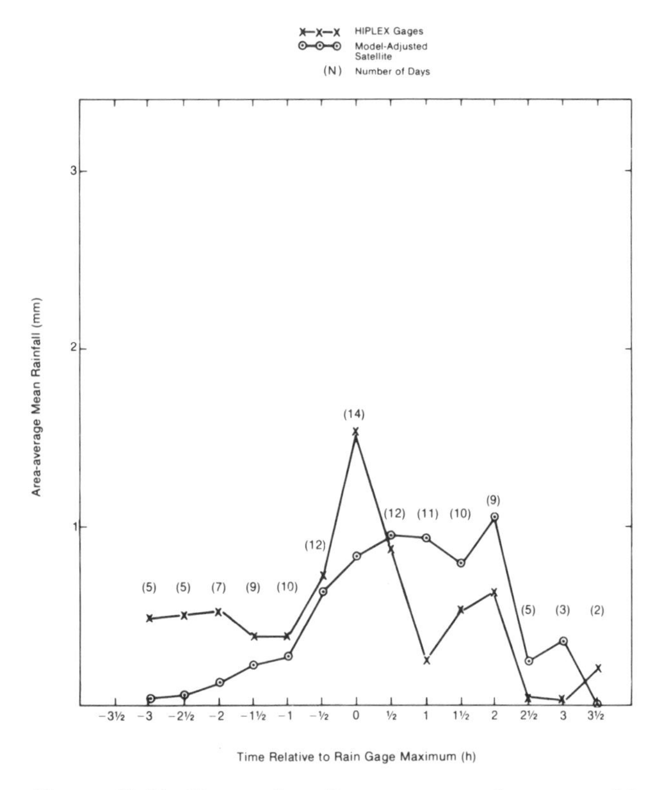

Figure 12.11. Time series of mean gage and mean model-adjusted satellite rainfalls. Data are composited to the time of maximum gage rainfall. Values are from the dense network rain-gage data (from Griffith et al., 1981).

trieved from satellite-measured radiances with sufficient accuracy to delineate major features of a severe-storm environment, and that the quality of retrieved soundings was improved by "conditioning" the retrieval with local weather statistics. Indeed, Smith et al. (1982) presented an entire case study of a severe thunderstorm episode based primarily on satellite-derived upper-air analyses. Figure 12.13, from this study, illustrates the three hourly variations of a satellite-based stability index and its relation to the convective storms.

Hillger and VonderHaar (1977, 1981) used satellite radiance data to retrieve soundings having a horizontal resolution much finer than that of the standard rawinsonde network. Resultant mesoscale analyses of satellite-derived 500-mb temperatures, integrated atmospheric water content, and surface dewpoint temperatures are shown in Figs. 12.14–12.16. Small-scale thermal waves and the dryline perceived in satellite data helped identify areas where the probability of convective development seemed high. The accuracy of detection of small-scale features and the importance of such features for thunderstorm development remain important research topics at this writing.

Along these lines Chesters et al. (1982) demonstrated by means of simulations of retrieved VAS data the potential value of quantitative satellite data in monitoring and diagnosing the severe-storm environment. Subsequently Petersen et al. (1984) showed that the VAS "split window" technique (i.e., combined analysis of middle-level dryness and low-level water-vapor concentration—see Chesters et al. [1983] for a description of the technique) can be employed to monitor the evolution of mesoscale features in the preconvective environment. In particular, they concluded that

> 1) Midlevel dryness indicated in the 6.7 μm VAS images can be effectively monitored and related to the positions of jet streaks. However, . . . bands of midlevel dryness can exist which are normal to the midlevel flow, yet exhibit characteristics similar to those of bands parallel to the flow.
>
> 2) Low-level moisture can be monitored effectively in an image format using either "split window" calculations or visible and infrared observations of low clouds.
>
> 3) Mesoscale regions of apparent convective instability can be delineated by overlaying the 6.7 μm and "split window" images to isolate regions where midlevel dry air overtakes low-level moisture.

Since their study relied almost exclusively on satellite data, the results are particularly encouraging and have immediate potential for forecast application. However, the full utilization of information from VAS can be realized only when these data are diagnosed in conjunction with all forms of conventional data (e.g., detailed analyses of surface and upper-air observations and derived parameters).

Grody (1983), in studying intense convective episodes, investigated the utility of microwave (50–58 GHz) radiances, to date available only from polar-orbiting satellites. He was able to document significant upper-tropospheric temperature changes in regions of mesoscale convective systems. It is likely that microwave data will become increasingly important in storm research once microwave sounders can be placed in geosynchronous orbit.

Finally, Mills and Hayden (1983) have utilized high-density, satellite-retrieved temperature and moisture profiles in initializing a mesoscale numerical weather prediction model. They studied a weather situation that produced widespread severe storms. They concluded that the high-resolution satellite data produced increased accuracy and detail in the situation simulated. It seems inevitable that satellite soundings will play an important role in future numerical studies of severe thunderstorms.

3. Operational Applications of Imagery

Applications and forecasting groups in the National Environmental Satellite, Data, and Information Services (NESDIS) and the National Weather Service (NWS) have documented many interesting cases illustrative of possible relationships between thunderstorms and their mesoscale environment. Examples include a tendency for afternoon storms to form in regions that were clear during the morn-

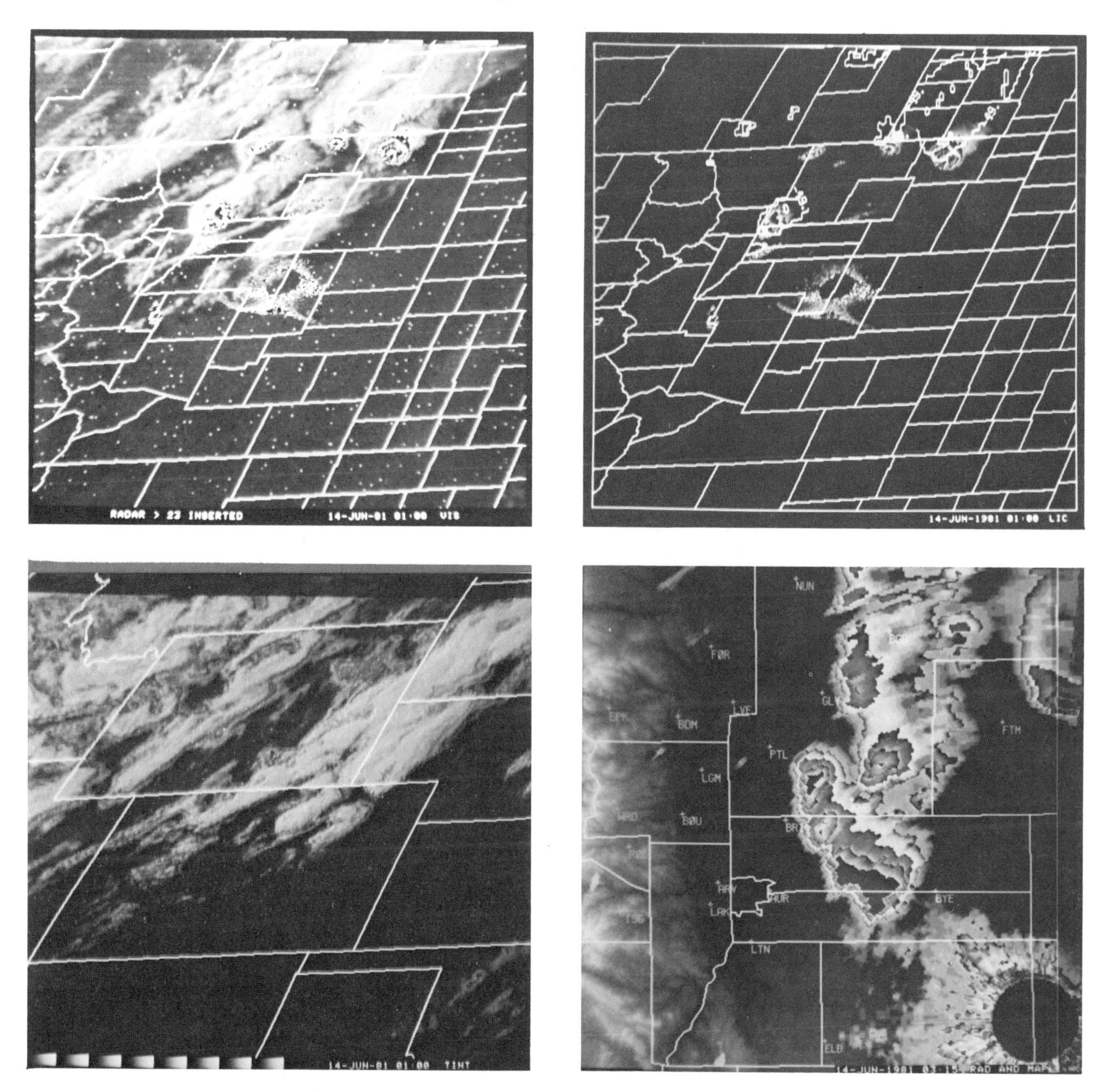

Figure 12.12. (a) Visible satellite image replaced by Limon radar low-level reflectivities above 23 dBZ for 0100 GMT, 14 June 1981. County outlines are superimposed. The data resolution is approximately 2km. (b) The same radar image as (a), overlaid by IR cloud-top temperature contours position-corrected for satellite viewing angle. The −49°C contour was chosen because this was the tropopause temperature as determined from the 0000 GMT, 14 June 1981, Denver sounding. Contours within the −49°C locate the minimum cloud-top temperature. (c) An infrared-visible satellite combination whereby the pseudocolor enhancement for the visible image is defined by the colocated IR temperature values; Gray scale, +55°C to 0°C; green, −30°C to −40°C; cyan, −40°C to −50°C; blue, −50°C to −63°C. (d) Limon low-level reflectivities superimposed on a gray-scale representation of topography derived from a high-resolution digitized map file. White indicates elevations above 4.7 km; black, elevations below 2.4 km (from Reynolds, 1983).

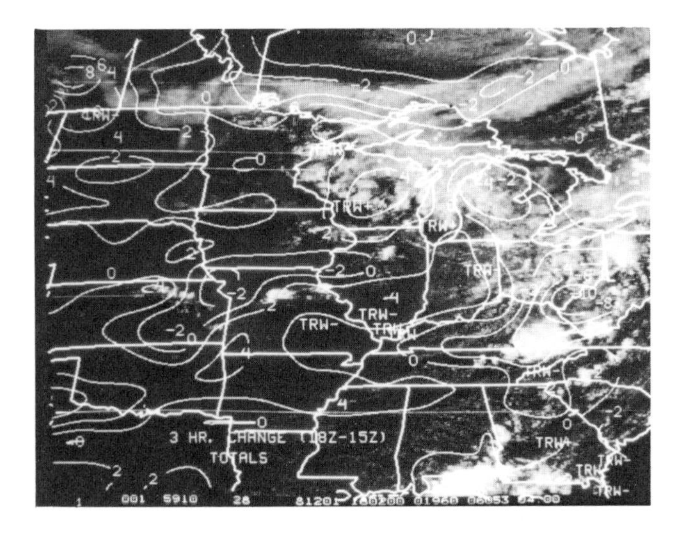

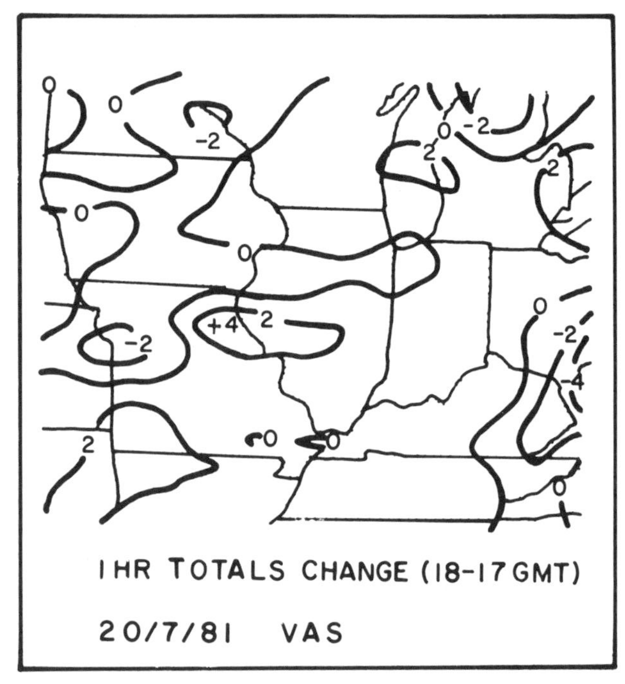

Figure 12.13. (a) Three-hour variation of satellite derived Total-Totals index (°C) between 1500 and 1800 GMT superimposed over the 1800 GMT VAS 11-μm image of cloudiness. Thunderstorms (TRW) that were observed between 2000 and 2300 GMT are also shown. (b) 1-h (1700 and 1800 GMT) variation of Total-Totals index (°C) showing that the maximum 1-h variation (4°C) observed by satellite soundings on 20 July 1981 occurred at the location of and just before the development of a severe convective storm over northeast Missouri (from Smith et al., 1982).

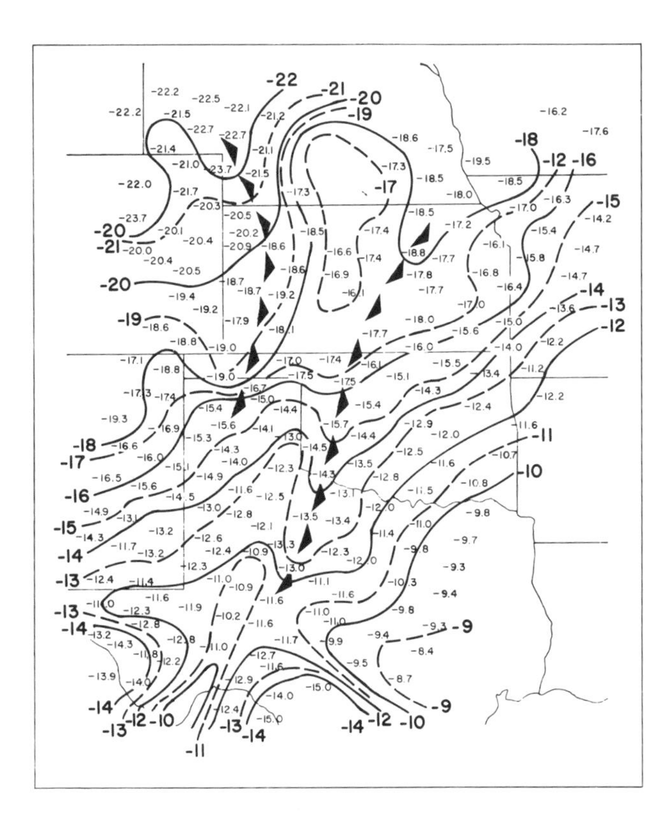

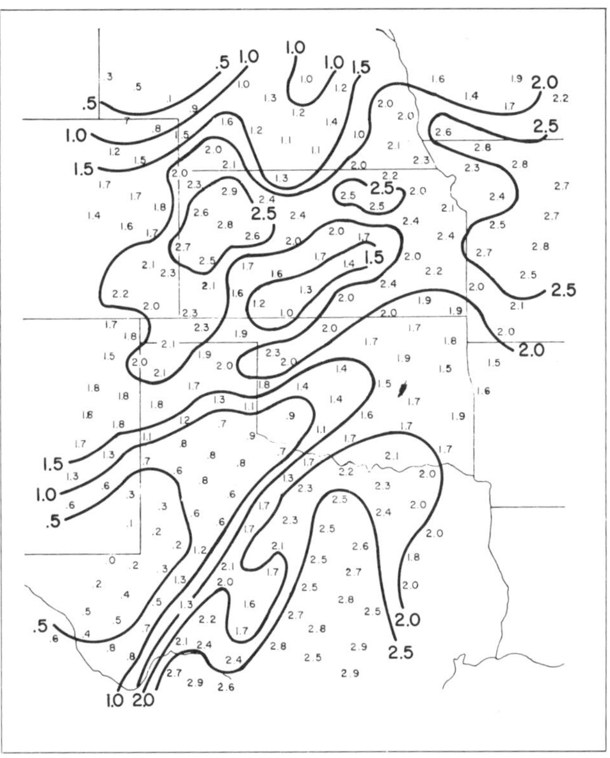

Figure 12.14. (top right) Mesoanalysis (°C) of a satellite-derived 500-mb temperature field obtained using an iterative retrieval algorithm with an initial-guess vertical temperature profile (from Hillger and VonderHaar, 1977).

Figure 12.15. (bottom right) Mesoanalysis (cm) of a satellite-derived precipitable water field obtained using a least-squares linear regression of radiosonde values against H_2O channel radiance residuals (from Hillger and VonderHaar, 1977).

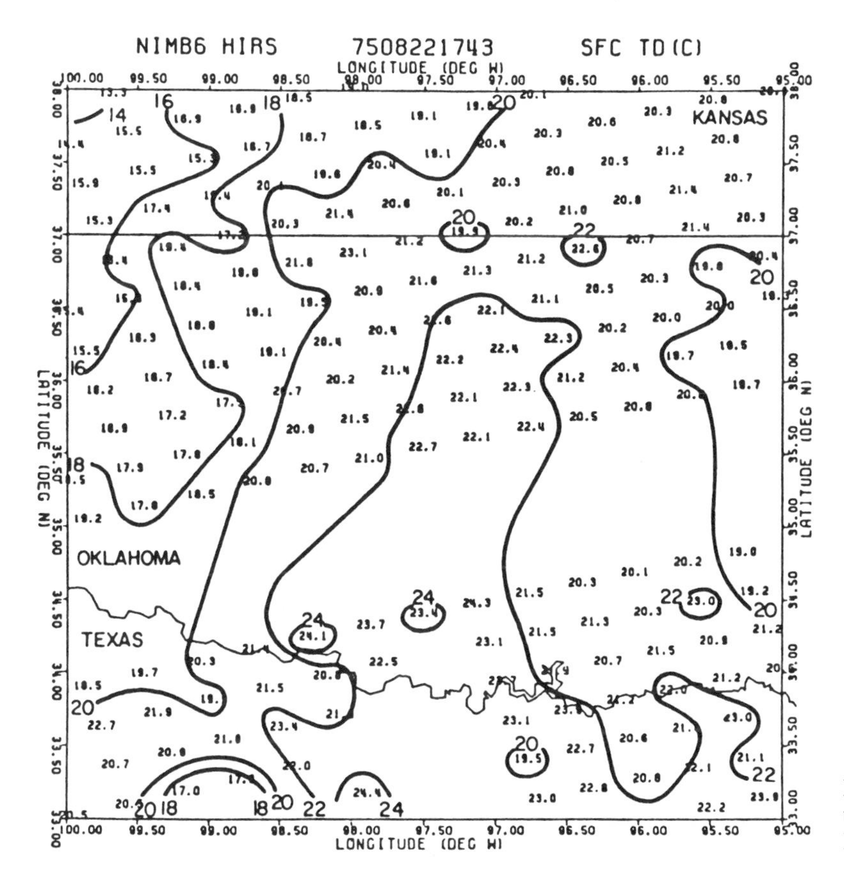

Figure 12.16. Satellite-derived, surface dew-point temperature analysis with contours every 2°C (from Hillger and VonderHaar, 1981).

ing or to form around the fringes of regions that were cloudy or foggy during the morning, and a tendency for intense storms to occur where and when cloud lines intersect (Purdom, 1976; Anderson et al., 1974; Scofield and Weiss, 1976). The following discussion demonstrates the use of satellite imagery in thunderstorm detection, monitoring, and forecasting.

a. Determination of Convective Rainfall

Meteorologists and hydrologists are called on to issue river forecasts and/or flood warnings when important precipitation events are occurring over regions that are relatively devoid of data. Naturally, geosynchronous satellite data have been considered as a potential source of valuable additional rainfall information. Scofield and Oliver (1977) developed an operational scheme for estimating amounts of middle-latitude convective rainfall. Their technique employs IR images that are produced routinely by NESDIS and enhanced according to the curve shown in Fig. 12.17 (top). This particular curve emphasizes cold cloud-top temperatures in the −30° to −70°C range. Scofield and Oliver noted that certain features, along with ongoing changes in the enhanced IR and visible imagery, have been associated with locally heavy convective rainstorms:

1. Quasi-stationary cumulonimbus (Cb) systems and/or Cb systems regenerating over the same location.
2. Cold, rapidly expanding Cb anvils.
3. Cb mergers.
4. Merging convective cloud lines.
5. Overshooting Cb tops (convective turrets that penetrate above the relatively flat anvil cloud).

Scofield (1982) described modifications that have been developed to deal with meteorologically unusual scenarios (e.g., heavy rains from thunderstorms with relatively low and warm cloud tops) and the current version of the technique used in operational rain estimations. Moses (1982) indicated that the scheme was being automated on the NOAA/NESDIS Interactive Flash Flood Analyzer (IFFA) system for easy, widespread application across the United States. At the current time, however, the procedure is not totally automated but is implemented by a satellite meteorologist working on the IFFA system in an interactive mode.

An illustration of operational use of the technique was presented by Scofield (1981) for the Bradys Bend, Pa., flash flood. Evolution of thunderstorms that produced the flash flood is shown in the IR images of Fig. 12.18a–f (up to 5.5 inches of rain fell in the Bradys Bend area, resulting in flooding that killed nine persons and caused more than $50

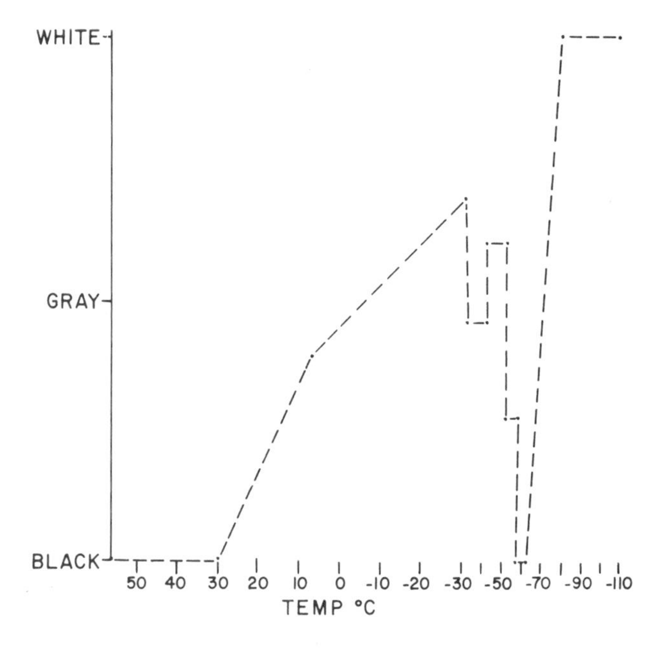

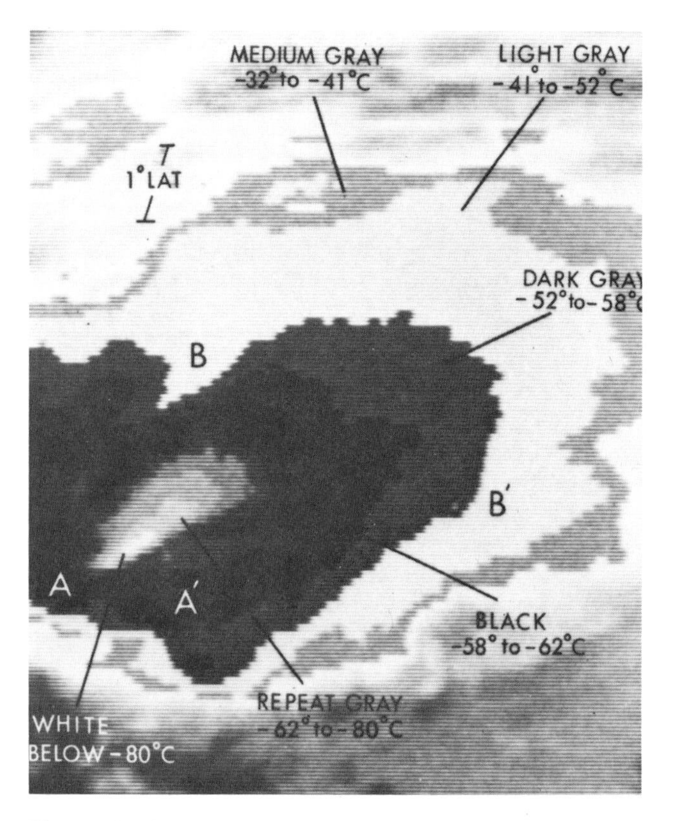

Figure 12.17. The MB enhancement curve for IR images (above) and an example (below).

million in damage). Figure 12.19 shows the official rain measurements in the area and the satellite-derived rainfall estimates for the same period. This technique has important operational utility, especially when it draws the forecaster's attention to an important event not revealed by casual perusal of data, as is illustrated here.

This particular method of interactive, satellite-derived rainfall estimation also indicates that considerable work remains to develop a totally automated and widely applicable scheme. Any such scheme will ultimately require local fine tuning to allow for particular terrain and climate variations and for integration of conventional data such as surface and radar rainfall measurements.

b. Monitoring Movement and Intensity of Thunderstorm Gust Fronts

The importance of gust fronts as a hazard to aviation has been emphasized by crashes that apparently resulted when aircraft penetrated gust fronts shortly before touchdown or after liftoff (Caracena et al., 1983). Gurka (1976) correlated satellite-observed characteristics of thunderstorm cloud arcs with concurrent surface-wind observations. Figure 12.20 shows the southward progression of such a cloud arc across the Baltimore–Washington, D.C., area. Selected surface-wind observations have been superposed on the images.

Gurka (1976) lists the following characteristics of thunderstorm gust fronts that can be identified from satellite imagery:

1. Wind gusts and wind shear occur very near to or at the leading edge of the cloud arc.
2. The strongest winds occur beneath the portion of the arc nearest the most vigorous convection.
3. The regions of vigorous convection are pinpointed by cloud-edge brightness gradients on enhanced IR imagery.
4. Cloud arcs bounding mesohighs associated with a cloudy area and active convection are characterized by stronger gusts than those of narrow cloud arcs trailed by clear skies.
5. Rapidly moving cloud arcs are generally associated with strong low-level winds.

c. Severe-Thunderstorm–Jet-Stream Relationships

Many researchers have presented both observational and theoretical studies that related positions of jet streams, or bands of maximum windspeeds, with the development, location, and orientation of severe thunderstorms. Whitney (1977), studying severe-storm occurrences on 16 active days, found that severe activity rarely developed south of the subtropical jet stream. The positions of jet-stream features were determined from both satellite photographs and conventional upper-air sounding data.

Figure 12.21 shows Whitney's analysis of the severe-thunderstorm situation of 30 April 1975. The subtropical

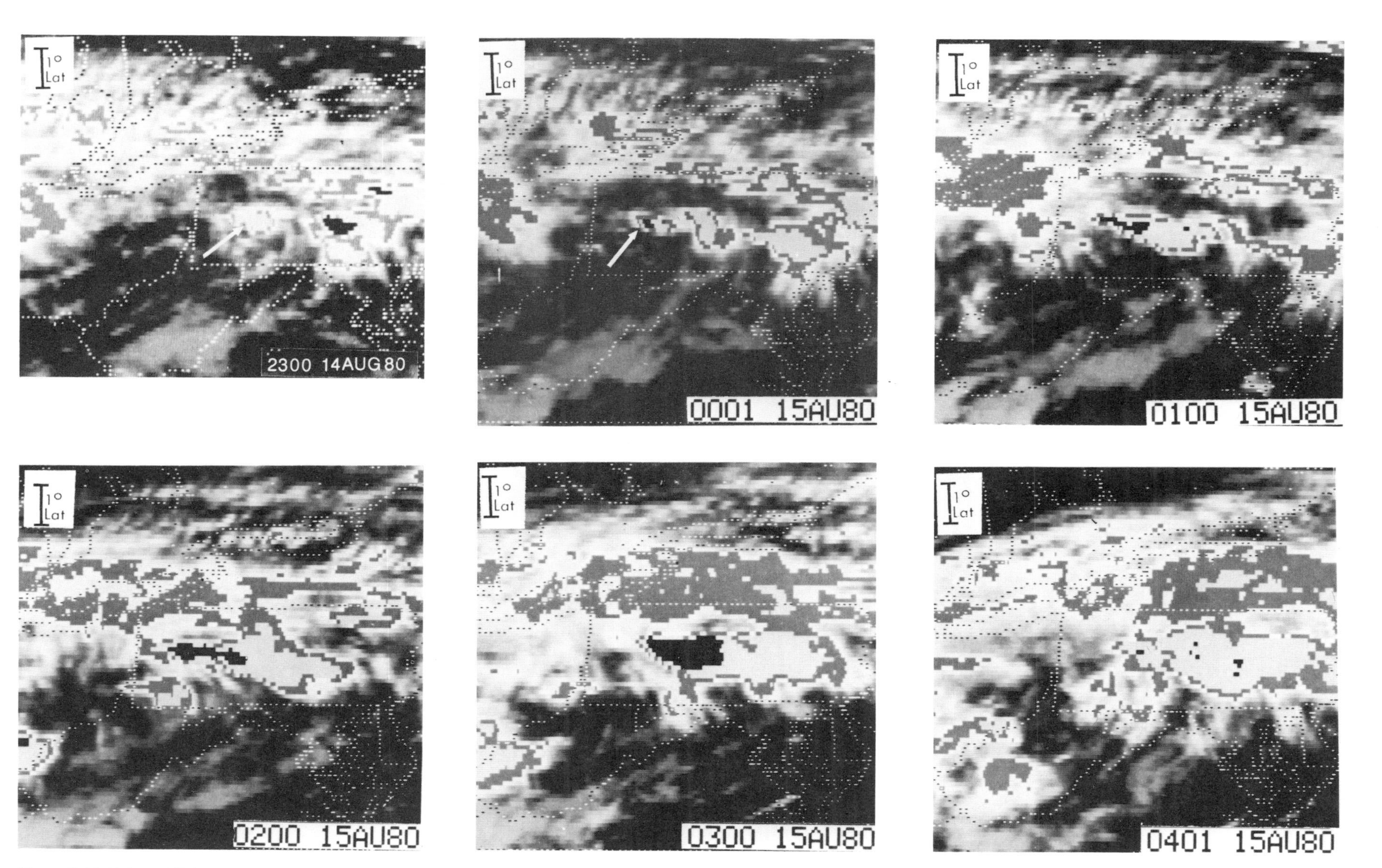

Figure 12.18. Enhanced IR images (MB curve; see Fig. 12.17) during the Bradys Bend, Pa., flash flood. Arrows in (a) and (b) show the storms of interest over Bradys Bend (from Scofield, 1981).

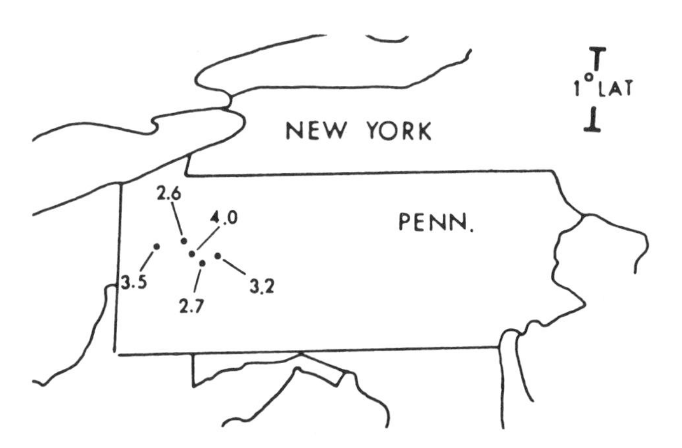

Figure 12.19a. Three-hour observed rainfall (in inches), 0000–0300 GMT, 14 August 1980 (from Scofield, 1981).

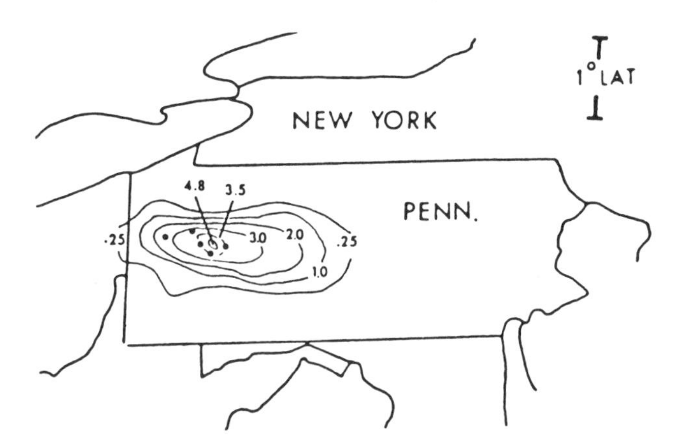

Figure 12.19b. Three-hour satellite-derived rainfall estimate (in inches), modified for warm tops, ending at 0300 GMT, 15 August 1980. The dots are the locations of rainfall observations in Fig. 12.19a (from Scofield, 1981).

jet stream sharply delineates the southern boundary of the large area of severe-storm occurrences. Whitney found that severe thunderstorms usually develop first along the surface front very near to the polar jet stream (note that a double polar-jet structure is analyzed in Fig. 12.21). Activity then moves eastward ahead of the surface front while developing southward across the zone between the diverging jet streams. Miller and McGinley (1978) proposed a technique for forecasting severe thunderstorms that integrates conventional upper-air data, surface analyses, and satellite imagery to identify zones or areas having storm potential. The detection and tracking of various jet streams and jet-stream branches with the use of both satellite imagery and upper-air reports play a crucial role in this innovative forecast scheme. Many of these ideas are being explored operationally at the National Severe Storms Forecast Center, since forecasters there have the interactive CSIS system available for real-time use (see Ostby, 1984; Anthony and Wade, 1983).

d. Thunderstorm Boundaries and Intersections

Making use of satellite imagery, Purdom (1976) showed that the leading boundary of a thunderstorm-associated mesohigh often appears as an arc-shaped line of convective clouds. Earlier he showed (Purdom, 1973) that the intersection of such an arc cloud with another boundary sometimes marks a local region with high potential for intense convec-

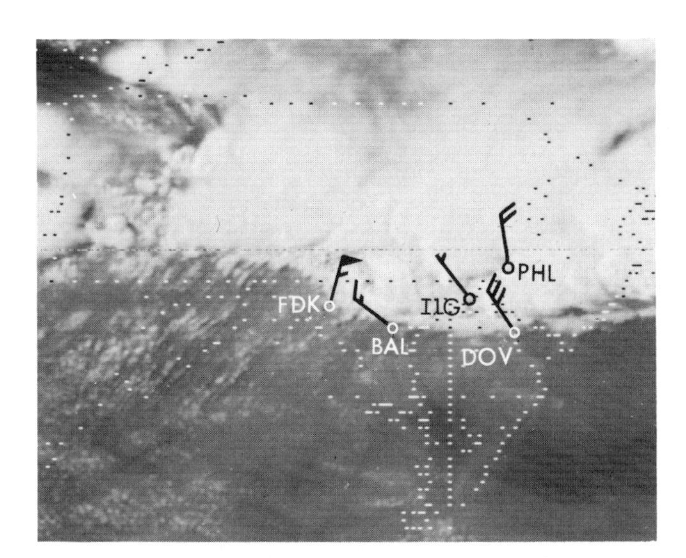

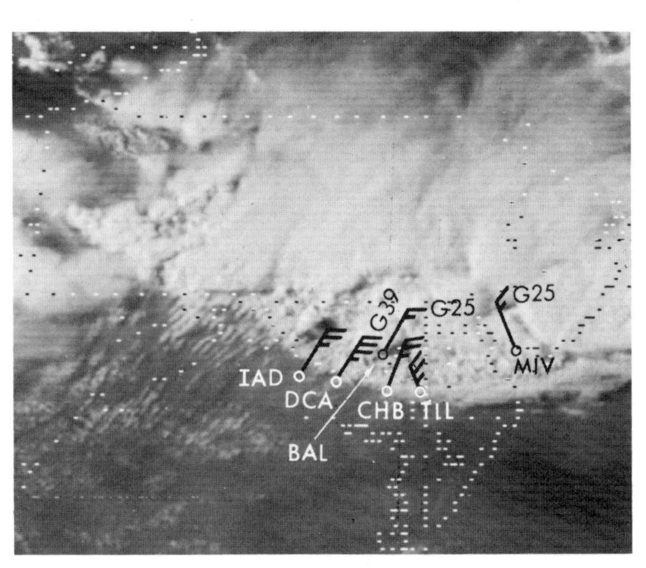

Figure 12.20. Visible 1-km-resolution satellite images over the Washington, D.C., area, with surface wind observations (above) 2100 GMT, 3 July 1975, and (below) 2200 GMT, 3 July 1975 (from Gurka, 1976).

Figure 12.21. Unenhanced infrared picture at 0000 CST, 30 April 1975, showing concurrent severe storms and the synoptic features of jet streams and surface fronts. Large ×'s are positions of wind maxima on the axes of the jet streams. Subtropical windspeed maximum is >100 kn. Diagonal-striped lines are surface fronts. Dotted lines are geographical and political boundaries (from Whitney, 1977).

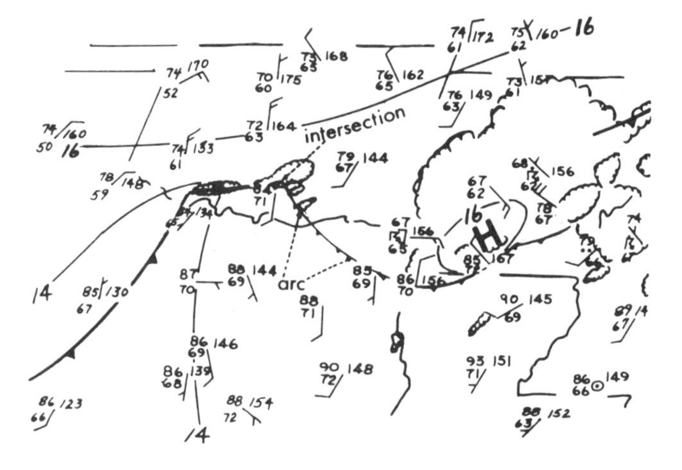

Figure 12.22. Top: GOES-1 1 km visible imagery for 2100 GMT, 26 May 1975. A cloud arc, produced by the storms over Arkansas, stretches from A to B to C. A second arc line is indicated at F. Center: Surface analysis from 2100 GMT, 26 May 1975, showing selected clouds and cloud boundaries. Bottom: GOES-1 1-km visible imagery for 2230 GMT, 26 May 1975 (from Purdom, 1976).

tive development and severe-storm occurrence. These findings provided "photographic" verification of well-known severe-storm characteristics presented earlier by Magor (1969) and Miller (1972). Such a convective development is presented in Fig. 12.22, which shows the 2100 GMT GOES-1 photograph on 26 May 1975, a corresponding mesoscale surface analysis, and a subsequent photograph for 2230 GMT. The impressive convective development in south-central Oklahoma occurred in the often-favored arc line–boundary intersection zone at point A.

Purdom (1982) also found that other types of mesoscale baroclinic zones can play important roles in the development of localized vertical motion that leads to thunderstorm formation. Interaction of baroclinic zones along the sea-breeze front and zones generated by early-morning cloud cover are illustrated in Fig. 12.23. Figure 12.24 shows a satellite-derived evaluation of convective storm generation mechanisms in the southeastern United States. The data in this figure, for example, indicate that large percentages (60%–90%) of afternoon storms develop at intersections of outflow boundaries from previous storms. Location of such boundaries precisely in GOES imagery and observation of their interaction with other boundaries provide the forecaster with information useful for short-range prediction of intense convective developments.

e. Signatures of Severe Storm Tops

McCann (1983) used enhanced IR imagery to examine severe thunderstorms that display a warm area over high storm tops and a signature he terms the "enhanced-V" (note that Negri [1982; see Sec. 2a], McCann [1983], and Heymsfield et al. [1983b] proposed various theoretical explanations of this phenomenon). Figure 12.25 is a satellite image illustrating the enhanced-V. McCann found that when a storm displays an enhanced-V the probability of subsequent severe weather is high. The median lead time from enhanced-V identification to the first severe weather report was 30 min. A low false-alarm ratio makes this identification technique a potential severe-storm warning tool; however, a relatively low probability of detection indicates that many severe storms do not show an enhanced-V. Purdom (1984) proposed improved ways to enhance and display IR data that would be easy to interpret in operational use. Figure 12.26 illustrates a pronounced enhanced-V in this new kind of display, along with a concurrent radar display.

The potential uses of VAS sounding data in the operational forecast environment (particularly that of the National Severe Storms Forecast Center) were reviewed by Petersen et al. (1983). They illustrate that VAS-derived data can fill in temporal and spatial gaps that currently exist within the conventional data base. Eventually the usefulness of VAS and other satellite-based products will be reflected by their acceptance as mesoscale nowcasting and forecasting tools. Thus the operational verdict awaits routine distri-

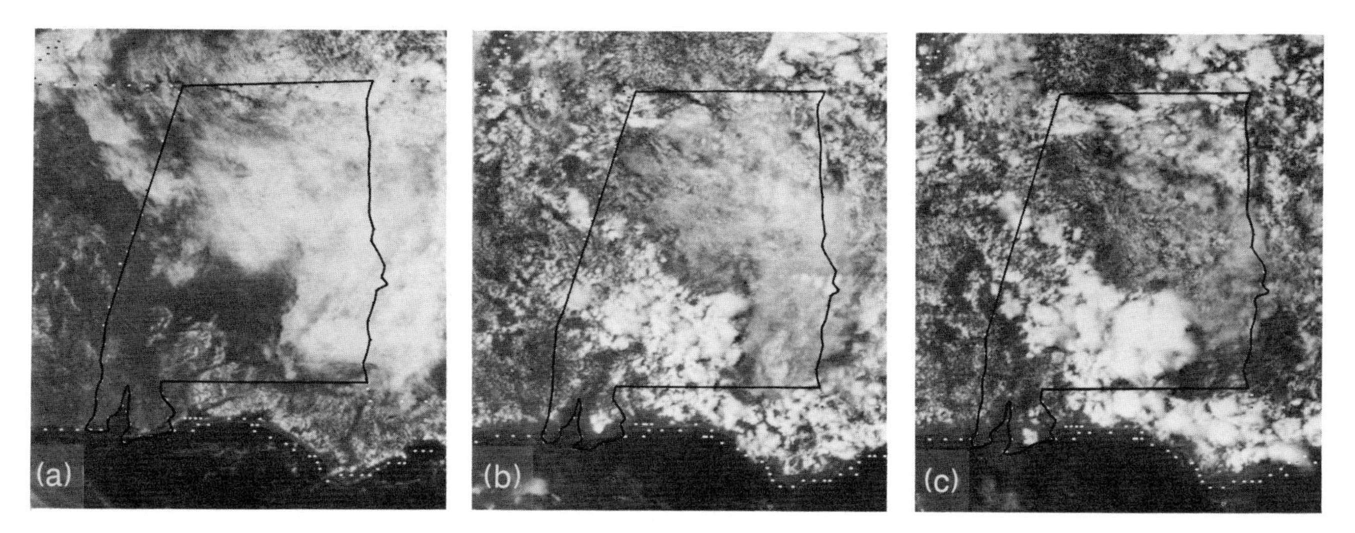

Figure 12.23. GOES visible images for 27 May 1977 for (a) 1530, (b) 1830, and (c) 1930 GMT. Boundaries indicated (Alabama) encompass an area of about 130,000 km². This series of images shows the dramatic effect early cloud cover can have on afternoon thunderstorm development. The clear region in southwestern Alabama during the morning becomes filled with strong convection during the day; conversely, the morning's cloudy region over the remainder of the state evolves into mostly clear skies. Note also how the strongest activity later in the day develops in the "notch" of the clear region in south-central Alabama, as one might expect from the merging of cloud breeze fronts (from Purdom, 1982).

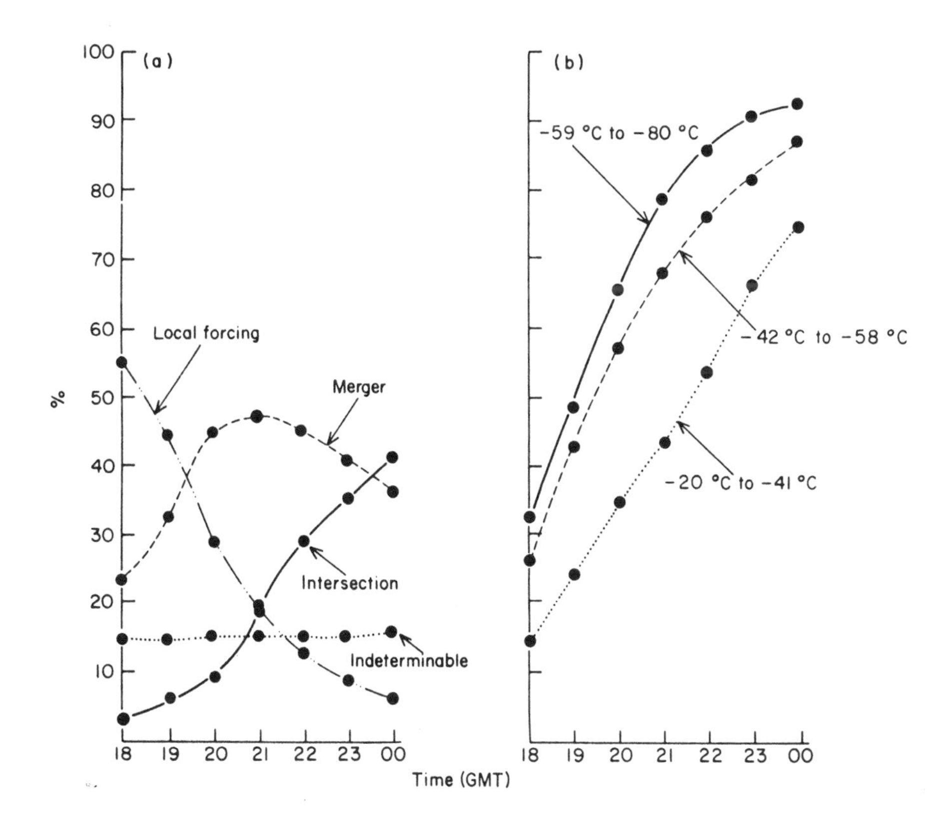

Figure 12.24. (a) Distribution of convective generation mechanisms for all storms with tops colder than −20°C versus time for the summer of 1979 over the southeastern United States. The sample contained 9,850 storms. (b) Percentage of storms, in various temperature ranges, owing to arc cloud intersections and mergers for a given hour, assuming that indeterminable storms in (a) are equally distributed (from Purdom, 1982).

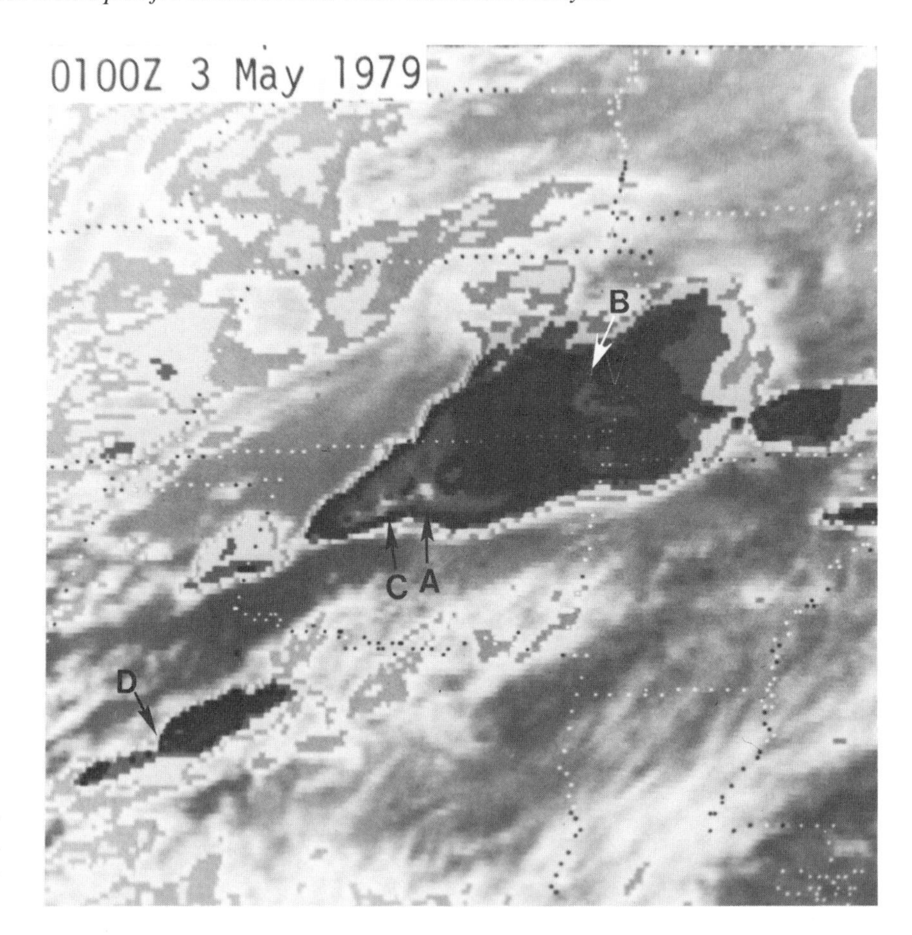

Figure 12.25. Satellite imagery of the lower Midwest processed with the MB curve. Letters A, B, C, and D identify four enhanced-V storms (from McCann, 1983).

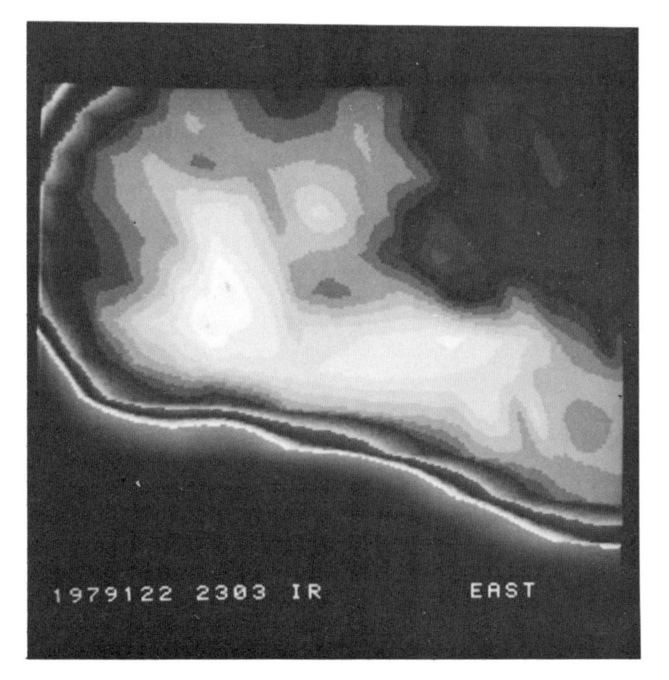

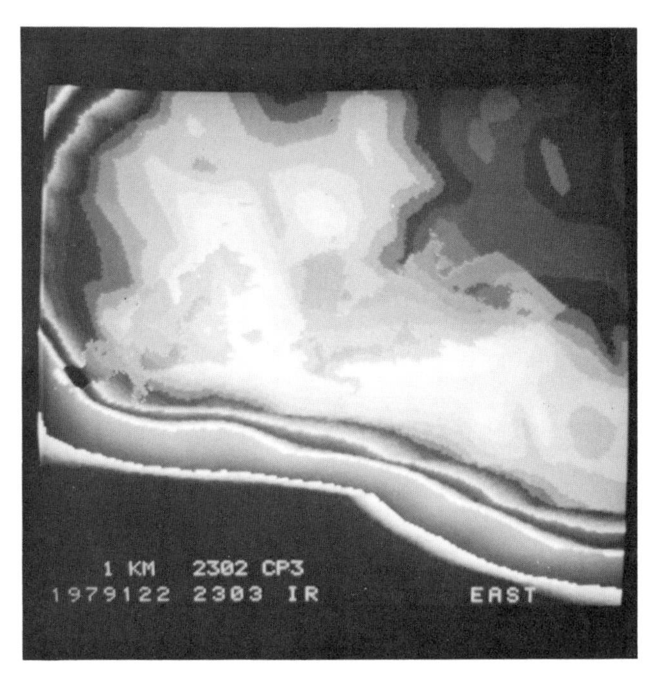

Figure 12.26. (a) GOES-East infrared image for 2 May 1979 at 2302 GMT, displayed in an interpolated format with an enhancement table showing 1°C temperature changes. The coldest temperatures are −69.2°C. (b) The same image, displayed with 2302 GMT 1-km Doppler radar reflectivity data. At the time of the image tornado activity was in progress in the eastern storm complex (from Purdom, 1984).

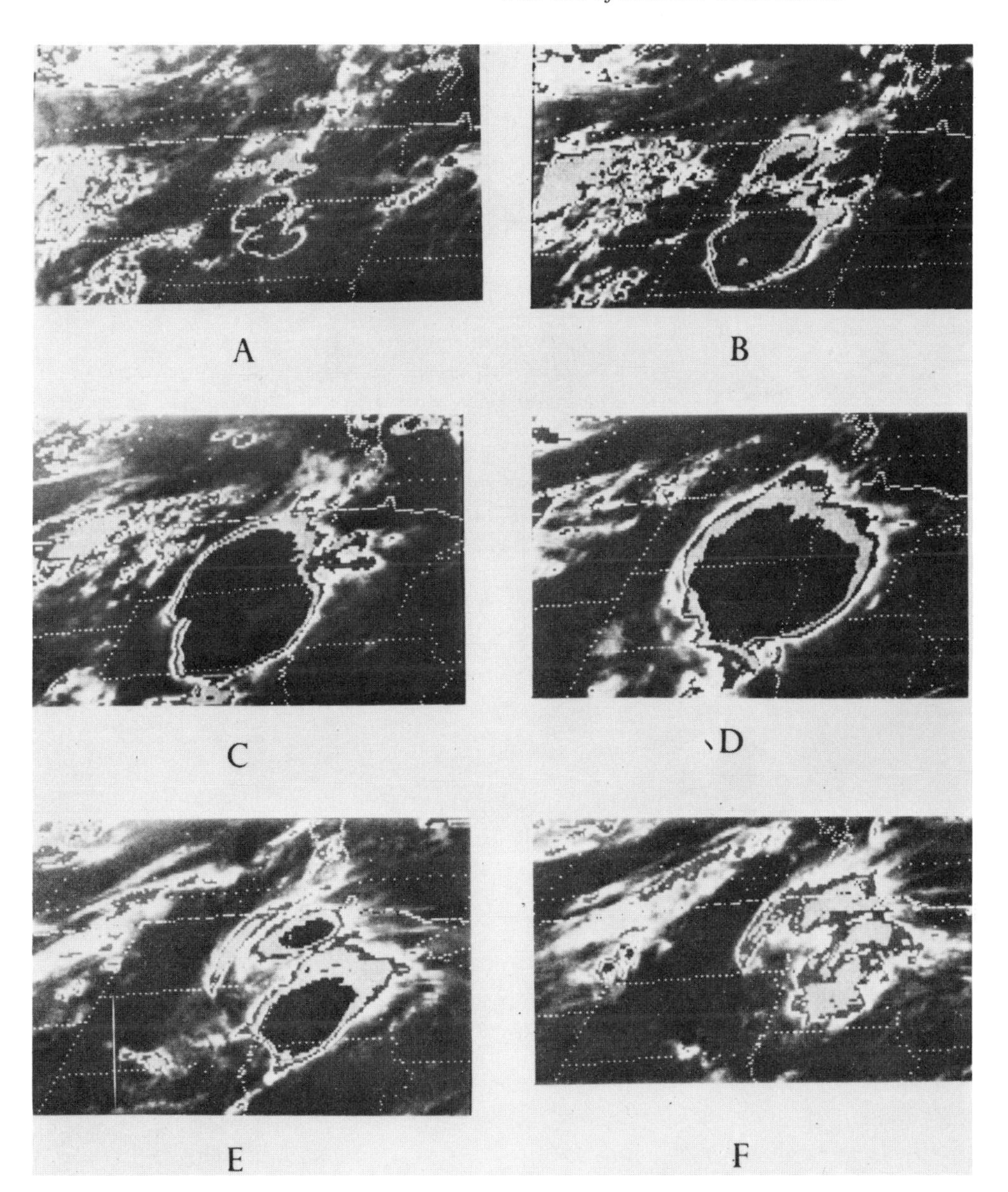

Figure 12.27. Enhanced infrared satellite images on 12 July 1979 for (A) 0030 GMT, (B) 0300 GMT, (C) 0600 GMT, (D) 0900 GMT, (E) 1430 GMT, and (F) 1630 GMT showing the life cycle of a Mesoscale Convective Complex (from Maddox, 1980).

bution of the VAS data and derived satellite products to National Weather Service forecast offices and other operational meteorological organizations.

f. Mesoscale Convective Weather Systems

Identification of large, long-lived convective weather systems, called Mesoscale Convective Complexes (MCC's) by Maddox (1980), was made possible by the operational availability of enhanced IR satellite images at half-hour intervals. The life cycle of a typical summertime MCC is illustrated in Fig. 12.27. MCC's produce a wide variety of significant convective weather phenomena including tornadoes, hail, winds, flash floods, and intense electrical storms, in addition to widespread beneficial rains. Substantial research efforts (e.g., the OK-PRE-STORM field program) are required to understand the physical mechanisms that produce MCC's; however, satellite imagery has already helped improve local weather forecasts by facilitating identifiction of these systems in real time.

4. Summary

Satellite data are being applied extensively in thunderstorm research and operational forecasting and warning. Uses have ranged from subjective inputs into mesoscale analyses to complex quantitative techniques to infer atmospheric vertical temperature profiles. The rapid development of human-machine-interactive computer systems that allow selective processing of digital satellite data has led to increased thunderstorm-related research. As the space program advances, restrictions on weight and size will be eased, and increasingly sophisticated infrared and microwave instruments will be placed in geosynchronous orbit. The concurrent design and flight of satellite systems specifically developed for severe-storm studies and applications should help provide the data crucially needed both for better understanding and for better forecasting of thunderstorms and thunderstorm systems.[1]

[1]The authors appreciate and acknowledge the help of the many researchers who made their figures available to us.

13. Use of Velocity Fields from Doppler Radars to Retrieve Other Variables in Thunderstorms

Carl E. Hane, Conrad L. Ziegler
and Peter S. Ray

1. Introduction

A knowledge of temperature, pressure, and water quantities throughout storm volumes should facilitate description and understanding of forces important to the internal structure and evolution of storms. Measurements have usually been made directly and in situ with airborne and ground-based instruments. In addition, important remote measurements of reflectivity by incoherent radar have for many years provided means for interpretation of storm structure and evolution. The advent of Doppler radar, and, more particularly, the use of two or more Doppler radars to probe the same storm, has allowed estimates of the three components of air motion within storms to be combined with reflectivity information for further storm analysis. Moreover, availability of detailed velocity data has stimulated an area of research known as "retrieval" in which fields of unseen variables are derived by use of the dynamical equations applicable to deep moist convection.

Retrieval is not new in the atmospheric sciences. Radiance measurements by satellite have been used to retrieve, by calculations, temperature and moisture profiles between the Earth's surface and the orbiting sensors. There are, in addition, many other examples in science in which deduction of a set of unknown variables from observation of a different set is accomplished through diagnostic or predictive relations. In this chapter "retrieval" denotes calculation of three-dimensional distributions of certain variables, based on velocity fields derived from Doppler radar measurements. The uniqueness of this particular application perhaps lies in the unusual completeness and accuracy of the observed fields. It should further be noted that, though basic knowledge of the three components of air motion is assumed here, it is in principle possible to derive the same solution quantities given even less velocity information. Although our main emphasis is on retrieval in thunderstorms, derivation of fields in the precipitation-free boundary layer from Doppler-based velocity fields also falls within the meaning of "retrieval" as used here. "Dynamic retrieval" denotes the derivation of pressure and buoyancy from Doppler-derived velocity by utilizing the momentum equations. "Microphysical retrieval" refers to derivation of water quantities and temperature from Doppler-derived winds by using equations for continuity of water substance and the thermodynamic equation.

Retrieval research began in earnest in 1978. Gal-Chen (1978) devised a method by which he calculated density and pressure perturbations using output from Deardorff's (1974) boundary layer model in place of observations. This work demonstrated the feasibility of the method and tested the sensitivity of solutions to errors in the velocity field and to other inadequacies potentially present in real data. Hane and Scott (1978) formulated a two-dimensional method for retrieving departures of pressure and temperature from their horizontal averages (deviation quantities) in convective clouds. Numerical model output was again used to examine the sensitivity of the solution to potentially lacking velocity and water information. This work demonstrated the feasibility of retrieval methods in thunderstorms. Using composited data from single Doppler radar observations in the mesocyclone region of storms and the horizontal equations of motion, Bonesteele and Lin (1978) derived pressure perturbations on horizontal planes. The derived horizontal and vertical pressure gradients were found to be consistent with various features of the flow field that exist in the mesocyclone region. Liese (1978) used analytic data to test a formulation for deriving pressure and thermal variables from the momentum equations. Plausible boundary layer fields were derived, and filtering techniques for minimizing the effect of random error in the velocity field were investigated.

Chong et al. (1980) were the first to apply a dynamic retrieval method using dual Doppler radar data. Their observations were made during an episode of weak convection. Their method is slightly different from that of Gal-Chen (1978) and specifically solves for horizontal gradients in pressure and potential temperature. Gradients of pressure derived in a convective cell were consistent with updraft and downdraft locations, and buoyancy perturbations in the updraft region were on the order of 2 K in the relatively weak circulation. Hane et al. (1981) applied Gal-Chen's (1978) method to thunderstorm flow fields obtained from a numerical simulation by Klemp and Wilhelmson (1978b). The solution fields were found to be virtually identical to the model output in a control case and significantly sensitive to omission of time derivatives ($\partial u/\partial t$, $\partial v/\partial t$, $\partial w/\partial t$) in the velocity fields, to omission of turbulence terms in the equations, and to errors in the input velocity and water fields in tests cases.

Brandes (1983, 1984) used a slightly different approach based on the momentum equations in analyzing pressure and buoyancy fields in two tornadic storms. He found that intensification of the mesocyclone circulation led to lowered pressure near the surface and subsequent increase in rear downdrafts. Roux et al. (1984) have applied a method similar to that of Gal-Chen (1978) in deriving pressure and temperature distributions in a West African squall line that was observed by dual Doppler radar. Encouraging results were obtained from comparisons of retrieved pressure and temperature fields with pressure and temperature data obtained from surface stations. Using boundary layer Doppler observations obtained in Project PHOENIX (a field experiment carried out in northeast Colorado during the summer of 1978), Gal-Chen and Kropfli (1984) derived pressure and temperature perturbations and computed heat-flux profiles for comparison with tower data; an encouraging amount of agreement with the tower data was achieved. Hane and Ray (1985) retrieved pressure and buoyancy fields in a tornadic storm observed by four Doppler radars. Pressure distributions showed good agreement with theoretical predictions through the depth of the storm, and structure of the mature low-level mesocyclone was in basic agreement with results of Brandes (1983) and with numerical simulation results of Klemp and Rotunno (1983).

Microphysical retrieval is a recent innovation, although the conservation principles forming the basis of this method have been widely applied since World War II to the study of thunderstorm structure. Evidence of the strong relationship between airflow and precipitation in thunderstorms dates to the Thunderstorm Project (Byers and Braham, 1949). Additional storm structure details such as weak echo regions and hook echoes have been studied since the 1960s (Browning, 1964; Marwitz, 1972; Chisholm and English, 1973; Browning and Foote, 1976). The classic conceptual models have implicitly relied on continuity principles to deduce qualitative thunderstorm structure from radar reflectivity. For example, vertical motion removes precipitation from the vault region at a rate that counterbalances the production of precipitation from cloud droplet coalescence and results in low net hydrometeor content and reflectivity. Recent investigations use winds derived from the data of multiple Doppler radars in place of inferred airflow (Foote and Frank, 1983), although the resulting inferences concerning the microphysical thunderstorm structure remain largely qualitative.

The quantitative study of precipitation formation follows from coordinated use of modeled or observed airflows, continuity principles for heat and water substance, and models of microphysical processes. Kessler (1969) studied the distribution of cloud water and rainwater in relation to modeled one- and two-dimensional flows, finding that the impact of airflow on spatial precipitation distribution and amount of precipitation accumulated aloft is governed by the ratio of precipitation fallspeed to updraft speed. Rutledge and Hobbs (1983) studied the dominant precipitation mechanisms in extratropical cyclone rainbands through the combined use of continuity principles, approximate representations of warm rain and ice microphysical processes, and two-dimensional motions inferred from a single Doppler radar. Condensate in weak or strong updraft zones (15 cm s^{-1} or 60 cm s^{-1}, respectively) is efficiently removed by either deposition (weak updraft) or riming (strong updraft) on ice particles falling into the cloud from a higher-level formation zone.

The use of multiple-Doppler-derived three-dimensional airflow and continuity principles may provide a more realistic diagnosis of natural precipitation fields, since important advection effects are expressed in full detail. Ziegler (1984) has developed and applied a retrieval method that proceeds from the three-dimensional air motion observed by Doppler radar, through accompanying models of thermodynamic and microphysical processes, to diagnosis of internally consistent fields of hydrometeor content and temperature. The accuracy of a simplified version of the method has been demonstrated by using dynamically simulated cloud velocity fields as model input. In applications to observed storms, good agreement is noted between observed and retrieved radar reflectivities, which are partitioned into separate contributions from rain, hail, and snow. Retrieved thermal and water substance buoyancies are qualitatively consistent with the observed airflow.

Retrieval brings better knowledge of interrelationships of all variable fields, which should, in turn facilitate parameterization[1] to account for the important effects of convection in larger-scale predictive models. Numerical simulations on the cloud scale also might benefit more directly through specification of initial conditions retrieved from observations. A simulated cloud might be allowed to evolve from one observation time to another, and retrieval methods might be applied at intervals to verify model predictions. Thus the much-needed task of verifying cloud models can be pursued.

The addition of microphysical information heretofore unavailable is an especially exciting development. Interpretation of radar reflectivity fields is greatly enhanced by a knowledge of the particle types and sizes that produce a given reflectivity. Realistic environments for Lagrangian models of hydrometeor (e.g., hail) growth are provided by detailed mixing ratio information for the various water substances. Weather-modification research efforts, including

[1]A method for characterizing gross aspects of the behavior of small-scale processes without specification of process details, based on available large scale variables. For example, the behavior of an ensemble of raindrops might be described reasonably well without tracking each particle. Of course, formulation of a parameterization scheme involves consideration of detailed physical processes to determine how those processes might be summarized.

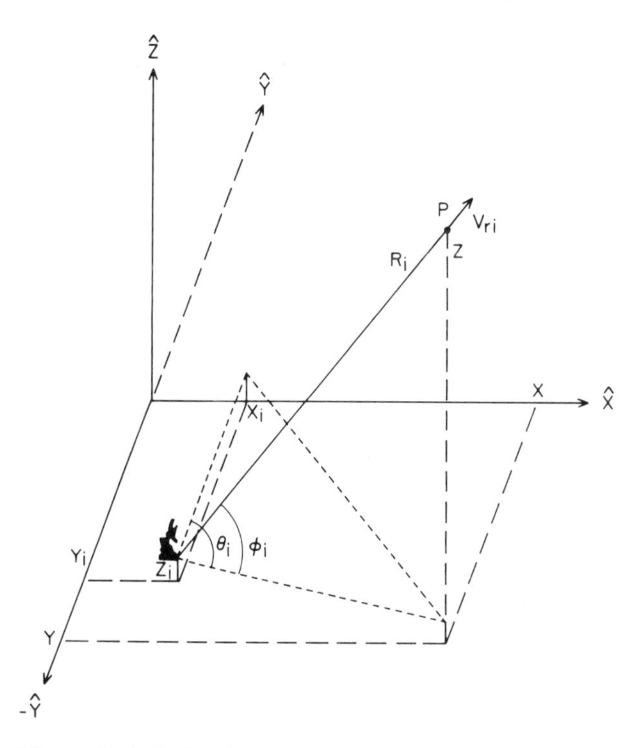

Figure 13.1. Radar observation of radial velocity. Radar located at (x_i, y_i, z_i) observes radial velocity V_{ri} at sampling volume $P(x,y,z)$. The orientation of P to the radar is defined by the elevation angle ϕ, the azimuth θ, and the range R_i.

both dynamic and microphysical seeding of clouds (see Vol. 2, Chap. 15), should benefit greatly from a complete specification of both dynamical and microphysical fields before and after seeding, for example. Lightning activity in storms (which depends upon separation of charge, which in turn is believed to depend upon the phase, size distribution, and motion of water substance) could be more fully understood, given such added comprehensive information.

2. Doppler Radar Analysis Techniques

Retrieval of dynamic and microphysical variables requires knowledge of the three components of air motion within the storm volume. Thus the analysis of radial velocity data from individual radars to produce air velocity components over the storm volume is a vital step toward retrieval.

With three independent (orthogonal) directions of air motion to be determined, at least three independent relationships (or equations) must be used to determine the wind uniquely at any point. Additional information results in an overdetermined problem, which allows the most nearly correct answer to be estimated through a method of least squares (Draper and Smith, 1966). In the application discussed here, the radar does not actually measure the wind motion at a point; it measures the average speed of the hydrometeors (liquid water or ice) toward or away from the radar in each sampling volume (labeled P in Fig. 13.1). The hydrometeor falls with respect to the air around it, and the radar measures a vertical component W, which is the sum of the vertical air motion w and the particle terminal fallspeed V_t:

$$W(x,y,z) = w + V_t. \tag{13.1}$$

The positive variable V_t requires that yet another equation be added to the three already identified to provide the basis for a unique solution. Normally this final equation is an empirical formula relating reflectivity and terminal fallspeed.

The relevant geometric relationships are illustrated in Fig. 13.1, where a radar located at (x_i, y_i, z_i) observes a point P at the position defined by coordinates (x,y,z). The point P is at a range from the radar given by

$$R_i = [(x - x_i)^2 + (y - y_i)^2 + (z - z_i)^2]^{\frac{1}{2}}. \tag{13.2}$$

In Fig. 13.1 the azimuth, usually defined as the clockwise direction of the antenna from north, is given by θ, and the elevation, defined as the antenna position different from horizontal, is given by ϕ. Usually the air motion components u, v, and w are in the directions of x, y, and z, respectively.

Many of the basic principles for deriving windfields from the combined use of separated Doppler radars were given by Armijo (1969). Most methods incorporate a terminal velocity relationship previously mentioned and the anelastic form of the equation of mass continuity

$$\frac{\partial u}{\partial x} + \frac{\partial v}{\partial y} + \frac{\partial w}{\partial z} = \kappa w, \tag{13.3}$$

where κ is a constant approximation to $-\partial(ln\rho)/\partial z$, and the wind components (when the measured particle motion is along the ith radar's beam axis) satisfy

$$u \cos \phi_i \sin \theta_i + v \cos \phi_i \cos \theta_i + W \sin \phi_i = V_i. \tag{13.4a}$$

Through standard trigonometric relationships (Eq. 13.4a) can be written

$$u(x - x_i) + v(y - y_i) + W(z - z_i) = R_i V_i. \tag{13.4b}$$

a. Methodology with Two Doppler Radars

Expressions for the three velocity components resulting from combination of Eqs. 13.1 and 13.4b for each of two radars are given in the Appendix, Sec. 1. Brandes (1977) and Ray et al. (1980) pointed out that Eq. 13.3 can be solved by inserting an initial estimate of w into Eq. 13.28 to derive horizontal wind components, followed by solution of Eq. 13.3 for a new value for w. This process is repeated until the adjustments are less than the presumed measurement error.

Air density decreasing with height suppresses error when

the integration proceeds downward from the storm top. This error reduction must be balanced against the possible effects of greater uncertainties with the upper boundary condition than with the lower boundary condition. Generally, for deep convection when the wind field is reasonably steady state, it is advantageous to integrate downward. Analysis of these effects can be found in Ray et al. (1980) and Nelson and Brown (1982). The spatial distribution of errors generated by the uncertainty in the estimates of V_{ri} is discussed in Ray et al. (1979). Since there is no measurement of the wind component perpendicular to the baseline joining the two radar locations, the uncertainty in that direction becomes unbounded. In general, the greater the sample of each wind component, the smaller the error. Error tends to be smallest when the intersection of the radar beams is near 90° and largest when the angle of intersection is less then 20° or greater than 160°.

After this initial analysis wind components are prescribed at all grid points sampled by at least two radars. However, the vertical velocities may depart substantially from reasonable values at locations far from where the integration begins. To mitigate these effects the winds are adjusted to new values. One means of adjustment is to require the integrated density-weighted horizontal divergence from the surface to Z_t to be a constant C, usually zero (Ziegler et al., 1983):

$$C = \int_0^{Z_t} \rho \left(\frac{\partial u}{\partial x} + \frac{\partial v}{\partial y}\right) dz = -\int_0^{Z_t} \frac{\partial w^*}{\partial z}\, dz. \quad (13.5)$$

where ρ is air density and $w^* = \rho w$. The result is that the horizontal wind components are adjusted so that when the components are integrated both new boundary conditions are satisfied.

Presented in variational form (Hildebrand, 1965; Courant and Hilbert, 1953), the functional to be minimized incorporating the integral constraint (Eq. 13.5) is

$$E = \int\int\{\int[\alpha^2 (u - u^0)^2 + \beta^2(v - v^0)^2]dz + \lambda[\int\rho\left(\frac{\partial u}{\partial x} + \frac{\partial v}{\partial y}\right)dz - C]\}\, dxdy, \quad (13.6)$$

where λ is the Lagrange multiplier, which varies only in the horizontal directions. The superscript 0 denotes an observed quantity, and the weights α and β are determined from an error analysis and are related to the Gauss-Precision Moduli, e.g., $\alpha^2 = (2\sigma_u^2)^{-1}$, where σ_u^2 is the variance of the u-component uncertainty. The result of minimization of Eq. 13.6 is that the adjusted values of the horizontal wind field (u and v) deviate as little as possible from the measured values (u^0 and v^0), and at the same time the integral constraint (Eq. 13.5) is satisfied. By using the techniques outlined in Hildebrand (1965), the value of λ that satisfies these conditions is found. Finally, the adjustment of each horizontal wind component is governed by the magnitude of its error compared with others in the same vertical column.

The adjustments are prescribed by

$$u = u^0 + \frac{\rho}{2\alpha^2}\frac{\partial\lambda}{\partial x}, \quad (13.7a)$$

and

$$v = v^0 + \frac{\rho}{2\beta^2}\frac{\partial\lambda}{\partial y}, \quad (13.7b)$$

where u^0 and v^0 are the initial grid-point estimation, u and v are the adjusted estimates, and α^2 and β^2 are proportional to the reciprocal of the error in u^0 and v^0. These errors are due to the geometrical relationship of the grid point to the radars, as given by Ray and Sangren (1983). The values of λ are obtained by solving the variational problem (Eq. 13.6), which incorporates the constraint given by Eq. 13.5. The vertical velocity is found by integrating Eq. 13.3 using the estimates found in Eq. 13.7.

However, a special case exists when α^2, β^2, and ρ do not vary with height. Here the adjustment of the horizontal wind components ($u - u^0$ and $v - v^0$) is constant and the corresponding adjustment of the horizontal divergence is

$$D - D^0 = \frac{w' - w^0}{Z_t}, \quad (13.8)$$

where w' is the specified lower (or upper) boundary condition and w^0 is the unadjusted value at that boundary. Here the adjustment of the divergence $D - D^0$ is constant for all heights. The adjustment in vertical velocity at any height l can be expressed as

$$w_l - w_l^0 = -\int_0^{Z_l} (D - D^0)\, dz,$$

or

$$w_l = w_l^0 - \frac{Z_l}{Z_t}(w' - w_l^0), \quad (13.9)$$

where t refers to boundary locations. Note that the adjustment in vertical velocity increases linearly with height.

b. Methodology with Three Doppler Radars

When more than two radars are available, an analogous equation set in the same form as Eq. 13.5 can be derived using standard least-squares methodology (Draper and Smith, 1966). Ray and Sangren (1983) discuss the resulting equation set and the distribution of error for combinations of two to nine radars. When, for example, a third radar is introduced, Eqs. 13.1 and 13.4 are used without need for the continuity relation (Eq. 13.3). The resulting wind components are given in the Appendix, Sec. 2.

Another approach was first described by Ray et al. (1978)

and expanded in Ray et al. (1980). In this approach the solution for three or more Doppler radars is found directly, but in such a way that the u,v,w wind components at a grid point are derived as the first term in a Taylor expansion of the wind field in the near vicinity of the grid point. The wind field is further required to satisfy constraints, among which mass continuity (Eq. 13.3) is prominent. This is done locally only in the region between grid points. After the field has been determined in this way, it is adjusted so that the mass continuity relationship holds (in its discretized form) when the grid-point values are used. These relationships are conveniently expressed by using techniques of variational calculus.

3. Dynamic Retrieval Methods

Three methods have been formulated for derivation of pressure and buoyancy distributions from the velocity distribution in deep convection. Each method is based solely on the three momentum equations written in a form applicable to deep convective clouds. Differences reside in the manner in which the momentum equations are utilized and in the form of solution variables. In addition, there are some differences in treatment of boundaries and in treatment of missing data in documented applications to date. Only methods that have been applied to observed thunderstorm data to derive both pressure and buoyancy are discussed here.

a. Dynamic Method A

The method originally formulated by Gal-Chen (1978) was applied by Hane et al. (1981) in the calculation of pressure and buoyancy distributions in thunderstorm circulations. The horizontal momentum equations can be written

$$\frac{\partial \pi}{\partial x} = \frac{1}{c_p \bar{\theta}_v}\left[D_u - \frac{\partial u}{\partial t} - \mathbf{V} \cdot \nabla u + fv\right] = F \qquad (13.10)$$

and

$$\frac{\partial \pi}{\partial y} = \frac{1}{c_p \bar{\theta}_v}\left[D_v - \frac{\partial v}{\partial t} - \mathbf{V} \cdot \nabla v - fu\right] = G, \qquad (13.11)$$

where π is nondimensional perturbation pressure, $\bar{\theta}_v$ is the base state virtual potential temperature, D_u and D_v are the turbulent terms in the u and v equations, respectively, f is the Coriolis parameter, and the overbar refers to the base state (condition in the storm environment). Expressions for D_u and D_v can be found in the description of the three-dimensional cloud model of Klemp and Wilhelmson (1978a).

All terms in F and G can, in principle, be calculated from the velocity components derived from Doppler wind data. As outlined by Gal-Chen (1978), if F and G are error-free, the condition $\partial F/\partial y = \partial G/\partial x$ is satisfied. However, in general F and G contain errors, and Eqs. 13.10 and 13.11 cannot be solved directly. Instead, the solution of these equations may be posed as a variational problem, where minimization of the functional

$$J = \iint \left[\left(\frac{\partial \pi}{\partial x} - F\right)^2 + \left(\frac{\partial \pi}{\partial y} - G\right)^2\right] dxdy$$

is sought. Carrying out the variation, an Euler equation in the form of a Poisson equation for π results:

$$\frac{\partial^2 \pi}{\partial x^2} + \frac{\partial^2 \pi}{\partial y^2} = \frac{\partial F}{\partial x} + \frac{\partial G}{\partial y}. \qquad (13.12)$$

A solution for π in at least-squares sense is found at each altitude from Eq. 13.12 by utilizing Neumann (first derivative) boundary conditions over a generally irregular domain $[(x_1,x_2), (y_1,y_2)]$. Specifically, with F and G given by Eqs. 13.10 and 13.11, boundary conditions are

$$\frac{\partial \pi}{\partial x} = F \text{ at } x = x_1 \text{ and } x = x_2,$$

and

$$\frac{\partial \pi}{\partial y} = G \text{ at } y = y_1 \text{ and } y = y_2.$$

The solution for π is not unique; only the horizontal gradients of π are unambiguously determined. Therefore, the horizontal average of the solution $\langle\pi\rangle$, is calculated and subtracted from π to yield unique values, $\pi - \langle\pi\rangle$ at every point.

The deviation of buoyancy from its horizontal average, $B - \langle B\rangle$, can be calculated from the third equation of motion. If the horizontal average of the third equation is subtracted from the third equation itself and the buoyancy term is isolated on the left-hand side, the resulting expression for deviation buoyancy is

$$\begin{aligned} B - \langle B\rangle &\equiv (\theta - \langle\theta\rangle)\bar{\theta}^{-1} + 0.61(q_v - \langle q_v\rangle) \\ &\quad - (q_c - \langle q_c\rangle) \qquad (13.13) \\ &= \frac{c_p\bar{\theta}_v}{g}\frac{\partial}{\partial z}(\pi - \langle\pi\rangle) + R - \langle R\rangle, \end{aligned}$$

where R contains the remaining terms in the vertical momentum equation:

$$R = \frac{1}{g}\left(\frac{\partial w}{\partial t} + \mathbf{V} \cdot \nabla w - D_w\right) + q_r.$$

Here θ is the potential temperature, q_v the water-vapor mixing ratio, q_c the cloud-water mixing ratio, g the acceleration of gravity, D_w the turbulence term, and q_r the rainwater mixing ratio. The water vapor and cloud-water mixing ratio deviations are included in the solution buoyancy expression, since they cannot be deduced from radar data. The vertical pressure gradient term can be evaluated following solution of Eq. 13.12 on horizontal planes at each altitude. Other terms on the right-hand side can be calculated or estimated from Doppler observations. In particular, the q_r term is approximated by use of an empircal formula relating reflectivity and rainwater concentration.

The solution of Eq. 13.12 is complicated by the fact that analyzed Doppler velocity components within storms occupy irregularly bounded regions and often even contain data voids. In applications of this method to date, the data have been accepted as given along the generally irregular surfaces that separate grid points with and without information. Therefore, the Neumann boundary conditions are calculated from Eqs. 13.10 and 13.11 according to the local orientation of surfaces. Two consequences of this treatment are (1) that it is necessary to use an iterative method for solution of Eq. 13.12 rather than a direct method (which might be used in the regularly bounded case) and (2) that there is loss of solution at grid points near boundaries, owing to the finite differencing.

The solution for buoyancy given by Eq. 13.13 is appropriate for some purposes, although the potential temperature itself seems more generally useful. The cloud-water component of buoyancy is usually small compared with the contribution from potential temperature, since 3.3 g/kg of cloud is equivalent to about 1 K in temperature. The vapor contribution is most important in low levels where the vapor deviation from average sometimes attains about 3 g/kg, contributing 0.5 K to the buoyancy deviation, $B - \langle B \rangle$. If $B - \langle B \rangle$ is approximated by the potential temperature term alone, then θ can be calculated from $B - \langle B \rangle$, given $\langle \theta \rangle$. In theory, $\langle \theta \rangle$ can be determined from a one-point independent measurement of θ, given the $B - \langle B \rangle$ solution, although in practice a statistical determination of $\langle \theta \rangle$ utilizing numerous measurements of θ is preferred.

The solution of Eq. 13.12 produces the quantity π from which $\pi - \langle \pi \rangle$ is subsequently calculated. It is necessary that the averaging to produce $\langle \pi \rangle$ be performed over the region enclosing points with data at all altitudes so that $\pi - \langle \pi \rangle$ can be related from level to level. If $\langle \pi \rangle$ were zero or the same value at all altitudes, then the averaging in Eq. 13.13 could be dropped and Eq. 13.13 solved for B itself rather than $B - \langle B \rangle$. In other words, the only reason for casting Eq. 13.13 in terms of deviations from horizontal averages is the z dependence of $\langle \pi \rangle$. In general, in the thunderstorm case $\langle \pi \rangle$ is nonzero and may vary from level to level because the storm volume defined by the region enclosing points with data at all altitudes is only a portion of the volume containing the complete storm-related π perturbation. In the boundary layer where the velocity-filled domain size may be considerably greater than the scale of perturbations in the flow, the assumption that $\langle \pi \rangle = 0$ is approximately accurate. In summary, in thunderstorms, horizontal gradients in π and B are accurately calculated, but vertical gradients in both quantities are affected by variations in $\langle \pi \rangle$ with height. Caution must be exercised so that solutions are interpreted as deviations from horizontal averages rather than perturbations from a base state.

b. Dynamic Method B

A second approach to solution of the three component momentum equations for pressure and buoyancy was developed by Brandes (1983, 1984). The momentum equations in vector form in this scheme are written

$$\frac{d\mathbf{V}}{dt} = -c_p\bar{\theta}_v\nabla\pi + \mathbf{D} + \mathbf{k}g\left(\frac{B}{\bar{\theta}} + q_c\right), \quad (13.14)$$

where $\bar{\theta}_v$ is the base-state virtual potential temperature, D represents subgrid-scale turbulent forces, q_c is mixing ratio of all condensate, and the buoyancy is given by

$$B = \theta' + a\bar{\theta}q_v', \quad (13.15)$$

where primed quantities represent deviations from base-state profiles (i.e., $\theta' = \theta - \bar{\theta}$) and $a = 0.61$.

A three-dimensional elliptic equation for perturbation pressure is derived by taking the divergence of Eq. 13.14 and combining with the anelastic continuity equation, given by

$$\nabla \cdot \rho\mathbf{V} = 0.$$

The resulting expression for calculation of perturbation pressure is

$$\nabla^2\pi + \frac{\partial ln(\bar{\rho}\,\bar{\theta}_v)}{\partial z}\frac{\partial\pi}{\partial z} = \frac{1}{c_p\,\bar{\theta}_v}\left\{-\frac{1}{\bar{\rho}}\nabla\cdot\bar{\rho}(\mathbf{V}\cdot\nabla)\mathbf{V} + \frac{g}{\bar{\rho}}\frac{\partial}{\partial z}\left[\bar{\rho}\left(\frac{B}{\bar{\theta}} - q_c\right)\right] + \frac{1}{\bar{\rho}}\nabla\cdot\bar{\rho}D\right\}. \quad (13.16)$$

In applications to date, Eq. 13.16 has been solved by relaxation with the use of Neumann boundary conditions at lateral, upper, and lower boundaries. Boundary values for $\partial\pi/\partial x$, $\partial\pi/\partial y$, and $\partial\pi/\partial z$ are calculated from the three components of the equation of motion, a procedure analogous to that used in method A above. There is a dependence on buoyancy and q_c in Eq. 13.17 and in the top and bottom boundary conditions; however, there is no dependence on local time derivatives, as in Eq. 13.11. Solution of Eq. 13.16 yields values of π that include an arbitrary constant over a three-dimensional volume. Deviations from a volume average can be calculated for comparison of solutions at different times. Pressure gradients are known from the solution in both the horizontal and the vertical, in contrast to solutions of Eq. 13.11, which produce horizontal pressure gradients and vertical gradients of deviation-from-average pressure.

Buoyancy values for the solution of Eq. 13.16 are provided by the solution of a two-dimensional elliptic equation over horizontal planes at needed levels in the storm volume. This elliptic equation is formed by dot multiplication of the curl of the vorticity equation by unit vector **k** (z direction),

$$\nabla_H^2 B = \frac{\bar{\theta}}{g}\{\mathbf{k}\cdot\nabla\times\left[\frac{d\boldsymbol{\omega}}{dt} - (\boldsymbol{\omega}\cdot\nabla)\mathbf{V} + (\nabla\cdot\mathbf{V})\boldsymbol{\omega}\right] + \mathbf{k}\cdot\nabla\times\nabla\times[\mathbf{D} - \mathbf{k}gq_c]\}, \quad (13.17)$$

where $\boldsymbol{\omega}$ is the vorticity vector. A pressure-dependent term in this equation has been found to be very small and is omitted. Therefore, buoyancy can be calculated without knowledge of the pressure field or in situations where pressure is not needed. The boundary condition $B = 0$ has been assumed in application of this method on each horizontal plane. This assumption is apparently appropriate except in areas where strong forcing extends outside the computational grid. In applications of this method boundaries of the computational volume have been extended outward from the Doppler-observed volume by a grid-filling procedure to produce a regularly bounded array. This procedure results in a reduction of boundary influences on the interior solution.

The condensate mixing ratio, q_c, is needed in both Eq. 13.16 and Eq. 13.17. An empirical formula relating q_c to reflectivity,

$$q_c = \frac{3.91 \times 10^{-9}}{\bar{\rho}} Z^{0.55}, \quad (13.18)$$

is used at all grid points where Z is measured. Once again, the presence of hail and of cloud droplets can result in erroneous values of q_c; however, errors in the π and B solutions are expected to be relatively small. The turbulence terms have been computed by the use of Schlesinger's (1978) anelastic adaptation of the Deardorff (1970) scheme. Estimates for the $d\boldsymbol{\omega}/dt$ term needed for the buoyancy solution have been obtained by calculating uncentered differences in time (by necessity) from consecutive data collections.

c. Dynamic Method C

A method very similar to that of Gal-Chen (1978) described in Sec. 3a above, has been employed by Roux et al. (1984). It uses analyzed velocity fields from dual Doppler radar, and the horizontal momentum equations are expressed in a form similar to Eqs. 13.10 and 13.11:

$$\nabla_H(\pi) = \mathbf{A}_H, \quad (13.19)$$

where

$$\mathbf{A}_H = \frac{-\left(\frac{d\mathbf{V}}{dt} - \mathbf{D}\right)}{c_p\bar{\theta}_v}.$$

A solution for π is obtained through minimization of the functional,

$$J = \iint_{D(z_0)} [\nabla_H(\pi) - \mathbf{A}_H]dxdy + \mu \iint_{D(z_0)} \left[\left(\frac{\partial^2\pi}{\partial x^2}\right)^2 + 2\left(\frac{\partial^2\pi}{\partial x\partial y}\right)^2 + \left(\frac{\partial^2\pi}{\partial y^2}\right)^2\right] dxdy, \quad (13.20)$$

where $\mathbf{A}_H$ has horizontal components A_x and A_y, $D(z_0)$ is the horizontal domain (at altitude z_0) where data $\mathbf{A}_H$ are available, and μ is a parameter controlling the cutoff wave number associated with the filtering brought about by the second term. The first term is identical to that producing the Euler equation (Eq. 13.12) in Gal-Chen's treatment. The purpose of the second term is to filter as far as possible the contribution of experimental errors in the determination of π. In practice Eq. 13.20 has been discretized by use of a finite-element approach (see, e.g., Strang and Fix, 1973). This avoids the need for integration boundary conditions and produces a particular solution π at each altitude whose mean value in the horizontal plane of interest is zero.

To obtain a solution for a temperature variable, the vertical component of the equation of motion is horizontally differentiated to produce

$$\nabla_H \theta_{cl} = \mathbf{B}_H, \quad (13.21)$$

where

$$\mathbf{B}_H = (c_p\bar{\theta}_v^2/g)\left[\frac{\partial}{\partial z}(\nabla_H\pi)\right] + \bar{\theta}_v\nabla_H\left(\frac{dw}{dt} + gq_r - D_w\right) \quad (13.22)$$

and

$$\theta_{cl} = \theta_v - q_c\bar{\theta}_v.$$

The quantity θ_{cl} is the "virtual cloud temperature" perturbation, which takes into account the unknown cloud-water content q_c; q_r is the precipitating water content, which can be estimated from reflectivity measurements; θ_v is the virtual potential temperature. $\mathbf{B}_H$ can therefore be calculated since $\nabla_H\pi$ is deduced from Eq. 13.19, and other terms can be computed from Doppler radar measurements. In practice the methods for solving Eqs. 13.21 and 13.19 are identical; both use a functional completely analogous to Eq. 13.20. Solutions for both pressure and temperature contain a constant that may vary from level to level, as was the case in dynamic method A; therefore, caution must be exercised in interpreting vertical gradients in these calculated fields.

d. Sensitivity of Solutions to Errors

The quality of retrieved pressure and buoyancy fields depends directly on the quality of analyzed velocity fields. In all the methods described, the order of differentiation involved in the buoyancy calculation is higher than in the pressure calculation. Therefore, since some noise is present in analyzed velocity fields, buoyancy solutions in general contain more variability than pressure solutions do. When retrieval techniques were first suggested, many researchers were pessimistic about obtaining even pressure solutions that are realistic, because of the combined effect of high-order derivatives and noisy data. Not often recognized was that differentiation in the procedure is followed by integration, and the latter smooths the fields.

Sources of error other than noisy velocity fields are also present in the application of these methods to observed convective circulations. In certain cases volume scans by the Doppler radars are not frequent enough to provide estimates of time-change terms. Omission of these terms can result in significant solution error, as demonstrated by Gal-Chen (1978) and Hane et al. (1981) in tests using numerical model output in place of observations, and by Gal-Chen and Kropfli (1984) using boundary layer data. In applications of methods B and C, estimates of time changes have been included in the calculations, but the sensitivity to inclusion of these terms has not been reported. Another source of error is uncertainty inherent in the turbulence parameterization used in a particular method. Tests by Hane et al. (1981) suggested that errors from this cause are significantly less than those owing to omission of local time changes or to noisiness and inaccuracies in the velocity fields. Errors also arise in buoyancy calculations, owing to inability to measure cloud-water mixing ratios and to estimate rainwater mixing ratios accurately. These errors are probably relatively small, in approximately the same range as turbulence-related errors.

The effect of velocity errors on retrieved pressure gradients has been assessed in the case of Gal-Chen's method by calculation of the nondimensional quantity

$$E_r = \frac{\iint \left[\left(\frac{\partial \pi}{\partial x} - F \right)^2 + \left(\frac{\partial \pi}{\partial y} - G \right)^2 \right] dxdy}{\iint (F^2 + G^2)\, dxdy}, \qquad (13.23)$$

where F and G are defined in Eqs. 13.10 and 13.11. The $\partial\pi/\partial x$ and $\partial\pi/\partial y$ in this expression are evaluated directly from the retrieved pressure. Thus E_r is a measure of how well the retrieved pressure gradients fit the individual momentum equations. If the velocity fields are error-free and complete, then $E_r = 0$ should result. An E_r value less than 0.25 is generally considered to represent a good fit. A rationale for setting an upper bound for E_r at 0.5 (corresponding to useful information) was presented by Gal-Chen and Kropfli (1984). At the very least, E_r can be considered a relative measure of the goodness of fit for the pressure solution. In applications of retrieval methods to boundary layer and thunderstorm data, calculated E_r values have generally been in the 0.10 to 0.30 range.

Roux et al. (1984) have approached the assessment of error by calculating uncertainties in each of the three components of velocity; this leads to estimates of uncertainty in the derived pressure and temperature fields. In application of their method to a West African squall line they deduced uncertainties of 0.1 to 0.2 mb in derived pressure at various altitudes and uncertainties of 0.3 to 0.5°C in "cloud" temperature gradient. Hane et al. (1981) inserted random noise in velocity fields produced by a numerical model and assessed the effect on solution fields. The correct solution (known from model output) was lost unless filtering was applied to solutions whose correponding velocity input contained noise of known wavelength. It is recognized that velocity errors are often probably in the form of bias errors rather than random errors.

4. Microphysical Retrieval Method

The microphysical retrieval method developed by Ziegler (1984) employs a warm-cold cloud numerical kinematic model based on continuity equations for heat and the water substance. It is thus fundamentally different from the dynamic retrieval method, which is based on the momentum equations. The model is time-dependent and three-dimensional and incorporates thermodynamic and transportive processes, parameterized microphysics, and Doppler radar observations of velocity. The current version of the model includes an expanded ice-phase parameterization (Ziegler, 1985). The system of continuity equations depicts variations of potential temperature θ, water-vapor mixing ratio q_v, cloud droplet total number concentration N_c and mixing ratio q_c, raindrop total number concentration N_r and mixing ratio q_r, hail total concentration N_h and mixing ratio q_h, ice mixing ratio q_i and (nonprecipitating) cloud-ice mixing ratio q_x. Units of temperature (K), concentration (cm^{-3}), and mixing ratio (gg^{-1}) are conventional. The model domain corresponds to the Doppler radar observation domain.

The cold-frontal rainband model developed by Rutledge and Hobbs (1983) is analogous to the microphysical retrieval scheme discussed in this section. The rainband model is two-dimensional and does not employ multiple Doppler-derived winds, though it would be straightforward to increase the number of dimensions.

a. The Kinematic Framework

The prognostic equations for θ, q_v, N_c, q_c, N_r, q_r, q_x, q_i, N_h, and q_h can be effectively represented in the form

$$\frac{\partial \phi}{\partial t} = -u\frac{\partial \phi}{\partial x} - v\frac{\partial \phi}{\partial y} - w\frac{\partial \phi}{\partial z} + M_\phi + D_\phi, \qquad (13.24)$$

where ϕ represents the temperature, mixing ratio, or concentration; M_ϕ the microphysical source or sink rates;

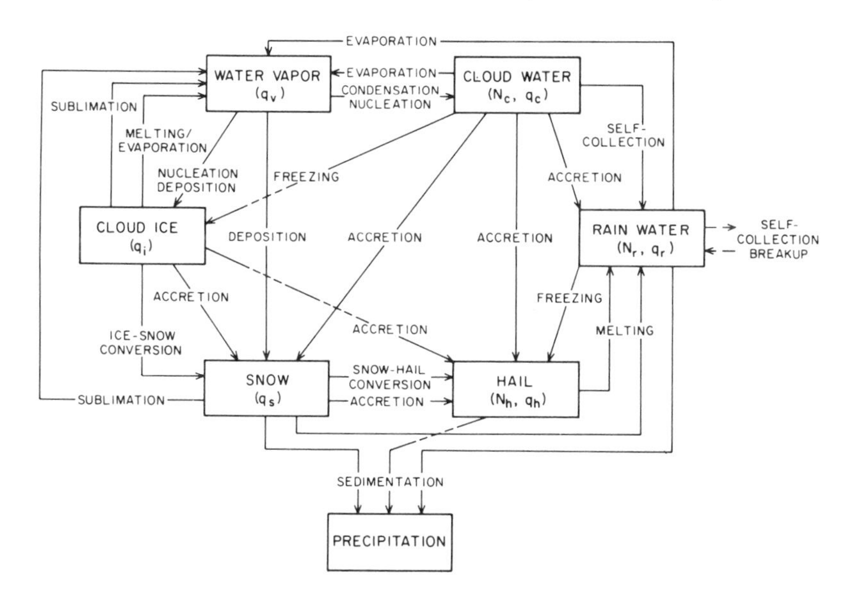

Figure 13.2. Major microphysical processes in the cloud model. Modeled species are in boxes, and arrows indicate change of species accompanying each microphysical process.

and D_ϕ the turbulent diffusion rates. The west-east, north-south, and vertical air motion components are denoted by u, v, and w, respectively. The diffusion terms are approximated by the expression

$$D\phi = \nabla_3 \cdot (K_d \nabla \phi),$$

where K_d is the eddy mixing coefficient. The value of K_d is parameterized in terms of the spatially variable deformation of the velocity field (Takahashi, 1981). In the kinematic approach to solution of Eq. 13.24 the airflow components appearing in the advection terms are specified from Doppler observations.

b. The Microphysical Parameterization

The cloud model parameterizes the water substance distribution as well as various transitions of form and phase. The parameterized scheme assumes a particular functional form of the continuous distribution, the variation of the distribution being defined by changes in power moments of the assumed distribution. The underlying physics is implicit in expressions for the power moment tendencies. Physical laws govern changes of individual particle mass or concentration; multiplication by the particle size distribution and subsequent integration over all particle sizes yield the total concentration and mass tendencies. In the alternative finite difference method, a discrete set of mass or size categories is used instead of continuous size distributions; the physical processes are applied directly to each category (Berry and Reinhardt, 1974). Figure 13.2 maps the various form and phase changes affecting water substance in the microphysical model.

The gamma function has been selected to represent model cloud and rain distributions (Williams and Wojtowicz, 1982; Willis, 1984); the inverse exponential function represents the model hail and ice distributions (Douglas, 1960; Houze et al., 1979). The functional form of the various distributions is assumed to be constant in space and time; only the distribution parameters vary. The gamma distribution used here has the form

$$n(v) = N(\nu + 1)^{\nu+1}(v/v_0)^{\nu} \exp[-(\nu + 1)v/v_0)]/\Gamma(\nu + 1)v_0, \quad (13.25)$$

where N is the total concentration (cm^{-3}), v_0 is the mean volume (cm^3), ν is the shape parameter, v is volume, and $n(v)$ is the concentration per unit volume interval. The inverse exponential distribution is given by

$$n(D) = N_0 \exp(-\lambda D) \quad (13.26)$$

where N_0 is the intercept parameter (concentration of particles per unit size range at zero size), λ is the slope parameter, and D is diameter. The value of ν is fixed in Eq. 13.25, while N and v_0 vary. The value of ν is chosen (in the range 0 to 3, typically) to maximize agreement between the simulated evolution of drop distributions in the parameterized model and a highly resolved finite difference model. The parameterization of hail allows independent variation of total concentration and mixing ratio, which is equivalent to independent variation of N_0 and λ. The ice parameterization fixes N_0 and varies λ only.

The more significant microphysical parameterizations and adjustments utilized in the model are summarized below:

1. Condensation-evaporation adjustment: Cloud condensation or evaporation occur as air parcels rise or sink moist-adiabatically. The Soong and Ogura (1973) moist adjustment procedure determines the variations of water vapor, cloud, and potential temperature under the constraint of exact saturation. The saturation ratio q_v/q_{vs} is computed

with a linearized form of Teten's formula after all variables have been advected and diffused. With supersaturation the decrease in q_v (with attendant increases in θ and q_c) necessary to maintain exact saturation is computed. In the event that cloud already exists, the explicit condensation rate associated with the known cloud droplet distribution is computed. The condensation rate is the minimum of the rate constrained by exact saturation and the explicit condensation rate. The principal advantage of this approximate adjustment scheme is that the exact saturation constraint is relaxed. Supersaturation is permitted in updraft regions, where intense scavenging so reduces cloud that it is incapable of consuming excess vapor as fast as it is supplied. The evaporation of cloud is assumed to be instantaneous. If cloud is consumed, rain evaporation may occur if $q_v < q_{vs}$. If saturation deficit exists with respect to ice, ice sublimation may occur.

2. Cloud nucleation parameterization: Cloud nucleation depends on the updraft strength (rate of supersaturation generation) and on thermodynamic properties at cloud inflow boundaries, or on supersaturation in the cloud interior (calculated by the condensation scheme). Concentration at inflow boundaries is determined from a power law relation with supersaturation and an estimate of peak supersaturation (Twomey, 1959).

3. Warm rain coalescence parameterization: The warm rain coalescence parameterization employs an analytic solution of the stochastic coalescence equation (SCE) that uses the gamma distribution and an accurate polynomial form of the collection kernel (Pruppacher and Klett, 1978). This analytic test function solution of the SCE offers a simple and straightforward conception of the underlying physics of the coalescence process. The Berry and Reinhardt (1974) concepts of cloud droplet self-collection or autoconversion, rain collection of cloud, and rain self-collection are used here. The drop breakup is parameterized by reducing the rate coefficient for self-collection of large hydrometeors, according to the coalescence efficiency for filament breakup (Brazier-Smith et al., 1972; Low and List, 1982). Breakup provides an asymptotically approached upper limit on drop size, reducing increase of large sizes as the frequency of breakup following coalescence increases.

4. Hail parameterization: The hail parameterization represents dry and wet hail growth by using the threshold diameter D_w for the onset of wet growth from Lagrangian hail growth theory (Nelson, 1983). The total hail distribution is subdivided by D_w into two parts that are characterized by either dry or wet growth. Approximate effects of low-density riming on the dry growth rate are included. Hail accretes ice during wet growth only. Hail melting considers the effects of conduction, forced convection, and evaporation. Frozen raindrops are transferred to the hail distribution with conservation of number and mass.

5. Ice parameterization: Ice particles form by nucleation and grow by vapor diffusion through the Bergeron process and by riming of cloud water. The mass residing in the ice distribution at diameters exceeding an assumed embryo size is transferred to the hail distribution by the ice-hail conversion process. The deposition growth parameterization models individual growth as a function of temperature and crystal habit, though individual ice-crystal habits are not considered. Melting and sublimation of ice are also modeled.

6. Cloud-ice parameterization: Small cloud-ice particles form by the spontaneous freezing of cloud water at or below $-40°C$. Sublimation of cloud ice occurs below ice saturation. Cloud ice is assumed to melt instantaneously at or above freezing and to evaporate in an environment below water saturation.

7. Hydrometeor sedimentation: The sedimentation rates for rain, hail, and ice mixing ratios are all in the well-known flux divergence form (e.g., Klemp and Wilhelmson, 1978a),

$$(M_\phi)_{\text{sed}} = \frac{1}{\bar{\rho}}\frac{\partial}{\partial z}(\bar{\rho}V\phi), \tag{13.27}$$

where V is the distribution-averaged fallspeed, ρ is air density, and ϕ represents the rain, hail, or ice mixing ratios. The overbars denote base-state quantities. Sedimentation rates of precipitation concentration are represented by terms analogous to Eq. 13.27, though average air density does not appear in those expressions.

The manner in which the individual microphysical processes contribute to the source-sink terms in Eq. 13.24 appears in the Appendix, Sec. 3.

c. Model Solution

The finite-difference form of the model is integrated forward in time from a specified initial state to produce the solution fields of temperature and water substance. Second-order leapfrog time and fourth-order space differencing is employed in the finite-difference equations (Klemp and Wilhelmson, 1978a). The model continuity equations are prognostic in the microphysical variables whose diagnosis is sought. This situation contrasts with the dynamic retrieval method, wherein π and B are diagnostically determined from the momentum equations.

Several solution methods are available. One technique involves fixing the observed storm-relative wind and integrating the model to the steady state. This is the simplest and most common approach, although the assumption of a fixed wind field results in varying degrees of error, owing to effects of probable transients in the true temperature and water substance fields. An alternative method allows wind nonsteadiness by inserting time-spaced wind analyses during model integration from a suitable initial condition. The initial fields might be the steady model output corresponding to the fixed initial wind analysis, perhaps modified by comprehensive in situ microphysical observations.

d. Retrieval Error Sources

Primary sources of retrieval error are faulty approximation of local time changes of derived quantities and the wind field, errors in wind analysis, and errors attributable to the parameterization of microphysical processes. The steadiness assumption precludes possible influence of storm history on storm structure at the retrieval time; the influence of such transient characteristics is highly variable. Analyzed winds possess varying amounts of random and/or bias errors. Retrieved quantities, particularly the thermal variables, are tracers whose variation in space is fundamentally constrained by characteristic curves of the solution of the advection equation and by the boundary values. Both microphysical and dynamical retrieval proceed from the velocity field, but microphysical retrieval is less sensitive to random errors than dynamical retrieval because, in the former case, solution variables evolve to a steady state through a time-dependent calculation. Thus, in microphysical retrieval, the dependent variables are globally, rather than locally, constrained by the wind field. The advection terms contain undifferentiated wind components, so that possible wind errors are not amplified by differencing, as in the dynamic method. On the other hand, wind bias errors may incorrectly displace characteristic curves and their boundary intersection points, producing significant retrieval errors. Incorrect microphysical parameterization may produce retrieval errors that follow the motion along characteristic lines.

A warm-cloud version of the retrieval scheme was tested by Ziegler (1984), who used a dynamically simulated cloud as input. Typical temperature and water-vapor mixing ratio errors in the cloud, traceable mainly to the steadiness assumption, were of order 0.1°C and 0.1 g kg^{-1}, respectively. Rainwater errors were typically 0.5 g kg^{-1}. In other tests retrieval errors were slightly amplified by the addition of a random error ($\sigma = 1$ m s^{-1}) to the input wind field, but no overall changes occurred. Errors in the thermal fields, introduced by the steadiness assumption, are relatively small because the advective time scale is much shorter than the period over which the wind field fluctuates. Precipitation fallout is characterized by a relatively long time scale, so that larger rainwater errors result. During model integration with a varying wind field, initial error owing to assumed steadiness propagates into the solution in time.

The retrieval scheme and dynamic simulation used identical microphysical formulations, though natural processes are imperfectly represented by the warm-cloud parameterization. Additional tests have employed observed wind fields to measure the sensitivity of model output to different formulations and/or exclusion of certain key microphysical processes. The thermal fields are rather insensitive to those effective changes of the form and phase of hydrometeors, since the vapor-liquid (condensation) phase transition predominates in the continuity equations. Precipitation content is conversely rather sensitive to the parameterization used. In particular, the hail distribution responds strongly to parameterization of riming density and efficiency. Because of nonlinear error propagation, retrieval verification with in situ observations may be the only feasible means of gagging the parameterization error. Although individual microphysical processes may be imperfectly represented, several case studies have nonetheless demonstrated the model's ability to explain precipitation structure within a variety of storm types.

5. Application of Dynamic and Microphysical Retrieval to a Common Data Set

To illustrate deductions from Doppler velocity fields, the microphysical method of Ziegler (1984) and the dynamic method of Hane et al. (1981) are here applied to the same analyzed wind fields from the Del City, Okla., tornadic storm of 20 May 1977. Application of the two methods results in three-dimensional fields of all the major thunderstorm variables. The merit of combining the two techniques thus lies in having the unprecedented ability to examine the interrelation of the various fields. This availability of all fields has implications in the initiation of cloud-scale numerical simulations and in other future areas of thunderstorm research. The velocity fields are taken from the analysis of Ray et al. (1981), who also examined other storms that occurred on the same day. Fields shown here are primarily at the 1833 CST analysis time, approximately 7 min before the start of the storm's tornadic stage. The wind field at this analysis time was derived with the methodology for two Doppler radars (as described in Sec. 2a above).

a. Presentation of Retrieved Fields

At 1833 CST the Del City storm contained a well-developed mesocyclone located generally 10 to 12 km south of the low-level maximum in reflectivity. Figure 13.3 shows the storm structure at this time, including horizontal distributions of reflectivity, storm relative wind vectors, and vertical wind speeds at 2 km height. The updraft maximum at this height is several kilometers northeast of the mesocyclone in the strong relative flow from the south-southeast. The location of maximum upward motion slopes toward the north-northwest with height. Maximum downdrafts are north-northwest of the mesocyclone in a rainy area. The characteristic "notch," north of the mesocyclone, is clearly visible in the reflectivity pattern.

The derived pressure field at 5.5-km height is shown in Fig. 13.4. The isopleths represent pressure deviation from horizontal average at 0.5-hPa (mb) intervals. The strong updraft region extends southward from the circled ×, which marks its maximum, and is located in the region of strong gradient between high and low pressure. The pressure difference across the updraft region is approximately 5 hPa. Linear theory (Rotunno and Klemp, 1982) predicts that the

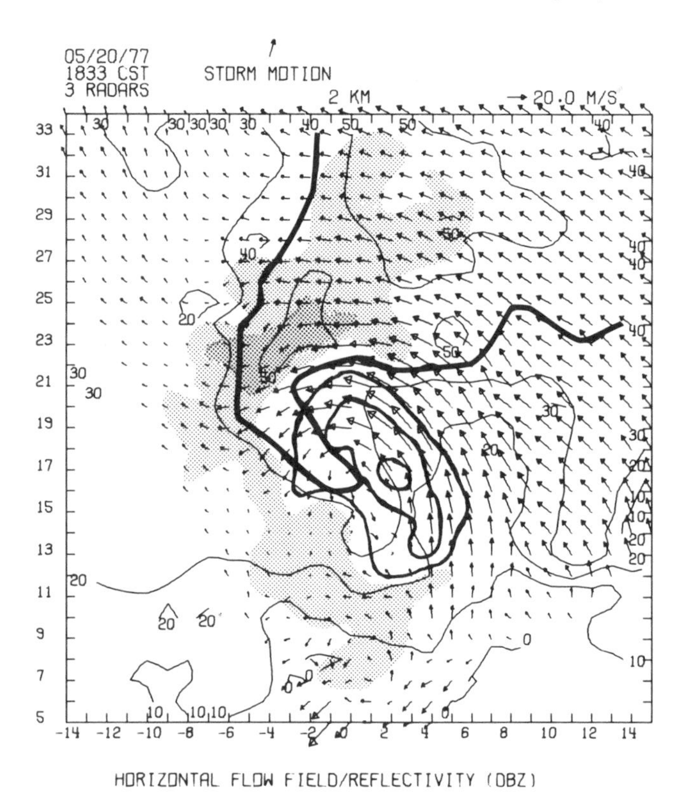

Figure 13.3. Storm relative Doppler-derived winds at 2 km for the Del City storm at 1833 CST on 20 May 1977. Areas with downdrafts of less than -1 m s^{-1} are stippled. Darker stippling indicates downdrafts less than -5 m s^{-1}. Updrafts of 5, 10, and 15 m s^{-1} are indicated by the medium-thick contours. Reflectivities are contoured from 0 to 50 at 10-dBZ intervals, the heavy contour represents 40 dBZ (after Ray et al., 1981).

maximum pressure gradient will be oriented along the direction of the environmental shear vector at each height in the storm, and that is found to be very nearly the case in this storm (Hane and Ray, 1985). The heavy solid line is the area of maximum vertical vorticity, which lies generally to the right of the environmental shear vector through the updraft maximum.

Figure 13.5 illustrates the pressure structure present in the tornadic stage of this storm at 1847 CST. The wind analysis was carried out with the methodology for three Doppler radars described in Sec. 2b above. The pressure field shown is at the lowest altitude available (500 m above the surface) and was chosen because it illustrates the very strong pressure lowering that has taken place in the mesocyclone region. This lowering results from intense positive vertical vorticity in this region; the maximum vorticity is indicated by the $\oplus$ near coordinates $x = 3$, $y = 30$. A secondary region of strong vorticity is indicated farther east, ahead of the gust front, which has rotated around the southeast portion of the strong circulation. The pressure lowering near the ground contributes to an increasing perturbation pressure with height in this region and therefore to downward acceleration of air, promoting a localized downdraft formation, noted by Klemp and Rotunno (1983) in numerical simulations and by Brandes (1983, 1984) in analyzed observations. The axis of maximum upward motion is indicated in Fig. 13.5 by the solid line that forms an arc generally in the northeast quadrant of the mesocyclonic circulation. Minimum pressure is also approximately configured in an arc about 1 km radially outward from the vertical velocity arc. Higher pressure is located north-northeast of the mesocyclone area in rain-cooled air and in the area behind the gust front south and southeast of the mesocyclone.

The internal consistency of the pressure field can be checked in a relative sense by calculating E_r, given in Eq. 13.23. Results of the E_r calculation at the 1833 and 1847 CST analysis times are presented in Fig. 13.6. Lower values of E_r indicate a better fit of the solution pressure field to the individual horizontal momentum equations. The best fit is obtained at 1833 in low altitudes, but low E_r values are obtained at high altitudes of the storm at both times. The higher E_r values at 1847 CST at low altitudes are attributed either to errors in the velocity field stemming from a location of the storm less favorable for radar observation or to neglect of time tendencies in the momentum equation in a region where significant changes may be occurring. A slight

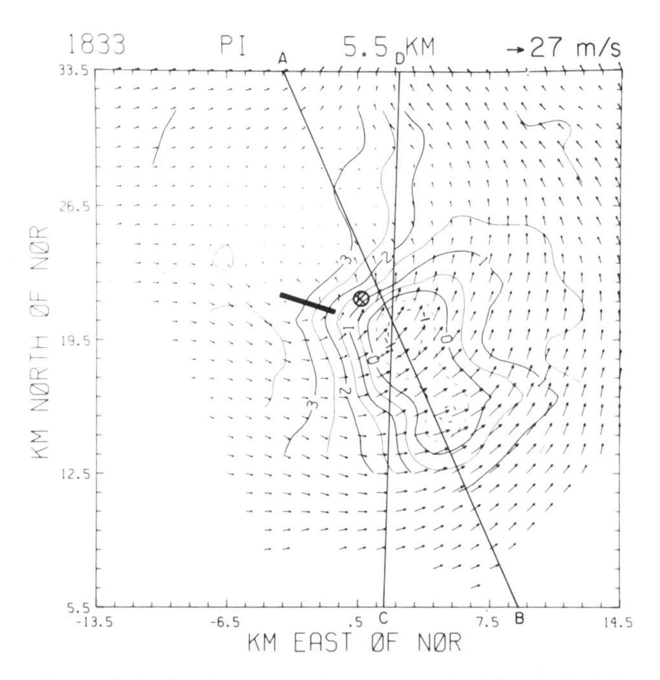

Figure 13.4. Retrieved deviation pressure field for the Del City, Okla., storm at 1833 CST. Contours are at 0.5 hPa; altitude is 5.5 km. Storm relative wind vectors are superimposed. The circled $\times$ marks the maximum updraft; the bar marks the area of maximum vertical vorticity. Lines A–B and C–D locate the vertical cross sections shown in Figs. 13.7–13.11.

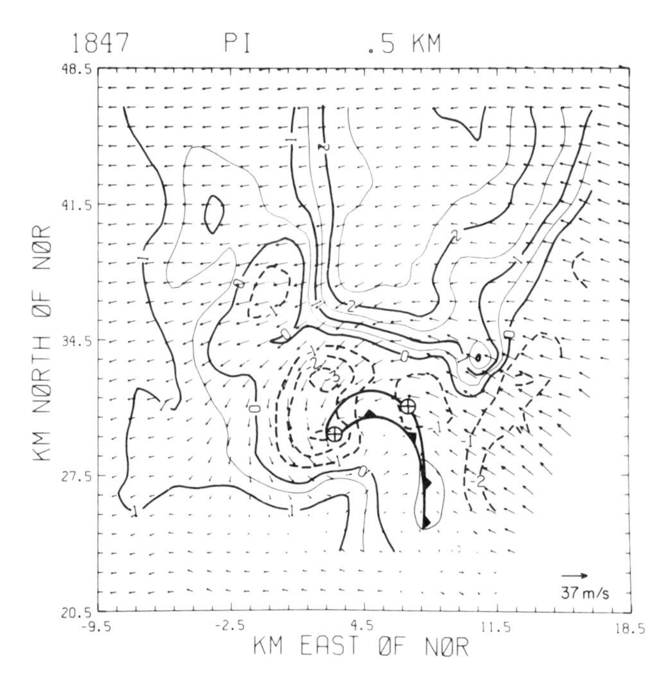

Figure 13.5. Retrieved deviation pressure field for the Del City storm at 1847 CST. Contours are at 0.5 hPa in positive areas and at 1 hPa in negative areas; altitude is 0.5 km. Approximate location of gust front is marked with standard frontal symbols. The circled + marks the locations of maximum vertical vorticity, and the heavy solid line marks the location of maximum updrafts. The approximate coordinates of the tornado at this time are $x = 3.5$, $y = 29.5$.

improvement in the pressure fit is also indicated with inclusion of turbulence terms in the solution equations, compared with neglect of turbulence.

A comparison of observed and microphysically retrieved radar reflectivity at 1833 CST is presented in the vertical cross section A–B in Fig. 13.7a, b. The A–B cross section is oriented roughly northwest-southeast, as illustrated in Fig. 13.4. There are striking similarities between the observations (Fig. 13.7a) and the retrieved reflectivity (Fig. 13.7b) obtained by summing the partial theoretical reflectivities of rain and hail. The precipitation core tilts slightly to the south with height from (x,y,z) coordinates $(-2, 26.2, 0)$ to $(1.4, 22.5, 12)$, aligning along the principal axis of flow deformation in the plane and illustrating the overwhelming control of advection in redistributing existing precipitation. Small-scale features such as the southward extension of the core to $(4.7, 15.2, 5)$, the shallow bounded weak-echo region beneath this overhang, and the indentation of contours near $(3, 18.9, 7)$ are well reproduced. Extensive weak stratiform precipitation impinging on the Del City storm core north of $(-0.2, 26.2)$ is not retrieved because the storms producing the associated hydrometeors are outside the microphysical retrieval domain.

Retrieved cloud-water and cloud-ice mixing ratios in the A–B vertical cross section are shown in Fig. 13.8. Cloud-water mixing ratios exceed 3.5 g kg^{-1} at 5 km; cloud-ice mixing ratios exceed 1.5 g kg^{-1} at 11 km. Ice crystal mixing ratios (not shown) attain maximum values of order 0.1 g kg^{-1}. Cloud-water mixing ratios outside the plane (not shown) exceed 4 g kg^{-1}. Retrieved cloud-water mixing ratios are roughly 50 percent of their adiabatic value in the middle-level cloud core. This reduction is due mainly to accretion by rain and hail. Updrafts on the northwest side of the storm above 6 km are cloud-free because dry upper tropospheric air is being ingested. Also note that the maximum cloud-water mixing ratios are not coincident with the strongest updraft as suggested by simple parcel theory.

The regions of precipitating water substance in the A–B vertical cross section are illustrated in Fig. 13.9. Retrieved mixing ratios for rain and hail and the region of cloud (cloud water plus cloud ice) exceeding 0.5 g kg^{-1} are presented in Fig. 13.9a; the mixing ratios for rain derived from measured reflectivity and a standard Z-q relation are illustrated in Fig. 13.9b. Retrieved rain and hail mixing ratios locally exceed 7 g kg^{-1} in the updraft core, and the rain core reaches the surface within the low-level downdraft. The precipitation core is northwest of the cloud center, the cloud base being rather precipitation-free and the low-level precipitation core being cloud-free. The peak retrieved mix-

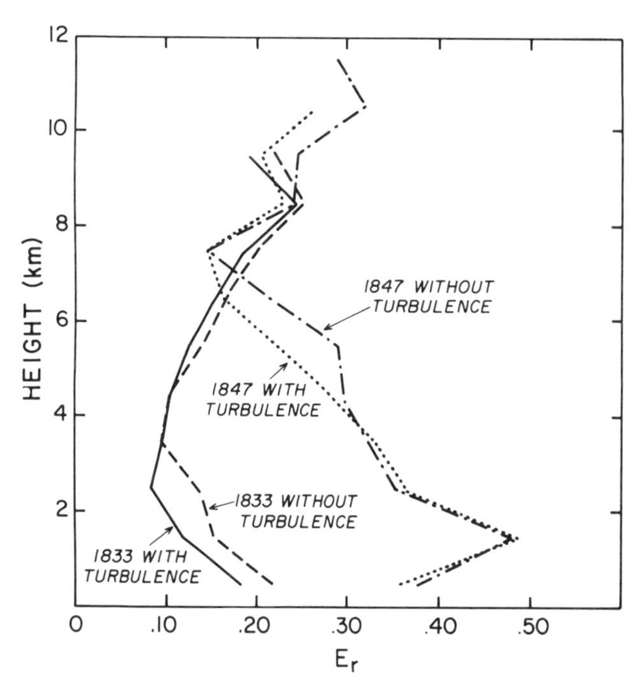

Figure 13.6. Height distribution of error parameter associated with the retrieved pressure. Smaller E_r values indicate a better "fit" for the pressure solution. Curves indicate distribution at both 1833 CST and 1847 CST for cases with and without turbulence (after Hane and Ray, 1985).

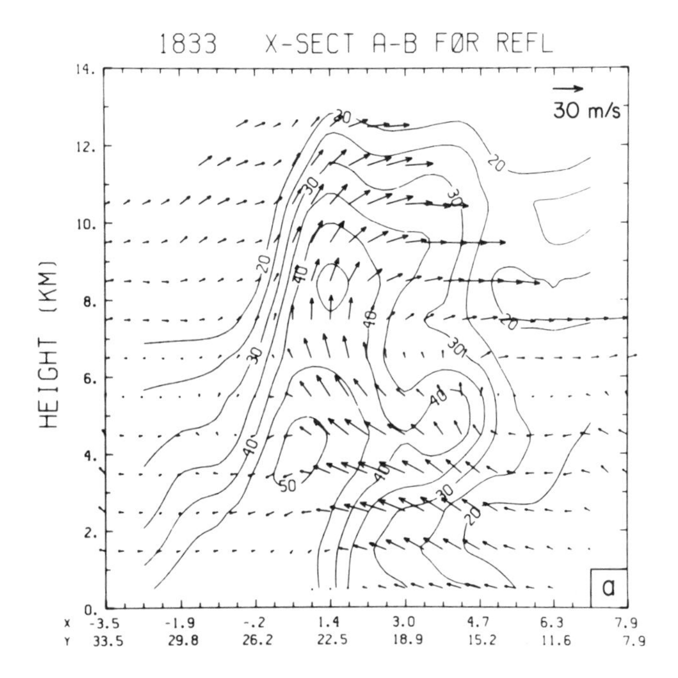

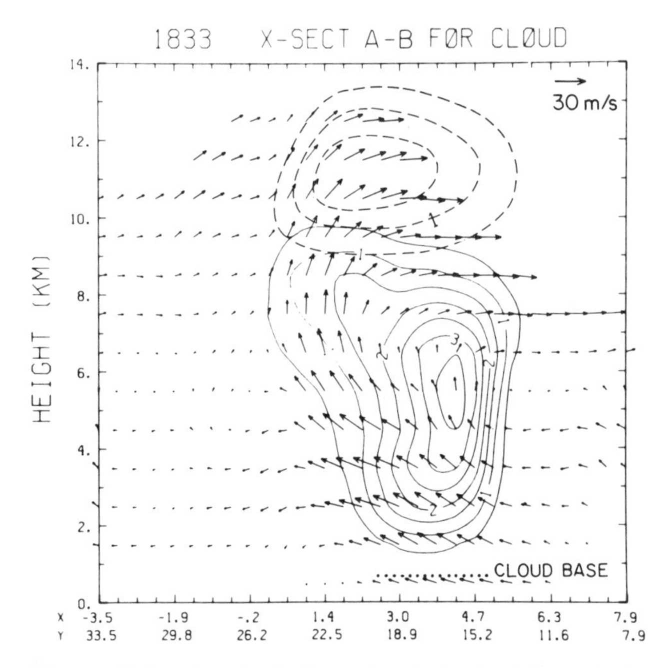

Figure 13.8. Microphysically retrieved cloud mixing ratio for cloud liquid water (solid contour) and for cloud ice (dashed contour) in the vertical cross section A–B. Contour interval and starting contour are 0.5 g kg^{-1} for each hydrometeor. The horizontal dotted line locates cloud base estimated from retrieval output.

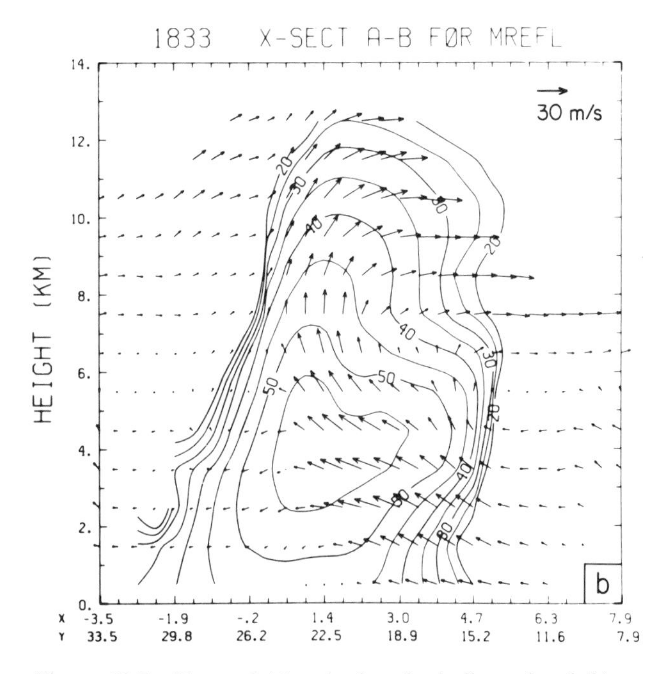

Figure 13.7. Observed (a) and microphysically retrieved (b) radar reflectivity (dBZ) in the vertical cross section A–B oriented as in Fig. 13.4. Horizontal axis indicates (x,y) distance (km) relative to Norman, Okla.; the (stretched) vertical axis indicates height (km AGL). Contour interval is 5 dBZ. Observed airflow is added for reference.

ing ratios for rain are roughly double the corresponding rain content determined from observed reflectivity and an assumed drop distribution. This difference in rainwater mixing ratio might result from a faulty assumption about drop size distribution, erroneous enhancement or retrieved values owing to changes in storm intensity, or uncertainties of radar calibration.

The distribution of retrieved thermal variables in the Del City storm is illustrated in Fig. 13.10 by the equivalent potential temperature (θ_e) field in vertical cross section C–D, oriented as in Fig. 13.4. The θ_e values were derived from retrieved potential temperatures and water-vapor mixing ratio by using Bolton's (1980) formula. Airflow vectors in the plane are also plotted in Fig. 13.10 for reference. The updraft core is essentially undiluted through most of its depth, and inflow air originates below 1 km. The peak θ_e value of approximately 340 K increases by about 1 K in the vicinity of the 8-km hail core (not shown), owing to combined effects of a strong water saturation tendency in the updraft and the fusion heating that accompanies rain freezing to hail and hail riming. Note that the middle-level environmental air is potentially cold with θ_e below 326 K; the thermally buoyant updraft core illustrates the dynamic consistency of results.

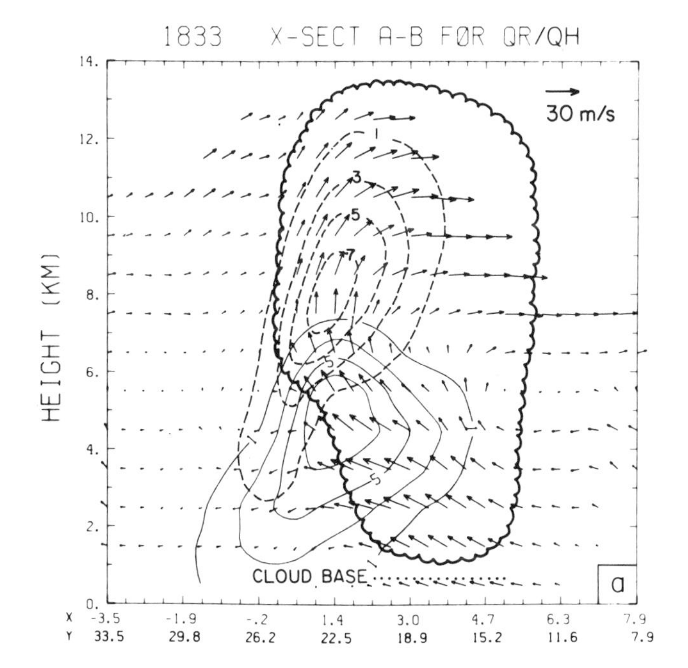

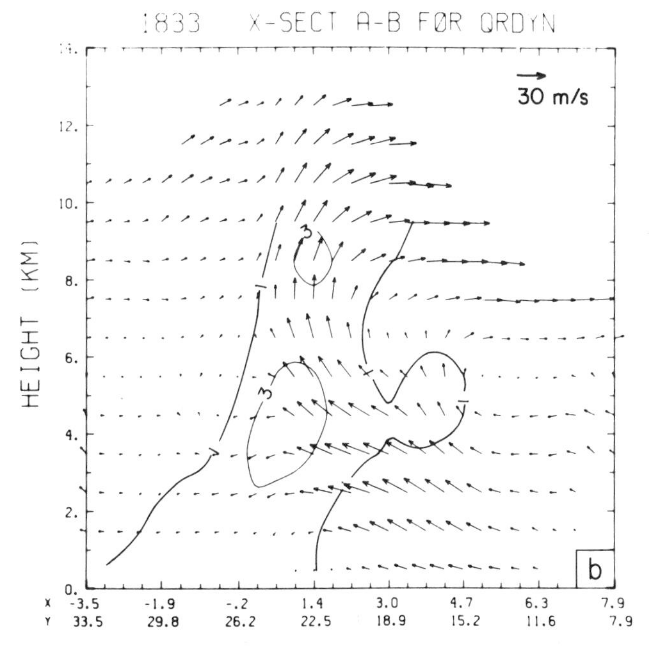

Figure 13.9. Precipitation contents from the microphysical retrieval (a) and derived from reflectivity and standard Z-q relation (b), in the vertical cross section A–B. Retrieval output (Fig. 13.9a) indicates mixing ratio for rain (solid contour, 2 g kg^{-1} interval), for hail (dashed contour, 2 g kg^{-1} interval), and for cloud water (0.5 g kg^{-1}, scalloped curve). Rain content derived from reflectivity (b) is indicated by solid contour having a 2 g kg^{-1} interval.

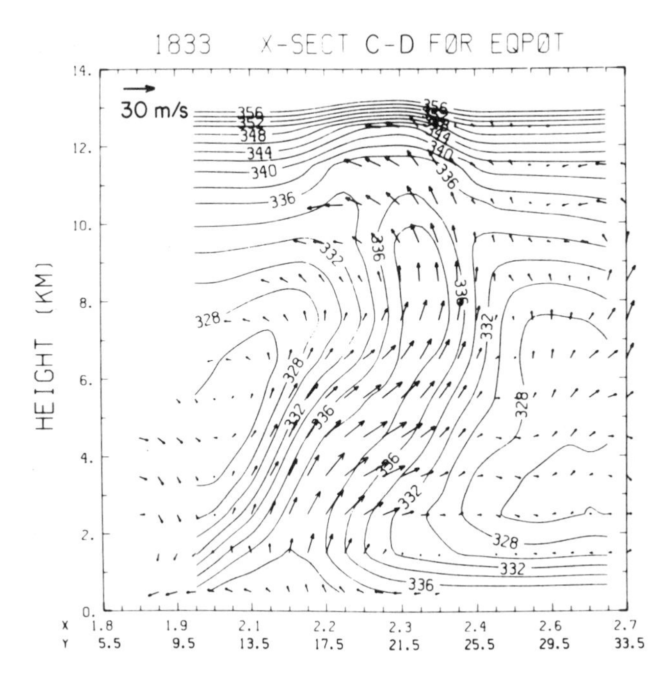

Figure 13.10. Equivalent potential temperature (θ_e) and airflow in the vertical cross section C–D, oriented as in Fig. 13.4. Solid θ_e contours have intervals of 2 K starting at 326 K.

b. Comparison of Dynamically and Microphysically Retrieved Fields

Temperatures and hydrometeor mixing ratios are related to buoyancy accelerations and airflow changes through the vertical equation of motion. Common buoyancy quantities of the dynamical and microphysical retrieval schemes illuminate the contrasting diagnoses resulting from application of independent constraining equations to a common wind field. The temperature-equivalent total thermal buoyancy and the potential temperature are selected as vehicles in the comparison.

The total thermal buoyancy may be derived by the dynamic retrieval method or computed directly from microphysical retrieval results. This quantity has dimensions of temperature and is defined by the expression $T = \theta - \langle\theta\rangle + 0.61\,\overline{\theta}\,(q_v - \langle q_v\rangle)$. (The dynamic relation of T to buoyancy is given by Eq. 13.13; θ and q_v are obtained through microphysical retrieval [Eq. 13.24].) Microphysical and dynamical derivations of T in the vertical cross section A–B are compared in Fig. 13.11. Vector velocities in the plane have been added for reference. The microphysical retrieval (Fig. 13.11a) deduces a warm, moist updraft core in middle levels where $T > 3$°C and a negatively buoyant, high-level, penetrative updraft. A cold region where $T <$

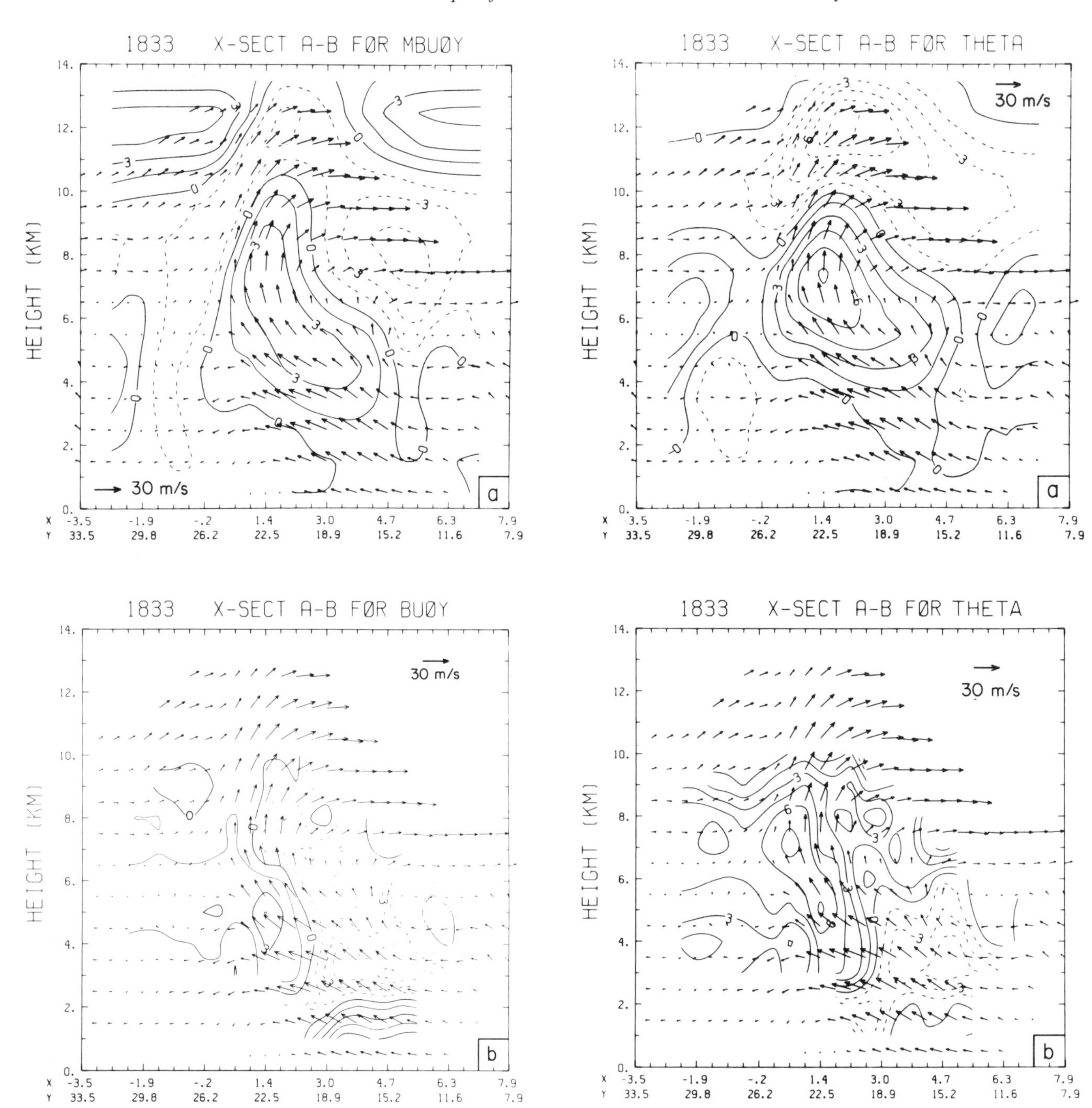

Figure 13.11. Total thermal buoyancy (deviation from horizontal average) derived from microphysical (a) and dynamic (b) retrievals, in the vertical cross section A–B. Contour interval is 1.5°C. Negative values are indicated by dashed contours; nonnegative values are indicated by solid contours.

Figure 13.12. Perturbation potential temperature, $\theta - \bar{\theta}$, in cross section A–B from the results of (a) microphysical retrieval and (b) dynamic retrieval aided by water quantities and a potential temperature profile from microphysical retrieval. Contour interval is 1.5°C; vector wind components in the vertical section are superimposed.

$-4.5°C$ is adjacent to the middle-level warm updraft core. The corresponding dynamic retrieval (Fig. 13.11b) has similar features, most notably the warm middle-level updraft core ($T > 4.5°C$) and the cold region south of the updraft core ($T < -4.5°C$). The most notable difference between the dynamic and the microphysical retrievals appears in Fig. 13.11b as an encroachment of cold air into the southern flank of the updraft core. Since research aircraft did not penetrate the Del City storm, it is now known whether this discrepancy is due to differing sensitivities to random and systematic wind-field errors of the retrieval schemes or to a possibly significant degree of nonsteadiness[2] of the 1833 CST wind field. The greatest significance of the comparison is that two independent diagnostic methods retrieve similar buoyant updraft cores, a condition vital to the sustenance of this storm that produced a large tornado at about 1840 CST.

A second variable available for comparing the dynamic and microphysical retrieval outputs is the potential temperature, which can be derived from dynamic buoyancy results by employing a point measurement of θ to render that solution unique. Figure 13.12 illustrates the retrieved fields of $\theta - \overline{\theta}$ in the vertical cross section A–B; the isopleths represent deviations from the base-state profile at each level. The microphysical retrieval (Fig. 13.12a) associates the middle-level updraft with temperature excess greater than 7.5°C; the high-level penetrative updraft possesses a thermal defect of equal magnitude. Cool temperature decrements in the rainy low-level downdraft are larger than $-1.5°C$. The middle-level updrafts are warmed primarily by condensation heating, and the penetrative high-level updraft processes air parcels above their equilibrium level in the stable upper troposphere. The low-level rainy downdraft is cooled primarily by evaporation.

An alternative derivation of $\theta - \overline{\theta}$ (Fig. 13.12b) is based on dynamically retrieved buoyancy aided by the microphysically retrieved vapor and condensate fields, as well as a vertical profile of θ in the updraft core. Specifically, Eq. 13.13 is solved for $\theta - \langle\theta\rangle$; water vapor and total condensed water are taken from the microphysically retrieved results, pressure from the dynamic retrieval field, and velocities from the Doppler radar as analyzed. Then, given θ from the microphysical results for one point in the updraft at each level, $\langle\theta\rangle$ is calculated at each level, and, therefore, θ can be calculated over the entire volume. Warming in the updraft region calculated by dynamical retrieval covers a narrower volume than that deduced with microphysical retrieval (Fig. 13.12a), although the surrounding thermal gradients generally agree. Greater variability is present, especially in low levels, in the dynamically retrieved results. The dynamically retrieved middle-level temperatures north of (−0.2, 26.2) are slightly warmer than those diagnosed with the microphysical retrieval, which relates weak subsidence to small diabatic heating rates. The particular choice of a θ values from the microphysical retrieval may introduce an uncertain degree of bias into the dynamically calculated θ. As with buoyancy (Fig. 13.11), it is again concluded that the two retrieval methods diagnose similar magnitudes of thermal buoyancy necessary to maintain the updraft.

6. Conclusion

Doppler radar data can be analyzed to define air-velocity fields within thunderstorms, and these in turn can be used to estimate fields of all the major variables within storm volumes. Retrieval therefore represents a major advance in thunderstorm studies, providing reasonably accurate estimates of wind, temperature, pressure, and water quantities over three-dimensional storm volumes at one time. This comprehensive information facilitates quantitative estimation of the forces at work within thunderstorms, and it provides potential for partial initialization and verification of three-dimensional, fully time-dependent numerical simulations. Significant impact should be realized in a diversity of microphysical problem areas, including cloud electrification, precipitation scavenging, and storm modification.

Much work remains to place the new techniques of dynamical and microphysical retrieval on a firm basis:

1. In situ measurements within storms observed by Doppler radar must be obtained for verification of all derived fields, including air velocity. Techniques for remote observation of the storm environment must also be improved.
2. Techniques for combining results of dynamic and microphysical retrieval (and perhaps the methods themselves) must be advanced to improve the quality of the results and the efficiency of the solution methods.
3. Errors in velocity fields result in errors in other derived quantities. Therefore, the nature of velocity errors and the paths by which these errors are propagated and transformed in various retrieval techniques must be investigated further. Eventually velocity analyses should account for balances in all the major governing equations, with minimal differences between solution quantities and those observed over whole storm volumes.

[2]The substantial agreement noted between observed and retrieved reflectivity suggests that the steadiness assumption is valid in this case.

Appendix
Mathematical Details

1. Equations for Velocity Components in Two-Radar Analysis

Combining Eqs. 13.1 and 13.4b for each of two radars we obtain:

$$u = \frac{R_1V_1(y-y_1) - R_2V_2(y-y_1)V_t[(z-z_1)(y-y_2) - (z-z_2)(y-y_1)]}{(x-x_1)(y-y_2) - (x-x_2)(y-y_1)} - w\frac{[(z-z_1)(y-y_1) - (z-z_2)(y-y_2)]}{(x-x_1)(y-y_2) - (x-x_2)(y-y_1)} = C_1 - wC_2, \tag{13.28}$$

and

$$v = \frac{R_2V_2(x-x_1) - R_1V_1(x-x_2) - V_t[(z-z_2)(x-x_1) - (z-z_1)(x-x_2)]}{(x-x_2)(y-y_1) - (x-x_1)(y-y_2)} - w\frac{[(z-z_2)(y-y_1) - (z-z_2)(y-y_1)]}{(x-x_2)(y-y_1) - (x-x_1)(y-y_2)} = C_3 - wC_4,$$

where C_2 and C_4 are functions of position only, and C_1 and C_3 are functions of distance, radial velocity as measured by the two radars, and terminal velocity.

The vertical air motion w can be obtained from the linear inhomogeneous partial differential equation (Armijo, 1969)

$$-C_2\frac{\partial w}{\partial x} - C_4\frac{\partial w}{\partial y} + \frac{\partial w}{\partial z} = w\left(\frac{\partial C_2}{\partial x} + \frac{\partial C_4}{\partial y} + \kappa\right) - \left(\frac{\partial C_1}{\partial x} + \frac{\partial C_3}{\partial y}\right). \tag{13.29}$$

The natural coordinate for the solution is in a cylindrical coordinate system. Techniques for solution have been documented by Ray et al. (1975), Miller and Strauch (1974), Doviak et al. (1976), Lhermitte and Miller (1970) and Doviak et al. (in Chap. 9 of this book).

2. Equations for Velocity Components in Three-Radar Analysis

Wind components resulting when data from three radars are combined are

$$u = \frac{(y_3 - y_2)R_1V_1 + (y_1 - y_3)R_2V_2 + (y_2 - y_1)R_3V_3}{x_1(y_2 - y_3) + x_2(y_3 - y_1) + x_3(y_1 - y_2)}, \tag{13.30a}$$

$$v = \frac{(x_2 - x_3)R_1V_1 + (x_3 - x_1)R_2V_2 + (x_1 - x_2)R_3V_3}{x_1(y_2 - y_3) + x_2(y_3 - y_1) + x_3(y_1 - y_2)}, \tag{13.30b}$$

and

$$W = \left\{\frac{1}{z[x_1(y_2 - y_3) + x_2(y_3 - y_1) + x_3(y_1 - y_3)]}\right\} \times \{[x(y_2 - y_3) + x_2(y_3 - y) + x_3(y - y_2)]R_1V_1 + [x(y_3 - y_1) + x_1(y - y_3) + x_3(y_1 - y)]R_2V_2 + [x(y_1 - y_2) + x_1(y_2 - y) + x_2(y - y_1)]R_3V_3\}. \tag{13.30c}$$

Note from the first term in Eq. 13.30c that the vertical wind component w $(= W - V_t)$ varies inversely with height. When errors owing to geometry are computed, this term is squared. Thus the errors in the vertical component are proportional to $(1/z)^2$, and many analysts use only the horizontal wind component part of the solution, subsequently integrating the continuity relation to obtain w. Note that in either approach the terminal velocity V_t is not part of the horizontal wind and V_t uncertainty effects are therefore absent.

3. Interrelation of Microphysical Processes

The expressions that follow illustrate the contribution of individual microphysical processes to the source-sink terms M_ϕ found in the model equations represented by Eq. 13.1. Symbolic names appearing in the equations are defined in Table 13.1; the references noted there describe the current formulations of the various processes.

$$M_\theta = \frac{L_v}{c_p\overline{\pi}}(\text{COND} - \text{RVAP}) + \frac{L_s}{c_p\overline{\pi}}(-\text{ISUB} - \text{XSUB} - \text{XMEV})$$

$$+ \frac{L_f}{c_p \bar{\pi}} (\text{RFRZ} + \text{HDRY} + \text{HWET} \quad (13.31)$$

$$+ \text{INUC} + \text{IDEP} + \text{ICOC} - \text{HMLT}$$

$$- \text{IMLT} + \text{CFRZ}),$$

$$M_{q_v} = \text{RVAP} + \text{ISUB} + \text{XSUB} + \text{XMEV} - \text{COND}, \quad (13.32)$$

$$M_{q_c} = \text{COND} - \text{CCOC} - \text{RCOC} - \text{HDRY} - \text{HWET} - \text{INUC} - \text{IDEP} - \text{ICOC} - \text{CFRZ}, \quad (13.33)$$

$$M_{q_r} = \text{CCOC} + \text{RCOC} + \text{RSED} + \text{HMLT} + \text{IMLT} - \text{RVAP} - \text{RFRZ}, \quad (13.34)$$

$$M_{q_h} = \text{HDRY} + \text{HWET} + \text{HCOI} + \text{HSED} + \text{ICNV} - \text{HMLT}, \quad (13.35)$$

$$M_{q_i} = \text{INUC} + \text{IDEP} + \text{ICOC} + \text{ISED} - \text{ISUB} - \text{IMLT} - \text{ICNV}, \quad (13.36)$$

$$M_{q_x} = \text{CFRZ} - \text{XSUB} - \text{XMEV}, \quad (13.37)$$

where L_v, L_f, and L_s are, respectively, the latent heat of vaporization, fusion, and sublimation and c_p is the specific heat of air at constant pressure. Analogous expressions define M_{N_c}, M_{N_r}, and M_{N_h}, but they are not shown here. The concentration continuity equations also contain a term accounting for the effect of expansion (contraction) of rising (sinking) air parcels. The Exner function π is defined by

$$\pi \equiv \left(\frac{p}{p_0}\right)^{\kappa} = \left(\frac{p}{p_0}\right)^{R/c_p},$$

where R is the gas constant for dry air and p is the air pressure. The reference pressure p_0 is set to 1000 mb.

Table 13.1. Microphysical Processes and Their Symbolic Representations

Process	Symbolic name	Reference
Condensation of vapor	COND	Ziegler (1985)
Evaporation of vapor	RVAP	Soong and Ogura (1973)
Sublimation of ice	ISUB	Lin et al. (1983)
Sublimation of cloud ice	XSUB	Rogers (1976)
Melting or evaporation of cloud ice	XMEV	Ziegler (1985)
Freezing of raindrops	RFRZ	Ziegler (1985)
Dry growth of hail	HDRY	Ziegler (1985)
Wet growth of hail	HWET	Ziegler (1985)
Nucleation of ice particles	INUC	Rutledge and Hobbs (1983)
Growth of ice particles by vapor diffusion	IDEP	Rutledge and Hobbs (1983)
Growth of ice particles by riming of cloud water	ICOC	Lin et al. (1983)
Melting of hail	HMLT	Orville and Kopp (1977)
Melting of ice	IMLT	Ogura and Takahashi (1971)
Freezing of cloud water	CFRZ	Ziegler (1985)
Cloud droplet self-collection	CCOC	Ziegler (1985)
Rain collection of cloud	RCOC	Ziegler (1985)
Rain sedimentation	RSED	Klemp and Wilhelmson (1978a); Wisner et al. (1972)
Accretion of ice by hail	HCOI	Ziegler (1985)
Hail sedimentation	HSED	Klemp and Wilhelmson (1978a); Wisner et al. (1972)
Ice-hail conversion	ICNV	Ziegler (1985)
Ice sedimentation	ISED	Klemp and Wilhelmson (1978a); Ogura and Takahashi (1971)

References

Chapter 1

Alberty, R. L., D. W. Burgess, C. E. Hane, and J. F. Weaver, 1979. Project Severe Environmental Storms and Mesoscale Experiment (SESAME). SESAME 1979 Operations Summary, NOAA Environmental Research Laboratories, Boulder, Colo.

Angell, J. K., W. H. Hoecker, C. R. Dickson, and D. H. Pack, 1973. Urban influence on a strong daytime airflow as determined from tetroon flights. *J. Appl. Meteorol.* 12:924–36.

Barnes, S. L., 1964. A technique for maximizing details in numerical weather map analysis. *J. Appl. Meteorol.* 3:396–409.

———, 1973. Mesoscale objective map analysis using weighted time-series observations. Tech. Memo. ERL NSSL-62, NOAA Environmental Research Laboratories, Boulder, Colo. (NTIS COM-73-10781).

———, 1978a. Oklahoma thunderstorms on 29–30 April 1970; part I: morphology of a tornadic storm. *Mon. Weather Rev.* 106(5):673–84.

———, 1978b. Oklahoma thunderstorms on 29–30 April 1970; part II: radar-observed merger of twin hook echoes. *Mon. Weather Rev.* 106(5): 685–96.

———, 1985. Omega diagnostics as a supplement to LFM/MOS guidance in weakly forced convection situations. *Mon. Weather Rev.* 113:2122–41.

———, and S. P. Nelson, 1978. Oklahoma thunderstorms on 29–30 April 1970; part IV: study of a dissipating severe storm. *Mon. Weather Rev.* 106(5): 704–12.

———, J. H. Henderson, and R. J. Ketchum, 1971. Rawinsonde observation and processing techniques at the National Severe Storms Laboratory. Tech. Memo. ERL NSSL-53, NOAA Environmental Research Laboratories, Boulder, Colo. (NTIS COM-71-00707).

Bengtsson, L., M. Ghil, and E. Kallen, eds. 1981. *Dynamic Meteorology: Data Assimilation Methods*. Springer-Verlag, New York.

Bleeker, W., and M. J. Andre, 1950. Convective phenomena in the atmosphere. *J. Meteorol.* 7:195–209.

Braham, R. R., Jr., 1952. The water and energy budgets of the thunderstorm and their relation to thunderstorm development. *J. Meteorol.* 9:227–42.

Brandes, E. A., 1977. Gust front evolution and tornado genesis as viewed by Doppler radar. *J. Appl. Meteorol.* 16:333–38.

Brock, F. V., 1974. Portable Automated Mesonet. *Atmos. Tech.* 6:21–25.

———, and P. K. Govind, 1977. Portable Automated Mesonet in operation. *J. Appl. Meteorol.* 16:299–310.

Brooks, E. M., 1949. The tornado cyclone. *Weatherwise* 2: 32–33.

———, 1951. Tornadoes and related phenomena. In *Compendium of Meteorology,* T. F. Malone (ed.), American Meteorological Society, Boston, pp. 673–80.

Brown, R. A. (ed.), 1976. The Union City, Oklahoma, tornado of 24 May 1973. Tech. Memo. ERL NSSL-80, NOAA Environmental Research Laboratories, Boulder, Colo.

Browning, K. A., and R. J. Donaldson, Jr., 1963. Airflow and structure of a tornadic storm. *J. Atmos. Sci.* 20:533–45.

Buell, C. E., 1960. The structure of two-point wind correlations in the atmosphere. *J. Geophys. Res.* 65:3353–66.

Byers, H. R., and R. R. Braham, Jr., 1948. Thunderstorm structure and circulation. *J. Meteorol.* 5:71–86.

———, and R. R. Braham, Jr., 1949. *The Thunderstorm*. U.S. Weather Bureau, Washington, D.C.

———, and H. R. Rodebush, 1948. Causes of thunderstorms of the Florida peninsula. *J. Meteorol.* 5:275–80.

Changnon, S. A., and R. G. Semonin, 1971. Metromex: an investigation of inadvertent weather modification. *Bull. Am. Meteorol. Soc.* 52:958–67.

Cressman, G. P., 1959. An operational objective analysis system. *Mon. Weather Rev.* 87:367–74.

Danielson, E., 1975. Predicting and modifying severe weather. In *NCAR Annual Report,* National Center for Atmospheric Research, Boulder, Colo., 89–95.

Doswell, C. A., III, 1977. Obtaining meteorologically significant surface divergence fields through the filtering property of objective analysis. *Mon. Weather Rev.* 105:885–92.

Eddy, A., 1964. The objective analysis of horizontal wind divergence fields. *Q. J. R. Meteorol. Soc.* 90:424–40.

Fankhauser, J. C., 1969. Convective processes resolved by a mesoscale rawinsonde network. *J. Appl. Meteorol.* 8:778–98.

Ferguson, S. P., 1933. The early history of aerology in the United States. *Bull. Am. Meteorol. Soc.* 14:252–56.

Foote, G. B., and C. A. Knight, 1977. *Hail: A Review of Hail Science and Hail Suppression*. Meteorol. Monogr. 16(38), American Meteorological Society, Boston.

Fujita, T. T., 1962. Index to the NSSP surface network. National Severe Storms Project Report No. 6, U.S. Weather Bureau, Washington, D.C.

———, 1963. Analytical mesometeorology: a review. In *Severe Local Storms,* Meteorol. Monogr. 5(27), American Meteorological Society, Boston, pp. 77–125.

———, and R. Stuhmer, 1963. Proposed mechanism of hook echo formation, with a preliminary mesosynoptic analysis of tornado cyclone case of 26 May 1963. Research Paper No. 27, Mesometeorology Project, Department of the Geophysical Sciences, University of Chicago.

———, H. Newstein, and M. Tepper, 1956. Mesoanalysis, an important scale in the analysis of weather data. Res. Paper No. 39, U.S. Weather Bureau, Washington, D.C.

Gandin, L. S., 1963. *Objective Analysis of Meteorological Fields.* Translated by Israel Program for Scientific Translations (National Technical Information Service, Springfield, Va., TT-65-50007).

Holloway, J. L., Jr., 1958. Smoothing and filtering of time series and space fields. In *Advances in Geophysics,* Vol. 4, Academic Press, New York, pp. 351–89.

Houghton, D. D. (ed.), 1985. Part 2, Measurements, pp. 283–503, in *Handbook of Applied Meteorology*. John Wiley & Sons, New York.

Kessler, E., 1965. Purposes and program of the U.S. Weather Bureau National Severe Storms Laboratory, Norman, Okla. *Trans. Am. Geophys. Union* 46(2):389–97.

———, 1977. National Severe Storms Laboratory: history and 1976 program. NSSL Special Report, NOAA Environmental Research Laboratories, Boulder, Colo.

Kreitzberg, C. W., and H. A. Brown, 1970. Weather systems within an occlusion. *J. Appl. Meteorol.* 9:417–32.

Lhermitte, R. M., and Kessler, E., 1965. A weather radar signal integrator. Preprints, International Conference on Cloud Physics, Tokyo and Sapporo, Japan, May 24–June 1, 302–308.

Maddox, R. A., 1980. An objective technique for separating macroscale and mesoscale features in meteorological data. *Mon. Weather Rev.* 108:1108–21.

Middleton, W. E. K., 1969. *Invention of the Meteorological Instruments.* Johns Hopkins Press, Baltimore, Md.

Newton, C. W., 1963. Dynamics of severe convective storms. In *Severe Local Storms,* Meteorol. Monogr. 5(27), American Meteorological Society, Boston, pp. 33–57.

NSSP (National Severe Storms Project) Staff, 1961. Objectives and basic design. National Severe Storms Project Report No. 1, U.S. Weather Bureau, Washington, D.C.

Sahashi, K., 1972. Preliminary report on the feasibility study of mesomicrometeorological interaction with sonic anemometer—thermometers installed at the WKY-TV tower. Report to Environmental Research Laboratories on NOAA grant N22-47-72(G).

Sasaki, Y., 1960. An objective analysis for determining initial conditions for the primitive equations. Tech. Report (Ref: 60-16T), Department of Oceanography and Meteorology, Texas A&M University, College Station.

Tepper, M., 1950. A proposed mechanism of squall lines: the pressure jump line. *J. Meteorol.* 7:21–29.

Williams, D. T., 1953. Pressure wave observations in the central Midwest, 1952. *Mon. Weather Rev.* 81:278–98.

Chapter 2

Bates, F. C., 1962. Tornadoes in the Central United States. *Trans. Kansas Acad. Sci.* 65:215–46.

———, 1963. An aerial observation of a tornado and its parent cloud. *Weather* 18:12–18.

Bedard, A. J., Jr., and C. Ramzy, 1983. Surface meteorological observations in severe thunderstorms. Part I: Design details of TOTO. *J. Climate Appl. Meteorol.* 22:911–18.

Bluestein, H. B., 1980. The University of Oklahoma Severe Storm Intercept Project—1979. *Bull. Amer. Meteorol. Soc.* 61: 560–67.

———, 1983a. Surface meteorological observations in severe thunderstorms. Part II: Field experiments with TOTO. *J. Climate Appl. Meteorol.* 22:919–30.

———, 1983b. Measurements in the vicinity of severe thunderstorms and tornadoes with TOTO: 1982–1983 results. Preprints, Thirteenth Conference on Severe Local Storms, Tulsa, Oklahoma. American Meteorological Society, Boston, pp. 89–92.

———, 1984a. Further examples of low-precipitation severe thunderstorms. *Mon. Weather Rev.* 112:1885–88.

———, 1984b. Photographs of the Canyon, Texas, storm on 26 May 1978. *Mon. Weather Rev.* 112:2521–23.

———, 1985. Wall clouds with eyes. *Mon. Weather Rev.* 113: 1081–85.

———, 1986. Visual aspects of the flanking line in severe thunderstorms. *Mon. Weather Rev.* 115:788–95.

———, and C. R. Parks, 1983. A synoptic and photographic climatology of low-precipitation severe thunderstorms in the Southern Plains. *Mon. Weather Rev.* 111:2034–46.

———, E. W. McCall, G. P. Byrd, G. R. Woodall, and R. L. Walko, 1987. Use of a portable radiosonde unit for probing severe thunderstorms and the dryline. Preprints, Sixth Symposium on Meteorological Observations and Instrumentation, New Orleans. American Meteorological Society, Boston, pp. 444–47.

Brady, R. H., J. Wilson, and E. Szoke, 1986. Boundary layer influences on the formation of the 26 July 1986 Erie, Colorado, tornado. Preprints, Twenty-third Conference on Radar Meteorology, Snowmass, Colo. American Meteorological Society, Boston, pp. JP178–81.

Brewster, K. A., 1986. Photographs of a funnel-producing indented cloud-base swirl. *Mon. Weather Rev.* 114:1771–74.

Brock, F. V., G. Lesins, and R. Walko, 1987. Measurement of pressure and air temperature near severe thunderstorms: An inexpensive and portable instrument. Preprints, Sixth Symposium on Meteorological Observations and Instrumentation, New Orleans. American Meteorological Society, Boston, pp. 320–23.

Brown, R. A. (ed.), 1976. The Union City, Oklahoma tornado of 24 May 1973. Tech. Memo. ERL NSSL-80, NOAA Environmental Research Laboratories, Boulder, Colo.

Burgess, D. W., 1976. Anticyclonic tornado. *Weatherwise* 29:167.

———, and R. P. Davies-Jones, 1979. Unusual tornadic storms in eastern Oklahoma on 5 December 1975. *Mon. Weather Rev.* 107:451–57.

———, J. D. Bonewitz, and D. R. Devore, 1978. Operational Doppler radar experiments: results year 1. Preprints, Eighteenth Conference on Radar Meteorology, Atlanta, Georgia. American Meteorological Society, Boston, pp. 442–48.

———, S. V. Vasiloff, R. P. Davies-Jones, D. S. Zrnić, and S. E. Frederickson, 1985. Recent NSSL work on windspeed measurements in tornadoes. Proceedings, Fifth U.S. National Conference on Wind Engineering, Texas Tech University, Lubbock (1A-53-1A-60).

———, V. T. Wood and R. A. Brown, 1982. Mesocyclone evolution statistics. Preprints, Twelfth Conference on Severe Local Storms, San Antonio, Texas. American Meteorological Society, Boston, pp. 422–424.

Campbell, B. D., E. N. Rasmussen, and R. E. Peterson, 1983. Kinematic analysis of the Lakeview tornado. Preprints, Thirteenth Conference on Severe Local Storms, Tulsa, Oklahoma. American Meteorological Society, Boston, pp. 62–65.

Colgate, S. A., 1982. Small rocket tornado probe. Preprints, Twelfth Conference on Severe Local Storms, San Antonio, Texas. American Meteorological Society, Boston, pp. 396–400.

Danielsen, E. F., 1975. A conceptual theory of tornadogenesis based on macro-, meso-, and microscale processes. Preprints, Ninth Conference on Severe Local Storms, Norman, Oklahoma. American Meteorological Society, Boston, pp. 376–83.

Davies-Jones, R. P., and J. H. Golden, 1976. On the relation of electrical activity to tornadoes. *J. Geophys. Res.* 80:1614–16.

———, D. W. Burgess, and L. R. Lemon, 1975. An atypical tornado-producing cumulonimbus. *Weather* 31:337–47.

Donaldson, R. J., and W. E. Lamkin, 1963. Visual observations beneath a developing tornado. *Mon. Weather Rev.* 92:326–29.

Doswell, C. A., 1985. The operational meteorology of convective weather. Vol. 2: Storm scale analysis. NOAA Tech. Memo. ERL ESG-15.

Doviak, R. J. (ed.), 1981. 1980 Spring Program Summary. Tech. Memo. ERL NSSL-91, NOAA Environmental Research Laboratories, Boulder, Colo.

Eskridge, R. E., and P. Das, 1976. Effect of a precipitation-driven downdraft on a rotating wind field: a possible trigger mechanism for tornadoes. *J. Atmos. Sci.* 33:70–84.

Gannon, R., 1973. Tornado! How science tracks down the dread twister. *Pop. Sci.* 203:64–66, 122–25.

Golden, J. H., 1974. The life cycle of Florida Keys' waterspouts. I. *J. Appl. Meteorol.* 13:676–92.

———, and B. J. Morgan, 1972. The NSSL/Notre Dame Tornado Intercept Program, spring 1972. *Bull. Am. Meteorol. Soc.* 53:1178–79.

Hauptman, W., 1984. On the dryline. *Atlantic* 253, No. 5, pp. 76–87.

Hoadley, D. K. (ed.), 1978–86. *Storm Track,* Vol. 1, No. 1–Vol. 9, No. 3 (bimonthly newsletter from 3415 Slade Court, Falls Church, Va. 22042).

Jensen, B., E. N. Rasmussen, T. P. Marshall, and M. A. Mabey, 1983. Storm scale structure of the Pampa storm. Preprints, Thirteenth Conference on Severe Local Storms, Tulsa, Oklahoma. American Meteorological Society, Boston, pp. 85–88.

Knight, C. A., and N. C. Knight, 1974. Drop freezing in clouds. *J. Atmos. Sci.* 31:1174–76.

———, and ———, 1976. Hail embryo studies. Preprints, International Cloud Physics Conference, Boulder, Colorado. American Meteorological Society, Boston, pp. 222–26.

Lee, J. T., R. P. Davies-Jones, D. S. Zrnić, and J. H. Golden, 1981. Summary of AEC-ERDA-NRC supported research at NSSL 1973–1979. Tech. Memo. ERL NSSL-90, NOAA Environmental Research Laboratories, Boulder, Colorado.

Lemon, L. R., and C. A. Doswell, 1979. Severe thunderstorm evolution and mesocyclone structure as related to tornadogenesis. *Mon. Weather Rev.* 107:1184–97.

Lilly, D. K., 1965. Experimental generation of convectively driven vortices. *Geofis. Intl.* 5:43–48.

———, 1975. Severe storms and storm systems: scientific background, methods and critical questions. *Pure Appl. Geophys.* 113:713–34.

Mach, D. M., D. R. McGorman, W. D. Rust, and R. T. Arnold, 1986. Site errors and detection efficiency in a magnetic direction-finder network for locating lightning strikes to ground. *J. Atmos. Oceanic Technol.* 3:67–74.

Marshall, T. C., W. D. Rust, and W. P. Winn, 1984. Screening layers at the surface of thunderstorm anvils. Preprints, Seventh International Conference on Atmospheric Electricity, Albany, N.Y. American Meteorological Society, Boston, pp. 246–47.

Marshall, T. P., 1984. Chasing tornadoes. *Weatherwise,* 36:184–87.

——— (ed)., 1986–. *Storm Track,* Vol. 9, No. 4– (bimonthly newsletter available from 1336 Brazos Blvd., Lewisville, Tex. 75067).

———, and E. N. Rasmussen, 1982. The mesocyclone evolution of the Warren, Oklahoma tornadoes. Preprints, Twelfth Conference on Severe Local Storms, San Antonio, Texas. American Meteorological Society, Boston, pp. 375–78.

Moller, A. R., 1978. The improved NWS storm spotters' training program at Ft. Worth, Texas. *Bull. Am. Meteorol. Soc.* 59: 1574–82.

———, and D. Boots, 1983: The role of community preparedness and the National Weather Service's watch/warning system in the Paris, Texas, tornado. Preprints, Thirteenth Conference on Severe Local Storms, Tulsa, Oklahoma. American Meteorological Society, Boston, pp. J12–J15.

———, C. Doswell, J. McGinley, S. Tegtmeier, and R. Zipser, 1974. Field observations of the Union City tornado in Oklahoma. *Weatherwise* 27:68–77.

Morton, B. R., 1966. Geophysical vortices. In *Progress in Aeronautical Sciences,* Vol. 7. Pergamon Press, New York, pp. 145–193.

NOAA (National Oceanic and Atmospheric Administration), 1977. Tornadoes: a spotter's guide. Film available from Disaster Preparedness Office, National Weather Service, Silver Spring, Md.

———, 1980. Tornado: a spotters guide. Slide series supplement to film available from Weather and Flood Warnings Coordination Program, National Weather Service, Silver Spring, Md.

Rasmussen, E. N., R. E. Peterson, J. E. Minor, and B. D. Campbell, 1982. Evolutionary characteristics and photogrammetric determination of wind speeds within the Tulia outbreak tornadoes 28 May 1980. Preprints, Twelfth Conference on Severe Local Storms, San Antonio, Texas. American Meteorological Society, Boston, pp. 301–304.

Rossmann, F. O., 1960. On the physics of tornadoes. In *Cumulus Dynamics,* Pergamon Press, New York, pp. 167–74.

Rossow, V. J., 1970. Observations of waterspouts and their parent clouds. Tech. Note D-5854, National Aeronautics and Space Administration, Washington, D.C.

Rust, W. D., D. R. MacGorman, and R. T. Arnold, 1981. Positive cloud-to-ground lightning flashes in severe storms. *Geophysical Research Letters* 8:791–94.

Schwiesow, R. L., 1981. Horizontal velocity structure in waterspouts. *J. Appl. Meteorol.* 20:349–60.

———, R. E. Cupp, P. C. Sinclair, and R. F. Abbey, 1981. Waterspout velocity measurement by airborne Doppler radar. *J. Appl. Meteorol.* 20:341–48.

Smith, M., 1974. Visual observations of a tornadic thunderstorm. *Weatherwise* 27:256–58.

Taylor, W. L. (ed.), 1982. 1981 Spring Program Summary. Tech. Memo. ERL NSSL-93, NOAA Environmental Research Laboratories, Boulder, Colo.

Tegtmeier, S. A., 1974. The role of the surface, subsynoptic system in severe weather forecasting. Master's thesis, University of Oklahoma, Norman.

Vonnegut, B., 1960. Electrical theory of tornadoes. *J. Geophys. Res.* 65:203–12.

Ward, N. B., 1961. Radar and surface observations of tornadoes on May 4, 1961. Preprints, Ninth Weather Radar Conference, Kansas City, Missouri. American Meteorological Society, Boston, pp. 175–80.

Wilson, J. W., 1986. Tornadogenesis by nonprecipitation-induced wind shear lines. *Mon. Weather Rev.* 114:270–84.

Ziegler, C. L., P. S. Ray, and N. C. Knight, 1983. Hail growth in an Oklahoma multicell storm. *J. Atmos. Sci.* 40:1768–91.

Chapter 3

aufm Kampe, H. A., and H. K. Weickmann, 1952. Trabert's formula and the determination of the water content in clouds. *J. Meteorol.* 9:167–71.

Baumgardner, D., 1983. An analysis and comparison of five water droplet measuring instruments. *J. Appl. Meteorol.* 22: 891–910.

———, and J. E. Dye, 1983. The 1982 cloud particle measurement symposium. *Bull. Amer. Meteorol. Soc.* 64:336–70.

———, J. W. Strapp, and J. E. Dye, 1985. Evaluation of the forward scattering spectrometer probe, part II: corrections for coincidence and dead-time losses. *J. Atmos. Oceanic Technol.* 3:67–74.

Bensch, R. R., and J. McCarthy, 1978. The low-level cloud features and airflow of an Oklahoma hailstorm. *Mon. Weather Rev.* 106:566–71.

Blau, H. H., Jr., M. L. Cohen, L. B. Lapson, P. von Thuna, R. T. Ryan, and D. Watson, 1970. A prototype cloud physics nephelometer. *Appl. Optics* 9:1798–1803.

Brown, E. N., 1973. A flowmeter to measure cloud liquid water content. National Center for Atmospheric Research, *Atmos. Tech.* 1:49–51.

———, C. A. Friehe, and D. M. Lenschow, 1983: The use of pressure fluctuations on the nose of aircraft for measuring air motion. *J. Climate Appl. Meteorol.* 22:171–80.

Buck, L., 1973. Development of an improved Lyman-alpha hygrometer. National Center for Atmospheric Research, *Atmos. Tech.* 2:43–46.

Byers, H. R., and R. R. Braham, 1949. *The Thunderstorm.* U.S. Government Printing Office, Washington, D.C.

Cannon, T. W., 1974. A camera for photography of atmospheric particles from aircraft. *Rev. Sci. Instrum.* 45:1448–55.

Davis, C. I., and D. L. Veal, 1974. Decelerator design and testing. Inter. Prog. Rep. No. 9, Department of Atmospheric Resources, University of Wyoming.

Elliott, H. W., 1947. Cloud droplet camera. Rep. MI-701, Division of Mechanical Engineering, National Research Council of Canada, Ottawa.

Foote, G. B., and J. C. Fankhauser, 1973. Airflow and moisture budget beneath a northeast Colorado hailstorm. *J. Appl. Meteorol.* 12:1330–53.

Fujita, T. T., 1985. *The Downburst: Microburst and Macroburst.* Satellite and Mesometeorology Research Project Research Paper No. 210, University of Chicago, Chicago, Ill. (NTIS PB-148880).

Hallett, J., R. W. Hanaway, and P. B. Wagner, 1972. Design and construction of a new cloud particle replicator for use on a pressurized aircraft. AFCRL-72-80410, Nevada University–Reno Laboratory of Atmospheric Physics.

Hildebrand, P., and C. Mueller, 1985: Evaluation of meteorological airborne Doppler radar, part I: dual-Doppler analyses of air motion. *J. Tech.* 2.

Hobbs, P. V., R. J. Harper, and R. G. Joppa, 1973. Collection of ice particles from aircraft using decelerators. *J. Appl. Meteorol.* 12:522–28.

Jessup, E. A., 1972. Interpretations of chaff trajectories near a severe thunderstorm. *Mon. Weather Rev.* 100:653–61.

Kayton, M., and W. R. Fried, 1970. Avionics Navigation Systems. John Wiley & Sons, New York.

King, W. D., D. A. Parkin, and R. J. Handsworth, 1978. A hot-wire liquid water device having fully calculable response characteristics. *J. Appl. Meteorol.* 17:1809–13.

Knight, C. A., N. C. Knight, W. W. Grotewold, and T. W. Cannon, 1977. Interpretation of foil impactor impressions of water and ice particles. *J. Appl. Meteorol.* 16:997–1002.

Knollenberg, R. G., 1970. The optical array: an alternative to scattering or extinction for airborne particle size determination. *J. Appl. Meteorol.* 9:86–103.

———, 1976. Three new instruments for cloud physics measurements: the 2-D spectrometer, the forward scattering spectrometer probe, and the active scattering aerosol spectrometer. Preprints, International Conference on Cloud Physics, Boulder, Colorado. American Meteorological Society, Boston, pp. 554–61.

Kyle, T. G., 1975. The measurement of water content by an evaporator. *J. Appl. Meteorol.* 14:327–32.

Laktionov, A. G., N. K. Nikiforova, V. V. Smirnov, G. T. Shchelchkov, and O. A. Volkovitsky, 1972. New automatic equipment for the investigation of drop and crystal microstructure in clouds. Abstract, International Cloud Physics Conference, Royal Meteorological Society, London.

Lavoie, R. L., J. A. Pena, R. de Pena, R. L. Riuth, R. P. Greiner, D. A. Corkum, J. L. Lee, and D. L. Hosler, 1970. Studies of the microphysics of clouds. Rep. No. 16, Department of Meteorology, Pennsylvania State University.

Lenschow, D. H., 1976. Estimating updraft velocity from an airplane response. *Mon. Weather Rev.* 104:618–27.

———, and W. T. Pennell, 1974. On the measurement of in-cloud and wet-bulb temperatures from an aircraft. *Mon. Weather Rev.* 102:447–54.

List, R. J., 1958. Smithsonian Meteorological Tables, Smithsonian Institution, Washington, D.C. 527 pp. [See, for example, pp. 365 in the sixth revised edition.]

McCarthy, J., 1974. Field verification of the relationship between entrainment rate and cumulus cloud diameter. *J. Atmos. Sci.* 31:1028–39.

———, G. H. Heymsfield, and S. P. Nelson, 1974. Experiment to deduce tornado cyclone inflow characteristics using chaff and NSSL dual Doppler radars. *Bull. Am. Meteorol. Soc.* 55: 1130–31.

McCullough, S., and P. J. Perkins, 1951. Flight camera for photographing cloud droplets in natural suspension in the atmosphere (NACA RM #50K01a).

Marwitz, J. D., 1973. Trajectories within the weak echo regions of hailstorms. *J. Appl. Meteorol.* 1:1174–82.

Meceret, F. J., and T. L. Schricker, 1975. A new hot-wire liquid cloud water meter. *J. Appl. Meteorol.* 14:319–26.

Mossop, S. C., and E. R. Wishart, 1970. Ice particles in maritime clouds near Tasmania. *Q. J. R. Meteorol. Soc.* 96:487–508.

———, A. Ono, and K. J. Hefferman, 1967. Studies of ice crystals in natural clouds. *J. Rech. Atmos.* 3:45–64.

Mueller, C., and P. Hildebrand, 1985. Evaluation of meteorological airborne Doppler radar, part 2: triple Doppler analyses of air motion. *J. Tech* 2.

Mullin, R. A., and W. G. Wolver, 1964. A dewpoint hygrometer for micrometeorological range applications. Presented at the Fifth Conference on Applied Meteorology, March 2–6, 1964, Atlantic City, New Jersey. American Meteorological Society, Boston.

Nicholls, S., 1982. An observational study of mid-latitude marine atmospheric boundary layer. Master's thesis, University of Southampton, United Kingdom, October, 1983.

Rodi, A., and P. Spyers-Duran, 1972. Analysis of time response of airborne temperature sensors. *J. Appl. Meteorol.* 11:554–56.

Ruskin, R. E., 1967. Measurements of water-ice budget changes at −5°C in AgI seeded tropical cumulus. *J. Appl. Meteorol.* 6:72–81.

———, and W. D. Scott, 1974. Weather modification instruments and their use. In *Weather and Climate Modification,* W. N. Hess (ed.), John Wiley & Sons, New York.

Sasyo, Y., 1968. Studies and developments of meteorological instruments for cloud physics and micrometeorology (I). *Pap. Meteorol. Geophys.* (Japan) 19:587–98.

Schreck, R. I., V. Toutenhoofd, and C. A. Knight, 1974. A simple, airborne, ice particle collector. *J. Appl. Meteorol.* 13:949–50.

Sheets, R. C., and F. K. Odencrantz, 1974. Response characteristics of two automatic ice particle counters. *J. Appl. Meteorol.* 12:1309–18.

Spyers-Duran, P., and D. Baumgardner, 1983. In flight estimation of the time response of airborne temperature sensors. Preprints, Fifth Symposium on Meteorological Observations and Instrumentation, April 11–15, 1983, Toronto, Ont., Canada. American Meteorological Society, Boston, pp. 352–57.

Telford, J. W., and J. Warner, 1962. On the measurement from an aircraft of buoyancy and vertical air velocity in cloud. *J. Atmos. Sci.* 19:415–23.

Turner, F. M., and L. F. Radke, 1973. The design and evaluation of an airborne optical ice particle counter. *J. Appl. Meteorol.* 12:1309–18.

———, ———, and P. V. Hobbs, 1975. Optical techniques for counting ice particles in mixed-phase clouds. National Center for Atmospheric Research, *Atmos. Tech.* 8:25–31.

Veal, D. L., W. A. Cooper, G. Vali, and J. D. Marwitz, 1978. Some aspects of aircraft instrumentation for storm research. In *Hail: A Review of Hail Science and Hail Suppression.* Meteorol. Monogr. 16, American Meteorological Society, Boston, pp. 237–55.

Walther, C., 1985. Aircraft data system (ADS), user's manual 002-41ADS-002, available from Research Aviation Facility, Atmospheric Technology Division, National Center for Atmospheric Research, Boulder, Colo.

Warner, J., and T. D. Newnham, 1952. A new method of measurement of cloud water content. *Q. J. R. Meteorol. Soc.* 78:46–52.

Weickmann, H. K., 1947. Die Eisphase in der Atmosphare. *Reps. Trans.* No. 716, Ministry of Supply (A), Volkenrode, 244–47.

Wilmot, R. A., C. E. Cisneros, and F. L. Guiberson, 1974. High cloud measurements applicable to ballistic missile systems testing. Preprints, Sixth Conference on Aerospace and Aeronautical Meteorology, November 12–15, 1974, El Paso, Texas. American Meteorological Society, Boston, pp. 194–99.

Chapter 4

Agee, E. M., J. T. Snow, F. S. Nickerson, P. R. Clare, C. R. Church, and L. A. Schaal, 1977. An observational study of the West Lafayette, Indiana, tornado of 20 March 1976. *Mon. Weather Rev.* 105:893–907.

Aloway, J., C. B. Moore, and B. Vonnegut, 1970. Cameras for time-lapse photography. *Appl. Optics* 9:1811–13.

American Society of Photogrammetry, 1980. *Manual of Photogrammetry.* 4th ed., C. C. Slamo (ed.), American Society of Photogrammetry, Falls Church, Va.

———, 1983. *Manual of Remote Sensing.* 2 vols. 2d ed., R. N. Colwell (ed.-in-chief), American Society of Photogrammetry, Falls Church, Va.

Bhumralkar, C. M., 1973. An observational and theoretical study of atmospheric flow over a heated island, parts I and II. *Mon. Weather Rev.* 101:719, 731.

Cantilo, L. M. H., and W. L. Woodley, 1970. Cloud photogrammetry from airborne time-lapse photography. *J. Soc. Motion Pict. Telev. Eng.* 79:604–606.

Fankhauser, J. C., G. M. Barnes, L. J. Miller, and P. M. Roskowski, 1983. Photographic documentation of some distinctive cloud forms observed beneath a large cumulonimbus. *Bull. Am. Meteorol. Soc.* 64:450–62.

Fraser, A. B., 1968. Stereoscopic cloud photography. *Weather* 23:505–14.

Fujita, T., 1960. A detailed analysis of the Fargo tornadoes of June 20, 1957. U.S. Weather Bureau Res. Paper No. 42, Washington, D.C.

———, 1974. Overshooting thunderheads observed from ATS and Learjet. SMRP Res. Paper 117, University of Chicago.

———, 1975. New evidence from April 3–4, 1974 tornadoes. Preprints, Ninth Conference on Severe Local Storms, Norman, Oklahoma. American Meteorological Society, Boston, pp. 248–55.

Golden, J. H., and B. J. Morgan, 1972. The NSSL–Notre Dame Tornado Intercept Program, Spring 1972. *Bull. Am. Meteorol. Soc.* 53:1178–80.

———, and D. Purcell, 1977. Photogrammetric velocities for the Great Bend, Kansas, tornado of 30 August 1974: accelerations and asymmetries. *Mon. Weather Rev.* 105:485–92.

Holle, R. L., 1968. Some aspects of tropical oceanic cloud populations. *J. Appl. Meteorol.* 7:173–83.

———, and M. W. Maier, 1980. Tornado formation from downdraft interaction in the FACE mesonetwork. *Mon. Weather Rev.* 108:1010–28.

———, J. Simpson, and S. W. Leavitt, 1979. GATE B-scale cloudiness from whole-sky cameras on four U.S. ships. *Mon. Weather Rev.* 107:874–95.

Lerfald, G., and H. Erickson, 1979. Time-lapse photography using low cost camera systems. NOAA Tech. Memo ERL WPL-43, Boulder, Colo.

McNeil, G. T., 1954. *Photographic Measurements: Problems and Solutions.* Pitman Publishing Co., New York.

Malkus, J. S., and H. Riehl, 1964. *Cloud Structure and Distributions over the Tropical Pacific Ocean.* University of California Press, Berkeley and Los Angeles.

Plank, V. G., 1969. The size distribution of cumulus clouds in representative Florida populations. *J. Appl. Meteorol.* 8:46–67.

———, 1974. A photoreconnaissance technique for conducting time-lapse studies of the development and motions of cumulus cloud populations and systems. Air Force Cambridge Res. Lab. Rep. AFCRL-TR-74-0250, Bedford, Mass.

Ronne, C., 1959. On a method of cloud measurement from aircraft motion picture films. Tech. Rep. No. 7, Reference No. 59-29, Woods Hole Oceanographic Institute, Woods Hole, Mass.

Scorer, R. S., 1972. *Clouds of the World*. Stackpole Books, Harrisburg, Pa.

Shaw, R. W., and J. S. Marshall, 1972. Showers observed by stereo cameras and radar. Stormy Weather Group, Sci. Rep. MW-53, McGill University, Montreal.

Simmon, P. D., 1978. Sensitivity analysis and application of an entraining plume model. *J. Appl. Meteorol.* 17:990–97.

Staff, Cumulus Group, 1976. 1975 Florida Area Cumulus Experiment (FACE): operational summary. Tech. Memo. ERL WMPO-28, NOAA Environmental Research Laboratories, Boulder, Colo.

Vonnegut, B., 1970. Time-lapse photography of clouds from high altitude balloons. *Appl. Optics* 9:1814–16.

Warner, C., 1977. Photos of seeded Alberta storms. Stormy Weather Group, Sci. Rep. MW-89, McGill University, Montreal.

———, 1978. Photogrammetry from aircraft nose camera movies. *J. Appl. Meteorol.* 17:1416–20.

———, 1981. Photogrammetry from aircraft side camera movies: Winter MONEX. *J. Appl. Meteorol.* 20:1516–26.

———, J. H. Renick, M. W. Balshaw, and R. H. Douglas, 1973. Stereo photogrammetry of cumulonimbus clouds. *Q. J. R. Meteorol. Soc.* 99:105–15.

———, J. Simpson, D. W. Martin, D. Suchman, F. R. Mosher, and R. F. Reinking, 1979. Shallow convection on day 261 of GATE: mesoscale arcs. *Mon. Weather Rev.* 107:1617–35.

Whitney, L. F., Jr., and E. P. McClain, 1967. Cloud measurements using aircraft time-lapse photography. ESSA Tech. Rep. NESC-40, Washington, D.C.

Wolf, P. R., 1974. *Elements of Photogrammetry*. McGraw-Hill Book Co., New York.

Chapter 5

Abdullah, A. J., 1966. The "musical" sound emitted by a tornado. *Mon. Weather Rev.* 94:213–20.

Aristotle, 384–22 B.C. *Meteorologica*. Trans. H. P. D. Lee. Loeb Classical Library, Harvard University Press, Cambridge, Mass., 1951.

Arnold, R. T., and H. E. Bass, 1977. Unusual storm acoustics. Res. Rep. 77-1, Department of Physics, University of Mississippi, University.

———, ———, and L. N. Bolen, 1976. Acoustic spectral analysis of three tornadoes. *J. Acoust. Soc. Am.* 60:584–93.

Balachandran, N. K., 1979. Infrasonic signals from thunder. *J. Geophys. Res.* 84:1735–45.

———, 1983. Acoustic and electric signals from lightning. *J. Geophys. Res.* 88:3879–84.

Bass, H. E., 1980. The propagation of thunder through the atmosphere. *J. Acoust. Soc. Am.* 67:1959–66.

———, and R. E. Losey, 1975. Effect of atmospheric absorption on the acoustic power spectrum of thunder. *J. Acoust. Soc. Am.* 57:822–23.

Bhartendu, 1968. A study of atmospheric pressure variations from lightning discharges. *Can. J. Phys.* 46:269–81.

———, 1969. Audio frequency pressure variations from lightning discharges. *J. Atmos. Terr. Phys.* 31:743–47.

Bohannon, J. L., A. A. Few, and A. J. Dessler, 1977. Detection of infrasonic pulses from thunderclouds. *Geophys. Res. Lett.* 4:49–52.

Brode, H. L., 1956. The blast wave in air resulting from a high temperature, high pressure sphere of air. Rand Corp. Rep. RM-1825-AEC.

Brown, E. H., and F. F. Hall, Jr., 1978. Advances in atmospheric acoustics. *Rev. Geophys. Space Phys.* 16:47–110.

Colgate, S. A., 1967. Enhanced drop coalescence by electric fields in equilibrium with turbulence. *J. Geophys. Res.* 72: 479–87.

———, and C. McKee, 1969. Electrostatic sounds in clouds and lightning. *J. Geophys. Res.* 74:5379–89.

Dawson, G. A., C. H. Richards, and E. P. Knider, 1968. Acoustic output of a long spark. *J. Geophys. Res.* 73:815.

Dessler, A. J., 1973. Infrasonic thunder. *J. Geophys. Res.* 78: 1889–96.

Evans, L. B., H. E. Bass, and L. C. Sutherland, 1972. Atmospheric absorption of sound: theoretical predictions. *J. Acoust. Soc. Am.* 51:1565–75.

Few, A. A., 1968. Thunder. Ph.D. diss., Rice University, Houston, Tex.

———, 1969. Power spectrum of thunder. *J. Geophys. Res.* 74:6926–34.

———, 1970. Lightning channel reconstruction from thunder measurements. *J. Geophys. Res.* 75:7517–23.

———, 1974a. Lightning sources in severe thunderstorms. Preprints, Conference on Cloud Physics, Tucson, Arizona, 21–24 October. American Meteorological Society, Boston, pp. 387–90.

———, 1974b. Thunder signatures. *EOS* 55:508–14.

———, 1975. Thunder. *Sci. Am.* 233:80–90.

———, 1982. Acoustic radiations from lightning. *Handbook of Atmospherics,* ed. Hans Volland, 257–90, CRC Press, Boca Raton, Fla.

———, and T. L. Teer, 1974. The accuracy of acoustic reconstructions of lightning channels. *J. Geophys. Res.* 79:5007–11.

———, A. J. Dessler, D. J. Latham, and M. Brook, 1967. A dominant 200-hertz peak in the acoustic spectrum of thunder. *J. Geophys. Res.* 72:6149–54.

———, H. B. Garrett, M. A. Uman, and L. E. Salanave, 1970. Comments on letter by W. W. Troutman. *J. Geophys. Res.* 75:4192–95.

Flora, S. D., 1954. *Tornadoes of the United States*. University of Oklahoma Press, Norman.

Georges, T. M., 1971. Acoustic ray paths through a model vortex with a viscous core. *J. Acoust. Soc. Am.* 51:206–209.

Golden, J. H., and B. J. Morgan, 1972. The NSSL/Notre Dame Tornado Intercept Project, spring 1972. *Bull. Am. Meteorol. Soc.* 53:1178–80.

Hill, R. D., 1968. Analysis of irregular paths of lightning channels. *J. Geophys. Res.* 73:1897–1906.

Holmes, C. R., M. Brook, P. Krehbiel, and R. McCroy, 1971. On the power spectrum and mechanism of thunder. *J. Geophys. Res.* 76:2106–15.

Jones, D. L., 1968. Comments on paper by A. A. Few, A. J. Dessler, D. J. Latham, and M. Brook, "A dominant 200-hertz peak in the acoustic spectrum of thunder." *J. Geophys. Res.* 74:5555.

———, G. G. Goyer, and M. N. Plooster, 1968. Shock wave from a lightning discharge. *J. Geophys. Res.* 73:3121–27.

Kinsler, L. E., and A. R. Frey, 1962. *Fundamentals of Acoustics*. John Wiley & Sons, New York.

Klinkowstein, R. E., 1974. A study of acoustic radiation from an electrical spark discharge in air. Master's thesis, Massachusetts Institute of Technology.

Krider, E. P., and C. Guo, 1983. The peak electromagnetic power radiated by lightning return strokes. *J. Geophys. Res.* 88: 8471–74.
Lighthill, M. J., 1952. On sound generated aerodynamically, I. *Proc. R. Soc. A.* 211:564–87.
———, 1954. On sound generated aerodynamically, II. *Proc. R. Soc. A.* 222:1–32.
Lucretius, T., 98–55 B.C. *On the Nature of Things.* Trans. H. A. J. Munro, Book 6, Great Books of the Western World, William Benton (ed.). Chicago, 1952.
Nakano, M., 1973. Lightning channel determined by thunder. *Proc. Res. Inst. Atmos. Nagoya Univ.* 20:1–7.
Ogawa, T., and M. Brook, 1969. Charge distribution in thunderstorm clouds. *Q. J. R. Meteorol. Soc.* 95:513–25.
Pierce, E. T., 1955. The development of lightning discharges. *Q. J. R. Meteorol. Soc.* 81:229–40.
Piercy, J. E., T. F. W. Embleton, and L. C. Sutherland, 1977. Review of noise propagation in the atmosphere. *J. Acoust. Soc. Am.* 61:1403–18.
Powell, A., 1964. Theory of vortex sound. *J. Acoust. Soc. Am.* 36:177–95.
Remillard, W. J., 1969. Comments on paper by A. A. Few, A. J. Dessler, D. J. Latham, and M. Brook, "A dominant 200-hertz peak in the acoustic spectrum of thunder." *J. Geophys. Res.* 74:5555.
Ribner, H. S., and D. Roy, 1982. Acoustics of thunder: a quasilinear model for tortuous lightning. *J. Acoust. Soc. Am.* 72:1911–25.
Sozou, C., 1968. Symmetrical normal modes in a Rankine vortex. *J. Acoust. Soc. Am.* 46:814–18.
Taylor, W. L., 1978. A VHF technique for space-time mapping of lightning discharge processes. *J. Geophys. Res.* 83:3575–83.
Teer, T. L., 1973. Lightning channel structure inside an Arizona thunderstorm. Ph.D. diss., Rice University, Houston, Tex.
———, and A. A. Few, 1974. Horizontal lightning. *J. Geophys. Res.* 79:3436–41.
Troutman, W. W., 1969. Numerical calculation of the pressure pulse from a lightning stroke. *J. Geophys. Res.* 74:4595–96.
Uman, M. A., 1969. *Lightning.* McGraw-Hill Book Co., New York.
———, A. H. Cookson, and J. B. Moreland, 1970. Shock wave from a 4-meter spark. *J. Appl. Phys.* 41:3148–55.
Wilson, C. T. R., 1920. Investigations on lightning discharges and on the electric field of thunderstorms. *Phil. Trans. R. Soc. London,* Ser. A. 221:73–115.
Workman, E. J., R. E. Holzer, and C. T. Pelson, 1942. The electric structure of thunderstorms. National Advisory Committee for Aeronautics, Tech. Note 864 [available from National Technical Information Service, Springfield, Va.].
Wright, W. M., and N. W. Mendendorp, 1967. Acoustic radiation from a finite line source with N-wave excitation. *J. Acoust. Soc. Am.* 43:966–71.
Zel'dovich, Y. B., and Y. P. Raizer, 1967. *Physics of Shock Waves and High-Temperature Hydrodynamic Phenomena.* Academic Press, New York.

Chapter 6

Agee, E., C. Church, C. Morris, and J. Snow, 1975. Some synoptic aspects and dynamic features of vortices associated with the tornado outbreak of 3 April 1974. *Mon. Weather Rev.* 103:318–33.
———, J. T. Snow, and P. R. Clarke, 1976. Multiple vortex features in the tornado cyclone and the occurrence of tornado families. *Mon. Weather Rev.* 104:552–63.
Baker, D. M., and K. Davies, 1969. F2-region acoustic waves from severe weather. *J. Atmos. Terr. Phys.* 31:1345–52.
Bedard, A. J., 1971. Seismic response of infrasonic microphones. *J. Res. NBS* 75C:41–45.
Bowman, H. S., and A. J. Bedard, 1971. Observations of infrasound and subsonic pressure disturbances related to severe weather. *Geophys. J. R. Astron. Soc.* 26:215–42.
Brown, R. A., D. W. Burgess, and K. C. Crawford, 1973. Twin tornado cyclones within a severe thunderstorm: single Doppler radar observations. *Weatherwise* 26:63–69.
Brown, R. F., Jr. 1963. An automatic multichannel correlator. *J. Res. NBS* 67C:33–38.
Burridge, R., 1971. The acoustics of pipe arrays. *Geophys. J. R. Astron. Soc.* 26:53–69.
Chimonas, G., and W. R. Peltier, 1974. On severe storm acoustic signals observed at ionospheric heights. *J. Atmos. Terr. Phys.* 36:821–28.
Chrzanowski, P., J. M. Young, and H. L. Marrett, 1960. Infrasonic pressure waves from tornadic storms. NBS Rep. 7035, National Bureau of Standards, Washington, D.C. (Unpublished.)
Cook, R. K., and A. J. Bedard, 1971. On the measurement of infrasound. *Geophys. J. R. Astron. Soc.* 26:5–11.
———, and J. M. Young, 1962. Strange sounds in the atmosphere, part 2. *Sound* 1:25–33.
Daniels, F. B., 1959. Noise-reducing line microphone for frequencies below 1 cps. *J. Acoust. Soc. Am.* 31:529–31.
Davies, K., and J. E. Jones, 1971. Ionospheric disturbances in the F2 region associated with severe thunderstorms. *J. Atmos. Sci.* 28:254–62.
———, and ———, 1972a. Ionospheric disturbances produced by severe thunderstorms. Prof. Paper No. 6, NOAA Environmental Research Laboratories, Boulder, Colo.
———, and ———, 1972b. Infrasound in the ionosphere generated by severe thunderstorms. *J. Acoust. Soc. Am.* 52: 1087–90.
Davy, A., R. C. DiPrima, and J. T. Stuart, 1968. On the instability of Taylor vortices. *J. Fluid Mech.* 31:17–52.
Detert, D. G., 1969. A study of the coupling of acoustic energy from the troposphere to the ionosphere. AVCO Corp. Final Rep. to Marshall Space Flight Center, Huntsville, Ala., 15 February 1969.
Elliott, J. A., 1972. Microscale pressure fluctuations measured within the lower atmospheric boundary layer. *J. Fluid Mech.* 53:351–83.
Francis, S. H., 1973. Acoustic-gravity modes and large-scale traveling ionospheric disturbances of a realistic, dissipative atmosphere. *J. Geophys. Res.* 78:2278–2301.
Fujita, T., 1970. The Lubbock tornadoes: a study of suction spots. *Weatherwise* 23:160–73.
———, 1971. Proposed mechanism of suction spots accompanied by tornadoes. Preprints, Seventh Conference on Severe Local Storms. American Meteorological Society, Boston, pp. 208–13.
Georges, T. M., 1967. Ionospheric effects of atmospheric waves. ESSA Tech. Rep. IER 57-ITSA 54, U.S. Government Printing Office, Washington, D.C.

———, 1968a. HF Doppler studies of traveling ionospheric irregularities. *J. Atmos. Terr. Phys.* 30:735–46.
———, 1968b. Short-period ionospheric oscillations associated with severe weather. Proceedings, Symposium on Acoustic-Gravity Waves in the Atmosphere. U.S. Government Printing Office, Washington, D.C., pp. 171–78.
———, 1973. Infrasound from convective storms: examining the evidence. *Rev. Geophys. Space Phys.* 11:571–94.
———, 1976. Infrasound from convective storms, part 2: a critique of source candidates. Tech. Rep. ERL 380-WPL 49, NOAA Environmental Research Laboratories, Boulder, Colo.
———, and G. E. Greene, 1975. Infrasound from convective storms, part IV: is it useful for storm warning? *J. Appl. Meteorol.* 25:1303–16.
———, and J. M. Young, 1972. Passive sensing of natural acoustic-gravity waves at the earth's surface. In *Remote Sensing of the Troposphere,* V. E. Derr (ed.), U.S. Government Printing Office, Washington, D.C., pp. 21–1 to 21–20.
Goerke, V. H., and M. W. Woodward, 1966. Infrasonic observations of a severe weather system. *Mon. Weather Rev.* 94:395.
Grover, F. H., 1971. Experimental noise reducers for an active microbarograph array. *Geophys. J. R. Astron. Soc.* 26:41–52.
Hubbard, E. K., and A. J. Bedard, 1969. A pressure transducer for use as a component of an infrasonic microphone. ESSA Tech. Memo. ERLTM-WPL 4, NOAA Environmental Research Laboratories, Boulder, Colo.
Hung, R. J., R. E. Smith, G. S. West, and B. B. Henson, 1975. Detection of upper atmospheric disturbances in northern Alabama during extreme tornado outbreak of April 3, 1974. Preprints, Ninth Conference on Severe Local Storms. American Meteorological Society, Boston, pp. 294–300.
Jones, R. M., and T. M. Georges, 1976. Infrasound from convective storms, part III: propagation to the ionosphere. *J. Acoust. Soc. Am.* 59:765–779.
Kelvin, Lord, 1910. Vibrations of a columnar vortex. In *Mathematical and Physical Papers,* Cambridge University Press, 152–165 (also in *Phil. Mag.* [1880] 5:155–68).
Kimball, B. A., and E. R. Lemon, 1970. Spectra of air pressure fluctuations at the soil surface. *J. Geophys. Res.* 75:6771–77.
Kraus, M. J., 1973. Doppler radar observations of the Brookline, Massachusetts tornado of 9 August 1972. *Bull. Am. Meteorol. Soc.* 54:519–24.
Kropfli, R. A., and L. J. Miller, 1976. Kinematic structure and flux quantities in a convective storm from dual-Doppler radar observations. *J. Atmos. Sci.* 33:520–29.
Larson, R. J., L. B. Craine, J. E. Thomas, and C. R. Wilson, 1971. Correlation of winds and geographic features with the production of certain infrasonic signals in the atmosphere. *Geophys. J. R. Astron. Soc.* 26:201–14.
Lemon, L. R., D. W. Burgess, and R. A. Brown, 1978. Tornadic storm airflow and morphology derived from single-Doppler radar measurements. *Mon. Weather Rev.* 106:48–61.
McDonald, J. A., 1974. Naturally occurring atmospheric acoustical signals. *J. Acoust. Soc. Am.* 56:338–51.
———, E. J. Douze, and E. Herrin, 1971. The structure of atmospheric turbulence and its application to the design of pipe arrays. *Geophys. J. R. Astron. Soc.* 26:99–109.
Matheson, H., 1964. Instructions for the operation of NBS infrasonic equipment. NBS Rep. 8519, National Bureau of Standards, Washington, D.C. (Unpublished.)
Meecham, W. C., 1971. On aerodynamic infrasound. *J. Atmos. Terr. Phys.* 33:149–55.
Moo, C. A., and A. D. Pierce, 1972. Generation of anomalous ionospheric oscillation by thunderstorms. Effects of atmospheric acoustic-gravity waves on electromagnetic wave propagation, Advisory Group for Aerospace Research and Development, Conference Proceedings No. 115, North Atlantic Treaty Organization, Geneva, Switzerland.
Powell, A., 1964. Theory of vortex sound. *J. Acoust. Soc. Am.* 36:177–95.
Prasad, S. S., L. J. Schneck, and K. Davies, 1975. Ionospheric disturbances by severe tropospheric weather storms. *J. Atmos. Terr. Phys.* 37:1357–63.
Priestley, J. T., 1966. Calculation of the effectiveness of infrasonic line microphones for reducing wind noise. NBS Rep. 9380, National Bureau of Standards, Washington, D.C. (Unpublished.)
Procunier, R. W., and G. W. Sharp, 1971. Optimum frequency for detection of acoustic sources in the upper atmosphere. *J. Acoust. Soc. Am.* 49:622–26.
Reed, J. W., 1972. Attenuation of blast waves by the atmosphere. *J. Geophys. Res.* 77:1616–22.
Smart, E., and E. A. Flinn, 1971. Fast frequency-wavenumber analysis and Fisher signal detection in real-time infrasonic array data processing. *Geophys. J. R. Astron. Soc.* 26:279–84.
Smith, A. M. O., and A. B. Bauer, 1970. Static-pressure probes that are theoretically insensitive to pitch, yaw and Mach number. *J. Fluid Mech.* 44:513–28.
Smith, R. E., and R. J. Hung, 1975. Observation of severe-weather activities by Doppler-sounder array. *J. Appl. Meteorol.* 14:1611–15.
Swedish Defence Materiel Administration, 1985. *Infrasound: A bibliography of articles up till April 1983.* 115–88. Stockholm.
Thomas, J. E., A. D. Pierce, E. A. Flinn, and L. B. Craine, 1971. Bibliography on infrasonic waves. *Geophys. J. R. Astron. Soc.* 26:399–426.
Ward, N. B., 1972. The exploration of certain features of tornado dynamics using a laboratory model. *J. Atmos. Sci.* 29: 1194–1204.
Watts, J. M., and K. Davies, 1959. Rapid frequency analysis of fading radio signals. *J. Geophys. Res.* 65:2295–2301.
Young, J. M., and W. A. Hoyle, 1975. Computer programs for multidimensional spectra array processing. Tech. Rep. ERL 345–WPL 43, NOAA Environmental Research Laboratories, Boulder, Colo.

Chapter 7

Biggs, W. G., and P. J. Waite, 1970. Can TV really detect tornadoes? *Weatherwise* 23:120–24.
Chalmers, J. A., 1967. *Atmospheric Electricity.* Pergamon Press, New York.
Harth, W., 1972. VLF-atmospherics: their measurement and interpretation. *Z. Geophys.* 38:815–49.
Heydt G., and H. Volland, 1964. A new method for locating thunderstorms and counting their lightning discharges from a single observing station. *J. Atmos. Terr. Phys.* 32:609–21.
Horner, F., 1964. Radio noise from thunderstorms. In *Advances in Radio Research,* Vol. 2, J. A. Saxton (ed.). Academic Press, New York, pp. 121–204.
Hughes, H. G., and R. J. Gallenberger, 1974. Propagation of extremely low-frequency (ELF) atmospherics over a mixed day-night path. *J. Atmos. Terr. Phys.* 36:1643–61.
Kinzer, G. D., 1974. Cloud-to-ground lightning versus radar

reflectivity in Oklahoma thunderstorms. *J. Atmos. Sci.* 31: 787–99.

Krider, E. P., R. C. Noggle, and M. A. Uman, 1976. A gated wideband magnetic direction finder for lightning return strokes. *J. Appl. Meteorol.* 15:301–306.

Oetzel, G. N., and E. T. Pierce, 1969. VHF technique for locating lightning. *Radio Science* 4:199–201.

Taylor, W. L., 1973. Electromagnetic radiation from severe storms in Oklahoma during April 29–30, 1970. *J. Geophys. Res.* 78:8761–77.

———, 1978. A VHF technique for space-time mapping of lightning discharge processes. *J. Geophys. Res.* 83:3575–83.

Wait, J. R., 1970. *Electromagnetic Waves in Stratified Media.* Pergamon Press, New York.

Chapter 8

Bendat, J. S., and A. G. Peirsol, 1971. *Random Data: Analysis and Measurement Procedures.* John Wiley & Sons, New York.

Bent, R. B., and W. A. Lyons, 1984. Theoretical evaluations and initial operational experiences of LPATS (Lightning Position and Tracking System) to monitor lightning ground strikes using a time of arrival (TOA) technique. Preprints, Seventh International Conference on Atmospheric Electricity, Albany, N.Y., American Meteorological Society, Boston, pp. 317–24.

Bevington, P. R., 1969. *Data Reduction and Error Analysis for the Physical Sciences.* McGraw-Hill Book Co., New York.

Blakeslee, R. J., and E. P. Krider, 1984. The electric currents under thunderstorms at the NASA Kennedy Space Center. Preprints, Seventh International Conference on Atmospheric Electricity, Albany, N.Y. American Meteorological Society, Boston, pp. 265–68.

Bohannon, J. L., 1978. Infrasonic pulses from thunderstorms. Master's thesis, Rice University, Houston, Tex.

———, A. A. Few, and A. J. Dessler, 1977. Detection of infrasonic pulses from thunderclouds. *Geophys. Res. Lett.* 4:49–52.

Brook, M., R. Tennis, C. Rhodes, P. Krehbiel, B. Vonnegut, and O. H. Vaughan, Jr., 1980. Simultaneous measurements of lightning radiations from above and below clouds. *Geophys. Res. Lett.* 7:267–70.

Brown, K. A., P. R. Krehbiel, C. B. Moore, and G. N. Sargent, 1971. Electrical screening layers around charged clouds. *J. Geophys. Res.* 76:2825–35.

Chalmers, J. A., 1967. *Atmospheric Electricity.* Pergamon Press, Oxford.

Christian, H. J., and A. A. Few, 1977. The measurement of atmospheric electric fields using a newly developed balloon-borne sensor. In *Electrical Processes in Atmospheres,* Dolezalek and Reiter (eds.). D. R. Dietrich Steinkopff Verlag, Darmstadt, FRG, pp. 231–36.

———, C. R. Holmes, J. W. Bullock, W. Gaskell, A. J. Illingworth, and J. Latham, 1980. Airborne and ground-based studies of thunderstorms in the vicinity of Langmuir Laboratory, *Q. J. Roy. Meteorol. Soc.* 106:159–74.

Cooray, V., and S. Lundquist, 1983. Effects of propagation on the rise times and the initial peaks of radiation fields from return strokes. *Radio Sci.* 18:409–15.

Davis, M. H., M. Brook, H. Christian, B. G. Heikes, R. E. Orville, C. G. Park, R. A. Roble, and B. Vonnegut, 1983. Some scientific objectives of a satellite-borne lightning mapper. *Bull. Am. Meteorol. Soc.* 64:114–19.

Few, A. A., 1970. Lightning channel reconstruction from thunder measurements. *J. Geophys. Res.* 75:7517–23.

———, 1985. The production of lightning-associated infrasonic acoustic sources in thunderclouds. *J. Geophys. Res.* 90: 6175–80.

———, and T. L. Teer, 1974. The accuracy of acoustic reconstructions of lightning channels. *J. Geophys. Res.* 79:5007–11.

Gaskell, W., A. J. Illingworth, J. Latham, and C. B. Moore, 1978. Airborne studies of electric fields and the charge and size of precipitation elements in thunderstorms. *Q. J. Roy. Meteorol. Soc.* 104:447–60.

Goodman, S. J., H. J. Christian, W. D. Rust, D. R. MacGorman, and R. T. Arnold, 1984. Simultaneous observations of cloud-to-ground lightning above and below cloud tops. Preprints, Seventh International Conference on Atmospheric Electricity, Albany, N.Y. American Meteorological Society, Boston, pp. 456–62.

Hayenga, C. O., 1984. Characteristics of lightning VHF radiation near the time of return strokes. *J. Geophys. Res.* 89:1403–10.

———, and J. W. Warwick, 1981. Two-dimensional interferometric positions of VHF lightning sources. *J. Geophys. Res.* 86:7451–62.

Holmes, C. R., E. W. Szymanski, S. J. Szymanski, and C. B. Moore, 1980. Radar and acoustic study of lightning. *J. Geophys. Res.* 85:7517–32.

Horner, F., 1954. The accuracy of the location of sources of atmospherics by radio direction-finding. *Proc. IEEE, Radio Section,* 101:383–90.

Jacobson, E. A., and E. P. Krider, 1976. Electrostatic field changes produced by Florida lightning. *J. Atmos. Sci.* 33: 103–17.

Kasemir, H. W., 1972. The cylindrical field mill. *Meteorol. Rundsch.* 25:33–38.

Krehbiel, P. R., 1981. An analysis of the electric field change produced by lightning. 2 vols. Ph.D. diss., Rep. T-11, New Mexico Institute of Mining and Technology, Socorro.

———, M. Brooks, and R. A. McCrory, 1979. An analysis of the charge structure of lightning discharges to ground. *J. Geophys. Res.* 84:2432–56.

———, R. Tennis, M. Brook, E. W. Holmes, and R. Comes, 1984. A comparative study of the initial sequence of lightning in a small Florida thunderstorm. Preprints, Seventh International Conference on Atmospheric Electricity, Albany, N.Y. American Meteorological Society, Boston, pp. 279–85.

Krider, E. P., and J. A. Musser, 1982. Maxwell currents under thunderstorms. *J. Geophys. Res.* 87:11, 171–11, 176.

———, and C. Noggle, 1976. A gated, wideband magnetic direction finder for lightning return strokes. *J. Appl. Meteorol.* 15:301–306.

———, ———, A. E. Pifer, and D. L. Vance, 1980. Lightning direction-finding for forest fire detection. *Bull. Am. Meteorol. Soc.* 61:980–86.

Lhermitte, R., and P. R. Krehbiel, 1979. Doppler radar and radio observations of thunderstorms. *IEEE Trans. Geosci. Electron.* GE17:162–71.

———, and E. Williams, 1985. Thunderstorm electrification: a case study. *J. Geophys. Res.* 90:6071–78.

Ligda, M. G. H., 1956. The radar observation of lightning. *J. Atmos. Terres. Phys.* 9(516):329–46.

Liu, X., and P. R. Krehbiel, 1985. The initial streamer of intracloud lightning flashes. *J. Geophys. Res.* 90:6211–18.

MacGorman, D. R., 1977. Lightning in a Colorado thunderstorm.

M.S. thesis, Rice University, Houston, Tex.
———, 1978. Lightning location in a storm with strong wind shear. Ph.D. diss., Rice University, Houston, Tex.
———, A. A. Few, and T. L. Teer, 1981. Layered lightning activity. *J. Geophys. Res.* 86:9900–10.
———, M. W. Maier, and W. D. Rust, 1984. Lightning strike density for the contiguous United States from thunderstorm duration records. Report to the U.S. Nuclear Regulatory Commission (NTIS# NUREG/CR-3759).
———, W. L. Taylor, and A. A. Few, 1983. Lightning location from acoustic and VHF techniques relative to storm structure from 10-cm radar. *Proceedings in Atmospheric Electricity*, L. H. Ruhnke and J. Latham (eds.). A. Deepak Publishing, Hampton, Va., pp. 377–80.
Mach, D. M., D. R. MacGorman, W. D. Rust, and R. T. Arnold, 1986. Site errors and detection efficiency in a magnetic direction-finder network for locating lightning strikes to ground. *J. Atmos. Ocean. Tech.* 3:67–74.
Marshall, T. C., and W. P. Winn, 1982. Measurements of charged precipitation in a New Mexico thunderstorm: lower positive charge centers. *J. Geophys. Res.* 87:7141–57.
Mazur, V., and W. D. Rust, 1983. Lightning propagation and flash density in squall lines as determined with radar. *J. Geophys. Res.* 88:1495–1502.
———, B. D. Fisher, and J. C. Gerlach, 1984a. Lightning strikes to an airplane in a thunderstorm. *J. Aircraft* 21:607–11.
———, J. C. Gerlach, and W. D. Rust, 1984b. Lightning flash density versus altitude and storm structure from observations with UHF- and S-band radars. *Geophys. Res. Lett.* 11:61–64.
Orville, R. E., 1977. Lightning spectroscopy. Chap. 8 in *Lightning*, Vol. 1, R. H. Golde (ed.). Academic Press, London, 281–308.
Proctor, D. E., 1971. A hyperbolic system for obtaining VHF radio pictures of lightning. *J. Geophys. Res.* 76:1478–89.
———, 1981. VHF radio pictures of cloud flashes. *J. Geophys. Res.* 86:4041–71.
———, 1983. Lightning and precipitation in a small multicellular thunderstorm. *J. Geophys. Res.* 88:5421–40.
Ray, P. S., D. R. MacGorman, W. D. Rust, W. L. Taylor, and L. W. Rasmussen, 1987. Lightning location relative to storm structure in a supercell storm and a multicell storm. *J. Geophys. Res.*, 92:5713–24.
Robertson, F. R., G. S. Wilson, H. J. Christian, Jr., S. J. Goodman, G. H. Fichtl, and W. W. Vaughan, 1984. Atmospheric science experiments applicable to space shuttle spacelab missions. *Bull. Am. Meteorol. Soc.* 5:692–700.
Rust, W. D., and R. J. Doviak, 1982. Radar research on thunderstorms and lightning. *Nature* 297:461–68.
———, and C. B. Moore, 1974. Electrical conditions near the bases of thunderclouds over New Mexico. *Q. J. Roy. Meteorol. Soc.* 100:450–68.
———, D. R. MacGorman, and R. T. Arnold, 1981. Positive cloud-to-ground lightning flashes in severe storms. *Geophys. Res. Lett.* 8:791–94.
———, W. L. Taylor, and D. R. MacGorman, 1982. Preliminary study of lightning location relative to storm structure, *AIAA J.* 20:404–409.
Taylor, W. L., 1978. A VHF technique for space-time mapping of lightning discharge processes. *J. Geophys. Res.* 83:3575–83.
———, 1983. Lightning location and progression using VHF space-time mapping technique. *Proceedings in Atmospheric Electricity*, L. H. Ruhnke and J. Latham (eds.). A. Deepak Publishing, Hampton, Va., pp. 381–84.
———, E. A. Brandes, W. D. Rust, and D. R. MacGorman, 1984. Lightning activity and severe storm structure. *Geophy. Res. Lett.* 11:545–48.
Teer, T. L., and A. A. Few, 1974. Horizontal lightning. *J. Geophys. Res.* 79:3436–41.
Uman, M. A., 1969. *Lightning*. McGraw-Hill Book Co., New York.
———, and E. P. Krider, 1982. A review of natural lightning: experimental data and modeling. *IEEE Trans. Electromag. Comp.*, EMC-24, pp. 79–112.
———, Y. T. Lin, and E. P. Krider, 1980. Errors in magnetic direction finding due to nonvertical lightning channels. *Radio Sci.* 15:35–39.
———, C. E. Swanberg, J. A. Tiller, Y. T. Lin, and E. P. Krider, 1976. Effects of 200-km propagation on Florida lightning return stroke electric fields, *Radio Sci.* 11:985–90.
Vonnegut, B., and R. E. Passarelli, Jr., 1978. Modified cine sound camera for photographing thunderstorms and recording lightning. *J. Appl. Meteorol.* 17:1079–81.
———, O. H. Vaughan, Jr., M. Brook, and P. Krehbiel, 1985. Mesoscale observations of lightning from Space Shuttle. *Bull. Am. Meteorol. Soc.* 66:20–29.
Warwick, J. W., C. O. Hayenga, and J. W. Brosnahan, 1979. Interferometric directions of lightning sources at 34 MHz. *J. Geophys. Res.* 84:2457–63.
Weber, M. E., H. J. Christian, A. A. Few, and M. F. Stewart, 1982. A thundercloud electric field sounding: charge distribution and lightning. *J. Geophys. Res.* 87:7158–69.
Williams, E. R., C. M. Cooke, and K. A. Wright, 1985. Electrical discharge propagation in and around space charge clouds. *J. Geophys. Res.* 90:6059–70.
Wilson, C. T. R., 1929. Some thundercloud problems. *J. Franklin Inst.* 208:1–12.
Winn, W. P., 1968. An electrostatic theory for instruments which measure the radii of water drops by detecting a change in capacity due to the presence of a drop. *J. Appl. Meteor.* 7:929–37.
———, 1984. Personal communication. New Mexico Institute of Mining and Technology, Socorro, N.Mex.
———, and C. B. Moore, 1971. Electric field measurements in thunderclouds using instrumented rockets. *J. Geophys. Res.* 76:5003–17.
———, T. V. Aldridge, and C. B. Moore, 1973. Video tape recordings of lightning flashes. *J. Geophys. Res.* 21:4515–19.
———, C. B. Moore, C. R. Holmes, and L. G. Byerly, III, 1978. Thunderstorm on July 16, 1975, over Langmuir Laboratory: a case study. *J. Geophys. Res.* 83:3079–92.
———, ———, and ———, 1981. Electric field structure in an active part of a small, isolated thundercloud. *J. Geophys. Res.* 86:1187–93.
Zrnić, D. S., W. D. Rust, and W. L. Taylor, 1982. Doppler radar echoes of lightning and precipitation at vertical incidence. *J. Geophys. Res.* 87:7179–91.

Chapter 9

Note: The Illinois State Water Survey bibliography on hail and hail preservation is available to bona fide researchers at a small cost. For information contact Head, Atmospheric Science Section, Illinois State Water Survey, P.O. Box 203, Urbana, Illinois 61801. The March 1978 issue of *Atmosphere-Ocean,* the largest single-volume collection of papers on hailfall measurement, can be obtained from the University of Toronto Press, Journals Dept., 5201 Dufferin St., Toronto, Ont., Canada M3H 5T8.

Atlas, D., and F. H. Ludlam, 1961. Multiwavelength radar reflectivity of hailstorms. *Q. J. R. Meteorol. Soc.* 87:523–34.

———, M. Kerker, and W. Hitschfeld, 1953. Scattering and attenuation by non-spherical atmospheric particles. *J. Atmos. Terr. Phys.* 3:108–19.

Aydin, K., T. A. Seliga, and V. Balaji, 1986. Remote sensing of hail with a dual linear polarization radar. *J. Climate Appl. Meteorol.* 25:2475–84.

Barge, B. L., 1970. Polarization observations in Alberta. Preprints, Fourteenth Radar Meteorology Conference, Tucson, Arizona. American Meteorological Society, Boston, pp. 221–24.

Beckwith, W. B., 1960. Analysis of hailstorms in the Denver network. *J. Meteorol.* 23:348–53.

Bringi, V. N., T. A. Seliga, and K. Aydin, 1984. Hail detection with a differential reflectivity radar. *Science* 225:1145–57.

Browning, K. A., and G. B. Foote, 1975. Airflow and hailgrowth in supercell storms and some implications for hail suppression. Tech. Rep. 75/1, National Center for Atmospheric Research, Boulder, Colo.

———, J. Hallett, T. W. Harrold, and D. Johnson, 1968. The collection and analysis of freshly fallen hailstones. *J. Appl. Meteorol.* 7:603–12.

Carte, A. E., and G. Held, 1978. Variability of hailstorms on the South African plateau. *J. Appl. Meteorol.* 17:365–73.

———, ———, C. East, K. L. S. Gunn, W. Hitschfeld, J. S. Marshall, E. J. Stansbury, 1961. Alberta hail studies, 1961. Sci. Rep. MW-35, McGill University, Montreal.

———, R. H. Douglas, R. C. Srivastava, and G. N. Williams, 1963. Alberta hail studies, 1962/63. Sci. Rep. MW-36, McGill University, Montreal.

Changnon, S. A., Jr. 1966. Note on recording hail occurrences. *J. Appl. Meteorol.* 5:899–901.

———, 1968. Effect of sampling density on areal extent of damaging hail. *J. Appl. Meteorol.* 7:518–21.

———, 1971a. Note on hailstone size distributions. *J. Appl. Meteorol.* 10:168–70.

———, 1971b. Means for estimating areal hail-day frequencies. *J. Weather Modif.* 3:154–59.

———, 1971c. Hailfall characteristics related to crop damage. *J. Appl. Meteorol.* 10:270–74.

———, 1972. Illinois radar research for hail suppression applications, 1967–1969. Invest. Rep. 71, Illinois State Water Survey, Urbana.

———, and N. A. Barron, 1971. Quantification of crop-hail losses by aerial photography. *J. Appl. Meteorol.* 10:86–96.

———, and G. M. Morgan, Jr., 1976. Design of an experiment to suppress hail in Illinois. Bull. 61, Illinois State Water Survey, Urbana.

———, and D. W. Staggs, 1969. Recording hailgage evaluation. Final Rep. NSF-GA-1520, Illinois State Water Survey, Urbana.

CHIAA, 1978. *Crop-Hail Insurance Statistics.* Crop Hail Insurance Actuarial Association, Chicago.

Crow, E. L., 1974. Confidence limits for digital error rates. OT Rep. 74–51, Department of Commerce, U.S. Government Printing Office, Washington, D.C.

Douglas, R. H., and W. Hitschfeld, 1958. Studies of Alberta hailstorms. Sci. Rep. MW-27, McGill University, Montreal.

Doviak, R. J., and Zrnić, D. S., 1984. *Doppler radar and weather observations.* Academic Press, New York (see esp. pp. 208ff.).

Eccles, P. J., and D. Atlas, 1969. A dual wavelength radar hail detector. Tech. Rep. 14, Lab. for Atmos. Probing, University of Chicago, Chicago.

Federer, B., and A. Waldvogel, 1975. Hail and raindrop size distributions from a Swiss multicell storm. *J. Appl. Meteorol.* 14:91–97.

———, A. Waldvogel, W. Schmid, H. H. Schiesser, F. Hampel, M. Schweingrubber, W. Stahel, J. Bader, F. J. Mezeix, N. Doras, G. d'Aubigny, G. DerMegreditchian, and D. Vento, 1986. Main results of Grossversuch IV. *J. Climate Appl. Meteorol.* 5:917–57.

Fremstad, P. G., 1968. Transducer for measuring hailstone momentum. Rep. 68-4, Institute for Atmospheric Sciences, South Dakota School of Mines and Technology, Rapid City.

Garcia, R. R., B. E. Weiss, and A. H. Murphy, 1976. Relationships between crop damage and hailfall parameters on the high plains. NHRE Tech. Rep. NCAR-7100-76/4. National Center for Atmospheric Research, Boulder, Colo.

Goyer, G. G., 1970. The testing of airborne infrared detection systems for mapping hailswaths and measuring hailfall coverage at the ground. The Joint Hail Research Project-Summer 1969-Summary Report. NCAR Internal Rep., February 1970, 24–29.

———, 1971. The Joint Hail Research Project-Summer, 1970. NCAR Internal Rep., March 1971, 44–53.

———, 1978. Meeting review: the First International Workshop on Hailfall Measurement, Banff, Alberta, Canada, 22–26 October 1979. *Bull. Am. Meteorol. Soc.* 59:297–98.

Hitschfeld, W. F., 1971. Hail research at McGill, 1956–1971. Sci. Rep. MW-68, Stormy Weather Group, McGill University, Montreal.

Jameson, H. R., and A. J. Heymsfeld, 1984. Comments on "antenna beam patterns and dual-wavelength processing." *J. Climate Appl. Meteorol.* 23:855–57.

Johnson, G. N., and P. L. Smith, Jr., 1978. Measurement of hailfall intensity with a self-contained hailstone momentum sensor. *Atmosphere-Ocean* 16:86–93.

Joss, J., and A Waldvogel, 1969. Raindrop size distribution and sampling size errors. *J. Atmos. Sci.* 26:566–69.

Knight, C. A., and N. C. Knight, 1968. The final freezing of spongy ice: hailstone collection techniques and interpretations of structures. *J. Appl. Meteorol.* 7:875–81.

———, and P. Squires (eds.), 1982. *Hailstorms of the Central High Plains.* Vol. 1: *The National Hail Research Experiment.* Vol. 2: *Case studies of the National Hail Research Experiment.* Colorado Associated University Press, Boulder.

Koren, D., 1969. Development of a moving aluminum foil hail recorder. TEC 711, Meteorological Branch Downsview, Toronto, Ont.

Kozminski, C., and S. Bac, 1964. Prototype de grêlimètre enregistreur. *J. Rech. Atmos.* 1:55–56.

Long, A. B., 1978. Design of hail measurement networks. *Atmosphere-Ocean* 16:35–48.

Lozowski, E. P., and G. S. Strong, 1978. On the calibration of hailpads. *J. Appl. Meteorol.* 17:521–28.

———, R. Erb, L. Wojtiw, M. Wong, G. S. Strong, R. Matson, A. Ong, D. Vento, and P. Admirat, 1978. The hail sensor intercomparison experiment. *Atmosphere-Ocean* 16:94–106.

McNeil, R. D., and J. E. Houston, 1966. A survey of methods of remote detection of hail. Project Hailswaths, Final Rep. 66-9, Institute for Atmospheric Sciences, South Dakota School of Mines and Technology, Rapid City.

Marriott, W., 1892. Report on the thunderstorms of 1888 and 1889. *Q. J. R. Meteorol. Soc.* 18:23–29.

Matson, R., A. Long, D. Vento, and P. Admirat, 1978. The hail sensor intercomparison experiment. *Atmosphere-Ocean* 16: 94–106.

Morgan, G. M., Jr., 1969. Thunderstorm studies at Verona, Italy. Preprints, Sixth Conference on Severe Local Storms, Chicago, Illinois. American Meteorological Society, Boston, pp. 332–37.

———, 1973. A general description of the hail problem in the Po Valley of northern Italy. *J. Appl. Meteorol.* 12:338–53.

———, 1982. Precipitation at the ground. Chap. 4 in *Hailstorms of the Central High Plains. Vol. 1: The National Hail Research Experiment,* C. A. Knight and P. Squires (eds.), Colorado Associated University Press, Boulder, Colo.

———, and N. G. Towery, 1975. Small scale variability of hail and its significance for hail prevention experiments. *J. Appl. Meteorol.* 14:763–70.

———, and ———, 1976a. Crop damage-hailpad parameter study in Illinois. Final Rep. NSF S-5015, Urbana.

———, and ———, 1976b. On the role of strong winds in damage to crops and its estimation with a simple instrument. *J. Appl. Meteorol.* 8:891–98.

Mueller, E. A., and S. A. Changnon, Jr., 1968. A recording hailgage for use in hail modification projects. Preprints, First National Conference on Weather Modification. American Meteorological Society, Boston, pp. 494–502.

Musil, D. J., E. L. May, P. L. Smith, Jr., and W. R. Sand, 1976. Structure of an evolving hailstorm; part IV: internal structure from penetrating aircraft. *Mon. Weather Rev.* 104(5):596–602.

Nelson, S. P., and S. K. Young, 1979. Characteristics of Oklahoma hailfalls and hailstorms. *J. Appl. Meteorol.* 18:339–47.

Nicholas, T. R., 1977. A review of surface hail measurement, pp. 257–67 in *Hail: A review of hail science and hail suppression.* Meteorol. Monogr. 16(38).

———, 1978. A review of surface hail measurement. In *Hail: A Review of Hail Science and Hail Suppression,* Meteorol. Monogr. 16, American Meteorological Society, Boston, pp. 257–67.

Pini, E., 1885. *Sui temporali osservati nell' Italia superiore durante l'anno 1879.* Publ. Reale Osserv. Brera di Milano, No. 18, Hoepli, Milano.

Prohaska, K., 1905. Prohaska uber Zugrichtung, Starke und Geschwindigkeit der Hagelwetter, Dauer des Hagelfalles 1902 und im Mittle. *Met. Zeit.* 22:519–23.

———, 1907. Die hagelfalle des 6 Juli 1905 in den Ostalpen. *Met. Zeit.* 24:13–201.

Rinehart, R. E., and J. D. Tuttle, 1982. Antenna beam patterns and dual wavelength processing. *J. Appl. Meteorol.* 21: 1865–80.

———, and ———, 1984. Reply (to Jameson and Heymsfeld). *J. Climate Appl. Meteorol.* 23:859–61.

Roads, J. O. 1973. A study of hailswaths by means of airborne infrared radiometry. *J. Appl. Meteorol.* 12:855–62.

Roos, D. v. d. S. 1978. Hailstone sizes inferred from dents in cold-rolled aluminum sheets. *J. Appl. Meteorol.* 17: 1234–1239.

Schickedanz, P. T., and S. A. Changnon, Jr., 1971. The design and evaluation of the NHRE, in northeast Colorado, *J. Weather Modif.* 3:160–76.

Schleusener, R. A., and P. C. Jennings, 1960. An energy method for relative estimates of hail intensity. *Bull. Am. Meteorol. Soc.* 41:372–76.

Seliga, T. A., and V. N. Bringi, 1976. Potential use of radar differential reflectivity measurement or orthogonal polarizations for measuring precipitation. *J. Appl. Meteorol.* 15:69–76.

Silverman, B., 1981. On the sampling variance of precipitation gage networks. *J. Appl. Meteorol.* 20(12).

Srivastava, R. C., and A. R. Jameson, 1977. Radar detection of hail. In *Hail: A Review of Hail Science and Hail Suppression,* Meteorol. Monogr. 16, American Meteorological Society, Boston, pp. 269–77.

Stoute, G. E., and S. A. Changnon, Jr., 1968. Climatology of hail in the central United States. CHIAA Res. Rep. No. 38, Crop Hail Insurance Actuarial Association, Chicago.

Strong, S., and E. P. Lozowski, 1977. An Alberta study to objectively measure hailfall intensity. *Atmosphere* 15:34–53.

Summers, P. W. (ed.), 1968. Alberta hail studies 1967 field program. Unpublished report, Research Council of Alberta, Edmonton.

Torlaschi, E. C., R. G. Humphrey, and B. L. Barge, 1984. Circular polarization for precipitation measurements. *Radio Sci.* 19:93–200.

Towery, N. G., S. A. Changnon, Jr., and G. M. Morgan, Jr., 1976. A review of hail measuring instruments. *Bull. Am. Meteorol. Soc.* 57:1132–40.

———, R. E. Eyton, C. L. Dailey, and D. E. Luman, 1976. Annual report of remote sensing of crop-hail damage. Report of Research for the Country Companies. Urbana, Ill.

Veal, D. L., 1970. Proposed programs of hailstorm research in association with the NHRE. A proposal by Natural Resources Research Institute, College of Engineering, University of Wyoming, Laramie.

Vento, D., 1972. La determinazione della energia di impatto della grandine. *Rev. Ital. Geofis.* 21:73–77.

———, 1976. The hailpad calibration for Italian hail damage documentation. *J. Appl. Meteorol.* 15:1018–22.

———, and G. M. Morgan, Jr., 1976. Statistical evaluation of energy imparted to hail by wind in Europe and United States. Second WMO Scientific Conference on Weather Modification, WMO-No. 443, World Meteorological Organization, Geneva, pp. 281–85.

Waldvogel, A., B. Federer, W. Schmid, and J. F. Mezeix, 1978. The kinetic energy of hailfalls, part II: radar and hailpads. *J. Appl. Meteorol.* 17:1680–93.

Weickmann, H. K., 1969. ESSA-1968-Colorado hail research. Preprints, Sixth Conference on Severe Local Storms, Chicago, Illinois. American Meteorological Society, Boston, pp. 314–19.

Wilk, K. E., 1961. Radar investigation of Illinois hailstorms. Sci. Rep. 1, USAF 19(604)-4940, Illinois State Water Survey, Urbana.

Wojtiw, L., and J. H. Renick, 1973. Hailfall and crop damage in Alberta. Preprints, Eighth Conference on Severe Local Storms, Denver, Colorado. American Meteorological Society, Boston, pp. 138–41.

Chapter 10

Abramowitz, M., and I. A. Stegun, 1964. Handbook of mathematical functions. National Bureau of Standards, Appl. Math. Ser. 55, U.S. Government Printing Office, Washington, D.C.

Ahnert, P. R., W. F. Krajewski, and E. B. Johnson, 1986. Kalman filter estimation of radar-rainfall field bias. Preprints, Twenty-third Conference on Radar Meteorology and Conference on Cloud Physics. American Meteorological Society, Boston, pp. 3:JP33–JP37.

Armijo, L., 1969. A theory for the determination of wind and precipitation velocities with Doppler radars. *J. Atmos. Sci.* 26:570–73.

Atlas, D., 1964. Advances in radar meteorology. In *Advances in Geophysics,* Landsberg and Mieghem (eds.), Academic Press, New York, pp. 317–478.

———, K. R. Hardy, K. M. Glover, I. Katz, and T. G. Konrad, 1966. Tropopause detected by radar. *Science* 153:1110–12.

———, R. C. Srivastava, and R. S. Sekhon, 1973. Doppler radar characteristics of precipitation at vertical incidence. *Rev. Geophys. Space Phys.* 2:1–35.

Balsley, B. B., 1978. The use of sensitive coherent radars to examine atmospheric parameters in the height range 1-100 km. Preprints, Eighteenth Conference on Radar Meteorology. American Meteorological Society, Boston, pp. 190–93.

Battan, L. J., 1973. *Radar Observation of the Atmosphere.* University of Chicago Press, Chicago.

Bean, B. R., and E. J. Dutton, 1966. *Radio Meteorology.* National Bureau of Standards Monogr. 92, U.S. Government Printing Office, Washington, D.C.

Berger, M. I., and R. J. Doviak, 1979. An analysis of the clear air planetary boundary layer wind synthesized from NSSL's dual Doppler-radar data. Tech. Memo. ERL NSSL-87, NOAA Environmental Research Laboratories, Boulder, Colo.

Blake, L. V., 1970. Prediction of radar range. In *Radar Handbook,* M. I. Skolnik (ed.), McGraw-Hill Book Co., New York, 2.1–2.73.

Born, M., and E. Wolf, 1964. *Principles of Optics.* 2d ed. Macmillan Co., New York.

Brandes, E. A., 1975. Optimizing rainfall estimates with the aid of radar. *J. Appl. Meteorol.* 14:1339–45.

Brewster, K. A., and D. S. Zrnić, 1986. Comparison of eddy dissipation rates from spatial spectra of Doppler velocities and Doppler spectrum widths. *J. Oceanic and Atmos. Tech.* 3: 440–52.

Brown, R. A., L. R. Lemon, and D. W. Burgess, 1978. Tornado detection by pulsed Doppler-radar. *Mon. Weather Rev.* 106:29–38.

Browning, K. A., and R. Wexler, 1968. The determination of kinematic properties of a wind field using Doppler radar. *J. Appl. Meteorol.* 7:105–13.

Burgess, D. W., 1976. Single Doppler radar vortex recognition: part I—mesocyclone signatures. Preprints, Seventeenth Conference on Radar Meteorology. American Meteorological Society, Boston, pp. 97–103.

———, J. D. Bonewitz, and D. R. Devore, 1978. Joint Doppler operational project: results year 1. Preprints, Eighteenth Conference on Radar Meteorology. American Meteorological Society, Boston, pp. 442–48.

———, L. D. Hennington, R. J. Doviak, and P. S. Ray, 1976. Multimoment Doppler display for severe storm identification. *J. Appl. Meteorol.* 15:1302–1306.

Caton, P. A. F., 1963. Wind measurement by Doppler radar. *Meteorol. Mag.* 92:213–22.

Chadwick, R. B., K. P. Moran, and G. E. Morrison, 1978. Measurements toward a C_n^2 climatology. Preprints, Eighteenth Conference on Radar Meteorology. American Meteorological Society, Boston, pp. 100–103.

Chernikov, A. A., Y. V. Mel'nichuk, N. Z. Pinus, S. M. Shmeter, and N. K. Vinnichenko, 1969. Investigations of the turbulence in convective atmosphere using radar and aircraft. *Radio Sci.* 4:1257–1259.

Crane, R. K., 1979. Automatic cell detection and tracking. *IEEE Trans. Geo.-sci. Electron.* GE-17(4):250–61.

Crawford, K. C., 1977. The design of a multivariate mesoscale field experiment. Ph.D. diss., University of Oklahoma, Norman.

Donaldson, R. J., Jr., 1970. Vortex signature recognition by a Doppler radar. *J. Appl. Meteorol.* 9:661–70.

Doviak, R. J., and M. Berger, 1980. Turbulence and waves in the optically clear planetary boundary layer resolved by dual-Doppler radars. *Radio Sci.* 15(2):297–318.

———, and C. T. Jobson, 1979. Dual Doppler radar observations of clear air wind perturbations in the planetary boundary layer. *J. Geophys. Res.* 84:697–702.

———, and J. T. Lee, 1985. Radar for storm forecasting and weather hazard warning. *J. Aircraft.* 22(12):1059–1064.

———, and D. S. Zrnić, 1979. Receiver bandwidth effect on reflectivity and Doppler velocity estimates. *J. Appl. Meteorol.* 18:69–76.

———, and D. S. Zrnić, 1984. *Doppler Radar and Weather Observations.* Academic Press, New York.

———, D. Burgess, L. Lemon, and D. Sirmans, 1974. Doppler velocity and reflectivity structure observed within a tornadic storm. *J. Rech. Atmos.* 8:235–43.

———, R. M. Rabin, and A. J. Koscielny, 1983. Doppler weather radar for profiling and mapping winds in the prestorm environment. *IEEE Trans. Geosci. Remote Sensing* GE-21(1): 25–33.

———, P. S. Ray, R. G. Strauch, and L. J. Miller, 1976. Error estimation in wind fields derived from dual-Doppler radar measurement. *J. Appl. Meteorol.* 15:868–78.

———, D. Sirmans, D. Zrnić, and G. B. Walker, 1978. Considerations for pulse-Doppler radar observations of severe thunderstorms. *J. Appl. Meteorol.* 17:189–205.

Eilts, M. D., 1986. Low altitude wind shear detection with Doppler radar. *J. Climate and Appl. Meteor.*, Jan. 1987.

———, and R. J. Doviak, 1987. Oklahoma downbursts and their asymmetry. *J. Climate and Appl. Meteor,* Jan. 1987.

Evans, J. E., D. Johnson, 1984. The FAA transportable Doppler weather radar. Reprints, Twenty-second Conference on Radar Meteorology, Zurich, American Meteorological Society, Boston, pp. 112–13.

Foote, G. B., and P. S. duToit, 1969. Terminal velocity of raindrops aloft. *J. Appl. Meteorol.* 8:249–53.

Frisch, A. S. and S. F. Clifford, 1974. A study of convection capped by a stable layer using Doppler radar and acoustic echo sounders. *J. Atmos. Sci.* 31:1622–28.

Fujita, T. T., 1981. Tornadoes and downbursts in the context of generalized planetary scales. *J. Atmos. Sci.* 38:1511–34.

———, 1985. *The Downburst, Microburst and Macroburst,* University of Chicago Press, Chicago.

Fukao, S., T. Sato, T. Tsuda, M. Yamamoto, and S. Kato, 1986. High resolution turbulence observations in the middle and lower atmosphere by the MU radar with fast beam steerability: preliminary results. *J. Atmos. Terres. Phys.* 48:1269–78.

———, K. Wakasugi, T. Sato, S. Morimoto, T. Tsuda, I. Hirota,

I. Kimura, and S. Kato, 1985. Direct measurement of air and precipitation particle motion by very high frequency Doppler radar. *Nature* 316:712.

Gage, K. S., and J. L. Green, 1978. Evidence for specular reflection from monostatic VHF radar observations of the stratosphere. *Radio Sci.* 13:991–1001.

———, W. P. Birkemeyer, and W. H. Jasperson, 1973. Atmospheric stability measurement at tropopause altitudes using forward-scatter CW radar. *J. Appl. Meteorol.* 12:1205–12.

Gossard, E. E., and A. S. Frisch, 1976. Kinematic models of a dry convective boundary layer compared with dual Doppler-radar observations of wind fields. *Boundary-Layer Meteorol.* 10:311–30.

———, and W. R. Moninger, 1975. The influence of a capping inversion on the dynamic and convective instability of a boundary layer model with shear. *J. Atmos. Sci.* 32:2111–24.

Green, J. L., J. M. Warnock, R. H. Winkler, and T. E. VanZandt, 1975. Studies of winds in the upper troposphere with a sensitive VHF radar. *Geophys. Res. Lett.* 2:19–21.

———, R. H. Winkler, J. M. Warnock, W. L. Clark, K. S. Gage, and T. E. VanZandt, 1978. Observations of enhanced clear air reflectivity associated with convective clouds. Preprints, Eighteenth Conference on Radar Meteorology. American Meteorological Society, Boston, pp. 88–93.

Gunn, K. L. S., and T. W. R. East, 1954. The microwave properties of precipitation particles. *Q. J. R. Meteorol. Soc.* 80: 522–45.

Hardy, K. R., and I. Katz, 1969. Probing the clear atmosphere with high power, high resolution radars. *Proc. IEEE* 57: 468–80.

———, D. Atlas, and K. M, Glover, 1966. Multiwavelength backscatter from the clear atmosphere. *J. Geophys. Res.* 71: 1537–52.

Harrold, T. W., and K. A. Browning, 1971. Identification of preferred areas of shower development by means of high power radar. *Q. J. R. Meteorol. Soc.* 97:330–39.

Hennington, L., R. J. Doviak, D. Sirmans, D. Zrnić, and R. G. Strauch, 1976. Measurements of winds in the optically clear air with microwave pulse-Doppler radar. Preprints, Seventeenth Conference on Radar Meteorology. American Meteorological Society, Boston, pp. 342–48.

Herman, B. M., S. R. Browning, and L. J. Battan, 1961. Tables of the radar cross sections of water sphere. Tech. Rep. No. 9, Institute of Atmospheric Physics, University of Arizona, Tucson.

Hess, S. L. *Introduction to Theoretical Meteorology*. Holt, Rinehart and Winston, New York.

Hicks, J. J., and J. K. Angell, 1968. Radar observations of breaking gravitational waves in the visually clear atmosphere. *J. Appl. Meteorol.* 7:114–21.

IEEE, (Institute of Electrical and Electronics Engineers), 1955. Scatter propagation issue. *Proc. IEEE* (formerly IRE) 43: 1173–1570.

Jordan, E. C., 1950. *Electromagnetic Waves and Radiating Systems*. Prentice-Hall, Englewood Cliffs, N.J.

Katz, I. 1966. Sea breeze structure as seen by radar. Memo. BPD 66U-25, Appl. Phys. Lab., John Hopkins University, Baltimore, Md.

Kessler, E., 1985. Wind shear and aviation safety. *Nature* 315(6061):179–80.

Konrad, T. G., 1968. The alignment of clear air convective cells. Preprints, International Conference on Cloud Physics, Toronto, Ontario. American Meteorological Society, Boston, pp. 539–43.

———, 1970. The dynamics of the convective process in the clear air as seen by radar. Preprints, Fourteenth Conference on Radar Meteorology, Tucson, Arizona. American Meteorological Society, Boston, pp. 57–60.

Kraus, J. D., 1966. *Radio Astronomy*. McGraw-Hill Book Co., New York.

Kropfli, R. A., and N. M. Kohn, 1978. Persistent horizontal rolls in the urban mixed layer as revealed by dual-Doppler radar. *J. Appl. Meteorol.* 17:669–76.

Kuettner, J. P., 1971. Cloud bands in the Earth's atmosphere. *Tellus* 23:404–25.

Lee, J. T., and C. Goff, 1976. Gust front wind shear and turbulence—concurrent aircraft and surface based observations. Preprints, Seventh Conference on Aerospace and Aeronautical Meteorology and Symposium on Remote Sensing from Satellites, Melbourne, Florida. American Meteorological Society, Boston, pp. 48–54.

Lhermitte, R. M., 1970. Dual-Doppler radar observations of convective storm circulation. Preprints, Fourteenth Radar Meteorological Conference, Tuscon, Arizona. American Meteorological Society, pp. 153–56.

———, and D. Atlas, 1961. Precipitation motion by pulse Doppler radar. Preprints, Ninth Weather Radar Conference, 23–26 October, Kansas City, Missouri. American Meteorological Society, Boston, pp. 218–223.

Linden, P. F., and J. E. Simpson, 1985. Microbursts: a hazard for aircraft *Nature* 317:601–602.

McCarthy, J., and R. Serafin, 1984. The microburst: hazard to aviation. *Weatherwise* 37:3:120–27.

———, and J. W. Wilson, 1984. The microburst as a hazard to aviation: structure, mechanisms, climatology and nowcasting. Preprints, Nowcasting II Symposium, Norrkoping, Sweden, pp. 21–30.

———, R. Roberts, and W. Schreiber, 1984. JAWS data collection, analysis highlights, and microburst statistics. Preprints, Twenty-first Conference on Radar Meteorology. American Meteorological Society, Boston, pp. 596–601.

Meyer, J. H., 1970. Radar observations of land breeze fronts at Wallops Island, Virginia. Preprints, Fourteenth Conference on Radar Meteorology, Tucson, Arizona. American Meteorological Society, Boston, pp. 61–67.

Nathanson, F. E., 1969. *Radar Design Principles*. McGraw-Hill Book Co., New York.

———, and P. L. Smith, 1972. A modified coefficient for the weather radar equation. Preprints, Fifteenth Radar Meteorological Conference, Champaign-Urbana, Illinois, 10–12 October. American Meteorological Society, Boston, pp. 228–30.

NSSL (National Severe Storms Laboratory), 1979. Final report on the Joint Doppler Operational Project (JDOP), 1976–78. Tech. Memo. ERL NSSL-86, NOAA Environmental Research Laboratories, Boulder, Colo.

O'Bannon, T., 1978. A study of dual-Doppler synthesized clear air wind fields. Preprints, Eighteenth Conference on Radar Meteorology. American Meteorological Society, Boston, pp. 65–69.

Papoulis, A., 1965. *Probability Random Variables and Stochastic Processes*. McGraw-Hill Co., New York.

Probert-Jones, J. R., 1962. The radar equation in meteorology. *Q. J. R. Meteorol. Soc.* 88:485–95.

Ray, P. S., R. J. Doviak, G. B. Walker, D. Sirmans, J. Carter, and

B. Bumgarner, 1975. Dual-Doppler observation of a tornadic storm. *J. Appl. Meteorol.* 14:1521–30.

Rice, P. L., A. G. Langley, K. A. Norton, and A. P. Barsis, 1967. Transmission loss predictions for tropospheric communication circuits. National Bureau of Standards Tech. Note 101, vols. 1–2. U.S. Government Printing Office, Washington, D.C.

Rideout, V. C., 1954. *Active Networks.* Prentice-Hall, Englewood Cliffs, N.J.

Rinehart, R. E., 1979. Internal storm motion from a single non-Doppler weather radar. Ph.D. diss., Colorado State University, Fort Collins.

———, and M. A. Isaminger, 1986. Radar characteristics of microbursts in the Mid-South. Preprints, Twenty-third Conference on Radar Meteorology and Conference on Cloud Physics. American Meteorological Society, Boston, 3:J116–J119.

Roberts, R. D., and J. W. Wilson, 1986. Nowcasting microburst events using single Doppler radar data. Preprints, Twenty-third Conference on Radar Meteorology. Snowmass, American Meteorological Society, Boston, pp. 14–17.

Röttger, J., and C. H. Liu, 1978. Partial reflection and scattering of VHF radar signals from the clear atmosphere. *Geophys. Res. Lett.* 5:357–60.

Saxton, A. (ed.), 1964. *Advances in Radio Research.* Academic Press, New York.

Sherman, J. W., III, 1970. Aperture-antenna analysis. In *Radar Handbook,* M. I. Skolnik (ed.). McGraw-Hill Book Co., New York, 9.1–9.40.

Shuman, F. G., 1957. Numerical methods and weather predictions, II. *Mon. Weather Rev.* 85:357–87.

Silver, S., 1949. *Microwave Antenna Theory and Design.* McGraw-Hill Book Co., New York.

Skolnik, M. I., 1970. *Radar Handbook.* McGraw Hill Book Co., New York.

Smith, R. L., and D. W. Holmes, 1961. Use of Doppler radar in meteorological observations. *Mon. Weather Rev.* 89:1–7.

Stackpole, J. D., 1961. The effectiveness of raindrops as turbulence sensors. Preprints, Ninth Weather Radar Conference, Kansas City, Missouri. American Meteorological Society, Boston, pp. 212–17.

Sychra, J. 1972. Relation between real and pulse volume averaged fields of reflectivity and velocity. Canadian Meteorol. Service Res. Report CMRR 5/72.

Taylor, J. W., and J. Mattern, 1970. Receivers. In *Radar Handbook,* M. I. Skolnik (ed.). McGraw-Hill Book Co., New York, 5.1–5.50.

Uyeda, H., and D. S. Zrnić, 1986. Automatic detection of gust fronts. *J. Atmos. and Oceanic Tech.* 3:36–50.

VanZandt, T. E., J. L. Green, K. S. Gage, and W. L. Clark, 1978. Vertical profiles of refractivity turbulence structure constant: comparison of observations by the sunset radar with new theoretical model. *Radio Sci.* 13:819–29.

Wilson, J. W., R. D. Roberts, C. Kessinger, and J. McCarthy, 1984. Microburst structure and evaluation of Doppler radar for airport wind shear detection. *J. Climate and Appl. Meteor.* 23:898–915.

Wood, V. T., and R. A. Brown, 1983. Single Doppler velocity signatures: an atlas of patterns in clear air/widespread precipitation and convective storms. Tech. Memo. ERL NSSL-95, NOAA Environmental Research Laboratories, Norman, Okla.

———, and ———, 1986. Single Doppler velocity signature interpretation of nondivergent environmental winds. *J. Atmos. and Oceanic Tech.* 3:114–28.

Woodman, R. F., and A. Guillen, 1974. Radar observations of winds and turbulence in the stratosphere and mesosphere. *J. Atmos. Sci.* 31:493–505.

Yaglom, AA. M., 1949. On the local structure of the temperature field in a turbulent flow. *Dokl. Akad. Nauk. SSSR* 69:743.

Zeoli, G. W., 1971. IF versus video limiting for two-channel coherent signal processors. *IEEE Trans. Inform. Theory* IT-17:579–587.

Zrnić, D. S., 1975a. Estimated tornado spectra and maximum velocity statistics. NOAA Final Rep. Grant No. 04-5-022-17.

———, 1975b. Signal-to-noise ratio in the output of nonlinear devices. *IEEE Trans. Inform. Theory* IT-21:662–663.

———, and B. Bumgarner, 1975. Receiver chain and signal processing effects on the Doppler spectrum. Preprints, Sixteenth Radar Meteorological Conference, Houston, Texas, 22–24 April. American Meteorological Society, Boston, pp. 163–68.

———, and R. J. Doviak, 1975. Velocity spectra of vortices scanned with a pulse Doppler radar. *J. Appl. Meteorol.* 14: 1531–39.

———, and ———, 1978. Matched filter criteria and range weighting for weather radar. *IEEE Trans. Aerosp. Electron. Syste.* AES-14:925–30.

———, ———, and D. W. Burgess, 1977. Probing tornadoes with pulse Doppler radar. *Q. J. R. Meteorol. Soc.* 103:707–20.

———, ———, J. T. Lee, and R. S. Ge, 1983. Characteristics of gust fronts and downdrafts from single Doppler radar data. Preprints, Twenty-first Radar Conference. American Meteorological Society, Boston, pp. 650–54.

Chapter 11

Atlas, D., 1964. Advances in radar meteorology. In *Advances in Geophysics,* Vol. 10. Academic Press, New York, pp. 317–478.

———, and A. C. Chmela, 1957. Physical-synoptic variations of raindrop size parameters. Preprints, Sixth Weather Radar Conference, Cambridge, Massachusetts. American Meteorological Society, Boston, pp. 21–29.

———, and R. Meneghini, 1983: Simultaneous ocean cross-section and rainfall measurements from space with a radar pointing radar. Preprints, Twenty-first Conference on Radar Meteorology, Edmonton, Alberta. American Meteorological Society, Boston, pp. 719–26.

———, and V. G. Plank, 1953. Drop-size history during a shower. *J. Meteorol.* 10:291–95.

Aoyagi, J., 1964. Areal rainfall obtained by a 3.2 cm radar and a raingage network. Preprints, Eleventh Weather Radar Conference, Boulder, Colorado. American Meteorological Society, Boston, pp. 116–19.

Barnston, A. G., and J. L. Thomas, 1983. Rainfall measurement accuracy in FACE: A comparison of gage and radar rainfalls. *J. Climate Appl. Meteor.* 22:2038–52.

Battan, L. J., 1973. *Radar Observation of the Atmosphere.* University of Chicago Press, Chicago.

Best, A. C., 1950. Empirical formulae for the terminal velocity of water drops falling through the atmosphere. *Q. J. R. Meteorol. Soc.* 76:302–11.

Blake, L. V., 1970. Prediction of radar range. In *Radar Handbook,* M. I. Skolnik (ed.), McGraw-Hill, New York, 2–51 to 2–55.

Blanchard, D. C., 1953. Raindrop size-distribution in Hawaiian rains. *J. Meteorol.* 10:457–73.

Borovikov, A. M., V. V. Kostarev, I. P. Mazin, V. I. Smirnov, and A. A. Charnikov, 1970. *Radar Measurements of Precipitation Rate.* Trans. Israel Program for Scientific Translations (National Technical Information Service, Springfield, Va.).

Brandes, E. A., 1974. Radar rainfall patterns optimizing technique. Tech. Memo. ERL NSSL-67, NOAA Environmental Research Laboratories, Boulder, Colo.

———, 1975. Optimizing rainfall estimates with the aid of radar. *J. Appl. Meteorol.* 14:1339–45.

———, and D. Sirmans, 1976. Convective rainfall estimation by radar: experimental results and proposed operational analysis technique. Preprints, Conference on Hydrometeorology, Fort Worth, Texas. American Meteorological Society, Boston, pp. 54–59.

Bruer, L., and R. K. Kreuels, 1976. Seasonal variations of precipitation parameters and a new hydrological field experiment in Western Germany. Preprints, Seventeenth Conference on Radar Meteorology, Seattle, Washington. American Meteorological Society, Boston, pp. 432–37.

Burrows, C. R., and S. S. Attwood, 1949. *Radio Wave Propagation: Consolidated Summary Technical Report, of the Committee on Propagation of the National Defense Research Committee.* Academic Press, New York.

Cain, D. W., and P. L. Smith, 1976. Operational adjustment of radar estimated rainfall with rain gage data: a statistical evaluation. Preprints, Nineteenth Conference on Radar Meteorology, Seattle, Washington. American Meteorological Society, Boston, pp. 533–38.

Carbone, R. E., and L. D. Nelson, 1978. The evolution of raindrop spectra in warm-based convective storms as observed and numerically modeled. *J. Atmos. Sci.* 12:2302–14.

Cavalli, R., 1984. The operational Swiss weather radar information distribution network. Preprints, Twenty-Second Conference on Radar Meteorology, Zurich, Switzerland. American Meteorological Society, Boston, pp. 21–24.

Clarke, S., T. A. Seliga, and K. Aydin, 1983: Estimates of rainfall rate using the differential reflectivity (Z_{DR}) radar technique: comparisons with a raingage network and Z-R relationships. Preprints, Twenty-First Conference on Radar Meteorology, Edmonton, Alberta. American Meteorological Society, Boston, pp. 479–84.

Cohen, A., and A. Smolski, 1966. The effect of rain on satellite communications earth terminal rigid radomes. *Microwave J.* 9:111–21.

Collier, C. G., 1984. Radar meteorology in the United Kingdom. Preprints, Twenty-Second Conference on Radar Meteorology, Zurich, Switzerland. American Meteorological Society, Boston, pp. 1–8.

———, T. W. Harrold, and C. A. Nicholass, 1975. A comparison of areal rainfall as measured by a raingauge-calibrated radar system and raingauge networks of various densities. Preprints, Sixteenth Radar Meteorological Conference, Houston, Texas. American Meteorological Society, Boston, pp. 467–72.

Crawford, K. C., 1978. On the bivariate objective analysis of surface rainfall using interpolation. Preprints, Eighteenth Conference on Radar Meteorology, Atlanta, Georgia. American Meteorological Society, Boston, pp. 336–41.

Dahlström, B., 1973. Investigation of errors in rainfall observations: a continued study. Rep. 34, Department of Meteorology, University of Uppsala, Sweden.

Desautels, G., and K. L. S. Gunn, 1970. Comparison of radar with network gauges. Preprints, Fourteenth Radar Meteorology Conference, Tucson, Arizona. American Meteorological Society, Boston, pp. 239–40.

Doviak, R. J., and D. Sirmans, 1973. Reflectivity equation for NSSL's WDS-71 10-cm Doppler radar. National Severe Storms Lab., Norman, Okla., internal memorandum.

Foote, G. B., 1966. A Z-R relation for mountain thunderstorms. *J. Appl. Meteorol.* 5:229–31.

Fujiwara, M., 1965. Raindrop-size distribution from individual storms. *J. Atmos. Sci.* 22:585–91.

Grayman, W. M., and P. S. Eagleson, 1970. A review of the accuracy of radar and raingages for precipitation measurement. H.I.T. Hydrodyn. Lab. Rep. 119, Cambridge, Mass.

Greene, D. R., 1975. Hydrologic application of digital radar data. Preprints, Sixteenth Conference on Radar Meteorology, Houston, Texas. American Meteorological Society, Boston, pp. 353–60.

Gunn, K. L. S., and J. S. Marshall, 1955. The effect of wind shear on falling precipitation. *J. Meteorol.* 12:339–49.

Gunn, R., and G. D. Kinzer, 1949. The terminal velocity of fall for water droplets in stagnant air. *J. Meteorol.* 6:243–48.

Hall, M. P. M., S. M. Cherry, J. F. W. Goddard, and G. R. Kennedy, 1980. Raindrop sizes and rainfall rate measured by dual-polarization radar. *Nature* 285:195–98.

Harrold, T. W., and P. G. Kitchingman, 1975. Measurement of surface rainfall using radar when the beam intersects the melting layer. Preprints, Sixteenth Radar Meteorological Conference, Houston, Texas. American Meteorological Society, Boston, pp. 473–78.

———, E. J. English, and C. A. Nicholass, 1974. The accuracy of radar-derived rainfall measurements in hilly terrain. *Q. J. R. Meteorol. Soc.* 100:331–50.

Herndon, A., W. L. Woodley, A. H. Miller, A. Samet, and H. Senn, 1973. Comparison of gage and radar methods of convective precipitation measurement. Tech. Memo. ERL OD-18, NOAA Environmental Research Laboratories, Boulder, Colo.

Hildebrand, P. H., N. Towery, and M. R. Snell, 1979. Measurement of convective mean rainfall over small areas using high density raingages and radar. *J. Appl. Meteorol.* 18:1316–26.

Hitschfeld, W., and J. Bordan, 1954. Errors inherent in the radar measurement of rainfall at attenuating wavelengths. *J. Appl. Meteorol.* 11:58–67.

Huff, F. A., 1955. Comparison between standard and small orifice raingages. *Trans. Am. Geophys. Union* 36:689–94.

———, 1971. Evaluation of precipitation records in weather modification experiments. In *Advances in Geophysics,* Vol. 15. Academic Press, New York, pp. 59–134.

———, and N. G. Towery, 1978. Utilization of radar in operation of urban hydrologic systems. Preprints, Eighteenth Conference on Radar Meteorology, Atlanta, Georgia. American Meteorological Society, Boston, pp. 437–441.

Imai, I., 1960. Raindrop size distributions and Z-R relationships. Preprints, Eighth Weather Radar Conference, San Francisco, California. American Meteorological Society, Boston, pp. 211–18.

Israelsen, C. G., 1967. Reliability of can-type precipitation gage measurements. Tech. Rep. 2, Utah Water Research Laboratory, Utah State University, Logan.

Jameson, A. R., and K. V. Beard, 1982: Raindrop axial ratios. *J. Appl. Meteorol.* 21:257–59.

Jatila, E., and T. Puhakka, 1973a. Experiments on the measure-

ment of areal precipitation by radar. *Geophysica* 12.
———, and T. Puhakka, 1973b. On the accuracy of radar rainfall measurements. *Geophysica* 12.
Jones, D. M. A., 1955. Three-cm and 10-cm wavelength radiation back scatter from rain. Preprints, Fifth Weather Radar Conference, Asbury Park, New Jersey. American Meteorological Society, Boston, pp. 281–85.
———, 1966. The correlation of raingage-network and radar-detected rainfall. Preprints, Twelfth Conference on Radar Meteorology, Norman, Oklahoma. American Meteorological Society, Boston, pp. 204–207.
Joss, J. K., and E. G. Gori, 1978. Shapes of raindrop size distributions. *J. Appl. Meteorol.* 17:1054–61.
———, K. Schram, J. C. Thams, and A. Waldvogel, 1970. On the quantitative determination of precipitation by radar. Wissenschaftliche Mitteilungen Nr. 63, Eidgenössischen Kommission Zum Studium der Hagelbildung und der Hagelawehr.
———, J. C. Thams, and A. Waldvogel, 1968. The accuracy of daily rainfall measurements by radar. Preprints, Thirteenth Radar Meteorological Conference, Montreal. American Meteorological Society, Boston, pp. 448–51.
Kessler, E., 1965. Use of radar measurements for assessment of areal rainfall. Paper presented at 17th Session of the WMO Executive Committee, Geneva, Switzerland, May 1965.
———, 1969. *On the Continuity of Water Substance in Atmospheric Circulations.* Meteorol. Monogr. 10, American Meteorological Society, Boston.
———, and J. A. Russo, Jr., 1963. Statistical properties of weather radar echoes. Preprints, Tenth Weather Radar Conference, Washington, D.C. American Meteorological Society, Boston, pp. 25–33.
Kinzer, G. D., and R. Gunn, 1951. The evaporation, temperature and thermal relaxation-time of freely falling waterdrops. *J. Meteorol.* 8:71–82.
Klazura, G. E., 1977. Changes in gage/radar ratios in high rain rate gradients by varying location and size of radar comparison areas. Preprints, Second Conference on Hydrometeorology, Toronto. American Meteorological Society, Boston, pp. 376–79.
Koschmieder, H., 1934. Methods and results of definite rain measurements: III. Danzig report (1). *Mon. Weather Rev.* 62:5–7.
Kurtyka, J. C., 1953. Precipitation measurement study. Rep. of Investigation No. 20, Ill. State Water Supv. Div., Urbana, Ill.
Larson, L. W., 1971. Precipitation and its measurement, a state of the art. Water Resour. Ser. 24, Water Resour. Res. Inst., University of Wyoming, Laramie.
———, and E. L. Peck, 1974. Accuracy of precipitation measurements for hydrologic modeling. *Water Resour. Res.* 10: 857–63.
Laws, J. O., and D. A. Parsons, 1943. The relation of raindrop-size to intensity. *Trans. Am. Geophys. Union* 24:452–60.
Linsley, R. K., and M. A. Kohler, 1951. Variations in storm rainfall over small areas. *Trans. Am. Geophys. Union* 32:245–50.
McGuinness, J. L., 1963. Accuracy of estimating watershed mean rainfall. *J. Geophys. Res.* 68:4763–67.
Marshall, J. S., and W. M. Palmer, 1948. The distribution of raindrops with size. *J. Meteorol.* 5:165–66.
Martner, B. E., 1975. Final report on the University of Wyoming's participation in NSSL Project Storm Intercept, 1975 (data analysis). Department of Atmospheric Science, University of Wyoming, Laramie.
———, 1977. A field experiment on the calibration of radars with raindrop disdrometers. *J. Appl. Meteorol.* 16:451–54.
Mason, B. J., and J. B. Andrews, 1960. Drop-size distributions from various types of rain. *Q. J. R. Meteorol. Soc.* 86:346–53.
Meneghini, R., J. Eckerman, and D. Atlas, 1983: Determination of rain rate from a spaceborne radar using measurements of total attenuation. *IEEE Trans., GeoScience and Remote Sensing* GE-21:34–43.
Muchnik, V. M., M. L. Markovich, and L. M. Volynetz, 1968. The results of radar measurements of the areal rainfall. Preprints, Thirteenth Radar Meteorological Conference, Montreal. American Meteorological Society, Boston, pp. 392–95.
Nicks, A. D., 1966. Field evaluation of rain gage network design principles. Symposium Design of Hydrological Networks, *Intl. Assoc. Sci. Hydrol. Pub.* 67:82—93.
Probert-Jones, J. R., 1962. The radar equation in meteorology. *Q. J. R. Meteorol. Soc.* 88:485–95.
Puhakka, T., 1974. On the variability of the Z-R relationships in rainfall related to radar echo pattern. *Geophysica* 13.
Rigby, E. C., J. S. Marshall, and W. Hitschfeld, 1954. The development of the size distribution of raindrops during their fall. *J. Meteorol.* 11:362–72.
Rogers, C. W. C., and R. Wexler, 1963. Rainfall determination from 0.86 and 1.82 cm radar measurements. Preprints, Tenth Weather Radar Conference, Washington, D.C. American Meteorological Society, Boston, pp. 260–70.
Rogers, R. R., 1971. The effect of variable target reflectivity on weather radar measurements. *Q. J. R. Meteorol. Soc.* 97: 154–67.
Ryde, J. W., and D. Ryde, 1945. Attenuation of centimetre and millimetre waves by rain, hail, fogs, and clouds. General Electric Company Report No. 8670.
Saffle, R. E., and D. R. Greene, 1978. The role of radar in the flash flood watch warning system: Johnstown examined. Preprints, Eighteenth Conference on Radar Meteorology, Atlanta, Georgia. American Meteorological Society, Boston, pp. 468–73.
Schaffner, M. R., 1976. On the characterization of weather radar echoes, II. Preprints, Eleventh Conference on Radar Meteorology, Seattle, Washington. American Meteorological Society, Boston, pp. 478–85.
Sekhon, R. S., and R. C. Srivastava, 1971. Doppler radar observations of drop-size distributions in a thunderstorm. *J. Atmos. Sci.* 28:983–94.
Seliga, T. A., and K. Aydin, 1983: Possible detection of wide-spread glaciation throughout upper regions of a storm after local sunset from differential reflectivity (Z_{DR}) measurements during CCOPE. Twenty-First Conference on Radar Meteorology, Edmonton, Alberta. American Meteorological Society, Boston, pp. 530–37.
———, and V. N. Bringi, 1976. Potential use of radar differential reflectivity measurements at orthogonal polarizations for measuring precipitation. *J. Appl. Meteorol.* 15:69–76.
———, and ———, 1978. Preliminary results of differential reflectivity measurements in rain. Preprints, Eighteenth Conference on Radar Meteorology, Atlanta, Georgia. American Meteorological Society, Boston, pp. 134–38.
———, ———, and H. H. Al-Khatib, 1981a: A preliminary study of comparative measurement of rainfall rate using the differential reflectivity radar technique and a raingage network. *J. Appl. Meteorol.* 20:1362–68.
———, J. R. Peterson, and V. N. Bringi, 1981b: Hydrometeor characteristics in the May 2, 1979 squall line in central Okla-

homa as obtained from radar differential reflectivity measurements during SESAME. Preprints, Twentieth Conference on Radar Meteorology, Boston, Massachusetts. American Meteorological Society, Boston, pp. 561–66.

Sirmans, D., and R. J. Doviak, 1973. Meteorological radar signal intensity estimation. Tech. Memo. ERL-NSSL-64, NOAA Environmental Research Laboratories, Boulder, Colo.

Spilhaus, A. F., 1948. Raindrop size, shape, and falling speed. *J. Meteorol.* 5:108–10.

Srivastava, R. C., 1971 Size distribution of raindrops generated by their breakup and coalescence. *J. Atmos. Sci.* 28:410–15.

Twomey, S., 1953. On the measurement of precipitation intensity by radar. *J. Meteorol.* 10:66—67.

Ulbrich, C. W., and D. Atlas, 1975. The use of radar reflectivity and microwave attenuation to obtain improved measurements of precipitation parameters. Preprints, Sixteenth Radar Meteorological Conference, Houston, Texas. American Meteorological Society, Boston, pp. 496–503.

Wexler, R., 1947. Radar detection of a frontal storm 18 June 1946. *J. Meteorol.* 4:38–44.

Wilson, J. W., 1966. Storm-to-storm variability in the radar reflectivity-rainfall rate relationship. Preprints, Twelfth Conference on Radar Meteorology, Norman, Oklahoma. American Meteorological Society, Boston, pp. 229–33.

———, 1968. Factors affecting the accuracy of radar rainfall measurement. Final Rep. 7488-331 to the National Severe Storms Laboratory, Travelers Research Center, Inc., Hartford, Conn.

———, 1970. Integration of radar and raingage data for improved rainfall measurement. *J. Appl. Meteorol.* 9:489–97.

———, 1975. Radar-gage precipitation measurements during the IFYGL. CEM Rep. 4177-4546, Center for the Environment and Man, Hartford, Conn.

———, 1978. Observations of radome transmission losses at 5 cm wavelengths. Preprints, Eighteenth Conference on Radar Meteorology, Atlanta, Georgia. American Meteorological Society, Boston, pp. 288–91.

———, and E. A. Brandes, 1979. Radar measurement of rainfall—a summary. *Bull. Am. Meteorol. Soc.* 60:1048–58.

WMO (World Meteorological Organization), 1973. *Annotated Bibliography on Precipitation Measurement Instruments.* Rep. 17, WMO-343, Geneva.

Woodley, W., and A. Herndon, 1970. A raingage evaluation of the Miami reflectivity-rainfall rate relation. *J. Appl. Meteorol.* 9:258–63.

———, A. Olsen, A. Herndon, and V. Wiggert, 1974. Optimizing the measurement of convective rainfall in Florida. Tech Memo ERL-WMPO-18, NOAA Environmental Research Laboratories, Boulder, Colo.

———, ———, ———, and ———, 1975. Comparison of gage and radar methods of convective rain measurement. *J. Appl. Meteorol.* 14:909–28.

Zawadski, I. I., 1975. On radar-raingage comparison. *J. Appl. Meteorol.* 14:1430–36.

Chapter 12

Adler, R. F., and D. D. Fenn, 1979a. Thunderstorm intensity as determined from satellite data. *J. Appl. Meteorol.* 18:502–17.

———, and ———, 1979b. Thunderstorm vertical velocities estimated from satellite data. *J. Atmos. Sci.* 36:1747–54.

———, and ———, 1981. Satellite-observed cloud-top height changes in tornadic thunderstorms. *J. Appl. Meteorol.* 20: 1369–75.

Anderson, R. K., et al., 1974. Application of meteorological satellite data in analysis and forecasting. *ESSA Tech. Rep.* NESS 51, Washington, D.C.

Anthony, R. W., and G. S. Wade, 1983. VAS operational assessment findings for spring 1982/83. Preprints, Thirteenth Conference on Severe Local Storms, Tulsa, Okla. American Meteorological Society, Boston, pp. J23–J28.

Billingsley, J. B., 1976. Interactive image processing for meteorological applications at NASA/Goddard Space Flight Center. Preprints, Seventh Conference on Aereospace and Aeronautical Meteorology and Symposium on Remote Sensing from Satellites, Melbourne, Fla., American Meteorological Society, Boston, pp. 268–75.

Boucher, R. J., 1967. Relationships between the size of satellite-observed cirrus shields and the severity of thunderstorm complexes. *J. Appl. Meteorol.* 6:564–72.

Caracena, F., R. A. Maddox, J. F. W. Purdom, J. F. Weaver, and R. N. Green, 1983. Multiscale analyses of meteorological conditions affecting Pan American World Airways Flight 759. NOAA Tech. Memo. ERL ESG-2, Boulder, Colo.

Chesters, D., L. W. Uccellini, and A. Mostek, 1982. VISSR atmospheric sounder (VAS) simulation experiment for a severe storm environment. *Mon. Weather Rev.* 110:198–216.

———, ———, and W. Robinson, 1983. Low-level moisture images from the VISSR atmospheric sounder (VAS). "Split-window" channels at 11 and 12 microns. *J. Climate Appl. Meteorol.* 22:725–43.

Epstein, E. S., W. M. Callicott, D. J. Cottes, and H. W. Yates, 1984. NOAA satellite programs. IEEE *AES* 20 (4):325–44.

Erickson, C. O., 1964. Satellite photographs of convective clouds and their relation to the vertical wind shear. *Mon. Weather Rev.* 92:283–96.

Fritz, S., 1977. Temperature retrievals from satellite radiance measurements—an empirical method. *J. Appl. Meteorol.* 16: 172–76.

———, D. Q. Wark, H. E. Fleming, W. L. Smith, H. Jacobwitz, D. T. Hilleary, and J. C. Alishouse, 1972. Temperature sounding from satellites. NOAA Tech. Report NESS 59, Washington, D.C.

Fujita, T. T., E. W. Pearl, and W. E. Shenk, 1975. Satellite-tracked cumulus velocities. *J. Appl. Meteorol.* 14:407–13.

Griffith, C. G., J. A. Augustine, and W. L. Woodley, 1981. Satellite rain estimation in the U.S. High Plains. *J. Appl. Meteorol.* 20:53–66.

———, W. L. Woodley, P. G. Grube, D. W. Martin, J. Stout, and D. N. Sikdar, 1978. Rain estimation from geosynchronous satellite imagery—visible and infrared studies. *Mon. Weather Rev.* 106:1153–71.

Grody, N. C., 1983. Severe storm observations using the microwave sounding unit. *J. Climate Appl. Meteorol.* 22:609–25.

Gurka, J. J., 1976. Satellite and surface observations of strong wind zones accompanying thunderstorms. *Mon. Weather Rev.* 104:1484–93.

Hasler, A. F., W. E. Shenk, and W. C. Skillman, 1976. Wind estimates from cloud motions: phase 1 of an *in situ* aircraft verification experiment. *J. Appl. Meteorol.* 15:10–15.

Heymsfield, G.M., K. K. Ghosh, and L. C. Chen, 1983a. An interactive system for compositing digital radar and satellite data. *J. Climate Appl. Meteorol.* 22:705–713.

———, G. Szejwach, S. Schotz, and R. H. Blackmer, 1983b. Upper-level structure of Oklahoma tornadic storms on 2 May 1979. II: Proposed explanation of V pattern and internal warm region in infrared observations. *J. Appl. Meteorol.* 22: 1756–67.

Hill, K., and R. E. Turner, 1977. NASA's atmospheric variability experiments (AVE). *Bull. Am. Meteorol. Soc.* 58:170–72.

Hillger, D. W., and T. H. VonderHaar, 1977. Deriving mesoscale temperature and moisture fields from satellite radiance measurements over the United States. *J. Appl. Meteorol.* 16:715–26.

———, and ———, 1981. Retrieval and use of high-resolution moisture and stability fields from Nimbus 6 HIRS radiances in pre-convective situations. *Mon. Weather Rev.* 109:1788–1806.

Hubert, L. F., and L. F. Whitney, Jr., 1971. Wind estimation from geostationary-satellite pictures. *Mon. Weather Rev.* 99: 665–72.

Koch, S. E., M. desJardins, and P. J. Kocin, 1983. An interactive Barnes objective map analysis scheme for use with satellite and conventional data. *J. Climate Appl. Meteorol.* 22:1487–1503.

Kreitzberg, C. W., 1976. Interactive applications of satellite observations and mesoscale numerical models. *Bull. Am. Meteorol. Soc.* 57:679–85.

McCann, D. W. 1983. The enhanced-V: a satellite observable severe storm signature. *Mon. Weather Rev.* 111:887–94.

Mack, R. A., and D. P. Wylie, 1982. An estimation of the condensation rates in three severe storm systems from satellite observations of the convective mass flux. *Mon. Weather Rev.* 110:725–44.

———, A. F. Hasler, and R. F. Adler, 1983. Thunderstorm cloud top observations using satellite stereoscopy. *Mon. Weather Rev.* 111:1949–64.

McMillin, L. M., et al., 1973. Satellite infrared soundings from NOAA spacecraft. NOAA Tech. Report NESS 65. Washington, D.C.

Maddox, R. A. 1980. Mesoscale convective complexes. *Bull. Am. Meteorol. Soc.* 61:1374–87.

———, and T. H. VonderHaar, 1979. Covariance analyses of satellite-derived mesoscale wind fields. *J. Appl. Meteorol.* 18: 1327–34.

Magor, B. W., 1969. Mesoanalysis: some operational analysis techniques utilized in tornado forecasting. *Bull. Am. Meteorol. Soc.* 40:499–511.

Miller, R. C., 1972. Notes on analysis and severe storm forecasting procedures of the Air Force Global Weather Center. Air Weather Service TR 200 (Rev.).

———, and J. A. McGinley, 1978. Using satellite imagery to detect and track comma clouds and the application of the zone technique in forecasting severe storms. NASA Project Report, GE Management and Technical Services Co., Beltsville, Md.

Mills, G. A., and C. M. Hayden, 1983. The use of high horizontal resolution satellite temperature and moisture profiles to initialize a mesoscale numerical weather prediction model-A severe weather event case study. *J. Climate Appl. Meteorol.* 22: 649–63.

Moses, J. F., 1982. Forecasting convective precipitation by tracking cloud tops in GOES imagery. Preprints, International Symposium on Hydrometeorology, American Water Resources Association, Denver, Colo., pp. 129–37.

Negri, A. J., 1982. Cloud-top structure of tornadic storms on 10 April 1979 from rapid scan and stereo satellite observations. *Bull. Am. Meteorol. Soc.* 63:1151–59.

———, and R. F. Adler, 1981. Relation of satellite-based thunderstorm intensity to radar-estimated rainfall. *J. Appl. Meteorol.* 20:288–300.

———, and T. H. VonderHaar, 1980. Moisture convergence using satellite-derived wind fields: a severe local storm case study. *Mon. Weather Rev.* 108:1170–82.

———, R. F. Adler, and P. J. Wetzel, 1984. Rain estimation from satellites: an examination of the Griffith-Woodley technique. *J. Climate Appl. Meteorol.* 23:102–16.

Ninomiya, K., 1971a. Dynamical analysis of outflow from tornado-producing thunderstorms as revealed by ATS III pictures. *J. Appl. Meteorol.* 10:275–94.

———, 1971b. Mesoscale modification of synoptic situations from thunderstorm development as revealed by ATS III and aerological data. *J. Appl. Meteorol.* 10:1103–21.

Ostby, F. P., 1984. Use of CSIS in severe weather prediction. *Society of Photo-optical Engineers* Vol. 481: Recent Advances in Civil Space Remote Sensing, pp. 78–83.

Peslen, C. A., 1980. Short-interval SMS wind vector determinations for a severe local storms area. *Mon. Weather Rev.* 108: 1407–18.

Petersen, R. A., L. W. Uccellini, D. Chesters, and A. Mostek, 1983. The use of satellite data in weather analysis, prediction, and diagnosis. *Nat. Weather Digest* 8(1):12–23.

———, ———, A. Mostek, and D. A. Keyser, 1984. Delineating mid-and low-level water vapor patterns in pre-convective environments using VAS moisture channels. *Mon. Weather Rev.* 112:2178–98.

Purdom, J. F. W., 1976. Some uses of high-resolution GOES imagery in the mesoscale forecasting of convection and its behavior. *Mon. Weather Rev.* 104:1474–83.

———, 1982. Subjective interpretation of geostationary satellite data for nowcasting. Chap. 3.1 in *Nowcasting*, K. A. Browning (ed.). Academic Press, London.

———, 1984. Use of satellite soundings and imagery for nowcasting and very-short-range forecasting. Preprint volume, Nowcasting Symposium, Sweden, pp. 1–13.

Reynolds, D. W., 1980. Observations of damaging hailstorms from geosynchronous satellite digital data. *Mon. Weather Rev.* 108:337–48.

———, 1983. Prototype workstation for mesoscale forecasting. *Bull. Am. Meteorol. Soc.* 64:264–73.

———, and E. A. Smith, 1979. Detailed analysis of composited digital radar and satellite data. *Bull. Am. Meteorol. Soc.* 60:1024–37.

Scofield, R. A. 1981. Satellite-derived rainfall estimates for the Bradys Bend, Pennsylvania, flash flood. Preprints, Fourth Conference on Hydrometeorology, Reno, Nev. American Meteorological Society, Boston, pp. 188–93.

———, 1982. A satellite technique for estimating rainfall from flash flood producing thunderstorms. Preprints, International Symposium on Hydrometeorology. American Water Resources Association, Denver, Colo., pp. 121–28.

———, and V. J. Oliver, 1977. A scheme for estimating convective rainfall from satellite imagery. NOAA Tech. Memo. NESS 86. Washington, D.C.

———, and C. E. Weiss, 1976. Application of synchronous meteorological satellite products and other data for short range forecasting in the Chesapeake Bay region. Preprints, Sixth Conference on Weather Forecasting and Analysis, Albany, N.Y., American Meteorological Society, Boston, pp. 67–73.

Sikdar, D. N., V. E. Suomi, and C. E. Anderson, 1970. Convective transports of mass and energy in severe storms over the

United States—an estimate from a geostationary altitudes. *Tellus* 22:521–32.
Smith, E. A., 1975. The McIDAS system. *IEEE Trans. Geosci. Elec.*, GE-13, pp. 123–36.
———, T. A. Brubaker, and T. H. VonderHaar, 1979. All-digital video imaging system for atmospheric research (ADVISAR). Tech. Rep., Dept. of Atmospheric Science, Colorado State University, Fort Collins.
Smith, W. L., 1968. An improved method for calculating tropospheric temperature and moisture from satellite radiometer measurements. *Mon. Weather Rev.* 96:387–96.
———, 1983: The retrieval of atmospheric profiles from VAS geostationary radiance observations. *J. Atmos. Sci.* 40: 2025–35.
———, 1985. Satellites. Observing systems—satellites. In Chap. 10 (pp. 380–472) in *The Handbook of Applied Meteorology*. John Wiley & Son, New York.
———, V. E. Suomi, W. P. Menzel, H. M. Woolf, L. A. Sromovsky, H. E. Revercomb, C. M. Hayden, D. N. Erickson, and F. R. Mosher, 1981. First sounding results from VAS-D. *Bull. Am. Meteorol. Soc.* 62:232–36.
———, ———, F. X. Zhou, and W. P. Menzel, 1982. Nowcasting applications of geostationary satellite atmospheric sounding data. Chaps. 2, 4 in *Nowcasting*, K. A. Browning (ed.). Academic Press, London.
Suomi, V. E., R. Fox, S. S. Limaye, and W. L. Smith, 1983. McIDAS III: a modern interactive data access and analysis system. *J. Climate Appl. Meteorol.* 22:766–78.
Tecson, J. J., T. A. Umenhofer, and T. T. Fujita, 1977. Thunderstorm-associated cloud motions as computed from 5-minute SMS pictures. Preprints, Tenth Conference on Severe Local Storms, Omaha, Nebr. American Meteorological Society, Boston, pp. 22–29.
VonderHaar, T. H., 1969. Meteorological applications of reflected radiance measurements from ATS 1 and ATS 3. *J. Geophys. Res.* 74:5404–12.
Wexler, R., and R. H. Blackmer, Jr., 1982. Radar reflectivities and satellite imagery of severe storms 20 May 1977. *Mon. Weather Rev.* 110:719–24.
Whitney, L. F., Jr., 1961. Another view from TIROSI of a severe weather situation May 16, 1960. *Mon. Weather Rev.* 89: 447–60.
———, 1963: Severe storm clouds as seen from TIROS. *J. Appl. Meteorol.* 2:501–507.
———, 1977. Relationship of the sub-tropical jet stream to severe local storms. *Mon. Weather Rev.* 105:398–412.
———, and S. Fritz, 1961. A tornado-producing cloud pattern seen from TIROS I. *Bull. Am. Meteorol. Soc.* 42:603–14.
Wilson, T. A., and D. D. Houghton, 1979. Mesoscale wind fields for a severe storm situation determined from SMS cloud observations. *Mon. Weather Rev.* 107:1198–1209.
Woodley, W. L., C. G. Griffith, J. S. Griffin, and S. C. Stromatt, 1980. The inference of GATE convective rain fall from SMS-1 imagery. *J. Appl. Meteorol.* 19:388–408.

Chapter 13

Armijo, L., 1969. A theory for the determination of wind and precipitation velocities with Doppler radars. *J. Atmos. Sci.* 26:566–69.
Berry, E. X., and R. L. Reinhardt, 1974. An analysis of cloud drop growth by collection: Part I. Double distributions. *J. Atmos. Sci.* 31:1814–24.
Bolton, D., 1980. The computation of equivalent potential temperature. *Mon. Weather Rev.* 108:1046–53.
Bonesteele, R. G., and Y. J. Lin, 1978. A study of updraft-downdraft interaction based upon perturbation pressure and single-Doppler radar data. *Mon. Weather Rev.* 106:62–68.
Brandes, E. A., 1977. Flow in severe thunderstorms observed by dual-Doppler radar. *Mon. Weather Rev.* 105:113–20.
———, 1983. Relationships between thunderstorm mesoscale circulation and tornadogenesis. Ph.D. diss., University of Oklahoma, Norman.
———, 1984. Relationships between radar-derived thermodynamic variables and tornadogenesis. *Mon. Weather Rev.* 112:1033–52.
Brazier-Smith, P. R., S. G. Jennings, and J. Latham, 1972. The interactions of falling water drops: Coalescence. *Proc. R. Soc.* A326:393–408.
Browning, K. A., 1964. Airflow and precipitation trajectories within severe local storms which travel to the right of the winds. *J. Atmos. Sci.* 21:634–39.
———, and G. B. Foote, 1976. Airflow and hail growth in supercell storms and some implications for hail suppression. *J. R. Meteorol. Soc.* 102:499–533.
Byers, H. R., and R. R. Braham, Jr., 1949. Thunderstorm structure and circulation. *J. Meteorol.* 5:71–86.
Chisolm, A. J., and M. English, 1973. Alberta hailstorms. *Meteorol. Monogr.* 14(36).
Chong, M., F. Roux, and J. Testud, 1980. A new filtering and interpolating method for processing dual Doppler radar data: Performance in three dimensional wind restitution, ability to derive pressure and temperature fields. Preprints, Nineteenth Conference on Radar Meteorology, Miami Beach. American Meteorological Society, Boston, pp. 286–93.
Courant, R., and D. Hilbert, 1953. *Methods of Mathematical Physics*. Interscience Publishers, New York.
Deardorff, J. W., 1970. A numerical study of three-dimensional turbulent channel flow at large Reynolds numbers. *J. Fluid Mech.* 41:453–80.
———, 1974. Three-dimensional numerical study of the height and mean structure of a heated planetary boundary layer. *Bound.-Layer Meteorol.* 7:81–106.
Douglas, R. H., 1960. Size distributions, ice contents and radar reflectivities of hail in Alberta. *Nubila* 3:5–11.
Doviak, R. J., P. S. Ray, R. G. Strauch, and L. J. Miller, 1976. Error estimation in wind fields derived from dual Doppler radar measurements. *J. Appl. Meteorol.* 15:868–78.
Draper, W. R., and H. Smith, 1966. *Applied Regression Analysis*. John Wiley & Sons, New York.
Foote, G. B., and H. W. Frank, 1983. Case study of a hailstorm in Colorado. Part III: Airflow from triple Doppler measurements. *J. Atmos. Sci.* 40:686–707.
Gal-Chen, T. 1978. A method for the initialization of the anelastic equations: Implications for matching models with observations. *Mon. Weather Rev.* 106:587–606.
———, and R. A. Kropfli, 1984. Buoyancy and pressure perturbations derived from dual-Doppler radar observations of the planetary boundary layer: applications for matching models with observations. *J. Atmos. Sci.* 41:3007–20.
Hane, C. E., and P. S. Ray, 1985. Pressure and buoyancy fields derived from Doppler radar data in a tornadic thunderstorm. *J. Atmos. Sci.* 42:18–35.

———, and B. C. Scott, 1978. Temperature and pressure perturbations within convective clouds derived from detailed air motion information: Preliminary testing. *Mon. Weather Rev.* 106:654–61.

———, R. B. Wilhelmson, and T. Gal-Chen, 1981. Retrieval of thermodynamic variables within deep convective clouds: Experiments in three dimensions. *Mon. Weather Rev.* 109: 564–76.

Hildebrand, F. B., 1965. *Methods of Applied Mathematics,* 2d ed. Prentice-Hall, Englewood Cliffs, N.J.

Houze, R. A., Jr., P. V. Hobbs, P. H. Herzegh, and D. W. Parsons, 1979. Size distributions of precipitation particles in frontal clouds. *J. Atmos. Sci.* 36:156–62.

Kessler, E., 1969. On the distribution and continuity of water substance in atmospheric circulations. *Meteorol. Monogr.* 10(32).

Klemp, J. B., and R. Rotunno, 1983. A study of the tornadic region within a supercell thunderstorm. *J. Atmos. Sci.* 40: 359–77.

———, and R. B. Wilhelmson, 1978a. The simulation of three-dimensional convective storm dynamics. *J. Atmos. Sci.* 35: 1070–96.

———, and ———, 1978b. Simulation of right- and left-moving storms produced through storm splitting. *J. Atmos. Sci.* 35: 1097–1110.

Lhermitte, R., and L. J. Miller, 1970. Doppler radar methodology for the observation of convective storms. Preprints, Fourteenth Radar Meteorological Conference, Tucson. American Meteorological Society, Boston, pp. 133–38.

Liese, J. A., 1978. Temperature retrieval from dual-Doppler radar wind field data. Preprints, Twelfth Conference on Radar Meteorology, Atlanta. American Meteorological Society, Boston, pp. 94–99.

Lin, Y.-L., R. D. Farley, and H. D. Orville, 1983. Bulk parameterization of the snow field in a cloud model. *J. Climate Appl. Meteorol.* 22:1065–92.

Low, T. B., and R. List, 1982. Collision, coalescence and breakup of raindrops. Part I: Experimentally established coalescence efficiencies and fragment size distributions in breakup. *J. Atmos. Sci.* 39:1591–1606.

Marwitz, J. D., 1972. The structure and motion of severe hailstorms. Part I: Supercell storms. *J. Appl. Meteorol.* 11: 166–79.

Miller, L. J., and R. G. Strauch, 1974. A dual Doppler radar method for the determination of wind velocities within precipitating weather systems. *Remote Sensing Environ.* 3:219–35.

Nelson, S. P., 1983. The influence of storm flow structure on hail growth. *J. Atmos. Sci.* 40:1965–83.

———, and R. A. Brown, 1982. Multiple Doppler radar derived vertical velocities in thunderstorms: Part 1—Error analysis and solution techniques; Part II—Maximizing areal extent of vertical velocities. NOAA Tech. Memo. ERL NSSL-94.

Ogura, Y., and T. Takahashi, 1971: Numerical simulation of the life cycle of a thunderstorm cell. *Mon. Weather Rev.* 99: 895–911.

Orville, H. D.,and F. J. Kopp, 1977. Numerical simulation of the life history of a hailstorm. *J. Atmos. Sci.* 34:1596–1618.

Pruppacher, H. R., and J. D. Klett, 1978. *Microphysics of Clouds and Precipitation.* D. Reidel, Boston.

Ray, P. S., and K. L. Sangren, 1983: Multiple-Doppler radar network design. *J. Climate and Appl. Meteorol.* 22:1444–54.

———, R. J. Doviak, G. B. Walker, D. Sirmans, J. Carter, and B. Bumgarner, 1975. Dual-Doppler observation of a tornadic storm. *J. Appl. Meteorol.* 14:1521–30.

———, B. C. Johnson, K. W. Johnson, J. S. Bradberry, J. J. Stephens, K. K. Wagner, R. B. Wilhelmson, and J. B. Klemp, 1981. The morphology of several tornadic storms on 20 May 1977. *J. Atmos. Sci.* 38:1643–63.

———, J. J. Stephens, and K. W. Johnson, 1979: Multiple Doppler radar network design. *J. Appl. Meteorol.* 15:706–10.

———, K. K. Wagner, K. W. Johnson, J. J. Stephens, W. C. Bumgarner, and E. A. Mueller, 1978. Triple-Doppler observations of a convective storm. *J. Appl. Meteorol.* 17:1201–12.

———, C. L. Ziegler, W. Bumgarner, and R. J. Serafin, 1980. Single- and multiple-Doppler radar observations of tornadic storms. *Mon. Weather Rev.* 108:1607–25.

Rogers, R. R., 1976: *A Short Course in Cloud Physics.* Pergamon Press, Oxford.

Rotunno, R., and J. B. Klemp, 1982. The influence of shear induced pressure gradient on thunderstorm motion. *Mon. Weather Rev.* 110:136–51.

Roux, F., J. Testud, M. Payen, and B. Pinty, 1984. Pressure and temperature perturbation fields retrieved from dual-Doppler radar data: an application to the observation of a West-African squall line. *J. Atmos. Sci.* 41:3104–21.

Rutledge, S. A., and P. V. Hobbs, 1983. The mesoscale and microscale structure and organization of clouds and precipitation in midlatitude cyclones. VIII: A model for the "seeder-feeder" process in warm-frontal rainbands. *J. Atmos. Sci.* 40: 1185–1206.

Schlesinger, R. E., 1978. A three-dimensional numerical model of an isolated thunderstorm: Part I. Comparative experiments for variable ambient wind shear. *J. Atmos. Sci.* 35:690–713.

Soong, S., and Y. Ogura, 1973. A comparison between axisymmetric and slab-symmetric cumulus cloud models. *J. Atmos. Sci.* 30:879–93.

Strang, G., and G. J. Fix, 1973. *An Analysis of the Finite Element Method.* Prentice-Hall, Englewood Cliffs, N.J.

Takahashi, T., 1981. Warm rain development in a three-dimensional cloud model, *J. Atmos. Sci.* 38:1991–2013.

Twomey, S., 1959. The nuclei of natural cloud formation. Part II: The supersaturation in natural clouds and the variation of cloud droplet concentration. *Geofis. Pura e Appl.* 43:243–49.

Williams, R., and P. J. Wojtowicz, 1982. A simple model for droplet size distribution in atmospheric clouds. *J. Appl. Meteorol.* 21:1042–44.

Willis, P. T., 1984. Functional fits to some observed size distributions and parameterization of rain. *J. Atmos. Sci.* 41: 1648–61.

Wisner, C., H. D. Orville, and C. Myers, 1972: A numerical model of a hailbearing cloud. *J. Atmos. Sci.* 29:1160–81.

Ziegler, C. L., 1984. Retrieval of thermal and microphysical variables in observed convective storms. Ph.D. diss., University of Oklahoma, Norman.

———, 1985. Retrieval of thermal and microphysical variables in observed convective storms. Part 1: Model development and preliminary testing. *J. Atmos. Sci.* 42:1487–1509.

———, P. S. Ray, and N. C. Knight, 1983. Hail growth in an Oklahoma multicell storm. *J. Atmos. Sci.* 40:1768–91.

The Authors

Roy T. Arnold, a native of Jackson, Mississippi, studied physics and received the B.S. degree from Millsaps College in 1954 and M.S and Ph.D. degrees (1954 and 1960) from Vanderbilt University. After four years as a research scientist at the Sperry Microwave Electronics Company, in Clearwater, Florida, he joined the faculty of the University of Mississippi in 1963, where until retirement he was Professor of Physics. His research interests are diversified, but from the early 1970s he worked principally in physical acoustics. After the tornado outbreak of 3–4 April 1974, he was engaged in storm acoustic research, with a particular interest in tornadic sounds, and participated with his students in the storm intercept program headquartered at the National Severe Storms Laboratory.

Stanley L. Barnes is a senior meteorologist for the Weather Research Program in NOAA's Environmental Research Laboratories, Boulder, Colorado. After four years as a naval weather observer he studied meteorology at Oklahoma State University and Texas A&M University, receiving the B.S. and M.S. degrees from the latter. His Ph.D. degree (1967) is from the University of Oklahoma, where he assisted in establishing the meteorology program. During the late 1960s and early 1970s he was Chief of the Severe Storms Morphology and Dynamics Project at the National Severe Storms Laboratory. From 1974 until 1981 he served as meteorologist and office manager for the Severe Environmental Storms and Mesoscale Experiment Program (Project SESAME) and, during its 1979 field program, as Deputy Director of Operations. His publications treat severe thunderstorm phenomena synthesized from surface, upper-air, and radar reflectivity data. His most recent research foci are development of objective analysis tools, mesoscale applications of quasi-geostrophic theory, and development of computer-based aids for use by operational forecasters.

Edward A. Brandes is a research meteorologist at NOAA's National Severe Storms Laboratory, Norman, Oklahoma. He holds a B.S. degree in mechanical engineering from the New Jersey Institute of Technology (Newark) and obtained his M.S. degree in meteorology from the New York University and the Ph.D. in meteorology from the University of Oklahoma. From 1964 to 1968 he served as a duty weather forecaster and chief weather forecaster in the U.S. Air Force. His interests are in the use of Doppler radar observations to study the kinematics and dynamics of precipitating weather systems.

Robert P. Davies-Jones completed undergraduate work in physics at Birmingham University, England, received the Ph.D. in astrogeophysics from the University of Colorado, and during 1969–70 held a postdoctoral research appointment at the National Center for Atmospheric Research. His specialties are the fluid dynamics of tornadoes and severe thunderstorms. Now at the National Severe Storms Laboratory, he investigates tornadoes theoretically and directs work of the Tornado Intercept Project to observe severe storms in the field. He is the author of numerous publications, has received awards from NOAA and from the American Meteorological Society, and holds an adjunct appointment in the University of Oklahoma.

Richard J. Doviak is a senior scientist at the National Severe Storms Laboratory and Adjunct Professor of Electrical Engineering and Meteorology in the University of Oklahoma. He received his B.S. degree in electrical engineering from Rensselaer Polytechnic Institute and his M.S. and Ph.D. degrees from the University of Pennsylvania, where he was a member of the faculty before joining NSSL in 1971. From 1965 to 1967, he was principal investigator on a project that developed the first microwave spark chamber for detection of atomic particles. From 1967 to 1971 he led development of a research facility for remote sensing at the University of Pennsylvania's Valley Forge Research Center. His work at NSSL has related mostly to Doppler radar measurements of atmospheric processes. He is a consultant to corporations, teacher, editor of professional journals, and author or coauthor of many professional articles and of the book *Doppler Radar and Weather Observations*. From January through March 1987 he was Visiting Professor of Electrical Engineering in the University of Kyoto, Kyoto, Japan, where he lectured on Doppler radar and performed research related to Kyoto University's phased-array VHF wind-profiling radar.

Thomas M. Georges received the B.S. degree in engineering from Loyola University of Los Angeles in 1961 and an M.S. from U.C.L.A., also in engineering. Since 1963 he has worked at the Boulder Laboratories of the U.S. Department of Commerce, first with the National Bureau of Standards, now with NOAA's Wave Propagation Laboratory. In 1967 he received a Ph.D. in engineering from the University of Colorado. He is author or coauthor of more than 60 publications on radio and acoustic-gravity wave propagation, ionospheric dynamics, meteorology, radar oceanography, and atmospheric and ocean acoustics. He now does research in ocean acoustic remote sensing and has recently published a book on business and technical writing.

Carl E. Hane is a research meteorologist at the National Severe Storms Laboratory in Norman, Oklahoma. He received B.A. and B.S. degrees in mathematics and meteorology, respectively, from the University of Kansas, and M.S. and Ph.D. degrees in meteorology from Florida State University. In 1972–73 he was a postdoctoral fellow at the National Center for Atmospheric Research and during 1973–76 was an atmospheric scientist at Battelle-Northwest Laboratories, Richland, Washington. He joined NSSL in 1976. His research treats numerical simulation of thunderstorms, precipitation scavenging, retrieval of pressure and buoyancy distributions in thunderstorms, and analysis of the structure and evolution of squall lines. He is currently Adjunct Associate Professor of Meteorology in the University of Oklahoma and is a member of the Council of Fellows of the NOAA-OU Cooperative Institute for Mesoscale Meteorological Studies.

Ronald L. Holle is Program Manager of the Thunderstorm Studies Group of the Weather Research Program in Boulder, Colorado, a component of the NOAA Environmental Research Laboratories. He obtained B.S. and M.S. degrees in meteorology from Florida State University and undertook additional course work at the University of Miami. Principal areas of his research have emphasized photogrammetry of convective clouds in the tropics, mesoscale cloud organization and interaction, and aspects of cloud-to-ground lightning. He has participated in the PRE-STORM and AIMCS projects studying Mesoscale Convective Systems, as well as the FACE, BOMEX, and GATE projects in tropical locations. He studied GATE photographs at the University of Virginia, has been rapporteur to the World Meteorological Organization on tropical cloud codes and photographs, and has served as primary consultant for revision of the WMO *International Cloud Atlas*.

John McCarthy obtained a B.A. degree in physics from Grinnell College, Iowa, in 1964, an M.S. in meteorology from University of Oklahoma in 1967, and a Ph.D. in atmospheric sciences from the University of Chicago, in 1973. He is Senior Manager and Scientist III in the Research Applications Program, Atmospheric Technology Division, at the National Center for Atmospheric Research. He directs the Terminal Doppler Weather Radar Project, the Joint Airport Weather Studies Project, and the LLWAS Project, the last named being addressed toward improvements to the Low-Level Wind Shear Alert System. In 1984 he directed the Classify, Locate, and Avoid Wind Shear Project. He is Chair of the AMS Committee on Aviation, Range, and Aerospace Meteorology; a member of the NASA Shuttle Weather Forecasting Advisory Panel; and past member or chairman of several other professional panels and committees. During 1973–80 he held appointments as Associate Professor of Meteorology in the University of Oklahoma. He received the 1987 Losey Atmospheric Sciences Award from the American Institute of Aeronautics and Astronautics, and the 1987 Edgar S. Gorrell Award from the Air Transport Association.

Donald R. MacGorman is a physicist with the Storm Electricity and Cloud Physics Group of the National Severe Storms Laboratory. He received a B.A. in physics and M.S. and Ph.D. degrees in space physics and astronomy from Rice University. He joined NSSL following two years as a postdoctoral research associate of the National Research Council, and he presently holds appointments as Adjunct Assistant Professor in the Departments of Meteorology and Physics at the University of Oklahoma. He serves on the Working Group on Lightning Detection Systems of the U.S. Interdepartmental Committee for Meteorological Services and Supporting Research. His chief recent research interests have been applications of lightning strike locating systems to operational meteorology and relationships between lightning and the evolving structure and kinematic properties of storms.

Robert A. Maddox received his B.S. degree in meteorology from Texas A&M University in 1967 and M.S. and Ph.D. degrees in atmospheric science from Colorado State University in 1973 and 1981. From 1967 to 1975 he was an aviation and severe-storms forecaster in the U.S. Air Force. After release from active duty he joined NOAA's Environmental Research Laboratories in Boulder, Colorado. In July 1986 he was appointed Director of the National Severe Storms Laboratory in Norman, Oklahoma, where his primary research concerns mesoscale weather systems, use of satellite observations to study severe storms, and prediction of heavy rains and flash floods. Since 1982 he has been Visiting Professor at the National Weather Service Training Center, Kansas City, Missouri, and during 1984–1986 he was co-editor of the *Monthly Weather Review*.

Griffith M. Morgan, Jr., attended Ripon College, Wisconsin, and received a B.A. degree in physics from Northwestern University in 1955 and an M.S. degree in meteorology from New York University in 1961. He served in the U.S. Air Force as a weather forecaster in Europe and North

Africa. In 1962 he studied hailstorms in Italy under a Fulbright grant. During 1967–70 he was Director of the Observatory for Hailstorm Research in Verona, Italy; subsequently he was Senior Meteorologist at the Illinois State Water Survey. During 1976–82 he specialized in hail investigations at the National Center for Atmospheric Research, where he became head of the Design and Evaluation Group in the Convective Storms Division. In 1983 he joined Simpson Weather Associates, Charlottesville, Virginia, and for that company was Field Director of the Program for Atmospheric Water Supply, sponsored by the Water Research Commission of South Africa. During 1987, while maintaining seasonal direction of the South African project, he became director of a hail-prevention experiment supported by governments of Yugoslavia and Italy, near Gorizia, Italy. He is the author of more than 60 papers and reports, mainly on hailstorms, cloud physics, and synoptic meteorology.

Edward T. Pierce was a graduate of the University of Wales and received the doctorate in atmospheric physics from Cambridge University in 1951. After more than a decade of teaching and research at Cambridge in solar and meteorological research, he was employed for high-voltage research by Vickers, Ltd., and for research in geophysics by AVCO Corporation. During 1960–76 he was employed at Stanford Research Institute and became widely known as an authority on atmospheric electricity and the hazardous effects of electric fields and lightning on aircraft, supertankers, and rockets. During 1976–77 he worked at NOAA's National Severe Storms Laboratory in Norman, Oklahoma. In his professional career he produced more than 150 reports and publications. He was honorary President of the International Commission on Atmospheric Electricity from 1975 until his untimely death in 1978.

Peter S. Ray graduated from Iowa State University and in 1973 received a Ph.D. from Florida State University. He is Professor in the Department of Meteorology and the Supercomputer Computations Research Institute in Florida State University. From 1974 to 1985 he was a meteorologist at the National Severe Storms Laboratory, and from 1980 to 1985 he was Chief of the Meteorological Research Group and, in 1985, Deputy Director of the Laboratory. His meteorological papers have focused on the theory of scattering by hydrometeors and on convective dynamics as revealed by techniques of synthesis of multiple Doppler radar data and by dynamic modeling. He is a Fellow of the American Meteorological Society.

W. David Rust received the B.S. degree in physics and mathematics from Southwestern University, Georgetown, Texas, in 1966, and M.S. and Ph.D. degrees in physics from the New Mexico Institute of Mining and Technology, Socorro, in 1969 and 1973. After a term of NOAA's Atmospheric Physics and Chemistry Laboratory in Boulder as a postdoctoral research associate of the National Research Council, he joined that laboratory as a physicist in 1975. In 1977 he joined the National Severe Storms Laboratory, where he is now Chief of the Storm Electricity and Cloud Physics Research Group. He is also Adjunct Associate Professor of Meteorology and Physics in the University of Oklahoma and a Fellow of the NOAA-OU Cooperative Institute for Mesoscale Meteorological Research. He has received several awards of professional societies and is active in the work of professional committees. His principal personal research interest since joining NSSL has been the study of lightning flashes that transfer positive charge to ground and of electrical fields in severe storms.

Dale Sirmans is Senior Engineer and Assistant Director of the National Severe Storms Laboratory. He studied at the Georgia Institute of Technology and holds a B.S. in electrical engineering from the University of Oklahoma. Before joining the National Severe Storms Project in 1962, he worked in aviation electronics with both military and commercial carriers and with manufacturers. His activities include design and development of meteorological radar systems and signal processing techniques and engineering consulting.

Donald L. Veal completed undergraduate work at South Dakota State University and received M.S. and Ph.D. degrees in civil engineering from the University of Wyoming. He was a pilot and instructor in the U.S. Air Force and held various academic posts before becoming Head of the Department of Atmospheric Science in the University of Wyoming's College of Engineering in 1971. During 1987 he was on leave from that post to direct the National Hail Research Experiment of the National Center for Atmospheric Research. He was appointed President of the University of Wyoming during 1981 and held that position until administrative retirement in June 1987, when he became President of Particle Measuring Systems, Inc., which manufactures and provides instrumentation to measure microcontamination in liquids, gases, and air and to monitor facilities in a wide range of industries. Dr. Veal's contributions to the professional literature emphasize instrumentation and use of aircraft in meteorological investigations.

Thomas H. VonderHaar attended Parks College, St. Louis, Missouri, and the University of Wisconsin, Madison, where he earned his doctoral degree in 1968. In 1970 he joined the Department of Atmospheric Sciences in Colorado State University, and was Professor and Head of that Department from 1974 to 1984. Since 1980 he has been Director of the NOAA-CSU Cooperative Institute for Research in the Atmosphere. He and his students have published extensively on meteorological applications of satellite data and have contributed substantially to development and use of satellite-borne radiation sensors. He is a member of several learned societies and scientific commissions.

Kenneth E. Wilk is Head of the Interim Operational Test Facility of the Joint Systems Program Office, National Weather Service; he holds key responsibilities for development of radar methodologies to be applied in a new national network of Doppler weather radars. He has B.A. and B.S. degrees in physics and meteorology from the University of Illinois and Pennsylvania State University, respectively. During 1952–56 he provided weather forecasts for U.S. Air Force units in England and Germany. In 1962 he joined the U.S. Weather Bureau's National Severe Storms Project to establish its Weather Radar Laboratory in Norman. After 1964, when the Norman group became the nucleus of NSSL, he led the operation of many of NSSL's observational facilities. He was manager of the Joint Doppler Operational Project, which demonstrated Doppler radar capabilities and thereby established bases for use of Doppler radars throughout the United States.

James W. Wilson is a scientist at the National Center for Atmospheric Research, in Boulder, Colorado. He obtained a B.S. degree in physics from the University of North Carolina in 1959 and an M.S. degree in meteorology from the University of Washington in 1961. From 1961 until 1978, when he joined NCAR, he worked as a research scientist for the Travelers Research Center in Hartford, Connecticut (renamed Center for the Environment and Man in 1970). His research specialties are use of Doppler radar and mesometeorological observation networks to study air motions and precipitation structure of convective and stratiform storms, with a special emphasis on research applied to operational problems. He has played a leading role in the development of radar techniques to measure precipitation.

Conrad L. Ziegler is a research meteorologist at the National Severe Storms Laboratory, Norman, Oklahoma. He holds a B.S. degree in environmental science from Rutgers University and M.S. and Ph.D. degrees in meteorology from the University of Oklahoma. He specializes in development and application of numerical cloud models and multiple Doppler radar analysis techniques. He serves in an adjunct faculty capacity in the University of Oklahoma.

Dusan S. Zrnić is Chief of the Doppler Radar Group at the National Severe Storms Laboratory, Norman, Oklahoma, and Adjunct Professor of Electrical Engineering and Meteorology in the University of Oklahoma. He graduated in 1965 from the University of Belgrade with a B.S. degree in electrical engineering, and he obtained M.S. and Ph.D. degrees in electrical engineering at the University of Illinois, Urbana. Following employment as a research and teaching assistant with the Charged Particle Research Laboratory at the University of Illinois, he joined the Electrical Engineering Department of California State University, Northridge, where he became Associate Professor in 1974 and Professor in 1978. During 1973–74 he was a Post-Doctoral Fellow of the National Research Council at NSSL, and in 1975–76 he was on sabbatical at the same laboratory. His research experience includes circuit design, applied mathematics, magnetohydrodynamics, and signal processing and radar systems.

Index